Modern Sedimentation in the Coastal and Nearshore Zones of China

Editor in Chief Ren Mei-e

With 228 Figures

China Ocean Press Beijing
Springer-Verlag Berlin Heidelberg New York Tokyo

Editor in Chief
Professor Dr. Ren Mei-e

Nanjing University
Department of Geology
Nanjing, China

Published by: China Ocean Press
Fuxingmenwai Street, Beijing, China

Distributed by:
Domestic— China Ocean Press
Foreign— Springer-Verlag Berlin Heidelberg New York Tokyo

Springer-Verlag Berlin Heidelberg New York Tokyo is the exclusive Distributor for all countries outside the People's Republic of China

ISBN 3-540 13149-3 Springer-Verlag Berlin Heidelberg New York Tokyo

ISBN 0-387 13149-3 Springer-Verlag New York Heidelberg Berlin Tokyo

Printed in The People's Republic of China

Preface

Since 1949, especially since early 1960's, great progress has been made in the study of coastal sedimentation of China. There is a large amount of literature on the subject, but unfortunately, they are almost all in Chinese, at most with short English abstracts. So, they are little available to readers of other countries. Therefore, it seems appropriate now that a concise English treatise on the subject be published both for geoscientists interested in the subject and for general readers. This book is the result of cooperative efforts of its authors and editor. Authors from many institutions on the coast of China have contributed to this volume. They have long been engaged in research on specific topics they contributed and the main results of research are included in respective chapters of this book. Consequently, it may be said that this book presents a preliminary summary of research by Chinese scientists on sedimentation of coastal zone of China since the establishment of the People's Republic of China.

Sediment dynamics is an important approach of modern sedimentology. Many contributors of this book treat characteristics of sediments as a function of oceanographical conditions and sediment input and some have presented laboratory results to elucidate processes of sedimentation. But different authors have their different scientific approaches. Chapter 2 discusses sedimentation on sandy coast chiefly from point of view of dynamic geomorphology. Chapter 3 on mud flat emphasises two case studies: Wanggang of Jiangsu Province and Taizhou Bay of Zhejiang Province. They are the results of efforts of colleagues of Nanjing University and Hangzhou University respectively and up to the present, represent two most detailed studies on sedimentation of mud flat in China. In Chapter 4 on estuarine sedimentation, detailed treatment is made on modern sedimentation in the mouth of the Changjiang River. This is the result of many years' research on the subject by colleagues of Hangzhou University. Process of flocculation is discussed in some detail and many SEM photographs are included. Section on the transport and diffusion of suspended discharge of the Changjiang River summarizes results of the collaborative study by the Chinese and American oceanographers on the subject in 1980–1982. Section on sedimentation in the Zhujiang River mouth comes from colleagues of Zhongshan University who, for many years, have been involved in this study. The main part of deltaic sedimentation (Chapter 5) discusses sedimentary systems of the Changjiang River and Luanhe River delta, which are results of colleagues of Tongji University in recent years. A large amount of valuable data are included. Coral reef has received increasing attention from Chinese scientists, and considerable research work has been done in recent years. Chapter 7 emphasizes chemical and mineralogical characteristics of coral reef sediments which constitute one of the main aspects of coral reef research in the world.

The unbalanced development of research on modern sedimentation in coastal zone of China can be readily seen in this book. For example, although the Huanghe River is one of the largest rivers of

China and the most heavily sediment laden river in the world, but its estuarine and deltaic sedimentation are only briefly described in this book. This clearly shows that detailed study on sedimentation of this great river is badly needed. Similarly, the chapter on submarine sand ridges (Chapter 6), compared with other chapters, is very short. Evidently, this is due to inadequate study on this subject which is a comparatively new field in China. But a good start in study of China's submarine sand ridges began in 1980–1982, when systematic survey of submarine sand ridge of northern Jiangsu, the largest submarine sand ridge field in China, was carried out by First Institute of Oceanography, National Bureau of Oceanography and part of the results of this survey is included in Chapter 6.

It must be pointed out that the editor has only made some necessary revisions on the original manuscripts of the contributors whose viewpoints and conclusions have been remained intact. Therefore, careful readers may find some discrepanies between different chapters in this book. But the editor believes this will be proved useful for the advancement of sedimentology in China.

The contributors of this book are all very busy in their respective daily work. The editor wishes to express his gratitude to their friendly and close cooperation in the preparation of this book, without which this book cannot be published in such a short time.

Nanjing University Ren Mei-e

Contents

Chapter One
Introduction

The mainland coastline of China is about 18,000 km long. In the coastel zone of China, sediments are most varied. The coastal zone in this book includes offshore sea less than 10–20 m deep bacause the shallow offshore sea forms a part of subtidal zone of tidal flat and therefore constitutes an integral part of the profile of the flat. Moreover, submarine sand bars and sand ridges in the offshore zone often have a close relation with the coast and coastal flat. For example, the large submarine sand ridge field of northern Jiangsu is formed in deposits of the Old Huanghe River (Yellow R.) and the Old Changjiang River (Yangtze R.); at the same time, it constitutes a part of the source of sediments of mud tidal flat along the coast of Jiangsu. Furthermore, in its historical evolution, part of the submarine ridge field has incorporated into the coast and formed a part of the tidal flat and coastal plain of Jiangsu. For this reason, it seems impossible to get a complete understanding of the genesis and evolution of the coast and beach if considerable attention is not paid to submarine depositional land forms and surficial sediments in the offshore shallow sea.

During the past few decades, coastal research in China mainly focussed on geomorphic evolution, coastline changes and the evolution of coastal depositional land forms, like sand bars, spits, etc., but little attention was paid to sediments themselves. Ancient depositional systems of the coastal zone like deltas, sand bars-lagoons, and coral reefs often form rich reservoirs for oil and gas. The study of sedimentary facies and models of modern depositional systems of the coastal zone is important in prospecting mineral resources of ancient depositional systems. Since the seventies', considerable research has been made on sedimentation of the coastal zone in the People's Republic of China, and a large amount of valuable data has been accumulated. Unfortunately, they are not easily accessible to foreign readers as they are mostly published in Chinese. The contributors of this book give a systematic treatise on sedimentation problems in which they have made considerable research. Up to the present, this is the only trestise on sedimentation of the coastal zone of China ever published in Chinese or English. But as the research made in different areas are not in the same level and as the insight of contributors to the sedimentation problems is varied, there are certain differences in the treatment of problems and in the style of writing and there are also some discrepancies in the nomenclature (including grain size parameters) of sediments in different chapters. The editor has tried his best to retain the main viewpoint and style of writing of different contributors.

I. Factors Influencing Depositional Environments of Coastal Zone

Physical factors are most important in determining depositional environments. The study of physical processes provides basic information for the recognition and reconstruction of depositional environments (Reineck, et al. 1980). Among physical factors, hydrodynamic conditions and provenance of sediments are of greater significance. Therefore, in the study of sedimentation of coastal zone, attention must be paid to hydrology and sediments and their mutual relationship. This is the approach of sediment dynamics in modern sedimentology.

1. Oceanographical conditions

Sea current, wave and tidal current have important effects on the distribution, texture and structure of sediments of the coastal zone of China.

Major currents in the Bohai Sea, Huanghai Sea (Yellow Sea) and East China Sea are the Kuroshio and its branches (i.e. Taiwan warm current and Huanghai Sea warm current) and littoral currents, the latter having greater influence on the distribution and characteristics of sediments of the coastal zone (Fig. 1-1).

In this area, the Huanghai Sea littoral current and the East China Sea littoral current are the two main littoral currents. The former is a diluted sea water mass moving from the mouth of the Huanghe River through southern part of the Bohai Straits and along the coastal zone of Shandong Province southward to the offshore of Jiangsu and the mouth of the Changjiang River. The direction of this current is rather fixed, with little change in different seasons. Part of suspended sediments of the Huanghe River discharging into the sea is transported by the Huanghai Sea littoral current into the Huanghai Sea and diffuses there. The northern Bohai Stratits, which are not affected by the Huanghe River mud stream, have entirely different surficial sediments from that of the southern Bohai Straits. The former is the main passage through which the Huanghai Sea warm current enters the Bohai Sea and the current velocity is great. For example, in the Laotieshan passage, the current velocity reaches 6–7 knots per hour. Whereas the southern Bohai Straits is the passage of the Huanghai Sea littoral current and the current is weak. The difference of surficial sediments of these two passages is shown in Table 1-1.

Table 1-1 Surficial sediments of the Laotieshan Passage and southern Bohai Straits

	Laotieshan Passage	Southern Bohai Straits
Grain size	fine sand	clayey silt
Sorting	well	moderate or poor
Content of stable minerals	high	low
Content of flaky minerals	low	high
Content of organic matter	low	high

(After Zhao Chuanji, 1981)

Summer Winter

Fig. 1-1. Current circulation in the Bohai Sea, Huanghai Sea and East China Sea. 1. Main stream of Kuroshio; 2. Tsushima warm current; 3. Huanghai Sea warm current; 4. Taiwan warm current; 5. Huanghai Sea littoral current; 6. East China Sea littoral current.

The data in the above table show that the suspended sediments of the Huanghe River are transported from the Bohai Sea by the Huanghai Sea littoral current and partly deposited in the southern Bohai Straits.

The East China Sea littoral current comes mainly from the diluted water of the Changjiang River and flows southward along the coast to the coast of northern Fujian. But during the summer monsoon, it flows northeastward. In years of heavy discharge of the Changjiang River (as flood year of 1954), it may extend to Chejudo. The fine sediments of the Changjiang River carried by the East China Sea littoral current is deposited in the offshore between the mouth of the Changjiang River and the coast of northern Fujian, forming a distinct mud zone of deposition in the inner shelf of the East China Sea.

The South China Sea littoral current flows from the mouth of the Zhujiang River westward to the Gulf of Guangzhou. Influenced by large discharge from the Zhujiang River, it is a littoral drift of low salinity, which is only half of that of the South China Sea warm current. The South China Sea littoral current is very conspicuous during the NE monsoon of the winter, attaining a width of 30 nautical miles. During the summer, it still flows westward, but with a samaller width. It is responsible for the diffusion of sediments of the Zhujiang River towards the coastal zone of western Guangdong. In the summer, the upwelling of high saline water of the South China Sea warm current in the near-shore affects sedimentation of the Zhujiang Estuary. For example, the presence of stenonaline foraminifera, with well-preserved tests, is obviously related to the intrusion of the South China Sea warm current.

The tide in the nearshore zone of China is of regular and irregular semi-diurnal type, except the coast of the Beibu Gulf which has a regular diurnal tide. The tidal range and tidal current exert considerable influence on the sedimentation of the coastal zone. According to J. L. Davis (1964), the world coastline may be divided into 3 types according to magnitude of tidal range, i. e. microtidal (mean tidal range < 2 m), mesotidal (mean tidal range 2–4 m) and macrotidal (mean tidal range > 4 m). In China, the coast of Guangdong is mainly microtidal; the coast of the Bohai Sea is also chiefly microtidal, except the heads of the Bohai Gulf and Liaodong Gulf which have greater tidal ranges; the major part of the coast of the Huanghai Sea is mesotidal; and the coast of Zhejiang and Fujian is largely macrotidal (Fig. 1-2).

The magnitude of tidal range affects the sedimentation of coastal zone. Generally speaking, the width of mud tidal flat varies with tidal range (Table 1-2).

Table 1-2 Relation between width of tidal mud flat and tidal range in some localities of the world

	Louisiana (U.S.A.)	Surinam (S.America)	Westcoast of S. Korea
Tidal range (m)	0.5	2	5–9
Width of intertidal mud flat (km)	0.1–0.3	2–5	5–30

(After J. T. Wells, 1983)

Since width of mud flat is influenced by many factors, such as sediment supply, historical evolution, etc., considerable care must be taken in the study of relationship between tidal range and width of mud flat. For example, although tidal range in the Taizhou Bay, Zhejiang Province is greater than that of Wanggang, Jiangsu Province, mud flat in Wanggang is considerably wider than that along the Taizhou Bay due to abundant supply of sediment.

Tidal current is one of the principal factors influencing sedimentation of mud tidal flat. Magnitude of current velocity, asymmetry of duration and velocity of the flood and ebb, etc. have important effect on sedimentation of mud flat. Strong tidal current scours a tidal inlet through mud bank or sand bar, for example the tidal inlet in Zhenhai Bay, Guangdong (through mud bank) and in the south coast of Hainan Island (through sand bar). Owing to scouring of strong tidal current, sediment in tidal inlet is generally coarse, consisting chiefly of coarse sand, gravels, shells and coral fragments. In submarine sand ridge field in the offshore zone of Jiangsu, some very deep tidal channels have been eroded by tidal current. Deeps of more than –40 or –50 m are formed in the place where two deep tidal channels meet. Rocks or old sedimentary layers are often exposed in the floor of these channels and deeps.

Wave is another important factor influencing sedimentation of coastal zone of China, especially in sandy beaches. It is well known that in a monsoonal region like China, change in wind direction between summer and winter is very marked. There is also a great seasonal variation of wave height which is greatest from mid-autumn to the end of winter, but is smaller in summer. Therefore, a corresponding conspicuous change of profile and sediments of sandy beaches has been observed in China (see Chapter 2).

The profound influence of storms and storm waves on sedimentation is well known. As pointed out by Kumar (1976), the storm deposits in surficial sediment of the Atlantic inner shelf off northeastern United States have entirely different textures and sedimentary structures compared with deposits in under ordinary weather. The coastal zone of China is frequented by typhoon, a strong tropical cyclone, in summer and autumn and attacked by strong wind associated with cold waves in winter and early spring. When these strong winds blow, wave height may reach 7.5–9.0 m along the coast of the Bohai Gulf and the Huanghai Sea and huge waves with wave height exceeding 10 m have been observed in the East China Sea and South China Sea. These huge waves not only cause serious damage to the coast, scour the surface of the beach and bring bout coarsening of beach sediments, but also result in storm deposits along sandy and gravelly coast which may be several meters high above the present sea level. The storm tide formed by typhoon and cold wave may cuase a great rise of water level along the coast, the amount of rise of water level reaching 1–3 m or even more, thus causing a shift of position of sedimentary zones of mud flat. The typhoon deposit in mud flat of China is markedly different from its overlying and underlying sediments and has a good potential of preservation. Therefore, rare events like storms are important in geological history. The study of modern storm deposits in coastal zone provides a key for the understanding of ancient storm deposits. Recent study (Sun et al., 1983)[1)] on Cambrian storm deposits in Yunnan Province shows that the lithological character, sedimentary structure and sequence of ancient storm deposits are remarkably similar to its modern analogs.

2. Sources of sediments

Sources of sediments of coastal zone come from: (1) Erosion of adjacent coast; this is important for sandy and gravel beach. (2) Sediment discharge of rivers into the sea; this is the most important

---

1) Sun Shu et al., Sedimentation in rift basins of China, Institute of Geology, Academia Sinica, 1983 (manuscript).

source of sediments of coastal zone. Large sandy depositional bodies such as coastal sand bars, offshore sand bars and spits derive their sediments directly or indirectly from the neighbouring rivers. Great rivers like the Changjiang River, Huanghe River and Zhujiang River, which carry huge amount of suspended solids to the sea every year, are particularly important in sedimentation of coastal zone. Besides building large delta in river mouth, they contribute large amount of fine sediments (mud) which are transported by littoral current along the coast for a great distance, and build wide tidal mud flat along the coast of China. Part of suspended sediments in littoral current is settled down and deposited on shallow sea of the inner shelf, but afterwards, they, stirred up by waves and carried by tidal current to nearshore, they contribute a considerable quantity of material to mud flat. If the position of river mouth changes, the balance between oceanographical factors and sediment supply is upset. Amont three great rivers of China, the Huanghe River is notable for its huge sediment discharge and frequent shifts of river mouth in the historical time. With shifts of its river mouth, tidal flat near the former river mouth changes from progradation to retrogradation, and cheniers are formed along the coast.

(3) Biologic debris. In coastal zone, dead bodies of micro-organisms (such as foraminifera, ect.), molluscs and plants often add into sediments, forming an integral part of sediments of coastal zone. In favourable localities, as in some bays and lagoons, molluscs reproduce in large quantity. Under the action of waves, their shells accumulate and form bio-depositional bodies–shell bars, as Caofidian in the Bohai Sea. In tropical coast of South China, corals and red algae flourish, forming coral reefs and related sediments. These are the most important biological sediments in coastal zone of China. Mud flat along the coast of South China is often covered with mangroves which make sediments of mud flat rich in organic matter and trunks of mangroves, forming a unique type of sediment–sediment of mangrove flat. In subtropical and temperate coast, it is well known that diatoms and other algae may play an important role in accretion of tidal flat (Frostick et al., 1979). From SEM photographs, it can be seen that various diatoms often grow on sediments of mud flat in Jiangsu. The role of organisms in tidal flat of China is worthy of careful study.

II. Chief Characteristics of Sediments of Coastal Zone of China

A major part of coast of China consists of mountains and hills, with rocky promontories and bays, and sediments are largely sandy or gravelly. It is estimated that more than 1/2 of the total coastline of China Mainland is fringed with sandy-gravelly sediments. In some regions, mountains project directly to the sea, forming sheer sea cliffs, as Chengshantou at the end of Shandong Peninsula. Here, the coast is only fringed with narrow rocky platforms cut by waves. Rocky coast of China is usually formed by the Mesozoic granites. In a few localites, limestones outcrop along the coast, giving rise to karst coast, as Heishizui, Dalian.

Under the action of waves and littoral current, sandy-gravelly sediments form subaerial and subaqueous depositional bodies of many types and large dimensions. Coastal sand bar near Shijiusuo, southern Shandong, 16 km long and 7–8 m high, is one of the longest coastal sand bars in China.

Diaolongzui compound spit and Zhifudao tombolo along the coast of northern Shandong have attracted the interest of geoscientists since 1930s'. As some reservoirs beds in oil fields of China are ancient coastal bars along the sea or lakes, study of modern depositional sand bodies in coastal zone of China has great economic significance. In favourable conditions, sand blown from beaches can accumulate on the coast, forming high coastal sand dunes, as coastal dunes near the famous beach Beidaihe, Hebei. Texture and sedimentary structure of dune sand are different from beaches. Since sandy-gravelly sediments in coastal zone mainly derive from adjacent coastal mountains and rivers, their lithologic character and mineral assemblage are intimately related to geology of local regions.

In the past, tidal sedimentation was often regarded by most marine geoscientists as limited to tidal flat. Since 1960s', rapid advancement of marine sedimentology has shown that in subtidal shallow sea, especially in shallow sea of broad continental shelf, tidal current plays an important role in the distribution and deposition of surficial sediment on sea floor, forming tidal-current sand bodies–submarine sand ridges which can be preserved in geological record. The East China Sea, 600 km wide in its broadest part, is one of the widest shelves in the world. Here, tidal-current sand bodies are well-developed, among which submarine sand ridge field of Jiangsu, with its large extent and complicated morphology, is perhaps unique in the world.

There are many types of lagoons along the coast of China. They are chiefly lagoons of temperate-humid and subtropical-humid regions and are usually associated with sand barriers, forming barrier-lagoon systems. Lagoons are also often associated with cheniers. Lagoons in tropical humid region of South China, particularly those in coral-reef coast and coral-reef islands, have different sedimentary facies and zones and are therefore discussed in Chapter on Coral-Reef Sedimentation. Sediments of lagoons, deposited under low-energyenvironment, are fine-grained and rich in organic matter. Formed in reducing environment, lagoonal sediments often form oil source bed. For this reason, ancient barrier-lagoon systems often offer a good prospect for the exploration of oil.

Delta systems are sites of many important oilfields in the world. There are abundant literature on sedimentation of modern and ancient deltas. The Changjiang River, Huanghe River and Zhujiang River, well known in the world as great rivers, all have large delta systems in their mouths. Owing to different geological background, hydrology and sediment supply, each delta system has its own characteristics, representing three different types of deltas. The Luanhe River, draining into the Bohai Sea, is a smaller river with coarse sediments. Its fan delta is distinctly different from deltas of the three great rivers of China. In this book, we have focussed our attention to deltaic sedimentation of the Changjiang River, Huanghe River, Zhujiang River and Luanhe River, which will represent main delta types of China.

Estuary is an important part of coastal zone. In estuarine sections of large rivers, river discharge, in meeting tidal current, intermingles with salt water, producing zones of turbidity maximum and giving rise to a series of physicochemical and biochemical processes, among which flocculation is especially important in estuarine sedimentation. Owing to the importance of estuaries in navigation, trade and industry, study of modern estuarine sedimentation is of practical significance. Besides estuaries of the Changjiang River, Huanghe River and Zhujiang River, estuarine sedimentation of the

Qiantang River is discussed in some detail in this book,because the estuary of the Qiantang River, with a large intertidal sand shoal, is one the most famous estuaries in the world, and its sedimentation pattern is rather different from estuaries of the three great rivers of China.

Tidal mud flat is widely distributed in China. It is estimated that the coastline fringed with mud flat accounts for nearly 1/4 of total length of coastline of China Mainland. Since 1960's, Chinese geoscientists have gradually paid attention to the study of tidal mud flat. But up to the present, the researches made are quite different for different regions. Here, we have chosen mud flat of Wanggang, Dafeng County , Jiangsu Province and Taizhou Bay, Zhejiang Province as representatives of mud flat in open coast and mud flat in embayment coast respectively and made detailed case studies. This will be useful for the study of tidal flat sedimentation in the world.

Coral reefs of China include fringing reefs along Hainan Island and Dongsha Islands, Xisha Islands, Zhongsha Islands and Nansha Islands in the South China Sea. These Islands are almost all reef islands. Sediments of reef coast, mainly derived from corals and other marine organisms, are carbonate sediments. During recent years, with progress of marine geology and practical need of petroleum exploration, large amount of work has been done on sediments and morphology of coral reefs of China. The Chapter on Reef Sedimentation (Chapter 7) is a preliminary summary of recent studies on sedimentation and geology of coral reefs of China.

In short, this book comprises major types of sediments in coastal zone of China and is a general summary of progress made in studies on sedimentation of coastal zone of China since the establishment of the Peoples' Republic of China. It may be pointed out that sedimentation of deltas, tidal mud flat and estuaries and submarine sand ridges in China possesses certain characteristics of their own. In this respect, some sedimentary models discussed in this book may supplement vast literature on this subject and may be helpful to geologists of other countries.

References

Davis, J. L, 1964, Zeit, Geomorphologie,V. 8, pp 27–42.

Frositick, L. E. et al., 1979, Estuarine and Coastal Marine Science, V. 9, pp 569–576.

Jones, B. et al., 1976, Jour. Sed. Petrol., V. 46, pp 393–401.

Kumar, N. et al., 1976, Jour. Sed. Petrol., V. 46, pp 145–162.

Reineck, H. E. et al., 1973, Depositional Sedimentary Environments, 2nd Ed. 1980, Springer-Verlag. Well, J. T., 1983, Canadian Jour. Fisheries and Aquatic Sciences. pp 40, 130–142.

Yuan Yaochu et al., 1982, Acta Oceanol. Sinica V. 4, pp 1–11. (in Chinese)

Zhao Huanting, 1980, Oceanol. et Limnol. Sinica V. 11, 121–133. (in Chinese)

by Ren Mei-e

Nanjing University, Nanjing

Chapter Two
Coastal Sand-Gravel Sediments

I. Introduction

The snad-gravel coast in China, as in most coastal areas in the world, ranks first in its length and widespread distribution. According to incomplete statistics, the length of sand-gravel coast (including medium- and small-sized sandy esturaine deltas) accounts for 55–57% of China's total coast line, of which 60% are barriers with lagoons, 12% dunes. Among barriers with lagoons a considerable part are abandoned ones after filling up by aeolian sands. Essentially, these buried lagoons are same as dunes, because both are the products of the aeolian sand actions. But the lagoons manifest themselves differently from dunes, so they are not included in the sandy coast with dunes.

Along the sand-gravel coast in China, there are many depositional bodies which are different in the composition of sediments and developing stages. Here we may give two examples. Firstly, most of the sand-gravel coast in China are discontinuous and short stretch in cape and bay check. Sediments are mainly sands and gravels with limited thickness, such as that along Liaodong Peninsula, Shandong Peninsula and most bay coast of Fujian and Guangdong Provinces. Secondly, straightly and continuously distributed coasts of barriers and lagoons are rich in river-input sands, well-developed in lagoons and barriers, and quick in siltation, such as the west coast of Taiwan, west coast of Liaodong Gulf and east coast of Laizhou Bay and so on, where sand-gravel layers are thick and barriers and lagoons are suffering from frequent erosions and depositions. Big barriers and lagoons derived from continental shelf sands moving landward in sea-level rises are widely distributed along the Hainan Island.

In China, lagoons that grow simultaneously with barriers and spits are of small size. They are not noticeable compared with the typical modern coastal lagoons in the world (e. g. that along the Atlantic coast of North America and Gulf of Mexico). Their localities are not marked out on the lagoon charts compiled and published in foreign countries. However, it is precisely these inconspicuous small lagoons that have many types and complicated sedimentary environments. They record those important events of coastal sedimentary environments and the changes of sea level in China since the Holocene Epoch, reflecting the whole history of coastal changes ever since.

There are many factors that have affected the development of modern sediments on the sand-gravel coast of China and undoubtedly the first factor is sediment discharge of rivers into the sea.

The amount of suspended matters carried from the Huanghe River and the Changjiang River into their esturies and coastal waters averages more than 1.6 billion tons annually. This has not only created large-scale mud-flats (See Chapter 3), but also played a restraining and remoulding role in the sediments of many sand-gravel coasts. For example, part of fine particles carried by the abandoned Huanghe River into north Jiangsu have made, after diffusing towards the Hangzhou Bay, the sandy coast and beaches of the northern coast of the Bay be covered by clay silt. The suspended sediments from the abandoned Haunghe River mouth in northern Jiangsu flow southward and converge with the fine suspended sediments from the Changjiang River, diffuse still southward and deposit in large quantity in offshore waters and bays of Zhejiang and Fujian Provinces. From Central Fujian to the Hangzhou Bay, the coastal zone has been buried or partially buried by the thick claysilt. New sand body can not develop; beaches have become claysilted; wave erosion has greatly reduced.

Wind plays an important role on coast, particularly on sandy coast. Monsoon prevailing in China exert a great influence on the barriers of foreshore and nearshore regions in a broad area. The changes of erosion-deposit in the sand-gravel coast, particularly the seasonal cycle of profiles of shores, have a close relationship with the coastal engineering. For instance, on the northern and southern coasts of the Shandong Peninsula, wave actions on its southern coast are strong in summer and beaches are eroded by wave actions. Therefore, structures on the beach cycle zones are easily to be attcked. But on its northern coast, sea waves and their erosive actions on beaches are strong in winter, frequently damaging the structures on the beaches. For straight shoreline extending from north to south, sands are transported to the north in summer, resulting in the retreat of southern coast. This kind of long-period damage to beaches should not be ignored in the protection of beaches and coasts.

Strong northeast wind prevails in late autumn and early winter, and this is the major cause for barriers retreating shorewards and lagoons extremely buried. Although alluvial deposits from small rivers are also contributing increasing amounts of sediments to the lagoons, it is only a second cause in comparison with that by the prevailing wind.

Dunes are extensively developed along the sand-gravel coast of China, which is the product of the north wind actions. A strong south wind sometimes can whip sands aloft. It is only because a strong south wind often occurs in summer, being associated with heavy rain and big wave, that large dunes cannot be formed in this season. Even when it has no rain, as fine sand is mainly located in wave field, little sand can be moved by wind, because waves are bigger and the sea level rises. Therefore, large dunes cannot be formed.

Most Bays in China are strongly subject to tidal actions. In the famous Jiaozhou Bay and Dongshan Bay, tidal currents determine the sedimentary characteristics of the bays. The maximum current velocity at the bay mouth is 1.5–2 cm/s. Water is deep and current is swift in the bay mouth. At its bottom are bedrocks and large rocks. With the current velocity gradually reducing towards the bay, sediments become gravel, sand-gravel, shell and sand, clay-sand or silt successively. In the inner bay, where current velocity reaches zero and wave action is relatively active, beach sands frequently appear. The bottom where tidal currents are frequently flowing is a good place for sessile organisms such as oysters and barnacles to grow. The shallow Jimjiao in the center of the Jiaozhou Bay is exactly a submarine tidal delta formed by the deposition of shells.

In the bay where waves are dominant and currents are secondary, the erosion and transport of sands are different from the bays where tides are strong. For example, in the Zhifu Bay and Longkou Bay in the northern coast of the Shandong Peninsula, the common characteristics are mainly longshore drifting of sands with relatively fast sedimentation rate. In the Zhifu Bay, beach ridges are formed by gravel which have been eroded from the coast.

The erosion and sedimentation of coasts as well as the changes of beaches are greatly affected by storms. Under the stormy conditions, direct or indirect observations of the processes of coastal erosion and sedimentation are in operation and some good results have been gained.

II. Beach Cycle in Coastal Sand-Gravel Sediments

Owing to the constant changes of wave intensity, the original profile of equilibrium is often re-created, thus the beach is also changed. The changed beach profile resulting from the actions of different wave intensities is called "Beach Cycle".

In stormy weather, shore face on the upper beach is eroded, a more steep slope formed and a considerable amount of sands brought to nearshore areas due to the intensified actions of wave currents. After the strong waves, the sands are restored to former position through the action of medium waves over a period of time. For example, on the north side of Longkou Tombolo in the Shangdong Peninsula, when north wind in winter of 1978 was 15–20 m/s, sled was used to accurately survey the Tombolo from top to -5 m of the bottom (See Fig. 2-1(A)). The accuracy of survey was 10 m. The largest erosion point, when a storm happened, was between the beach slope and the first subaqueous bar. The range of erosion and deposit was 1.1 m.

Since beach cycle depends on the intensity of wave actions, not only a cycle over a short period of time is manifested by an apparent strong wind in some coastal zones, but also a long-term cycle from winter to summer is manifested. In this aspect, changes on the southern and northern coasts of the Shangdong Peninsula are typical ones.

Yantai and Longkou represent the northern coast of the Peninsula, and the vicinity of Qingdao the southern one. The southern beaches are eroded in summer bacause of big surfs, but sediments in the northern coast are created and broadened because the wave is weak. Reversely, the wave in the southern coast is small in winter and the accumulation of beaches is created; while the wave in the northern coast is strong and the beaches are much decreased. The curve as shown in Fig 2-1 (B) is the result of repeated surveys. The survey results show that on both the southern and northern coasts of the Shangdong Peninsula, the transfomation of beaches in summer and winter in the northern and southern coasts is numerically larger than that in a storm and the pattern of transformation is also more apparant.

There are only gravel and pebble on the southeastern beach of Shilaoren, a village 15 km to the east of Qingdao, whose slope being 21°–23° in summer. When winter comes, the gravel beach is covered by a layer of medium coarse sands, 50–70 cm in thickness (Fig. 2-1 (C)). These apparent

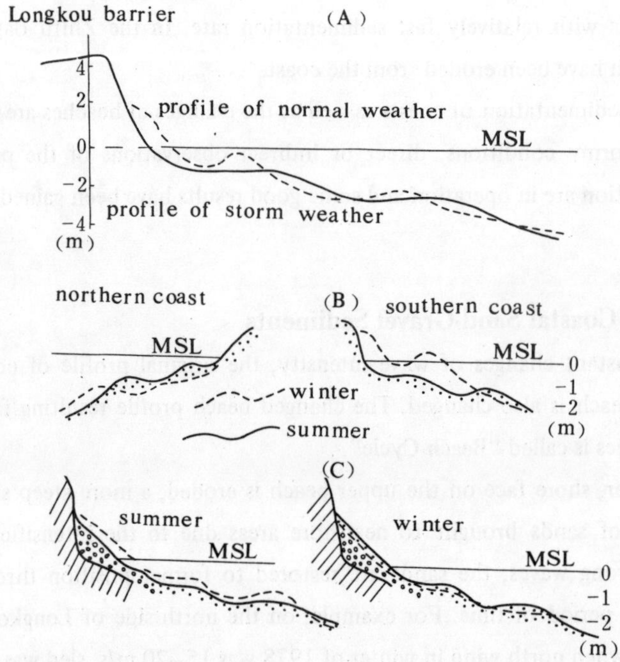

Fig. 2-1. Some case profiles of beach cycle in the Shangdong Peninsula.

(A) north coast of Longkou Tombolo; (B) southern and northern coasts of Shandong Peninsula; (C) silouren beach near Qingdao.

changes are due to strong wind in summer, when parts of sands are carried to nearshore areas of low tidal lines. The once down-carried sands return to the berm when wave is weak in winter.

III. Bay Sedimentation

Along the Chinese coast, bays of bedrock coast composed of sand-gravel in the beach and nearshore areas are rather common. Most coasts of Liaodong Peninsula, Shangdong Peninsula, Fujian and Guangdong Provinces are bay-shaped sand-gravel coasts.

There are many types of bays along the Chinese coast. They can at least be classified into three categories: 1) semi-closed bays dominated by the action of tidal currents with a mixture of sand and clay sediments, such as the Jiaozhou Bay, the Dongshan Bay and the Wutong Bay near Xiamen; 2) arc-shaped bays of sand-gravel dominated by the action of waves, such as the Longkou Bay, Zhifu Bay, the Guanghai Bay (along the eastern coast of Guangdong) and the Shanya Bay (along the Hainan Island); 3) clay bays as those along the middle coast of Zhejiang Province.

1. Bay sedimentation dominated by tidal current action

(1) *Jiaozhou Bay*

Situated in the middle section of the southern coast of the Shangdong Peninsula, it is a natural deepwater bay of medium size with a mouth nearly enclosed.

The water area of the Jiaozhou Bay is 496 km^2 when calculated in accordance to high tide line, or 296 km^2 when calculated in accordance to low tide line. The intertidal flat is 200 km^2. At the mouth of the bay, the deepest water is 53 m, decreasing gradually towards the inner bay. On average it is 10 m deep and the central area is 20 m deep. The area deeper than 5 m accounts for half of the bay. The environment of the bay are dominated by the Mesozoic Laoshan granite and the transgression of the Holocene, indicating that the Jiaozhou Bay was a valley depression developing along multiple faults. The narrowest in baymouth (3300 m) is from Tuandao to Xuejiadao.

According to the long-term records of the Qingdao Tidal Station, the average tidal range of the bay is 2.8 m with the maximum attaining 5 m . It is regular semi-diurnal tide with tidal currents weakening from bay mouth inwards. The maximum value of the observed tidal current velocity is 1.5 m/s, and the velocity of ebb tide, a slightly larger than that of flood tide. It is calculated that the maximum velocity can be close to 2 m/s as a result of the combination of spring tide and storm surge. The distribution of tidal currents are closely corrulated to the composition of sediments (see below).

There exist coastal sediments of different sources around the bay, the eastern shore being a beach and barrier at the mouth of the Lichuan River, and the western shore a sand-gravel beach of eroded bedrock. The sand-clay beach on the northern coast is an accumulation site of an annual input of more than 200 tons of sand-clay from the Daguhe River.

The distribution of the sedimentary types are rather regular in the Jiaozhou Bay though they are various types. Fig 2-2 is a distribution map of sediments based on the data of a score of surveys.

The major types of sedimentation in the Jiaozhou Bay are as follows:

1) Bedrock bottom without sediments. It is distributed in the area off Tuandao directing to Xuejiadao, extending 4–5 km on both sides, tapering convergently from 3 km onward in a shape of long rhombus. A greater part of the area is more than 30 m deep.

2) Gravel bottom. Sampling is hard to perform. We got several block of granite (2–15 cm in diameter) which were attached with various organisms, indicating that this kind of bottom gravel is seldom moved.

3) The sands that are dsitributed around the gravel bottom and identical to those in troughs. They are mainly composed of medium-fine sands, being mixed up with a small quantity of coarese sands and shells. Both the degree of sorting and psephicity are bad. Sometimes, much clay can be found mixed in the sands temperarily. Undoubtedly, the clay on the surface of sands is the outcome of neap tide during floods.

4) The mixture of shell fragments and sands, called as "Zianjiao" by local fishermen. It is situated

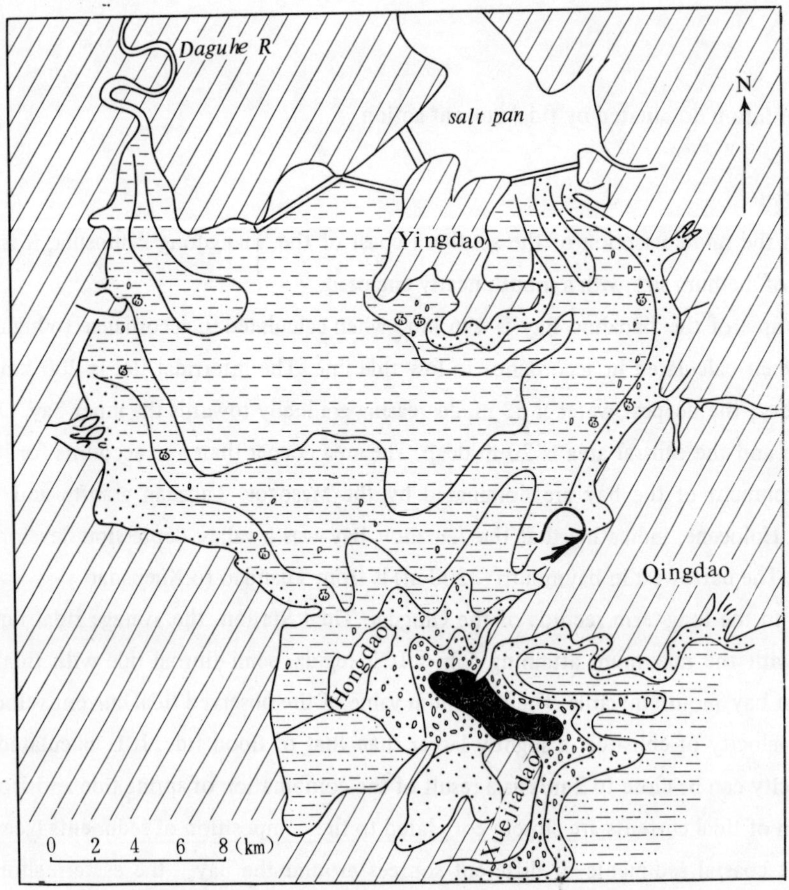

Fig. 2-2. A sketch map of tidal current sedimentation in the Jiaozhou Bay.

in the same area of sandy bottom but at the joint of tidal forks. According to the survey of Nanjing University [1)], at the top of the shallow, the depth is 5.2–6.8 m, and around the "Zianjiao", it is more than 12 m deep, bottom being composed of shells and sands. Analyses of bored materials from the loose accumulation layers at the "Zianjiao" show that from the surface to –16 m are coarse sands, shells mixed with small gravels (diameter, 4–5 mm). It is composed of quartz, feldspar and small granite with less psephicity. Shells account for over 50%. Much more shells are in the upper five meters, accounting for over 80%. Shells and fragments are varied in size, intact shells being 10–15 mm, small ones being fine sand and silt. Most of the shells are clams and oysters growing on nearby rocks or hard bottom. When extraordinary spring tide occurs, shells are crashed to death in the moving and collision of the rocks. Tidal currents bring the shells to accumulate on the convergent point. This is called current shallows or tidal ridges in the bay. Under the shell layers, there is dark grey silt and clay, being soft and with plant remains.

5) Muddy sands, averagely fine sands. Sorting is poor (3) because of much silt and medium sand

1) Zhu Dakui and Chui Chengqi, 1974. The formation and regulation of "Jianjao" Huangdao Oil Harbour (unpublished report).

contained. Skewness is variable. It is distributed in a larger extent in the outer than the inner bay where shell content is higher and sorting is poorer.

6) Sandy clay. They are mainly sand and clay, belonging to silt grade, with a small amount of fine sands and shell fragments and a poor sorting.

In view of composition and distribution, the sediments in the Jiaozhou Bay are Principally controlled by tidal currents. It is a typical bay tidal sediment characterized as follows: a) The grain size is controlled by tidal currents. There are gravel or non-sediment bed rock in areas where current velocity is larger, sandy sediments and shell fragment where current velocity is medium, and areas covered by sandy clay where current velocity is weak; b) The grain size of sediments within the bay mouth turns fining gradually and regularly. In the outer bay mouth, the sediments are sands owing to the wave action from the open sea, sandy bottom without clay is observed in this area. The reason is that there is no supply of clay which is not easy to remain at the bottom; c) In the bay, there is apparently a tidal delta or tidal ridge, such as "Zianjaio" and shallows. Out of the bay, there is no tidal delta because it is washed away by wave from the open sea.

In the middle of the bay, sandy layer of marine facies is not thick, mostly 1–3 m thick and underlain by weathering. The thickness of the sandy layer increases towards the inner bay. The thickest facies is on the area of the old river valley. Before transgression, there had been two main river valleys converging at the bay mouth. Drilling near Cankou reveals that marine facies at the old river valley is 5–7 m thicker than nearby area.

Following the old valleys where marine facies layer is thick, the Haixi Bay is the second in thickness (see Fig. 2-3). All the drill holes penetrated through sea layer in the central bay, being perpendicular to the coastal line. From the drilling data, the following characteristics can be reached. a) The clayey sand layer of the upper sea facies is greengrey in colour and soft in construction. The size of sand particle is from coarse to fine. The degree of sorting is below medium (2.6–4.8), which is better in shallow water and poorer in deep water. Psephicity is poor both in deep and shallow waters. The amount of sand increases from sea to land, indicating that the sediments are the outcome of the erosion of hill and coast. In the clayey sand layer many marine shell fragments are found. This layer is distinct from the underlying bedding face. b) Alluvial sand clay layer. It is yellow or light yellow in colour and hard in construction. The ratio between sand and clay is irregular. No fossil was found. This is a devolutionary layer before transgression with river alluvial sands. Under this layer, it is weathering granite.

According to the sedimentary characteristics of marine facies, the current of the Haixi Bay plays an important role in the sedimentation. The area is by the side of the dominant current and eddy current presents. Current velocity is low and wave action is weak. The area has a thicker layer of marine facies because it is situated in the diffusion area behind the narrow bay mouth.

(2) *Wutong Bay*

It is situated to the north of the Xiamen Island, with bay mouth racing east. As a small tidal bay, it has a water area of nearly 100 km^2, long and narrow. It is 72 km^2 in area in low tide.

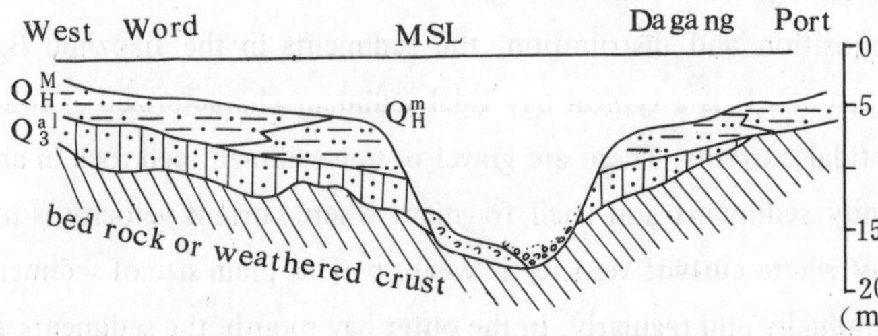

Fig. 2-3. Geological section across the bottom of the Jiaozhou Bay.

Exchange of terregenous fresh water and sea water always takes place in the bay, which is constantly filled with terregenous sediments and organisms. The maximum tidal range is 5.5 m . The maximum current velocity is over 1.2 m/s. In the sourroundings of the bay, there are knolls consisting sand stones of the Mesozoic Era and the Jurassic Period. The land surface is a bit uneven. Original topography in the bay bottom is relatively smooth. Layer is not deep. The area where it is over 5 m deep only accounts for one third of the total area.

Sediments in the Wutong Bay are from the Tong'an River runoff and coastal erosion, the latter accounts for a larger proportion. Around the bay is thick soft weathering crust which is easily to be eroded. Along a section of the coastal line 8–9 km long, there is freshly eroded cliff, more than 10 m high. In this bay, water is not deep and wave is not strong, the sandy clay cliff is very much developed which is seldom seen in other bays. It is worth noticing that under such conditions, modern sedimentary layer in the bay has been thin, indicating that the Wutong Bay was a smooth and even river depression before the transgression in the Holocene Epoch (Fig. 2-4).

Sediments in the bay is characterized by linear distribution, of which the main axis is wholely in accordance with the current direction, showing the flood and ebb currents are the major factors controlling the sediment distribution in the bay (Fig. 2-5). The composition of sediment particle size is the same with the current field.

Artificial factors excert great influences on modern sedimentation in the Wutong Bay. It has been proved by the preliminary investigations that: a) Since the causeway from Jimei to Gaiqi was built, the natural channel between Xiamen and Wutong has been basically blocked and the dynamic cycle system between Xiamen and Wutong Bay has been changed. The change has brought about heavey sedimentation on both sides of the causeway in the past twenty years (Fig. 2-6); b) The building of a series of causeways such as Pantu-Binzhou Causeway has made the bay water area reduced more than 20 km^2 (Fig. 2-5). As a result, the tidal prism of the bay is decreased, weakening the water circle in the bay, evidently influencing the sedimentary situation of the nearby projects, and also weakening to a certain extent the current system in the Xiamen Bay.

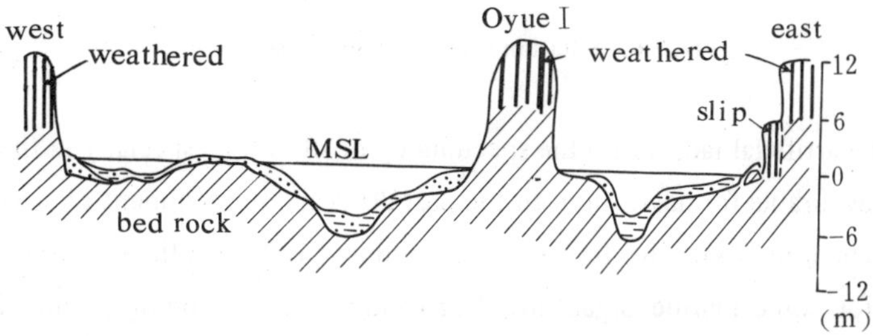

Fig. 2-4. Sediment profile of the Wutong Bay, Xiamen.

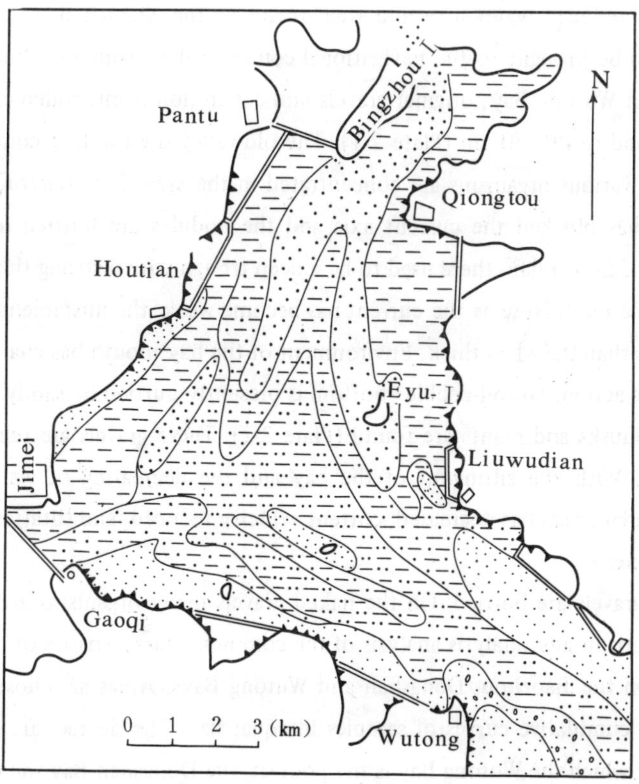

Fig. 2-5. Map of tidal deposits of the Wutong Bay, Xiamen.

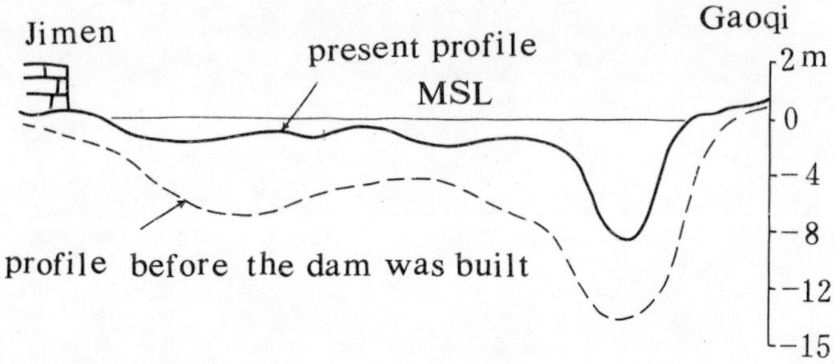

Fig. 2-6. Siltation near Jimei dike, Xiamen.

Results from the artificial factors are the speeding up of siltation, especially in the southwest part of the bay, and unfavourable for navigation. In the past 30 years, the sediment of coastal shallow near the causeways has changed from sand to mud. The sediment is 2–3 m thick, thus destructing breeding environments for some marine organisms. For example, the breeding ground for lancelet has nearly been destructed. Oyster nursery is also under severe threats of siltation. The area of muddy sediment has enlarged due to the reduction of the current velocity. The navigational course without muddy deposit in the past has now a thick layer of silt. The construction of Jimei Causeway and Xinlin Causeway has reduced water area and tidal prism of the Xiamen Bay. Unfavourable consequences are bound to be brought to the navigational course of the Xiamen Harbour. In the old river valley to the northern Wutong Bay, original gravels and concretion from eroded coastal cliff are now burried by muddy sand of 20–30 cm (Plate 2-1). The old valley used to be a course with large flow. Nodules attached by various organisms are concentrated in the area. The construction of the Pantu-Bingzhou causeway has blocked the current axis and the nodules are burried in a 30 cm layer of sands. To the north of Liuwudian, there used to be a deep trough with a strong tidal flow, and a layer of 10–20 cm unstable mud. Now as the current has become weak, the unstable mud has turned into stable sediment more than 0.5–1 m thick. Environment of the bay mouth has changed because of the weakening of current action. Gravel-pebble bottom is presently burried by sandy clay except a few , on which various mollusks and plants are found (Plate 2-2). These gravels are relict sediment at the Wutong Bay mouth. With the silting-up of the bay and the decreasing of water area as a result of reclamation, the tidal prism has decreased, current velocity reduced, and siltation occurs at the bay mouth to a certain extent.

A large part of gravels are burried. On the naked gravels many organisms are found. In order to make a better comparison and analysis analysis of the common characteristics of bay sediment under strong tidal current in the Jiaozhou, Dongshan and Wutong Bays. Areas are chosen from these three bays to compare the cumulative curves of samples in equal size. The degrees are very poor, all being no more than 3.8, of which the Wutong Bay is the poorest, the Dongshan Bay the second. Among the three bays, the Wutong Bay changes mostly in its sedimentary environments.

The reason for poor sorting in the Wutong Bay is that the current system is complex and the influence of river is more conspicous than that of the other two bays. The degree of sorting in the Jiaozhou Bay is the best. The main reason is that the sediment particles from the Dagu River are finer.

2. Bay sedimentation dominated by wave action

The coastal bays in China are mostly of arc or semi-circular ones facing the open seas with wave action being their main controlling factors. The transportation and accumulation of sediments from shore to shallow in this type of bays are mainly controlled by wave action while tidal current action ranks the second place. In contrast with the sedimentation of the above-mentioned three bays, the size of sediments of this type of bay is determined by wave strength. The degree of sorting and psephicity of sediments are relatively good. From shore to the centre of the bay, sediments are mainly sands reduced regularly as water becomes deeper.

(1) ***Longkou Bay***

It is located in the southeastern part of the Bohai Sea, northwestern of the Shandong Peninsula. The bay is separated by a big tombolo aproximately in east–west direction, 8 km long and 1.2–4 km wide. It is an irregular semi-circular bay. The bay mouth faces southwest. At the northern part of the bay mouth is an islet of metamorphos of the Precambrian period consisting of quartzite and gneiss. At the lower layer of the tombolo is yellow sandy soil of the late Pleistocene Epoch, which exists extensively in the northern part of the Shandong Peninsula.

There are two sorts of the Holocene Epoch layers. Roughly speaking, it is bound at five meter isohypse. This isohypse is a maximum limit of the transgression since the Holocene Epoch, namely the oldest coastline . The development of the coastline is on the yellow sandy layer of fluvial-alluvial. Marine sediments of the beach, lagoon and barrier lie under the old coastline.

As a weak tidal bay, the Longkou Bay has an irregular semi-diurnal tide with mean tidal range of about 1.2 m. The water level during storm and spring tide can reach 3.4 m because of strong wind action. The tidal current in the northern part of the bay is obviously to-and-flow tidal current. The maximum current velocity is found near the cape of the Jimu Isle. Velocity of flood tide is 1.2 m/s, ebb tide 1.0 m/s. Current in the inner bay has no apparent reciprocal characteristics. In general, the velocity is not faster than 20 m/s. Flood current comes from the north, ebb current flows westward. Bottom current velocity is 3–5 cm/s, being bigger than surface current velocity.The strongest wave in the bay is caused by the northeast wind. The constant wind is southerly which excerts little effects on the coastal sediments since it is an offshore wind. The observed maximum wind velocity is 34 m/s in NNE direction.

On the northern side of the Jimu Tombolo, the maximum wave height is 4.9 m, more frequently 3–4 m. So long as the north wind is 15–20 m/s, there will be high wave of 3.5–4 m. On the southern side of the Jimu Tombolo, owing to the diffraction of big waves passing through the Jimu Isle, energy is diffused and wave energy is greatly reduced. Wave is only 1–1.2 m in height. Under the same wind conditions, the wave can reach 2 m when west wind is 15–17 m/s.

The main sand source for the coastal sediment is the nearby sandy rivers originated in granite and metamorphic hilly areas. Under heavy weathering and erosion coupled with the riverbed gradient, washing action is strong and content of suspension is high. According to the data of the hydrologic station of Huangxian County, the amount of suspended sediments carried by the Huangshui River into the sea is 200,000 tons. Based on the composition of sand height and the thickness of the sand on the river bed at their lower reaches, the amount of medium and coarse sands carried into the sea in suspension can be estimated.

At the Longkou Bay and its nearby area, bay lagoon layers are widely distributed. According to the differences of its sedimentary environment, the bay lagoon facies can be further divided into the following three kinds.

1) *Shallow lagoon layer.* It is light green-grey to grey, loose sand and clay-sand layer, of which a large part is coarse and medium sands, with a small quantity of silt and clay mixed by many species of shells. On the upper part of this layer, the quantity of shells graduly becomes less. The lower layer is sandy terrestrial alluvial facies.

This kind of lagoon sediment lies on the inner side of the spit, i. e. the depression to the east of the Longkou Town. The size of sedimentary layer is $M_z = 0.5–2.8\ \phi$ fining-downward. The degree of sorting is good. This layer is distributed at the depression below the old coastline to the east of the Longkou Town. Its area is about 5 km^2 and thickness 1–1.5 m (Fig. 2-7).

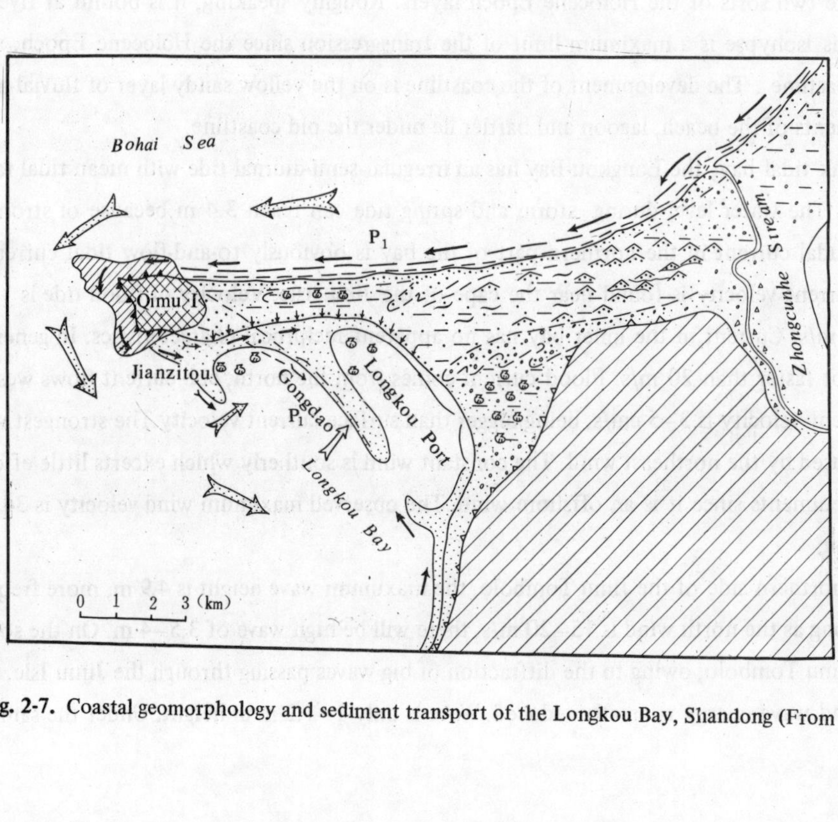

Fig. 2-7. Coastal geomorphology and sediment transport of the Longkou Bay, Shandong (From [27]).

2) *Lagoon peat bed.* It is black or dark grey sandy-clay layer. In which there are abundant organisms. Sediments are largely clay, with little quantity of silt-fine sand, about 1.5 m thick. the lower layer is terretrial yellowish sandy-clay layer. This lagoon peat bed is M_z= 2.5–3 ϕ. The degree of sorting is moderate. The top part is burried by wind sand (fine sand). This layer does not contain any shell. Its distribution area is at the northern part and in a form of east-west direction. It is more than 5 km in length, 300–400 m in width. At the southern coast of the Longkou Bay, peat bed can also be found.

3) *Bay sedimentary layer.* It is green-grey sandy clay, loose and stinking. The main component is clay, with silt and fine sand. the degree of sorting is poor, and M_z=3.3–5.9 ϕ. This is the sediment of modern sequence in calm water environment containing many species of shells. This layer is 2–4 m thick. Its lower limit is often 5.5–3.5 m. The lower layer is burried by marine sand or by alluvial layer of terrestrial yellow sand-clay. It is situated in the inner harbour of the Longkou Bay. At the inner side of the Jianzhitou, this layer also exists locally. As the large tombolo is gradually increasing, part of it has been burried by barrier sand (Fig. 2-8).

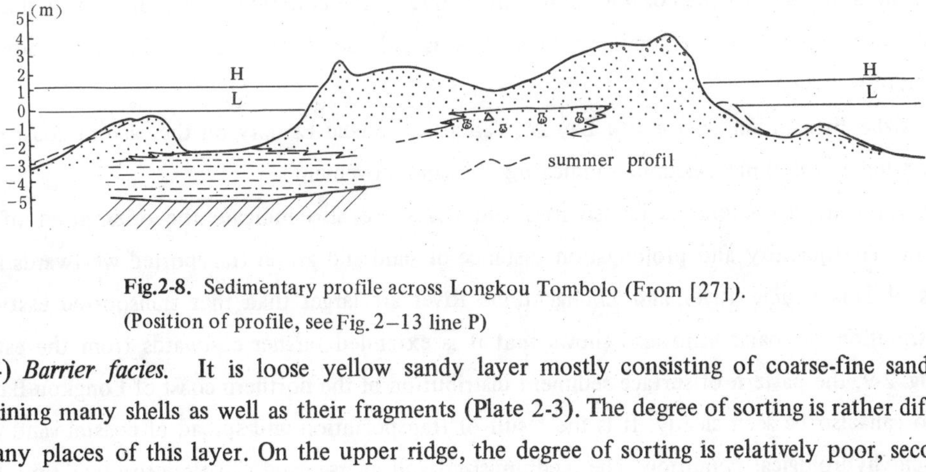

Fig.2-8. Sedimentary profile across Longkou Tombolo (From [27]).
(Position of profile, see Fig. 2–13 line P)

4) *Barrier facies.* It is loose yellow sandy layer mostly consisting of coarse-fine sands and containing many shells as well as their fragments (Plate 2-3). The degree of sorting is rather different in many places of this layer. On the upper ridge, the degree of sorting is relatively poor, second in the beach. The best degree of sorting is at the ridge, where the particle of sediment is comparatively fine and M_z is equal to 2.0–3.0 ϕ.

The size of barrier sediment is closely related to wave strength. Under the general conditions, the particle diameter is in direct proportion to wave strength. For example, at the northern coast of the tombolo, the M_z is equal to 0–0.5 ϕ, while at the southern coast, it falls to 0.5–1.2 ϕ. Near the high water line, the M_z is bigger than that of the low water line. The size becomes correspondingly smaller with the deepening of water because the wave action gradually becomes weaker.

The study of other coastal areas has proved that the height of coastal deposit geomorphology and material composition (particle diameter) are the indicators of wave strength.

In different bays, on both sides of a tombolo in the sea areas, the wave strength varies greatly, because of the different coastal contour and the distribution of islands. The differences of wave lead to the differences in the height and gradient of accumulation forms and size ot particles, such as the Luhuitou Tombolo in the Hainan Island with a reef flat of 500–800 m wide at its western coast playing the role of wave breakers, while the eastern coast facing the open sea with a discontinuous reef flat of less than 100 m wide. So the accumulation ridge in the eastern coast is 1.5 m higher than that of the western coastal ridgo. The coral fragments of the eastern coast is composed of much coarser ones than those of the western coast.

The above data provide us with an important revelation to the recognition of the height and size differences on the southern and northern coasts of the Jimu Tombolo. The height difference between the southern and northern coast is over 1.3 m. The grain size in the southern coast is also smaller by one ϕ number. This is the result of wave strength difference of the two coasts.

The sand sources of the Longkou Tombolo is the longshore drift from northeast. On the northern side of the tombolo the longshore drift flows from east to west when northeast wind blows. On the eastern part of the tombolo, the coastline is in the northeast orientation. As for this shoreline, sand is transported westwards not only during northeast wind, but also when the wave crest line is oblique to the coastline in an acute angel. The sand is moved southwestward into the tombolo. There is further evidence proving that the tombolo sand is from northeast coast. The estuary shape indicates the movement of the coastal sands. This is one of the commonly-used analysis method. No matter the estuary is large or small, so long as it is oblique to one side, it shows the dominant movement of sediment.

The Jiahe Estuary on west Zhifu Bay of Yantai and Shahe Estuary on the eastern coast of the Laizhou Bay are both typical examples indicating the sand movement.

The sediment characteristics in estuaries and nearshore also indicate the orientation of sand movement. The quantity and prolongation distance of sand and gravel transported westwards in the estuaries of Huangxuihe River and Zhongchunhe River are larger than that transported eastwards. The distribution of coarse sand also shows that it is extended farther eastwards from the estuary. From Fig. 2-9, the pattern of surface sediment distribution of the northern coast of Longkou Bay and Tombolo can also be seen clearly. It is the result of transportation and spread of coastal sand under the present hydrological condition. The medium-sized and coarse sand can flow through Jimu beach without being obstructed. As to fine sand, it is even distributed continuously from the northern coast to the inner bay. While to the south of the Longkou Town, little amount of coarse and medium sands have entered the bay. The finest sediment of the Longkou Bay is behind the Guandao, north of the inner bay. It is newly silted, soft and fine (silty clay), containing a little shell organisms.

Under storm wave, turbidity water mass is readily observed and its movement direction is an important symbol for judging the transportation of fine sand. The turbidity water area at the northern coast clearly shows two zones. Sub-turbid zone II is immediately close to the coast. The most turbid area is turbid zone I, which is 500 m offshore and parallel to the coastline. This zone is 2–3 km wide. Because the nearshore zone is composed of medium-fine sand, being rarely in suspension state, the turbidity is not great, while the silt-clay at 500 m beyond the coastline is easily stirred by storm wave and becomes suspension. It is also hard to settle down and is carried into the Longkou

Bay bypassing the Jimu Isle along with the current.

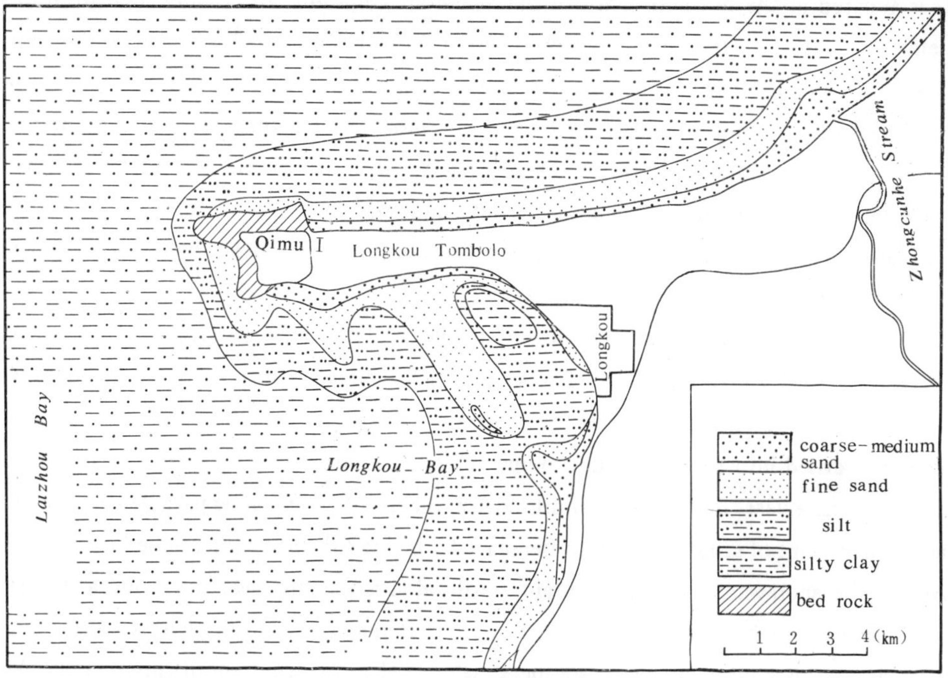

Fig. 2-9. Distribution of sediments in the Longkou Bay (From[27]).

To the east of the Jimu Isle, the coastline is straight. There isn't any obstacle protruding into the sea and cutting the sandy drift. Under the action of the predominant northeast wind in the zone, the sand on the northern coast can flow westwards along the coastline and enter the western coast of the Jimu Isle. After the northeast wave is diffracted, the energy is greatly reduced, but is can still carry part of sands southwestwards or southwards. Partibularly, under the action of north and northwest wave, the sand in the west Jimu Isle is transported towards the Longkou Bay. Now, the southern coast of the Jimu Tombolo is continuously accumulated. The mud in the Bay is coverred with sand ridges. This is the most powerful evidence for the sand at the northern coast by-passing the Jimu Isle and enters the southern coast (Plate 2–4, Fig. 2–8).

By drilling in the Longkou Bay and at the top of subaqueous bar (Guandao), we got the following results: a) the bar is moving towards the Bay, of which the clear proof is that the lagoonal sediment is burried by sand bar; b) In the inner side of Guandao sediment is a mixture of sand and clay. It is sediment in a calm water environment containing many organisms; c) At the outer Guandao, the slope of the bar is gentle (1:500), the inner slope is steep (1:10–20). There is a great disparity in the slope on both sides, showing that Guandao is moving towards the Bay from its slope and layer sequence. In winter, namely when the north wind is strong, on the northern side of the Longkou Tombolo, sand is transported westwards, but transverse moving of sand is also generated. During this period, the movement of rolling sediment on the northern coast is a zone of 500 m wide.

(2) ***Zhifu Bay***

It is situated at the northern coast of the Shandong Peninsula. An accumulated form of three kilometers long, consisted of sand and gravel, is developed between the Zhifu Island and the land. It is called Zhifu Tombolo. The Zhifu Island is a bed rock island, being approximately in N 60° W direction, like a long rhombus. Its long axis is 9.2 km, short axis 1.5 km. The Zhifu mountain rock is composed of "Zhifu System" of the Archaeozoic Era, mainly of gneiss, quartzite and schist. The rocks are comparatively hard, but 0–joint is multiple and fragile.

The land linking the Tombolo is a triangular sandy accumulation plain. Behind it are low hills in the Yantai City, their heights not exceeding 200 m.

The Jiahe River is the biggest river flowing to the sea in this coastal area, originating from the Chixia County, with a drainage area of 2340 km^2. In the drainage area, there are mainly low hills. According to surveys and calculations of data from the hydrological stations in the Yantai Prefecture, the materials carried into the sea amount to one million tons annually. The maximum amount can be 1.53 million tons a year. Because the sand is coarse, the quantity of rolling particle is not calculated. So the actual amount of sand flowing into the sea is far beyond the above-mentioned figures.

Among the shoreward winds in Yantain, north wind is the strongest, east wind and northwest wind the second. When wind velocity is >12 m/s, the frequency of north wind is much more than south wind. At this section of coast, southeast wind and south wind are offshore winds. Wind is weak, and the action of shore erosion and sand transportation is little.

The main dynamic element in this section of coast is wave. The north coast of the Zhifu Island is just facing the north Huanghai Sea. There is no obstruction on the nearshore, so big waves can reach the coast directly. According to local fishermen and boatment and surveying results by using wave meter, at the place of 200–400 m off the north coast of the island, wave length is around 35–45 m, sometimes being 50 m, wave height being 2.5–3 m when north wind is 15–20 m/s. The east Zhifu Bay has become a small semi-closed Bay, because of the obstruction by the isles at the outer bay mouth, such as Kongdongdao, Dandao, Madao Isles. Under the similar climate condition as the wave length at the southern Bay is 20–25 m, wave height is 1.5–1.6 m. In the middle section of the Bay, the wave length is 15–20 m, wave height 1.3 m. In the northern section of the Bay (near the Zhifu Island), the wave length is 13–16 m, wave height around 1 m.

West Zhifu Bay is much wider than East Zhifu Bay. Its arc degree is little and without obstructions by isles. Under the same climate conditions, the wave length can be 30–40 m, wave height 2–2.5 m. On both sides of East and West Bays, usually, the wave crest line is diffracting forward from the Bay mouth to the inner Bay. The wave crest line is oblique to the coastline in an acute angle. When the wave crest line reaches the inner Bay, it is almost parallel to the coastline.

In the Zhifu Bay, the tidal current velocity is small. During spring tide and when velocity of north wind is 5–8 m/s, the maximum current velocity is 18–20 cm/s. Generally the current velocity is only 6–8 cm/s. Such a small velocity in the Bay, plays little role on the transportation of sediments.

Zhifu Tombolo is very famous in China. The width of the Tombolo varies in different places, with both ends wide and middle narrow. By calculating the upper part of MSL, the lagoon on the northern side is 860 m wide. At the narrowest, it is only 520 m wide. From here southward, the Tombolo is gradually widened and linked to the Xishawang deltaic plain of sandy accumulation of Yantain.

The Tombolo is mainly composed of medium and fine sands (on the average its size is 2–3 ϕ). It is a sandy accumulation form higher than MHL. There are four gravel and sandy gravel ridges on it. The western two ridges are higher and longer. They are 3.8–4.0 m higher than MSL, and are entirely composed of coarse gravels. The eastern two sandy gravel ridges are lower and shorter, the height of which is 2.8–3.0 m higher than MSL, being composed of small gravel and sand.

At the juncture of the northern part of the Tombolo and the Zhifu Island, there is an arc-shaped gravel accumulation ridge. Within the arc-shaped ridge, there is an old lagoon. It is composed of green-grey or dark grey silt-clay and clay-silt, containing abundant organisms and shells. The shells are:

1. *Venerupis Philippinarum* (Adams & Reeve);
2. *Mya arearia japonica* (Linnaous);
3. *Macoma incongrua* (Martens);
4. *Ostrea* sp.;
5. *Batillaria zonalis* (Bruguiere).

All these organisms are living on the clay-silt (sand) intertidal flat and shallow. Thus it can be seen that, before the formation of the arc-shaped gravel ridge, it used to be a shallow here.

For the convenience of statement, we arrange the four gravel and sandy gravel ridges on the Zhifu Tombolo from west to east in an alphabetical order as A, B, C, and D. A, B ridges belong to West Group; C, D ridges belong to East Group. The data of the two groups are shown in Table 2-1. From Table 2-1, one can see that the dimension of the two groups are quite different. As for volume, the East Group is one third less than that of the West Group. Regarding height, the East Group is one meter lower than the West Group. In size and psephicity of gravel, the East Group is also inferior to the West Group. The cheif cause is that the wave parameters in East Bay is weaker than that in the West Bay. The erosion and accumulation actions in the East Bay are also inferior to the West Bay and shape of chenier ridge are basically the same with Jiahe River mouth.

Table 2-1 Data on the two groups of sand-gravel ridges on the Zhifu Tombolo

Items / Groups	Height (m) M.S.L.	Length (km)	Volume ($10^6 m^3$)	Characteristics		Gradient outside of the ridges	Shape
				Size (cm)	Psephicity		
West group	3.8–4.0	3.5	1.20	5–10	0.6–0.7	>15°	forked
East group	2.8–3.0	2.1	0.35	1–2	0.2–0.3	<12°	parallel

(i) *Sedimentary Characteristics of the Bay*

The composition of sediment sizes on the shallow of East and West bays of Zhifu Island is similar. Both are composed of silt and fine sands. However, the sediment is coarse in West Bay, containing more fine sands. From the beach extending to the subaqueous slope, the depth of sedimentary distribution of the two Bays are quite different from each other, judged from the range of the same size of sediment.

In East Bay of Zhifu, the distribution of fine sand in the southern section extends to one meter below the low tide line. While in the middle and northern section, the fine sand only reaches the low tide line. In East Bay, the silt is distributed to two meters below the low tide line in the south and one meter below the low tide line in the north. The lower part is clay silt (Fig. 2-10).

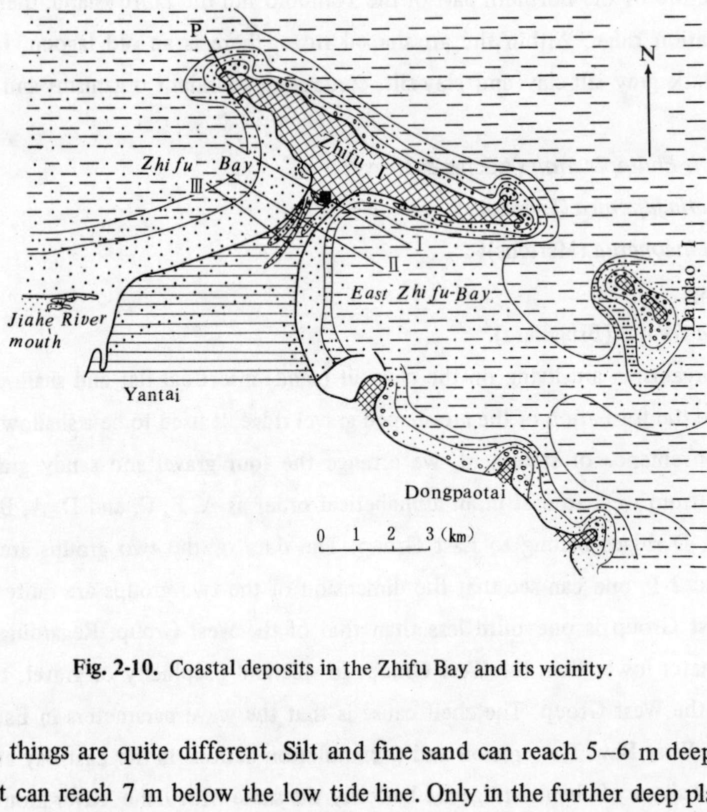

Fig. 2-10. Coastal deposits in the Zhifu Bay and its vicinity.

In West Bay, things are quite different. Silt and fine sand can reach 5–6 m deep below the low tide line. Clay silt can reach 7 m below the low tide line. Only in the further deep place can silt clay be found.

Obviously, the distribution of sands is 4–5 m deeper in West Bay than in East Bay of Zhifu. This phenomenon must be analysed in accordance with wave parameters.

The silt and fine or medium sands on the subaqueous slope are frequently sorted by waves. Its charactristic is that size grading is even and degree of sorting is below 1. The lower limit of silt distribution with good degree of sorting indicates that it is a place where wave action often takes place and sand is stirred up. It is only when the north wind velocity is over 15 m/s and water depth is close to 1/2–1/3 wave length that the clay-silt and silt-clay are stirred up and thus turbid water occurs.

The above-mentioned distribution charactristic of sediments fully shows that the wave strength in West Bay is stronger than that of East Bay. The depth of wave action is also bigger than that of East Bay. The main reason is that, at the mouth of East Bay, there are a series of isles such as Dangdao, Madao, Kongdongdao, etc. These isles play the role of wave breakers.

West Bay is an open bay. The parameters of wave are bigger than that of East Bay. The subaqueous slope is bigger. So, the distribution of silt, whose degree of dispersion is good, is also deeper. The conclusion can be drawn that, at the sea floor slope of loose sediment, the wave strength is the chief dynamic factor determining the sea floor gradient and the distribution of sediment size.

The second characteristic of nearshore sediment distribution is that, at both the promontories and other capes, there apparently exists an irregular subaqueous oval trough. We call it "nearshore trough".

On the nearshore trough, the sediment is sand, sand-gravel and gravel. At the some places the bottom is of colossal stone and bed rock. The diameters of gravels are quite different from each other. Shapes are various and psephicity is poor. The gravel features in the nearshore trough are quite different from those on the beach and gravel ridge. The reason for this specific forms and sediments under the water of the cape is that, when strong wave occurs, the huge wave crest line coming from the open sea converges when pushied towards the cape. Wave energy is concentrated. Nearshore drift and back current are produced here. At the same time, the tidal current is pressed, together with confluence of nearshore drift and back current, forming the strongest current far surpassing the current velocity of normal coast.

Under the strong wave condition, the strongest confluent current dashes the steep cliff and subaqueous slope of cape, washing fine particles in the nearshore trough away to distant places and the coarse gravels remain. At the same time, the gravels dashed by the strongest confluent current were rarely moved. But there is sand surrounding the gravel. Therefore, the extent of the gravel transportation is not beyond the limit of the sand distribution. The sand-gravel in the trough can not be transported to the Zhifu Bay to form the gravel ridge of Zhifu Tombolo. Meanwhile, by analysing the surface charactristic of the gravels collected from the trough, we often found some mollusks attached on the surface of the gravels, such as *Spirobranchus tricornigerus* (Grube) etc. This shows that the gravel here is not always in a state of being stirred.

In 1958, we made an investigation on a coastal project in the southern coast of the Shandong Peninsula and collected some gravels at the nearshore trough by the cape. On the uneven parts of the gravel surface, there are some fine sands and clays. This shows that fine materials can still deposit in the nearshore trough when sea water is calm. Once strong wind and wave occur, these temporary fine materials on the surface of the trough are stirred up by strong confluent currents and carried to some other places (Fig. 2-11).

(ii) *The main source and dynamic of sediments*

The Jiahe River is the main river along this coast. It empties into the sea over one million tons of materials annually, among which coarse sands are accumulated at the beach and the upper part of the subaqueous slope. Fine sands and clay are carried to deep water. The materials of the Jiahe River are sufficiently sorted by wave action with the deepening of water. At the same time they are longshore transported. The above conclusion can be drawn from the depth of sediment distribution in the profiles of Fig. 2-12, and from the topography and dynamic process of coastal geomorphology. By analysing the bay wave data it is proved that the wavecrest line of the Jiahe River is slant to the coast line in an acute angle (≤ 15°), with its arrow pointing to the Bay. The wavecrest line of the Jiahe River mouth is slant to the coast line in an angle of 30°–40°.When northwest wind or west wind prevails, the arrow is pointing to the east. When the northwest wind velocity is 14–17 m/s, the height of nearshore wave is 2 m, wave length is about 20 m before the nearshore wave is broken. There are three breaker zones at the nearshore. In front of the first breaker zone, the wave height is 2.3 m; at the second breaker zone, the wave height is reduced to 1.6 m; the third breaker zone is close to the shore, its wave height is only 1–0.8 m. In summer, wave is weak, the discharge of the river is large. The third breaker zone near the shore is intercepted at the river mouth. In winter, the discharge of the river is low, and wave is big. The third breaker zone continuously extends eastwards.

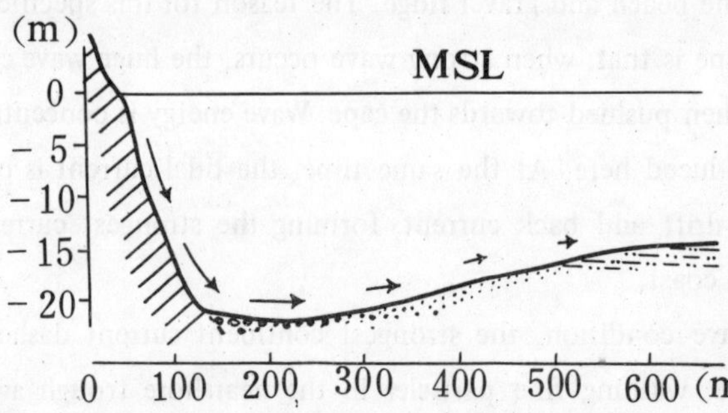

Fig. 2-11. Section of a nearshore trough at the western end of Zhifu Island (From[23]). (Position of section see Fig. 2-10)

The above information has strongly shown that the three subaqueous sandy bars are the reflection of continuous distribution of the three breaker zones from the Jiahe River mouth towards west Zhifu Bay, and has also shown that the sand at the West of the Jiahe River mouth is transported eastwards, when west wind or north wind blows. Just as many researches have proved, the most active movement of coastal sand is at the subaqueous ridges. It has been found out that breaker zones are the subaqueous sandy bars. On the northern coast of West Zhifu Bay, there is also a certain quantity of sand sources. This drift is chiefly gravel, its sources are three: a) The loose proluvial and eluvium on the hill slope is eroded and different sizes of materials are brought to the coast through several gullies. At present, there is a minimized triangular accumulation fan at the mouth of the largest gully. It was retransported and reaccumulated under the action of stormy wave to become two small feathery gravel spites; b) The bed rock at the southern coast in west Zhifu Island is eroded and become fragments, and the outcome of friction between the gravel and bed rock on the bench; c) The soft soil cliff is ceaselessly eroded backwards. There are considerable amount of gravel and sand cillaped from the cliff by the erosion of wave, thus also adding to the sources of drift to the northern tributary in west Bay. This sand-gravel drift is the main source for the west group gravel ridges of the Tombolo.

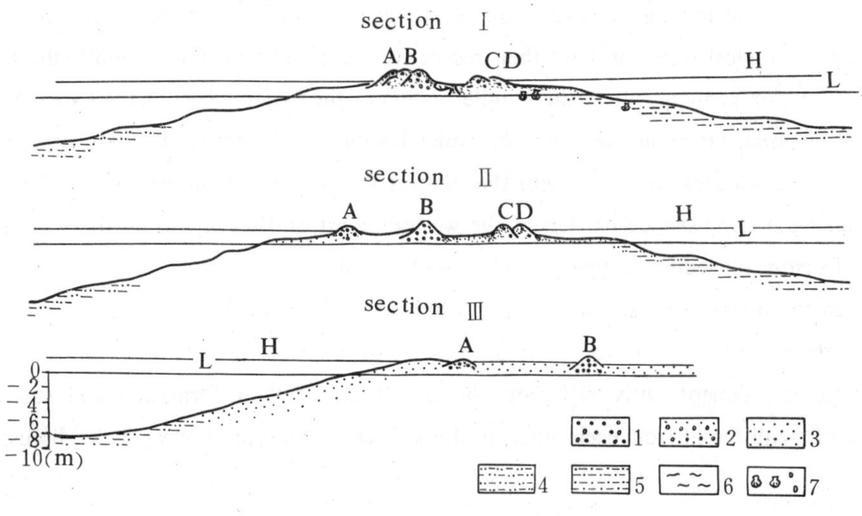

Fig. 2-12. Cross-sections of Zhifu Tombolo (From [23]).

(Position of sections see Fig. 2-10 I, II, III)

1. Gravel; 2. Sand-gravel; 3. Medium sand; 4. Silt; 5. Silt-clay; 6. Organic deposit; 7. Shell and detritus.

The gravels in West Bay of Zhifu are transported from north to south. While sand is transported from the Jiahe River mouth towards the inner bay. The two drifts move in an opposite direction. Such a case can be found both in China and in other countries. For instance, at Dungeness in the south-eastern coast of England, sand and gravel drift moves in an opposite direction. Small gravels are moved westwards under the action of normall wind, while large gravels are moved eastwards under the action of strong southwest wind with a long fetch.

IV. Coastal Sand Bodies and Their Origin

Coastal spit, ridge, barrier and tombolo are accumulation bodies of sand-gravel coasts commonly and widely distributed in the world. Most of them are on gentle coastal plains. For instance, at the eastern coast and southern coast of the United States, such accumulated sand bodies are well developed. The barrier coast in the United States accounts for 47% of its total coastline. At the coasts of Alaska, Holland, West Africa, Southern part of Brazil, Southeastern part of Australia, Mexico, the Black Sea and the Caspian Sea, there are considerable-sized sand bodies.

The sand-gravel accumulation bodies of China are not so well developed and widely distributed as that in North America. In particular, its dimension is relatively small. In addition to the western coast of Taiwan, western coast of Liaodong Bay and the coasts of the Hainan Island where there are large-sized, long-extended sand-gravel accumulation bodies. There are various small-sized sand-gravel accumulation bodies distributed along various sections of the coast, owing to the complexity of geological structure of China's coast, uneven topography, numerous bays and capes. Among them, the various sand-gravel accumulation bodies of the eastern coast of Laizhou Bay, eastern section of the northern coast in Shandong Peninsula, the northern coast of Haizhou Bay, southern Fujian and Guangdong are typical ones and have their respective characteristics. For example, the Diaolongzui system of complex pennate spit at the middle section of the eastern coast of the Laizhou Bay, Zhifu Tombolo of Yantai, Longkou Tombolo, the Gulei Tombolo of Dongshan Bay in southern Fujian are of distinquishing features compared with the same type of coastal accumulation bodies in the world. In addition , the chain-shaped barriers at the western coast of Taiwan and northern coast of Bohai Gulf are of significant sediment type in modern sedimentology.

Though the origin of coastal spit, barrier, tombolo and barrier island are approximately the same way under the same dynamic process, because their forms are extremely complex and varied, with the composition of sediment fairly different, the conditions of their formation and stages of their development are different from each other. In this section, some typical ones will be discussed.

1. Coastal spit

(1) *Sentougu pennate shell spit (chenier)*

The Sentougu pennate shell spit in the western coast of Bohai Bay is developed on the clay intertidal flat. It has four shell spits slanting in order and forming a system of shell spits. The pennate shell spit is mainly composed of materials of shell and its fragments, mixing with a small quantity of silt. The height of the shell spit from the foot to top is about 2 m.

The gradient of the ridge facing the sea is 15°–19°, the gradient facing the land is 11°–14°. There are species of salt-tolerant plant such as jujube and weed growing on the ridge and back slope. In the past eighty years, the chenier spit retreated landwards. At this period of time, Jiyunhe estuary was in the process of being gradually silted up. Therefore, when the shell fragments are transported forward and accumulated as a new spit on the already silted clay beach, the new spit is at the outside of the old spit, thus a fork is formed. As there have been multiple forks, a pennate spit is formed.

It is proved by interviews and profiles of beach sediments that the retreat of Sentogu village began in 1920's at 500 m east of the present village. The Chenier ridge was high and thick, with its width being more than 50 m . The quantity of silt was subsequently increased in the bay, shell organisms decreased. It is also because the slit layer of tidalflat and nearshore area has become thick, so the gradient of shallow area reduced, thus greatly weakening the wave energy. The greatly reduced shell organisms on the tidal flat and nearshore areas can not be transported to the coast, unable to supply the Chenier ridge with materials.

At the time of the retreat of Sentougu shell ridge, the sediment at the Jiyunhe River mouth has become higher because of the growth of silt in the bay. Under the predomimant wave caused by the action of northeast wind, the shell fragments eroded from the northern section of Sentougu shell ridge are transported southwards. The shell materials can only be accumulated at a certain position on the tidal flat corresponding to the wave energy. Because the tidal flat of the Jiyunhe mouth has ceaselessly become higher, the later-transported shell materials are accumulated in front of the old shell spit, where the tidal flat is being heightened, thus forming a new spit (Fig. 2-13).

The process of the heightening of the tidal flat is gradually and slowly carried out under calm climatic conditions[1)], while the fork of the shell spit is formed in a storm. It is revealed from the profile that the new fork of shell spit is developed when the originally heightened tidal flat has been eroded by the storm. A zone of relatively big slope is created. At this time the shell from the north is accumulated and forms a new ridge under the action of northeast wind wave.

The pennate spit of Sentougu shell ridge was formed under the condition of erosion in the northern part and shell transportation southwestwards. Besides the old ridge eroded and transported by waves the direct action of northeast wind is also an unneglectable factor. It brought shell fragments along the spit and did not transport the shell towards the depression behind the ridge. At present, there is basically no shell ridge at the Sentougu village. Therefore there is no natural fresh

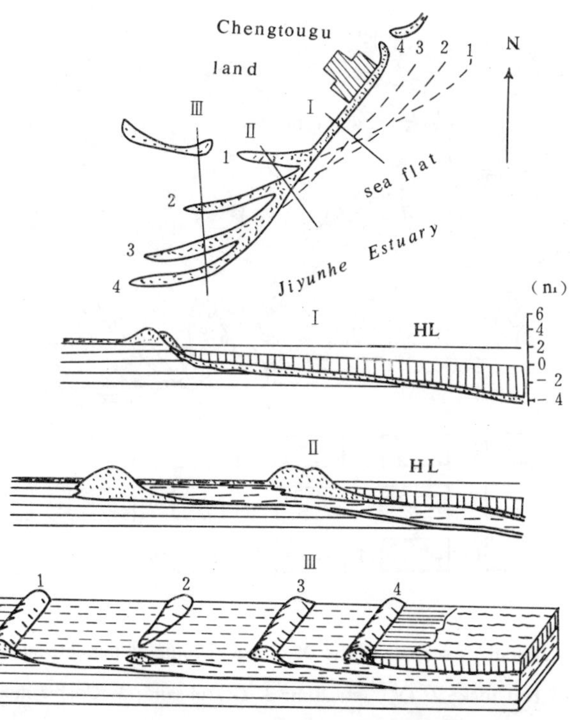

Fig. 2-13. The develpment patterns of pennate chenier at Sentougu along the western coast of the Bohai Gulf (From [25]).

1) Cai Aizhi, The dynamic process of erosion deposit of tidal flat. 1979 (unpublished).

water supply, the village is gradually.

(2) *Diaolongzui system of complex Pennate accumulation*

Diaolongzui is located at the middle section of the eastern coast of the Laighou Bay (37°25′N, 119°50′E).

The accumulation system at Diaolongzui consists of four pennate spits, their corresponding lagoons and some sandy ridges, together forming a complex accumulation body (Fig. 2-14). It can be divided into two parts: (i) Spit, barrier, ridge and their adjacent beaches. The composing materials of these accumulation bodies are chiefly coarse and medium sands in the northern part, along with small gravels. In the southern part, there are mainly fine or medium sands. Their dividing line is at the protruding place of Diaolongzui. (ii) Modern lagoon and old dry lagoon. In the northern lagoon, there are clay sand and a small quantity of coarse sand. In the southern lagoons, there are chiefly fine sands and silt clay. Their degree of sorting is very poor, all mixed with shells and abundant organisms.

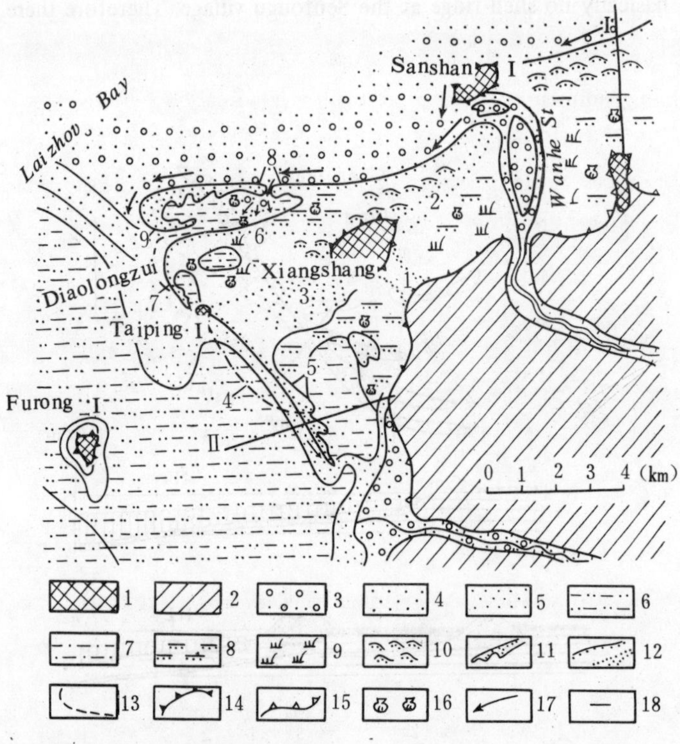

Fig. 2-14. Systematic scheme of the Diaolongzui complex pennate spits (From [24]). 1. Island of bed rock; 2. Q_3^{pdl} stratum; 3. Gravel-sand; 4. Coarse medium; 5. Fine sand; 6. Silt; 7. Silt-clay; 8. Lagoon-swamp; 9. Reed; 10. Dune; 11. Feathery spits; 12. Coastal and abandoned barrier; 13 Original position of spits; 14. Active cliff; 15. Relict cliff; 16. Sampling point of shells; 17. Direction of drift; 18. Sequence of morphological forms.

The subaqueous sediment is a gentle shallow composed of coarse sand and gravel at the northern part. The Laizhou shallow extending in NNW direction from Diaolongzui is composed of medium and fine sands. The degree of sorting is quite good. The top of the shallow is at about –1 m. At the southwestern side of the shallow is silt of very good sorting. In the middle, there is a narrow zone of fine sand. Obviously, the shallow itself is an abruptly changing zone of sediments.

The coastal sand ridge north of the Sanshan Island is composed of medium and fine sands, with a small quautity of coarse sand. The lagoon depression behind the ridge is filled up by fine sands of even size and it is almost one metre thick. The lower layer is composed of silt and fine sands, with small quantity of clay, and large quantity of shells. We collected a sample from this shell-rich layer below the surface of 1.2 m at 100 m west of Sanshan. The measurement by ^{14}C is 3780±210BP.

The accumulation body at the southern coast of the Sanshan Island is complex and various. Fig. 2-14 concisely shows the interrelations between these accumulation bodies and the process of coastal sediment since the Holocene Epoch. The accumulation bodies are developed in numerical sequence from ①–⑨.

According to the characteristic of sedimentary layer and marks of geomorphological form, there exists an old tombolo ① between Chanshan and the land. The accumulation bodies ②, ③ is composed of the same sediments. They are all mainly coarse and medium sands, with small quantity of gravels. When the ③ was formed, the longshore drift was carried to southwest in large quantity along the underwater slope of the ridge, and, bypassing the Taiping Isle, entered the shallow with fine sand and silt bottom. Thus the spit ④ was developed. Later on, the development of ⑥ had cut the sand supply, making ④ retreat to the present position of ⑤. At the northern shallow of ③, the size is –1 ϕ. Under the wave action, the sand particle bypassed the Taiping Isle and entered ④, its size is 2.5 ϕ. Besides, the height of ④ is 0.5–0.7 m lower than that of ③. The difference of the above two factors proved: (i) Wave energy is weakened after by-passing the Taiping Isle. The capability of transporting sand is also weakened; (ii) The height of coastal ridge is accordingly reduced after the wave energy is weakened. The height of coastal ridge depends on the wave strength, so the height of nearby accumulation body can be quite different. It is the same case with the heights of the various spits at Diaolongzui.

The accumulation body ⑥ began developing, making the drift hyphen originally moving westwards along ③ change its movement westwards to be along ⑥ and began building up of new accumulation bodies. But under the action of north wind wave, there is still partial sand transported southwards continuously, because sand source is little, leading to gradual extension of ④ and erosion backwards. On the slope of the ridge, a lagoon layer has appeared. The dynamic principle of this development process of pennate spit is the same with that of the formation of the pennate shell ridge of Sentougu in the Bohai Gulf.

As for the pennate spit ⑦, under the action of north wind, there is still a part of sand transported southwards and, bypassing the Taiping Isle, entered the accumulation body ⑤. In the development process of this small pennate spit, the input sand is now less than output sand, so that ⑦ is now in a state of retreating backwards. In Febuary 1979, under the action of strong north wind, the middle section of the ⑦ was washed to form a tidal inlet, and turned the section into a small barrier. In autumn of 1939 a big gap in the middle section of the ⑧ was opened in an extraordinary wind wave. Later on, in the lagoon at the mouth of this inlet, a small-size tidal delta composed of medium, coarse sands and small gravel was formed. Its origin as expounded by Swift (1975) is that both the flood tidal current and ebb tidal current can form this type of delta. Under the wave action, the ebbing tidal delta at the outside of the inlet is un-apparently removieded by waves. Hoyt pointed out that this type of delta caused by flood current is moved with the movement of sand barrier under the action of predominant longshore waves. Although this longshore movement is quite slow, the small tidal delta has, in the process of movement, made the lagoon accumulated with an irregular sand layer of different thickness.

In the middle section of the pennate spit ⑧, the ebb current in the inlet is weak, the water quantity is far less than that of flood tide. The strength and opportunity for transporting sand out of the inlet is small. In addition to the strong action of wave, the tidal delta out of the inlet is not developed.

The erosion and accumulation at the terminal end of the pennate spit changed rather rapidly. According to our repeated observations and comparison of land marks, the terminal end of ⑥ retreated 50 m from 1952 to 1976. The retreat averaged two meters annually. In recent years, because of the increasing accumulation of ⑧ and ⑨, the retreat of ⑥ has slowed down. In the same period, the western terminal end of ⑧ has extended 110 m. The yearly increase averaged 4.4 m. This figure can only show the modern sedimentation. The speed of the past sedimentation can not be estimated from it.

In treatise on the studying of coastal sedimentation and geomorphology, many typical forms and sediment process of pennate spit have been disussed. For example, Dungeness[61] in southern England, the southern and eastern coast of Australia, the Azove sea and the Caspian sea of the Soviet Union all have typical pennate spits. It should be pointed out that the Diaolongzui pennate spit is comparatively complicated. Its forming condition and developing process are quite different from other spits. So it needs to be discussed separately.

The development of accumulation system of Diaolongzui depends on the following conditions:

1. Stable longshore sand supply for development of the spit. At the eastern coast of the Laizhou Bay, there exists a longshore drift moving southwestwards from Longkou, but the main sand supply is from the Wanhe River.

Medium and fine sands dominate in the sediments of the eastern coast of the Laizhou Bay. We may conclude from this and the main direction of wave action that the sand current flowing from the north of the Sanshan Island towards west and east is relatively active. Its width is 2.5 km, reaching –8 m underwater. Although the Sanshan Island protrudes like a cape, it does not block the sand entering the Diaolongzui coast from north.

The nearshore wave at the north of Diaolongzui Spit is big. The maxiumu wave height exceeds 3.2 m. Waves of 3 m often occur in autumn, the direction of strong waves is N, NE. The waves from the west are weak. Under the same north wind of 20 m/s, the wave height is less than 2.4 m. This proves that the coastal sand north of the Sanshan Island is transported from northeast to soughwest. Only a small quantity of sand is transported from west to east. At the coast south of Diaolongzui, wave strength is even smaller. Under the same condition of above wind strength, the nearshore wave height is only 1.5 – 1.8 m. The wind from northwest, west and southwest can not induce big waves since the wind is weak and its fetch is short.

At the nearshore north and south of Diaolongzui Spit, obvious weakening of wave strength is observed from the following facts:

1) There is the Laizhou shallow extending from the accumulation body ⑧ at the north of Diaolongzui towards northwest. When the north wind waves reach the shallow, part of the wave energy is consumed on the shallow because the water is only one metre deep there. 2) By analysing the verified wave height at the two places south of Diaolongzui Spit, one can see that the waves, having beer broken with its energy partially consumed at the shallow, meet at the turning direction of the coastline, when the wave crest line undergoes a diffraction over 90° in which wave energy is reduced about a half.

The wave strength is corresponding to size distribution of nearshore sediments (Fig.2-14). Coarse and medium sandy bottom mixed with small gravel is distributed on the subaqueous ridge and, at its northern strong wave zone, wave energy is partly consumed after passing through the subaqueous ridge. Then medium and fine sands appear. At the southern part where waves are greatly weakened after diffraction, the silt bottom is corresponding to the weak wave zone. A thin layer of clay exists only on the suhaqueous bottom surface, –3 m at the southern part of the silt. More over, it must be two to three days after big wind wave is over that a layer of 2–5 cm yellowish thin and dilute clay, or "suspended clay" began emerging. At the southern part of the Laizhou Bay, and at the shallow south of Diaolongzui this sort of thin and dilute clay is temporary, disappearing at subaqueous –3 m and reappearing at –5 m[24] in several days.

Current velocity in the study area is weak. According to our survey, the velocity within five metre water, near the shore, does not exceed 20 cm/s, even when it is in spring tide. Such a small current velocity can only stir up clay and fine silt when it is under the action of big windwaves.

Spit shape is a reliable indicator showing the direction of sand transportation. According to the sedimentary distribution and geomorphology development as well as wave parameter of Diaolongzui coast, sediment supply in this coast is fairly rich. It is carried from east to west and transported southwards around Diaolongzui. The coastal sediment in this area is chiefly longshore movement. Transverse sand source is scarce.

2. The forming of pennate spit requires the action of waves from two different directions with rather large intensity range. Under the action of strong wave, sand is transported towards the terminal end of the spit, making the spit extend. The secondary strong wave, or the wave after diffraction, changes its direction, weakens and carries sand from the terminal end of the spit to the lagoon, thus creating a small hook-shaped accumulation spit meandering into the lagoon. The repeated emergences of hook-shaped accumulation spit result in a pennate spit.

3. The gradually increased underwater shallow and its corresponding sediment in accordance with specific order are maintained. The increase of shallow accumulation is the foundation of new pennate spit development. The development of a new group of spits indicates that the shallow has increased to a new width or height. It has been found that the gradient of nearshore shallow at a specific coast is the result of adptation to particle size of sediment and wave strength. It is shown that the shallow slope, when ③ is formed, is basically the same as that when ⑥ and ⑧ when they are formed, only with their positions moving outwards.

The gravel pennate spit A pennate accumulation spit of gravels, which is not big, has emerged at the northern part of West Zhifu Bay in Yantai. There is a groove from Zhifu hill slope, showing clearly the main source of gravel (Fig.2-15).

Most of the diameters of the gravel, of which the two pennate gravel spits are composed, are around 10 cm. Its main content is silicilith and silicilith plate. Its psephicity is close to medium. They originates from the bed rock on Zhifu Island. Gravels are eroded and carried into nearhsore zone through the big groove. After a big windwave, a small hook-shaped spit is formed. After actions of many times, a pennate spit is formed. At present, this big groove is still continuously carrying gravels to the beach of the groove mouth. It can be said that a third pennate gravel spit is now being formed.

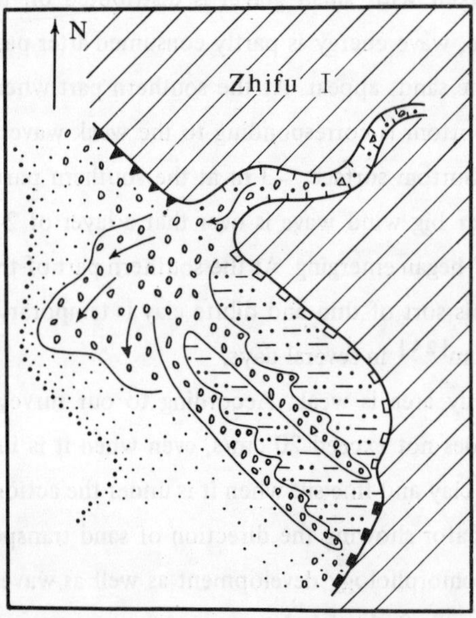

Fig. 2-15. Pennate gravel spits, Western Zhifu Bay, Yantai (From [23]).

There are also many pennate spits at other coasts of the Shandong Peninsula such as the Dingzi Bay mouth and Rizhao coast, as well as the coasts of Fujian and Taiwan, Nandujiang mouth and Basuo coast of the Hainan Island. They are mostly formed by longshore sand supply.

2. Coastal sand ridge

The sand ridge, spit and barrier, namely the sand-gravel accumulations have no substantial difference in their origin and material composition. They can all be composed of sand or gravel, and can all be formed by longshore sand supply or transverse sand supply. The main difference between them is that there is no lagoon behind the coastal ridge or water, though there are depressions. This type of accumulation form is often found in China's bigger gradient coast. It represents a sand-gravel accumulation body under a sort of environment, belonging to B.C type according to beach classification by Shepada.

Under the condition where the gradient at the lower part of coastal shallow is not big and loose sand layer is not thich, the bottom is eroded by waves, the coarse sand in the layer is pushed to the shore and fine sediment is moved to deep water. Because the gradient of primitive coastal topography is big and the beach is eroded by waves, so there is no lagoon water behind the sand ridge, though there exists the depression. The most typical examples are the Bay at the east of Penglai in the northern coast of the Shandong Peninsula (Fig. 2-16 A) and the gravel-sand ridge at the eastern coast of Dalian in the Liaodong Peninsula (Fig. 2-16 B).

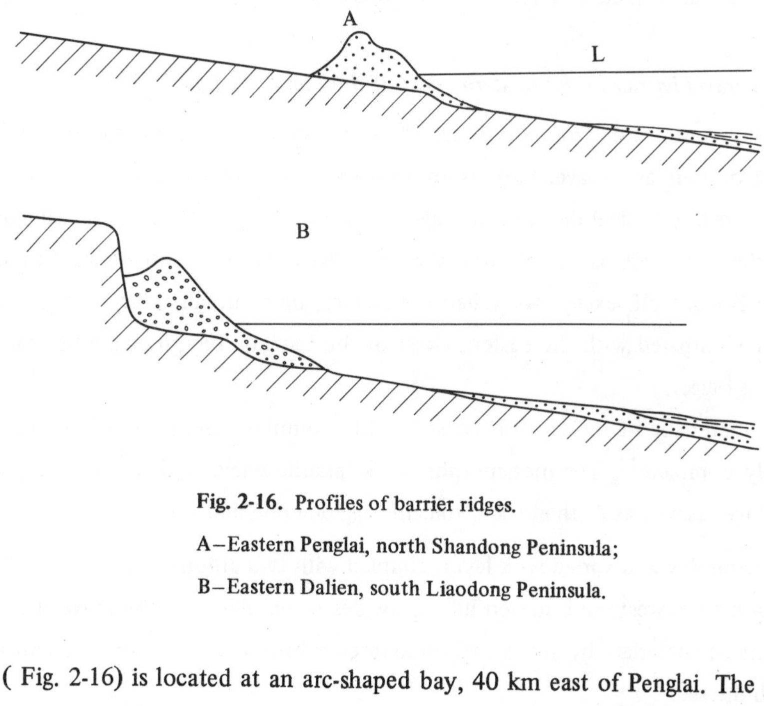

Fig. 2-16. Profiles of barrier ridges.

A–Eastern Penglai, north Shandong Peninsula;

B–Eastern Dalien, south Liaodong Peninsula.

The area A (Fig. 2-16) is located at an arc-shaped bay, 40 km east of Penglai. The gradient of the primitive coastal topography and of the shallow is between 1:100 to 1:150. The sediment, of which the sand ridge is composed, is medium, fine sands, containing a small quantity of shell fragments. Sands on the foreshore is relatively fine, being pure fine sands. The degree of sorting is good. Wind power has certain influence on the sand ridge, but can not make sand dune. According to the sequence of the sand ridge, the ridge is still slowly retreating. Between the sand ridge and the underlying yellow sand clay, there is only a thin and discontinuous fine sand-clay sediment. At the front slope of the snad ridge, particularly at the nearshore zone under the low tidal line, the fine sand is very thin. When there is storm, this thin layer of marine fine sand is transported to deep water zone, then the terrestrial yellow sand layer is exposed. At the same time, the yellow sand layer is continuously eroded.

About 60 km east of Dalian in the Liaodong Peninsula, the gravel ridge that piled on the bed rock keeps directly close to the edge of the 4 m eroded cliff (Fig. 2-16B). The gravel, of which the ridge is composed, is medium in diametre, flat and in good psephicity.

As water is shallow and slope is gentle, the bottom is effectively eroded and the bed rock at the bottom is exposed. By comparing the content of the gravel on the gravel ridge with the bed rock, the source of gravel is estimated to be from the nearby shallow waters.

3. Coastal sand barrier and barrier island

Most of the sand-gravel coasts in China are bed-rock coasts with many winding bays and isles. The gradient of primitive topography of inner continental shelf is rather steep and sand source is not rich. So the sand bodies at coastal zone are small, but the shapes of sand barriers and barrier islands are complex. Several typical examples are discussed as follows:

(1) *The sand gravel barrier at the western coast of the Liaodong Bay*

Stretching southwards from the Liugu River mouth to the northern coast of the Bohai Gulf, coastal sand barriers and gravel barriers are extensively developed. The lagoons coexisting with the sand barriers are also typical though most lagoons are at present silted by wind sand and river sand.

Along the over 3000 km coast from the Liugu River to the Caofeidian sand isle at the northern coast of the Bohai Gulf, except some bed rock capes, there are all rather straight sand gravel ridges and barriers. Compared with the eastern coast of the Laizhou Bay, they contain more gravel and the gravel body is bigger.

The characteristic of rock at this coast is fairly complex. The strata of various geological period are relatively complete[1)]. The metamorphic rock, granite gneiss and schist gravel, sand stone limestone sand shiver as well as Anshan rock, volcanic agglomerate and tuff.

Such a complex and varied rock layer, coupled with two groups of structural fault of northeast–southwest and east–west, and uneven hills, provides favourable conditions for the transportation of large amount of materials by rivers and undoubtedly brings forth complicate and variable minerals flowing into the sea.

Among many sandy rivers at the western coast of the Liaodong Bay, the amount of sea-flowing sand of the Liugu River is only secondary to that of the Luanhe River. 70% drainage area of the Liugu River is in hilly area, its river bed gradient is big, with a strong sand-carrying capacity. Most of the sands are of coarse particles. The runoff quantity in August averages 64.3 m/s. In dry seasons, it is only 3.18–2.71 m/s. According to the survey of Suizhong hydrologic station in 1959, the annual amount of transported sand was 4404 tons. Sand transportation concentrates in June, July and

---

1) Geological data are based on the specification of geological map by Institute of Geolagy, Liaoning Province.

August.

The distribution of major heavy minerals near the river mouth and offshore has a certain regularity. For example, near the river mouth, gregorite and magnetite account for 27.4%, while at the offshore (water depth, 5 m) they are only 10.38%.

Limonite and augite account for 2.55% and 6.51% at the river mouth but 1.30% and 3.57% offshore [1]. Onlv the green hornblende is much more in nearshore than in inner river mouth.

The above mineral composition shows that the sand comes from the metamorphic rock hilly area of the upper reaches of the Liugu River, and that the gradient of the river bed has become small near the river mouth. 1ts current velocity slows down, with heavy minerals abundantly concentrated and suddenly reduced when the river flows to the subaqueous deltaic plain.

There develop sand barriers or barrier isles at both sides of the Liugu River mouth. The river mouth gradually extends outwards, making gravel-sand barriers distribute in ranks. The sediment, of which the sand barrier (ridge)is composed, is mainly sand, mixed with some small gravels. The psephicity is below medium. The diameter of gravel is mostly 2–4 cm. The contents of gravel are fairly complex. In the metamorphic rocks, there are quartzite, gneiss, granite gneiss and various schists. In the sedimentary rocks, there are limestone, slate, gravel rock and flint, etc.

What is worth noticing is that some small gravels are collected from over 10 m underwater in the Liaodong Bay where there are sandy clay and clayey silt. Their psephicity is very poor and the diameter range of gravel is 7–9 cm, its content being mainly quatzite and flint, occasionally granite. Such big gravels are seldom found at the coastal ridge in the bay mouth.

The gravel at the nearshore sand-clay bottom in the Liaodong Bay, which is of poor psephicity, is not brought in by runoff current, but is frozen in the river bed in winter, and carried to a certain distance offshore. The gravel drops on the sandy clay bottom when ice melts in spring. Even in flood seasons, it is impossible for the Liugu River to carry the gravel with a diameter of 10 cm to the 10 m deep bottom 8 km off the shore.

From the sediment distribution map (Fig. 2-17), it can also be made out that the distribution of fine sand zone only reaches about 1000 m off the shore. In the vicinity of the bay mouth, the diametre of gravel does not exceed 3 cm. There is no gravel in the area of the fine sand distribution. This is another evidence to the origin of gravel in deep water zone.

The second feature of river mouth sediment is the asymmetrical distribution of coarse and fine sands. The extension area of fine sands is fairly big, demonstrating that the sediment flowing into the sea has a tendency of transporting southwards under the action of wave sorting.

The lagoon behind the barrier near the estuary is not apparent. This is because a part of sand on the barrier under the wind action is blown to fill the depression.

Qinhuangdao protrudes into the sea. It is a natural obstacle for the gravel of the Shihe River to

---

1) The information about heavy minerals is based on the literature by Cheng Lirong & Xu Wenqhiang.

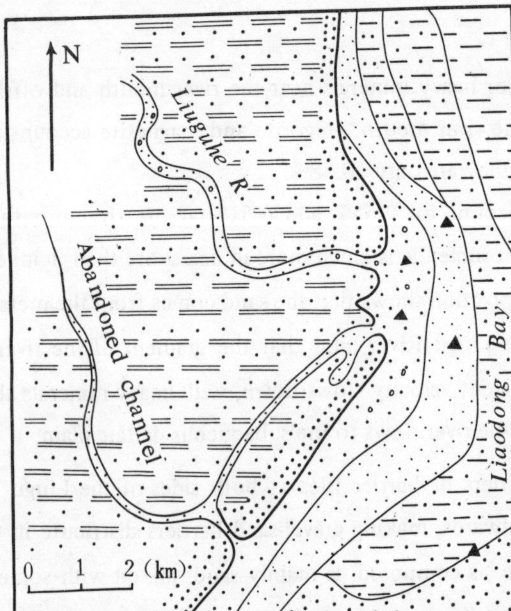

Fig. 2-17. Distribution of coastal sediment and barrier in the Estuary of the Liugu River, Western Liaodong Bay.

move southwards. So the southern coast of Qinhuangdao has been stable over a long period of time. The diameter of gravel at the beach is small, mostly being 1–3 cm. They are also in small quantity. Sands are mainly medium or fine ones. Lagoon layer is sandy clay, its underlying is terrestrial hard and yellowish sandsoil layer. Sand dunes are developed in the southern coast because it faces northeast. The sand dunes not only filled the lagoons behind the barrier, but also piled up towards land over 20 m above sea level.

In the coast within 10 km south of Beidaihe there are only two small naibourne streams carrying a small quantity of silts to the sea. Gravels are very few, 1–3 cm in diameter with poor psephicity. As the coastal slope used to be gentle here, the width of lagoon is wider than that in the northern part, drilling at the area inside the 6 cm water line has revealed that the sand layer is very thin within the 1.5 km shallow from low tidal line to 5 m deep water. In some places, there is even no sand and yellowish and hard terrestrial clay layer is readily exposed.

At the near shore, where the gradient of bottom is less than 1:300, sedimentary layer is absent owing to the action of wave. During big wind wave, waves exert certain eroding action on the bottom of terrestrial facies (Fig. 2-18). The loose marine sand bed emerges at the foreshore near the barrier. Its thickness is quite limited. The grey dark sandy clay of the lagoon is exposed in the front slope of the barrier, indicating the retreating of this section of coast.

There are two factors controlling the erosion of the coast and nearshore areas in Baidaihe. Waves have deformed and carried sand to the coast and built up sand barrier after eroding the bottom. The northeast wind blows the fine sand on the barrier towards southwest, almost in longshore direction. It provides the coastal dunes of the Dapu River with sands. The sand dunes south of this section are big

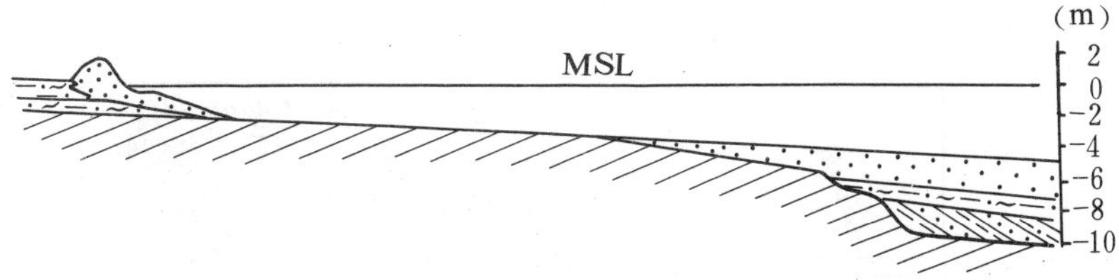

Fig. 2-18. Profile of south coast of Beidaihe.

in dimension, being the biggest dunes in the country. This will be discussed in section VI of this Chapter.

The section of coast from the south of the Luanhe River estuary to Nanbao is an example of modern estuarine spit developing into Chain barriers. The 18 million tons annually of sea-flowing sand, under the action of strong northeast wind, are transported southwestwards in autumn and winter, becoming the main sand sources for a chain of barriers south of the estuary.

According to the historical record and survey data on modern sedimentation and geomorphology, the Luanghe River has over a long period of time emptied into the sea around 1567 A.D. at the present estuary of the Dazhuanghe River or the Daqinhe River mouths, building sand isles at Caofeidian, Getuo, Shijutuo, Yuetuo etc. In 1811, the Luanhe River turned its channel eastwards, flowing into Chanli County, and emptied into the sea through the present abandoned Laomigou Channal. In 1833, the Luanhe River breached at the east of Chaijiazhuan Village, 20 km south of Luanxian County, and was divided into two streams, with the south one again flowing into the Daqinhe River and the east stream into Laomigou Channel. In 1886, the south stream was silted up and no longer emptied into the sea from the Daqinhe River. In 1915, the Luanhe River began flowing along the present channel. In 1938, the Laomigou tributary was silted up, and flowed entirely along the northern tributary. Everytime the river changed its course, a big estuarine spit would leave outside the estuary. After the spit was extended to a certain length, it would be broken up by the action of wave and current, becoming a series of chain sandy islands.

The materials emptying into the sea from the Luanhe River were transported southwards. But the amount of sand supplying for the spit and distance are quite limited. The gradually decline of transporting coarse sand southwestwards is a manifestation of limited distance under the wave action. For example, Dagan spit was formed when the Luanhe River emptied into the sea from Laomigou. But when this river was silted up, the original estuary was eroded for lacking sand supply, showing that there is neither enough time nor sufficient amount of supply for the seaflowing sand here. Of course, with the Luanhe River estuary increasingly protruding seawards and the quantity of sand at the estuary growing, the sand quantity supplied southwards is also increased. But it is absolutely impossible that the sand flowing into the sea from the Luanhe River mouth can be transported onto Getuo and Caofeidian isles. So, since the eastward shift of the Luanhe Estuary, Caofeidian and Getuo entered the period of retreating backwards. At present, these old estuarine sand isles are reducing gradually, in a stage close to disappearing and becoming a real remanent sand barrier. (Fig. 2-19).

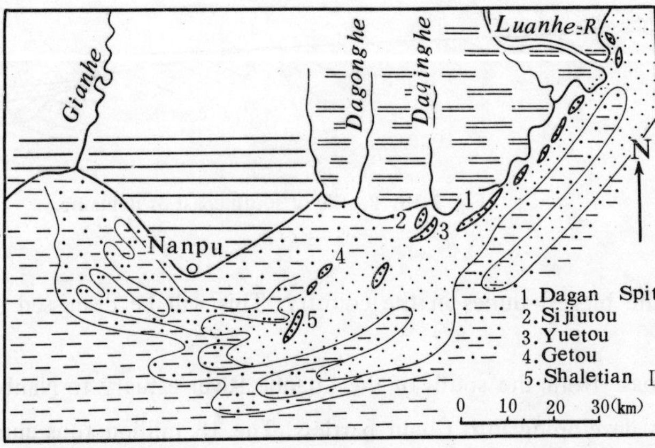

Fig. 2-19. Sediments and barrier chain from the Luanhe River mouth to Nanbao.

From the profiles from the beach to the bottom of Caofeidian sands on the isle brought southwestwards under the action of wave and wind power (Fig. 2-20 A). A part of sand is blown southwest. The range of their action is shown in the map. The 30 m deep trough is the outcome of washing by strong tidal current. The maxium current velocity here is 102 cm/s. If the storm is coupled with spring tide, then the maximum current velocity can be 150 cm/s. The velocity can exert washing action on the sea bottom of slightly compressed sand-clay layer, and roll the washed clay block at the bottom to balls.

The three functional zones on the profile are fairly clear. Wind blows the fine sand and small shell fragments on the sand isle and brings them southwestwards. The lower limit of wave action can reach 10 m deep beneath the surface of the water. The gradient is comparatively small in this range. There is a thin layer of fine sand, but it is discontinuous. The tidal current here strongly washes the clay bottom. Its most effective action zone is under the turning point of the slope. The tidal current plays a dominant role on the formation of this steep slope from –10 to –30 m. There is a short section of repetition of wave field and wind field. There is also a section of action zone with the tidal current. From Caofeidian westwards, this strong tidal current can reach Nanbao, entering the Jianhe River mouth and Sengtougu at the northwest of the Bohai Gulf. It still plays a dominant role on the coastal sediment there. The nearshore 13 m deep trough and the forked tidal ridge at the south of Nanbao is the extension of this strong current. The formation of Baimagan has a strong bearing on this strong current (Fig. 2-20 B).

The composition of heavy minerals of the coastal sediments south of the Luanhe River estuary has also illustrated the scope of sediment from the Luanhe River and the ups and downs of estuarine sand isle.

By analysing the sands on river bed near the estuary of modern Luanhe, on Dagan spit, on Caofeidian sandy isle, and on the bottom near Nanbao, we came to the following conclustions: a) They are of the same mineral sources, being similar in species and contents. Their common characteristic is high contents of extrusive rock debris, showing they are all from hilly areas of the Yanshan complex rock system. The content of hornblende is 20% in the Luanhe Estuary, 25% in Nanbao, garnet accounts for 18.5% of heavey minerals in the Luanhe River, 10.5% in Nanbao; gregorite accounts for 20.7% of heavy minerals in the Luanhe Estuary, 10.4% at Nanbou; b) The contents of heavy minerals change graetly owing to different sampling positions, different sedimentary environments, particularly the different dynamic strength. By simply using the content of heavy minerals to testify the orientation of sand transportation under such conditions, sometimes good results can not be obtained. For example, the extrusive rock debris at Caofeidian is only 2.0%. This can only explain that after being washed and dashed for a long period, this kind of minerals on the remänent isle can be easily carried away, and the content of the extrusive rock debris is low.

There exist two sandy zones of modern sedimentation in the nearshore bottom at the south of the Luanhe Estuary. The fine sand zone that is close to the edge of barrier, is mainly fine sand with 1–1.2 degree of dispersion, under the chief action of waves. The outside fine sand zone at subaqueous 10 m is a mixture of claysand under the action of longshore current, mixed with fine and medium sands. The contents of clay and fine silt can reach 15–31%, particle diameter being M_z= 3.1–3.9ϕ, with one or two reaching 4.3ϕ. Degree of dispersion in general is about 2.

The sedimentary profile of Nanpu coast (Fig. 2-20B) passes across several tidal ridges and troughs. They are the evident marks of the tidal current strength westwards and the range of their effects from Caofeidian at the northern coast of the Bohai Gulf. The sands from Caofeidian Isle, under the action of tidal current and wave, are brought mainly by tidal current and silted here. Controlled by tidal current, the Baimagan tidal ridge was formed. The sediment of Baimagan ridge is composed of fairly fine sand, mixed with silt and a small quantity of clay as well as many kinds of shell fragments.

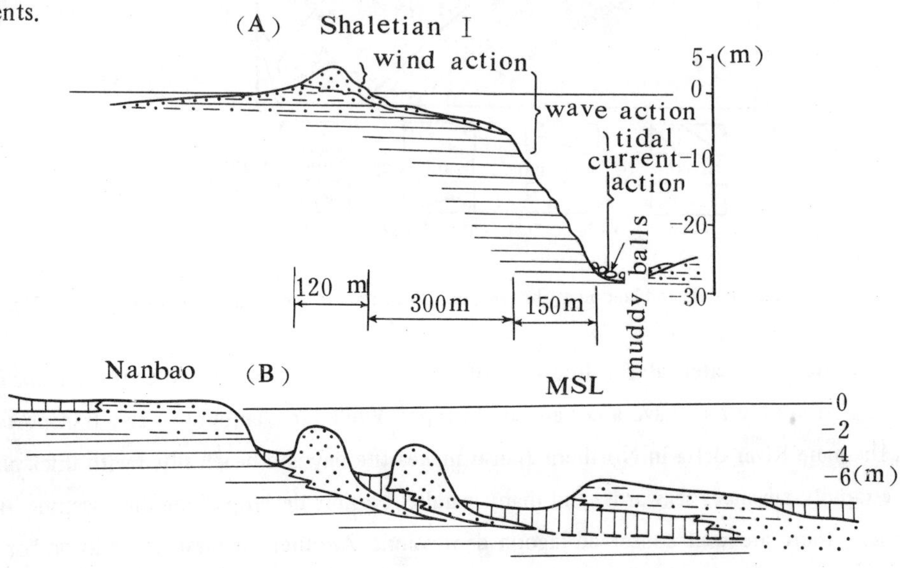

Fig. 2-20. Profiles of northern coast of the Bohai Gulf.
(A) Regression of barrier of Shaletian Islet;
(B) Nearshore tidal ridge of Nanbao.

As is shown in Fig 2-20B, these subaqueous tidal ridges have the same dynamic process as Caofeidian Isle i. e. moving landwards. So the clay layer is underlying the ridge.

Among the remanent sand isles formed at the Luanhe River estuary in different historical periods, Yuetuo and Shijiutuo are basically stable as they are protected by new sand isles. While the unprotected Dagan spite, Getuo, Caofeidian are eroded in varying degress and are in a state of retreating. The mineral composition of all these sand isles in different periods are the same, all belonging to the Luanhe River system, no mater they are in a stable or unstable state.

(2) *The barriers at the southern coast of the Shandong Peninsula*

The sand barrier at Fenjia Estuary in the southern coast of Shandong is a represent of the characteristic of barriers at the southern coast of the Peninsula. Not only the barrier does not retreat, but an old barrier is found at the lagoon behind the barrier (Fig. 2-21). Sand supply here comes from longshore drift. New spit is reconstructed outside the old barrier,. The sediment composition of both new and old spits is the same. Because there is a fairly big tidal inlet linking up the lagoon and the sea, the tidal delta inside the inlet can be quite clearly seen from Plate 2-5.

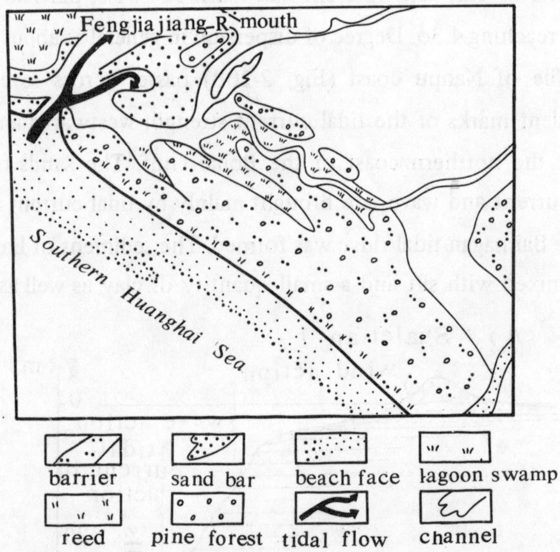

Fig. 2-21. Barrier and lagoon at the river mouth of Fenjia, Southern Shandong (From [17]).

Rizhao coast is located at the low hills of littoral area of the northern coast of Haizhou Bay. Small rivers from hilly area have a certain sand supply. While large amount of clay transported from the old Huanghe River delta in Northern Jiansu makes the bay and beach silted with thick clay. The coastline rapidly advances seawards, and many parallel barriers develop along the coastline. Between the barriers, there are well conserved lagoon depressions. Another favourable condition for the development of Rizhao coastal barrier is a longshore current from northeast, supplying continuously sand for the building up of the barriers in this zone. This sand drift has been proved by hydrologic and mineral composition data. Because there are various kinds of coarse and fine sand supplies, there is one, two, or even four barriers along the coast, and the interspace between the barriers is big. Most of them are 300 –500 m wide southwestwards.

(3) *Coastal barriers in the Hainan Island*

If the barriers and lagoons along Taiwan's western coast rank first in sedimentary rate and scope in China, then the Hainan Island tops the country in its extensive development of large ridges and barriers nearly around the whole Island (Fig. 2-22).

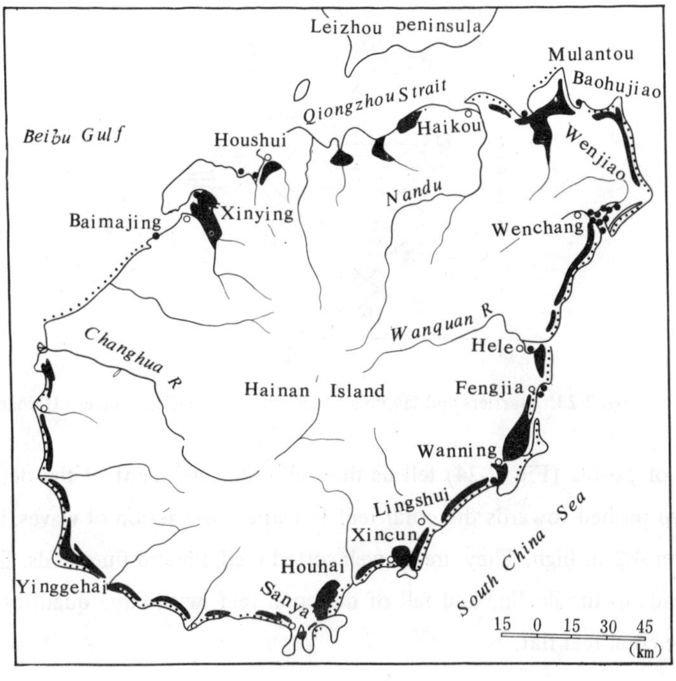

Fig. 2-22. Lagoons and barriers of Hainan Island.

Fig. 2-22 shows that most sections of the Hainan coast are surrounded by consecutive coastal sand bodies, interrupted locally at the capes, such as Mulantou, Dahuajue, Yuling, Luhuitou and Tian-yahaijiao etc. According to the preliminary statistic, the sand bodies account for 90% of the whole 850 km of the Hainan coastline.

The size of sand barriers in both the eastern and southern coasts of the Island is the biggest, with their height of 8–10 m. Their width is different in various places. Behind most of the barriers are long and narrow lagoons with shallow water. Some lagoons disappear as they are filled up by wind sands. The composition materials of barriers in the eastern and southern coasts are mainly coarse and medium sands. This is the major reason for the gradient of the sand barrier is big (13°–15°) and high.

The size of barriers and ridges at the western coast is not as big as that of the eastern and southern coasts. Their content of fine sand is more, mostly in a stable state. In the vicinity of the Changhuajiang Estuary., the barrier is well developed. From the estuarine delta towards south, it can be seen that north wind wave is dominant here (Fig. 2-23). Most evidently are three rows of estuarine spits indicating the sequence of the delta's advance seaward. But studying the sedimentary environment we have noticed the great changes happened at the Changhuajiang Estuary. At the present delta, particularly at the lagoon depression behind the southern barrier, there is an original coral reef flat less than half meter underground. This is sufficient to show that it was an environment with normal water quality. Sea water is so free from being diluted by river water and the sediment content is so low, that the growth of corals is not affected.

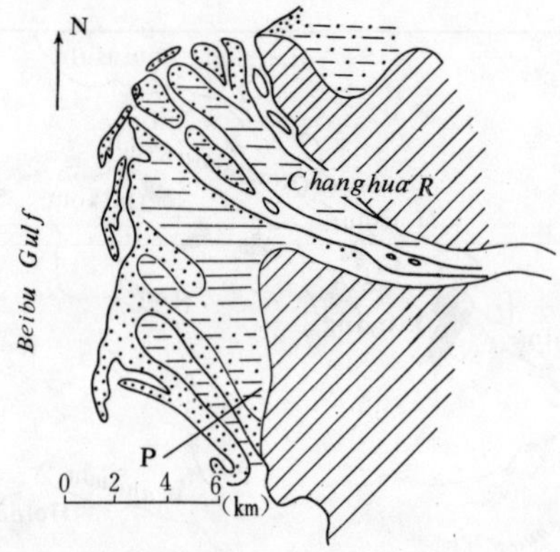

Fig. 2-23. Barriers and lagoons in Chanhuajiang Delta, western Hainan.

The sequences of profile (Fig. 2-24) tell us that while the sediment of the delta is transported southwards, it is also pushed towards the coral reef flat under the action of waves. It is accumulated into a ridge, not over 4.2 m high. They are all well sorted medium and fine sands. The accumulation of estuarine sand leads to the decline and fall of the coral reef, with large quantity of the estuarine sands covered on the coral reef flat.

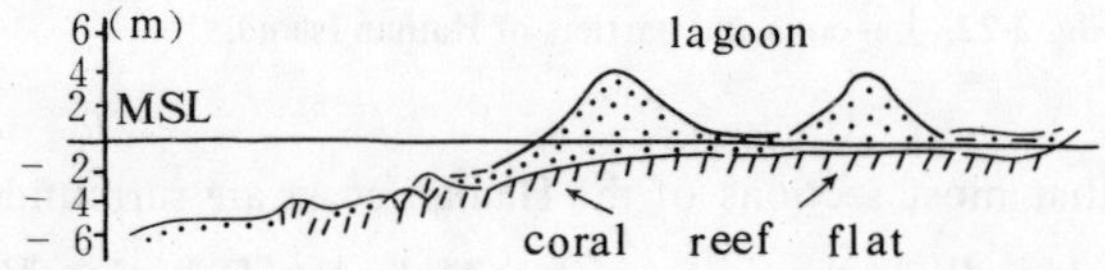

Fig. 2-24. Profile of sediment, south of the Changhuajiang River mouth, Hainan Island (for the position of section, see Fig. 2-23 P line).

The Houhai between Xincun and Yuling is the most developed barrier at the southern coast. Many longitudinal barriers parallel to the coastline are 8–10 m high, with individual section reaching 13 m high. The sediments, of which the big barrier is composed, are well sorted medium sand containing shell fragments. Fine and coarse sands are little.

There is a long and narrow lagoon of less than 1 km in width behind this 15 km long barrier. In the lagoon, there is little clay. Mangroves are seldom, and the trees are not high (Fig. 2-25). From the fact that some mangroves are burried by barrier, we know wind power edivently played a remolding role on the barrier.

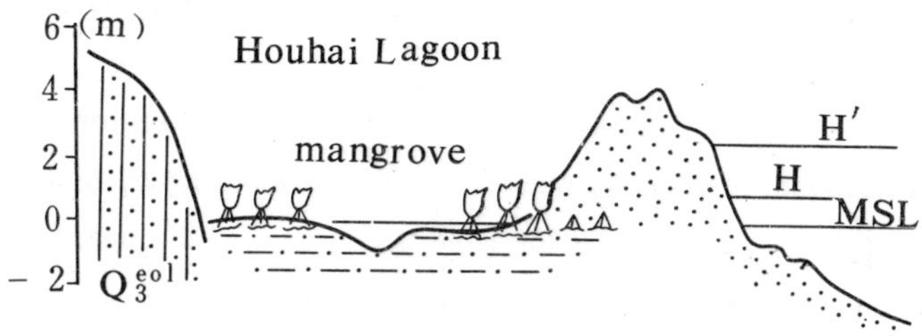

Fig. 2-25. Sediments of Houhai barrier and lagoon on the coast of southern Hainan.

There is little clay in the lagoon, mainly being fine sands. This phenomenon does not show that there is no source of clay sediment but no dynanic environment for settlement of suitable clay. As the lagoon is linked to the land by a small stream, the sediments in the lagoon are "cleaned" by flood and the component clay is flushed into the sea.

4. Tombolo

Tomboloes with various shapes, composition of different sediments and origin are commonly found along China's sand-gravel coast. This is a characteristic in the sandy coasts of the bay with hills and numerous islands. In the bay coast where sand source is abundant, tomboloes are especially developed. China's most developed areas of tomboloes are in Shandong, Southern Fujian and Eastern Guandong.

Tomboloes with different origins have a close connecttion with the studying of the coastal structure movement, eustasy of sea level and changes of sedimentary environments. For example, there have been different views on the problem whether tombolo is formed at the section where coast rises or at the section where coast falls as well as on the problem of sand sources. There have been various views about the origin of the well-known Zhifu Tombolo at the northern coast of Shandong.

The origin of the Zhifu Tombolo

We have discussed the modern sedimentary characteristic of the Zhifu Bay in Section III of this Chapter, and given a brief account of the world-known Zhifu Tombolo (see Fig. 2-10 to Fig. 2-12). Now, we would like to further present out arguments on the origin of the Zhifu Tombolo, for which much has been contended.

In this map, Leontyev of the Soviet Union declears that the formtion of the Zhifu Tombolo is an outcome of the advancement of the barrier from the bay mouth to inner bay. This point of view does not tally with the autual processes of the formation of the Zhifu Tombolo.

We hold that the accumulation shape of the Zhifu Tombolo i. e. the formation of the four gravel ridges is developed under the present condition of basically stable sea level.

The large amount of sediment flowing from the Jiahe River into the sea has gradually accumulated into sandy tombolo shallow. The eroded loose accumulation layer and the bed rock at the southern coast of the Zhifu Island produces large amount of gravel, a major source for the creation of the four gravel ridges. Under exceptionally big storm conditions, the gravel and sand-gravel re-transport to re-accumulate into ridges.

Fig. 2-26 reveals that, under the condition that the sea level was lower than the present, and that large sea area still land, the bed set was terrestrial yellow soil layer, mixed with little amount of small gravels. The Q_{2-3}^{dpl} at the bed set is the same as that on the hill slope. The overlying third layer is semiclosed calm environmental sediment, containing large amount of organisms and clay. On the upper part of this layer, a layer of *Ostrea denselamellosa* Lischke is found. Thereafter, in the rising process of sea level, if the gravel barrier at the bay mouth is to move to the inner bay, it is impossible to surmount the upper part of this layer. Moreover, judged from layer sequence, the second layer of marine silt-clay is also deposited. When the bay is relatively calm, the gravel barrier is also impossible to cross the upper part of this layer, with no remanent layer left. Facts have fully proved, Leontyev's inference is lack of foundation.

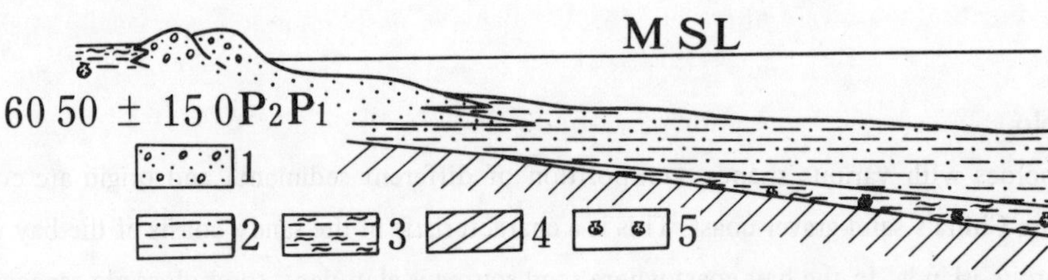

Fig. 2-26. Sedimentary profile of Zhifu Bay (From [23]).

1. Sand-gravel; 2. Silty clay; 3. Clay with organic matter;

4. Yellow sandy soil; 5. Shell.

The formation of the four gravel ridges on the Zhifu Tombolo is the outcome of longshore drift. The gravels come from the Zhifu Island. Energy is greatly consumed on the beach. In short, it is the result of the widening of beach southwards, leading to the decrease of the wave height at the ridge foot and the acute angle of the wave crestline slanting to the coast.

The height relation between the two ridges in the east group shows no apparent sign of emergence of the coast. The height of ridge C and ridge D is almost the same at the northern end, the middle section, and the southern terminal end of the Tombolo. This height can not show the emergence or submergence of the coast. In 1947, there was storm wave, in which sea water surmounted the ridge top, overflowed into the lagoon and inundated a vast land of the plain.

Thus it can be seen, that analysing in detail the material composition and characteristic of the subaqueous slope, strength of wave action and the relationship between coastal factors is significant to coastal sand movement, to the changing process of coastal sediment, to the determining of sea level change and new strutural movement.

Expounding in detail the formation process is extermely necessary for understanding the coastal characteristics and development in the eastern part of Shandong Peninsula.

At the present sea level, the Jiahe River supplys sediments to the area between the Zhifu Island and the land from the east until the subaqueous sandy shallow-subaqueous shallow tombolo are above the low and medium water levels. In this manner, the delta plain extends northward. The quantity of sands transported eastward bypassing the top of the deltaic accumulation reduces with the increasing height of the shallows.

At the same time of the shallow development, at the middle section by west in the southern coast of the Zhifu Island, two gravel spits are developed slanting to the island coast and in the opposite direction. The result of the development is that the two spits are linked up into a fully-closed lagoon. This is the oldest lagoon on the Zhifu Tombolo, marsh depression which has never been dried up (Fig. 2-27a).

We have analysed and compared the lithological characters of the gravel on the gravel ridge and bed rock of the Zhifu Island and of the mixed gravel in loose accumulation layer, we found they are the same.

Ridges A and B are formed by big gravels of quartzite with a diameter of 5–10 cm. Along the extending direction southward of gravel ridge, the diameter of gravel becomes gradually smaller.

Ridges C and D are formed by gravels in diameter of 2–5 cm. In the gravels are mixed large amount of medium and coarse sands, together with fairly much quartzite and gneiss. Along the ridge southwards, both the diameter and content of the gravel are gradually reduced.

By analysing profile I of Fig. 2-12, we found that ridge A is 20 cm lower than ridge B: while ridge B is over one meter higher than ridge C. This has proved: a) From the time of forming ridge B to the time of forming ridge A in the West Zhifu Bay, the height and width of the beach in profile I were not very much changed; b) Ridges B and C are symmetrical and were formed respectively by the West Bay and the East Bay over the same period of time. The height difference of these two ridges is over one meter. This is resulted by the fact that the wave in the East Bay is weaker than the West Bay.

Again, we analysed the middle section of the tombolo, i.e. the profile II of Fig. 2-12. Here ridge A is almost one meter lower than ridge B; while ridge B is about one meter higher than ridge C. The difference is the same with profile I. The increase of height difference between ridge A and ridge B has proved that after the formation of ridge B, sand in the West Bay greatly increased, the beach was rapidly heightened and widened. When waves pushed towards the shore, factors of wave are evidently weakened than the original ones.

On the profile III, ridge A is 1.3 m lower than ridge B. This strongly proves that the reason for increasing of height difference of ridge A and ridge B southwards is no other than the direct outcome of that the extent of the West Bay beach is increasing bigger and bigger southwards, and the wave factors gradually weakened.

The bottom of this lagoon is silt sediment of beach facies mixed with small amount of clay and molluscs of *Venerupes Philippinarum* (Adams & Reeve) and *Mya* sp. The height of beach face is higher than the low tidal level before the formation of lagoon. The time of turning form a semi-closed lagoon into a fully-closed lagoon is at 5000 BP. Since then the sedimentary environment has greatly changed. The sedimentary layer suddenly changed into clay-silt aboundant of organisms. The shell organism were substituted by semi-salty and low salty plants. The amount of organic matters has greatly increased. The silt layer containing marine shells was entirely substituted by the layer of organic matters rich in clay.

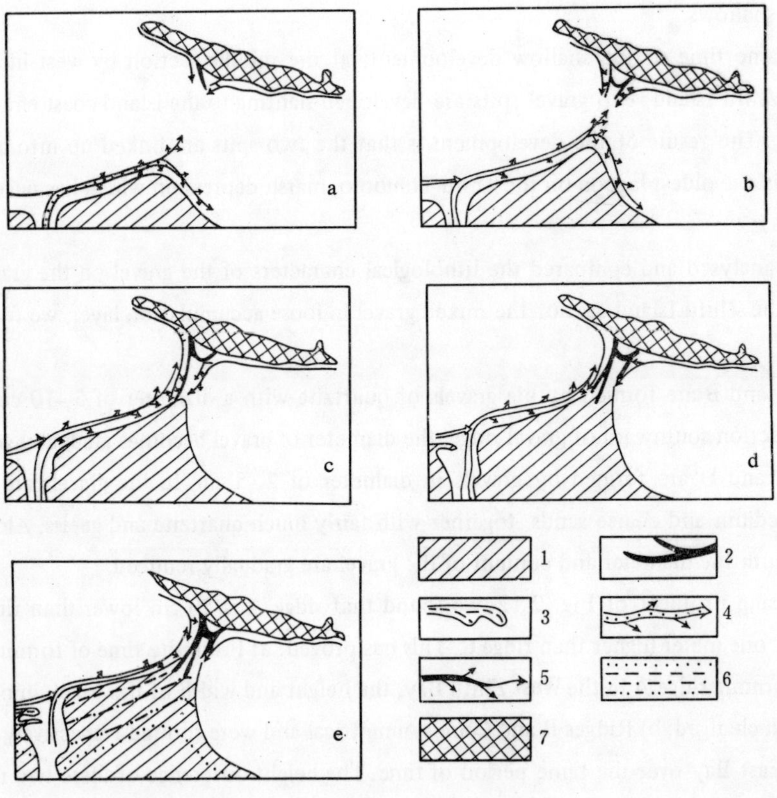

Fig. 2-27. Development of Zhifu Tombolo (From [23]).

1. Plain of accumulation; 2. Sandy-gravel ridge; 3. Spit; 4. Sandy-drift;

5. Gravel drift; 6. Stages of shoreling development; 7. Bed rock.

The formation of the first lagoon on the Zhifu Tombolo marks the first developing stage of the tombolo. In this stage, the movement of water and sand in both the East Bay and West Bay of Zhifu was interlinked when it was in low tide.

In stage II, the shallow-type tombolo between the island and land gradually became higher in the ceaseless supplying of sediments from the Jiahe River and with the triangular accumulation plain of the land developed northwards, ridge B and ridge C began their developing. By extending southwards, the two ridges became a fork against the old gravel ridge. This shows that the height of the shallow tombolo has exceeded low tidal line and close to medium tidal level. The shallow tombolo has thus been formed. At this time, the opportunity and quantity of sediments transported eastwards from the Jiahe River estuary were also greatly decreased. Ridge B and ridge C extended almost in parallel. That is a strong evidence of the forming of shallow tombolo as well as the main characteristic of the second stage (Fig. 2-27b).

In stage III, when ridge B and ridge C extended southwards, the eastern side of the triangular accumulation plain almost stopped growing because the amount of the sand transported eastwards from the Jiahe River mouth had been greatly reduced. While the growth in the western side of the plain accelerated, and the speed of the accumulation plain extending northwards also quikened. This process promoted the linking up of accumulation plain to ridge B as well as ridge C. Hence, the way of sediments entering the East Bay from Jiahe River mouth is completely cut off, with the growing of beach in East Bay basically holted. On the other side, the West Bay grows quickly.

The joining time of the accumulation plain and ridges B and C is the forming stage of Zhifu Tombolo. The lagoon between ridge B and ridge C is simultaneously formed. This lagoon is not connected to the hill slope. So, it is lack of clay source. Since the lagoon is formed, the thickness of sedimentary layer of lagoon phase has always been about one meter. (Fig. 2-27c).

In stage IV, there is a drift form the bay mouth to inner bay at the southern coast of the Zhifu Island, namely the norhtern coast of the East Bay and West Bay. These two drifts wash respectively the northern parts of ridge B and ridge C, making them retreated.

On the other hand, south of the converging point between ridges A and B, the beach increases, and extends southwards, the space between them being widened. It is worth noticing that when ridge D develops,. the northern end of ridge C is also eroded; but when ridge D extends southwards, the two ridges are close to each other, without being forked,. This proves that since the beginning of stage III, i.e. the sediment from Jiahe River no longer enters the East Bay, the beach of the East Bay has basically remained stable. Only the beach of the West Bay is widened southwards. The space between ridge A and ridge B is widened southwards, that is the increasing ratio of the beach (Fig. 2-27d).

From the formation of ridges A and D up to the present is the fifth stage. In this stage, the in the West Bay and coastline are continuously increased. The terminal end of ridge A is now at the land 200 m off the coast because the coast of the West Zhifu Bay has increased (Fig. 2-27e).

Another characteristic in stage V is two newly-developed pennate gravel spits at the big groove mouth in the West Bay. The position and shape of the pennate gravel spits have proved that the northern loose Q_{2-3}^{dpl} is a main supply source for the gravel spits on the Zhifu Tombolo under the present conditions of geology and physical geography. And the division between ridge A and ridge B is caused by the action of storm under the condition of continuous increase (in different speeds) of the beach. Thus, it can be inferred that once an extraordinarily big storm happened, the form of beach and subaqueous coastal slope would be greatly changed inevitably. The fan-shaped gravel cone and the two pennate gravel spits at the mouth of the big groove are likely to be remolded into a third gravel ridge in the West Zhifu Bay and to be forked at ridge A. The interspace between them will also be inevitably widened southwards.

5. The origin of coastal barrier

The various sand bodies and lagoonal sediments along the coasts of the bohai Gulf and the Shandong Peninsula are good areas for studying the theory of the origin of barriers. We have analysed the drilling data on more than 200 coastal barriers, lagoons, and nearshore areas. We have also analysed the characteristics of sedimentary facies in cores at various layers and studied the sequences of barriers and lagoons. We conclude that barrier are formed in various processes.

The barriers along the Bohai coast are formed by sands carried from rivers, and in longshore transportation and transverse accumulation, under the wave action after the transgression in the Holocene Epoch. The erosion of coast happened after the transgression was in full bloom, showing that sea water reached the gentle hill slope. Only at deltas of some sandy estuaries, the old marine cliffs are not obvious. The most parts, underlying the modern barrier, are not directly bounded by transgression sands. This type of sequence is very clear at the eastern coast of the Laizhou Bay and the western coast of the Liaodong Bay. Along the originally gentle coast, terrestrial facies is exposed at the bottom. For example, the profile near Baidaihe (Fig. 2-18) shows that no transgression sea facies remain everywhere. No matter how special the things may be, there does exist transgression sand under the barriers of the Laizhou Bay. Most barriers are neither linked to transgression sand nor are they in an underlying contact. There is an old marine eroded cliff behind the barrier, and under the lagoon facies, there is sediment of beach facies. All the above-mentioned phenomena are quite general (Fig. 2-28). What is worth noticing is that the sand source in the western coast of the Liaodong Bay is fairly abundant, obvious sea-eroded cliff as that in the eastern coast of the Laizhou Bay is produced.

A part of sediments in the Luanhe Delta are transported southwards by action of longshore dynamic factors, forming a series of chain-shaped barriers parallel to the coast. The forming of this kind of barriers has nothing to do with the changes of sea level structral movement. The formation of the chain-shaped barrier group in the western coast of Taiwan is the same as that of the above-mentioned formation.

Fig. 2-28 shows an area located at the place 50 km south of the present Luanhe Estuary, crossing a barrier and a lagoon. It is developed on a shallow at the outer edge of an alluvial plain. The sediment carried southwards from the present Luanhe Estuary is directly affected by waves and accumulated to become a chain barrier.The original sandclay layer of the beach was burried by lagoons behind the barrier. As can be seen from the profile (Fig. 2-28) that the sequence is clear. The sediment compositions on different positions are also obviously different.

The sand barrier can be formed by remanent sand of continental shelf, which is transversly transported to the coast, in the rising process of the sea level. The big barrier in Moupin at the northern coast, the big barrier in Rongcheng at the eastern coast and Chudao Tombolo of the Shandong Peninsula are good examples. All these big barriers are not formed by sands from nearby small streams, but are pushed upwards from continental shelf in the rising process of sea level. The transgression sand layer that is linked to the barrier is extended towards sea bottom (inner continental shelf). The barrier outside the Luanhe chain-lagoon is a typical example formed in this pattern. Fig. 2-29 is a profile showing the sedimentary relation between the barrier and lagoon at Songbuzui Longxudao. Transgression does not reach the hill slope directly. So there is no sea erosion trace at the juncture between lagoon and coast.

Underlying layer of the lagoon is terrestrial soil layer and regolith, and there is no beach sediment or nearshore sediment. All these facts are strong evidences that the barriers are formed by transverse sand supply in the rising process of sea level.

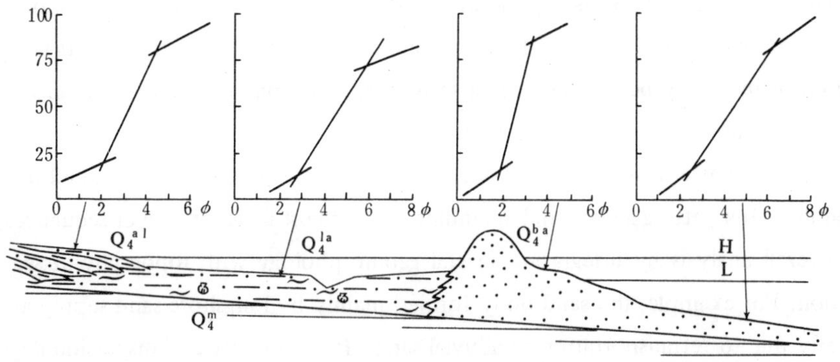

Fig. 2-28. Sedimentory sequence and grain-size in barriers south of the Luanhe River mouth.

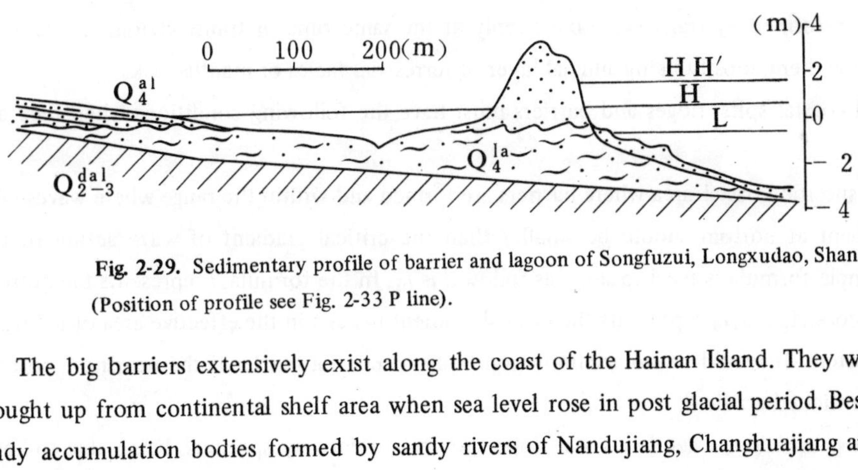

Fig. 2-29. Sedimentary profile of barrier and lagoon of Songfuzui, Longxudao, Shandong. (Position of profile see Fig. 2-33 P line).

The big barriers extensively exist along the coast of the Hainan Island. They were doubtlessly brought up from continental shelf area when sea level rose in post glacial period. Besides the deltaic sandy accumulation bodies formed by sandy rivers of Nandujiang, Changhuajiang and Wanquanhe, most parts of the coast have hardly any terregenous sand sources to build such big and widely-distributed coastal sandy bodies. Even the estuary of the second largest river-Changhuajiang River in the Hainan Island can only build a small size delta, affecting directly the coastal sediments of no more than 20 km. The sand barrier at the eastern and southern coasts of the Hainan Island is several times bigger than that at Changhuajiang. But the rivers here are much smaller than Changhuajiang River, being impossible to supply such large amount of sands.

Judging from the distribution of barriers and corals, we found that most of the barriers are not bound to original reefs, besides the original reef (which is burried in small-scale by southward extending of Changhuajiang delta at the north of Basugan) and other coasts where the barriers are partially linked to original reefs. This fully shows that the positions of barriers were fixed long before the sea level was basically stable after transgression. If the coastal barriers of the Hainan Island were formed mainly by sand supplies from the rivers, then there should have been two outcomes: a) The original reefs should have been occupied and burried owing to the growth of barriers and the enlargement of sand bodies; b) Below the lagoon layer and behind the present barriers, there should have existed a sedimentary layer of marine facies. In fact, these two phenomena do not exist. The case in the Changhuajiang delta is, after all, an isolated and rare one.

The forms of sand-gravel accumulation bodies are various, such as coastal barrier, ridge, spit and tombolo. Their formation patterns are also different. They may be formed by longshore sand supply; or by transverse sand supply; or by pushing the sand of continental shelf to shore at the same time of transgression; or they may be formed at nearshore area in changing a sedimentary environment after transgression and the increasing of sand sources.

The formation pattern of sand-gravel accumulation body should be determined mainly through sedimentology. After getting a clear understanding of sediment facies and layer sequences, one should carry out overall analysis by integrating coastal geomorphology with topogrphy at sea bottom and sedimentation. For example, the sand body formed mainly by longshore sand supply after transgression is chiefly due to retransportation of alluvial sands. Its psephicity and dispersion degree are poor. The underlying sand body is marine phase. Below the lagoon layer behind the ridge, it is bound to be a marine layer. On the land side of the lagoon, old eroded marine traces are often found. While the barrier and lagoon formed by transverse sand supply at the same time of transgression,. There is no marine layer below them, mostly being alluvial layer of terrestrial facies or mantle rocks.

In short, all coastal spits, ridges and barriers must have the following conditions when they are to be formed:

(A) At the specific coastal area where barriers are formed and within the range where waves take effect, the gradient at bottom should be smaller than the critical gradient of wave action to the bottom. The simple formula is used to show as follows: $i<i_0$. In the formula, i represents the bottom gradient of the coastal area; i_0 represents the critical gradient needed in the effective area of the wave action at the bottom. It would be no accumulation body of coastal barrier if the nearshore bottom had no above-mentioned gradient.

(B) The above-mentioned gentle coastal slope should be relatively specious, being able to keep the sand body on the gentle slope and not to be carried back to deep water area and dispersed under conditions of different wind waves.

(C) At the sea bottom, there should have sufficient sand sources for the building of barriers, or have sufficient quantity of longshore sand supply. In this aspect, the fluctuations of sea level are favourable for the formation and enlargement of the accumulation body of barriers. When the sea level falls, all the sand bodies are kept where they are. Waves are unable to carry them back to the lower place. Therefore, when the sea level falls, the sea bottom is eroded by wave. The eroded sand is pushed to high tidal line and to build sand body. When the sea level re-rises, the remanent sand body is pushed towards the shore, enlargement of the size of sand body occurs.

The growth and decline of coastal sand-gravel accumulation body depends on many factors. The most important one is the increase or decrease of the quantity of sand supply. Just as we expounded in discussing the forming of the pennate spit at Diaolongzui, the spit fork is due to the increase of sand sources at nearshore and the growth of beach accumulation; the retreat of the spit is the result of the decrease of sand supply, with sand output more than input, the nearshore is eroded backwards.

The vanishing phenomenon of coastal barrier is often found in China. The retreat of barrier near Beidaihe is owing to large quantity of coastal sand blown southwestward, supplying Big Dune of East Hebei with sand sources. The similar case also exists very seriously at the Shenfu Bay of southern Fujian. Particularly, at the northern coast of the Bay, the barrier retreats to the extent that there is no lagoon at all, and only a small ridge remains.

Another reason for the retreat of barrier is the strengthening of wind waves. In accordance with the principle of beach cycle, the slope foot of ridge and foreshore of the coast are eroded by big wind waves then the sand was carried to nearshore and below it. Strong wind can also push the sand to lagoon area behind the ridge (Plates 2-6,7), this also causes permenant lose of the beach sand. Moreover, the rising of sea level is, without doubt, one of the important reasons that make most barriers retreat.

V. Lagoon Sediments

Lagoonal sediment is inseparable from barrier or spit. To be more exact, the sediment of lagoon facies is often associated with barrier. So, naturally the study of lagoonal sediment is carried out simultaneously with the study of barriers and spits. In addition, because the lagoonal sedimentary layer is to be eassily preserved and contains abundant organism skeletons, therefore it is convenient to study the sedimentary environments, age dating and sedimentary rate.

The distribution of coastal lagoons in China is not so many as that in North America. The sizes are not so big as those in North America. But the lagoonal types and complexity of sedimentary environments have their own characteristics. In the past twenty years and more, the study of China's coastal lagoon sediment has been done in fairly details in the Shandong Peninsula and some lagoons in the Bohai Sea area. lagoonal sediments at the coast of the Shandong Peninsula

Lagoonal sediments at the coast of the Shandong Peninsula

There are 60 lagoons of different types along the Shandong Peninsula (excluding small lagoons in the Chanshan Archipelago and some buried, undiscovered lagoons), (Fig. 2-30). Total coastline of lagoon is 360 km, accounting for 32% of the whole coastline (1050 km) of the Shandong Peninsula.

Lagoons of the Shandong Peninsula are of many shapes, different sizes and complicated sedimentary composition. All the above-mentioned factors can give rise to changes of sedimentary environment of lagoons, resulting in sedimentary layers of different characteristics.

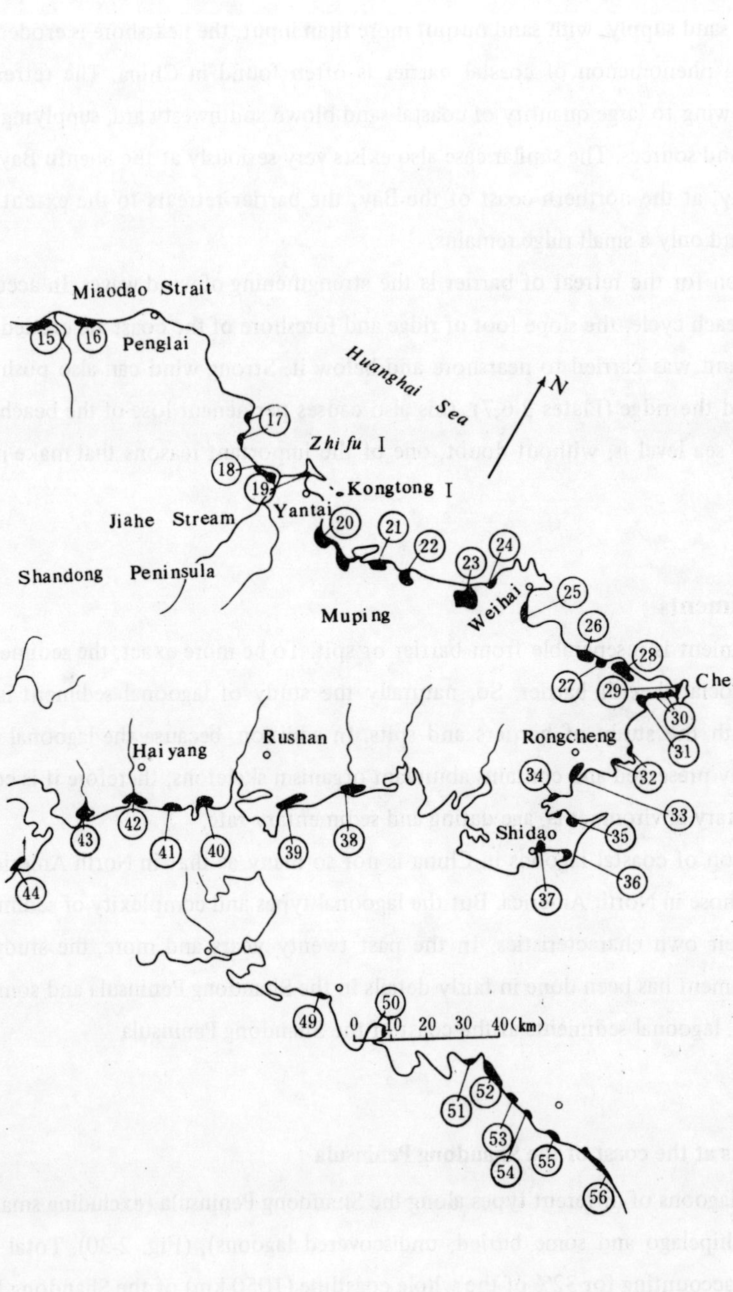

Miaodao Strait
Penglai
Huanghai Sea
N
Zhi fu I
Kongtong I
Yantai
Jiahe Stream
Shandong
Peninsula
Weihai
Muping
Chengshanjia
Rongcheng
Rushan
Hai yang
Shidao
10 20 30 40 (km)
15
16
17
18
19
20
21
22
23
24
25
26
27
28
29
30
31
32
33
34
35
36
37
38
39
40
41
42
43
44
49
50
51
52
53
54
55
56

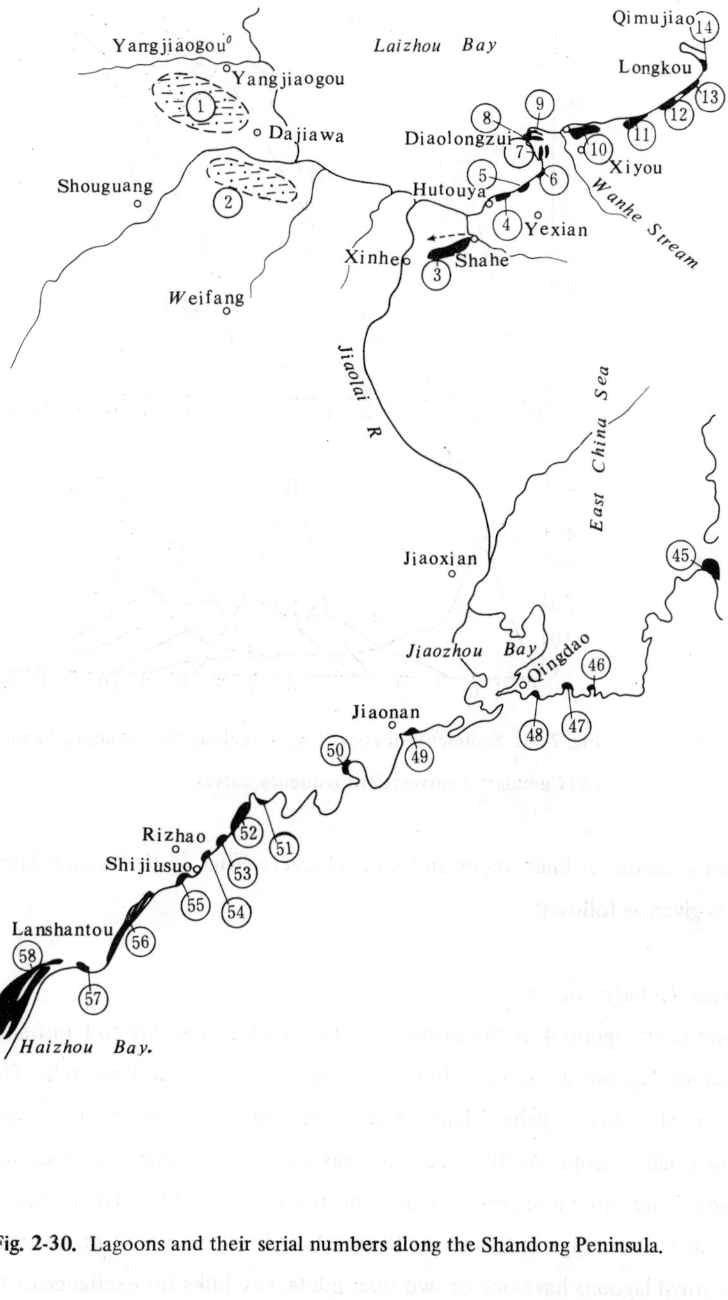

Fig. 2-30. Lagoons and their serial numbers along the Shandong Peninsula.

According to different geomorphologies and sedimentary characteristics of the coastal lagoons in the Shandong Peninsula, and taking into consideration the coastal geological structure and biological function, we have classified the lagoons in Table 2-2. Fig. 2-31 states briefly the five representative sedimentary types. Most lagoonal sediments in Shandong can be shown when compared with the Table 2-3.

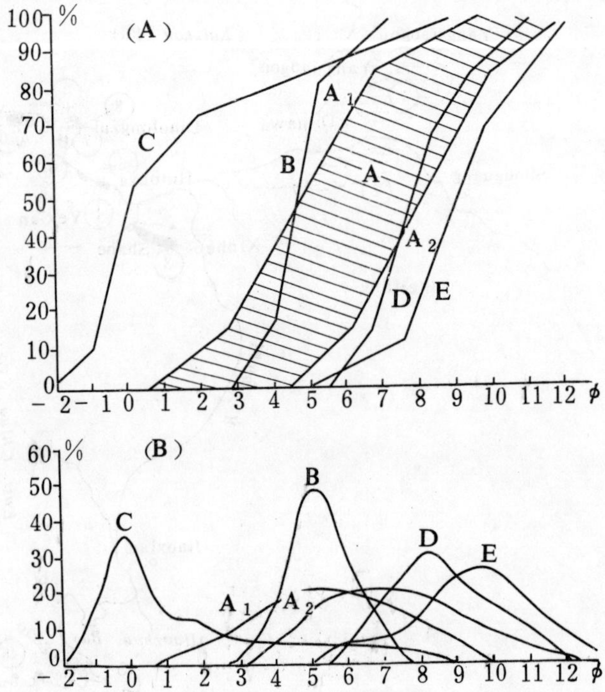

Fig. 2-31. Sediments of coastal lagoons along the Shandong Peninsula.
(A) Cumulative curves; (B) Frequency curves.

A brief account of main types and their characteristics of the coastal lagoons in the Shandong Peninsula is given as follows:

Type I, Coastal Sandy Lagoon

Starting from lagoon 4 at the eastern coast of the Laizhou Bay to Lanshantou of Rizhao County there are sandy lagoons along the whole hilly coastal area of the Peninsula. They are depressions or lagoons of small estuary, behind barriers and spits. Most lagoons are long, parallel to the coastline, narrow and small in scope. All the lagoonal areas are within several square kilometers. Most sediments are clay-sand. There are more eolian sands at northern coast, with a part of peat layers.

The coastal tide range at Shandong Peninsula is between 1.5–3 m, increasing from east to west. Therefore, most lagoons have one or two tidal inlets, key links for exchange of sea water and lagoonal water. When the lagoonal area is greatly reduced and the action of wave is stronger than the washing of current, the bay mouth is closed or very few water enters the lagoons. With the diluting of river water, salinity is below 5–8‰ in the lagoon and the saline-tolerant plants can thus thickly grow. A peat layer of 1–2 m thick is gradually accumulated. This type of lagoons can be subdivided into the

Table 2-2 The classification of coastal lagoons of the Shandong Peninsula

lagoonal type		Category & characteristic	Sediment Type	Biology & utilization
Type I Sandy coastal lagoons	1) Semi-closed lagoons behind barriers	Lagoons ⑦–⑨ ㉑–㉘, ㉛–㊴, ㊶–㊹, ㊼–㊾, (51)–(54). Lagoons with spits and barriers of tidalinlets.	Mainly cay-sand with a certain amount of eolian sand type (A) & (D) Table 2–4 in Fig. 2–55. Sediment in iagoon (8), (9) is type (C)	Having many species of shell orgainsm, some lagoons have been reclaimed into farmland, partial water area become marine culfures
	2) closed lagoons · barriers	Lagoons ⑭ ⑰ ⑲ ㉚ ㊵ ㊺ ㊻, water is entirely separated from the sea	Most sediments belong to type (A), mixed with type (B), lagoon (14) belonging to type (C)	No shell organism with plants growing, partially reclaimed into farm land
	3) Buried lagoons	Lagoons ④–⑥, ⑬ ⑮ ⑯ ⑳ ㉙ ㊿ and a few unnumbered small lagoons are all buried by eolian sand	The upper layer is type (B) lower part is type (A). Individual lagoons has peat layer or sediment (C)	Reclaimed into farm land or nearshore forest zone
Type II silty lagoons	1) nearshore plain depression lagoons	Lagoons ①–③, without obvious coastal barrier, turbid sea water enter the lagoons, when exceptionally big storm occurs	Mainly sediment of types (D) and (E), the underlying layer coarse silt of marine facies	Animals and plants seldom found, the underlying layer hypersaline as good salt fields
	2) Lagoons behind chenier ridges of clay beach in the bays	Lagoons (55) – (58) with discontinuous chenier ridge sea water often overflow into the lagoon during springtide	Thick sand-clay and fine silt-clay, mainly sediment of type (E)	Being able to be remolded into salt-field when ridges are repaired, rice can grow if there is water source
Type III small sized estuarine lagoons		Lagoons ⑩ ⑮ ⑱ ㊲ ㊳ depression inside estuarine barriers or spits	Mainly terrigenous river sands, belonging to type (A)-1. The underlying layer is often of 1–2 m thick a peat layer	Water quality changable with season, partilly reclamed into farmland.

following three types:

(1) *Semi-closed saline lagoon.* There is a tidal inlet linking the lagoon to the sea. Water quality in the lagoon is similar to sea water, suitable for breeding of benthonic organisms. The sediments are from clay to sand-gravel, generally being in mixture, lacking of micro-laminats. Grain size varies greatly with vertical and horizontal direction. In lagoons, into which fresh water empties, shells are unable to live when salinity lowers to a certain extent. At present, most lagoons in the world belong to this type. Many lagoons at the Atlantic coast or Northern America, the Gulf of Mexico and Australia are semi-closed saline lagoons.

Regarding the sources of lagoonal sediments, generally speaking, there are three sources: a) Suspended particles (including fine sand-clay) carried into calm lagoonal areas by tidal currents; b) Terrigenous sand carried into lagoons from small rivers and nearby waterlogged areas; c) The accumulation of biological remains. This kind of sediment layer accounts for a considerable proportion in some lagoons.

Many organisms are very sensitive to environmental factors. By making use of the distribution of biology in lagoons and changes of their sequences. We can realize the water quality in the lagoon and changes of sedimentary environments. For example, for an originally semi-closed lagoon with inter-flowing sea water and lagoonal water, it is suitable for the breeding of many species of benthonic organism. However, because of the quickening of lagoonal sediment, the water area and tidal prism are greatly reduced. The bay mouth is correspondingly narrowed. Water quality is also changed, the lagoonal floor has been turned into a beach. The time for inundating of seawater is too short. All these have changed the conditions for the breeding of shell organisms, many shells die or change their species, or even become extinct.

In most cases, semi-closed lagoonal environments with adequate fresh water are favourable for growing reeds. In the lagoon, there is abundantly accumulated thick peat or humus layer. In the lagoon where reeds can not grow because of high salinity, low algal plants can still be abundantly accumulated to form layers

Table 2-3 Size parameters of five sedimentary types of coastal lagoons

Parameters / Sedimentary types		$M_{z(\phi)}$	σ_1	S_k	K_g
(A)	1	5.1	1.64	0.04	1.1
	2	7.9	1.92	0.03	0.8
(B)		4.6	0.82	0.01	1.5
(C)		1.3	2.85	0.97	1.47
(D)		7.9	1.85	0.20	1.05
(E)		9.2	1.80	0.06	1.25

(2) *Sandy coastal lagoon of closed type.* In this type of lagoons, plants grow luxuriantly, for small supply of sea water and low salinity. Many buried lagoons behind the coastal ridges at the eastern coast of the Laizhou Bay are covered by fine sand layer of over one meter. Underlying this layer, it is black clay-peat. Overlying a thick layer of alluvial sand or yellow sand layer of terrestrial facies Q_3. This type of closed lagoon is caused by too much filling materials and weak current at the lagoon mouth.

(3) *Abandoned lagoon.* In the late period of the Holocene Epoch, the weak action or no action of eolian sand and the slow sediment rate were changed in the original lagoon with relative calm environment, owing to changing climate and strengthening of wind power. Because of the activities of wind sand, sediments in the lagoon have obviously become coarse. Sediment rate has been speeded up. At present, there are many lagoons at the northern coast, which are filled by eolian sand and turned into plains (see Table 2-2).

Type II, Silt Coastal Lagoons

They are distributed at the coastal depression, namely the littoral area and offshore zone where the topography is gentle. This type of lagoons has large lagoonal area and not too distinctive boundary line. The form separating the lagoon from the sea is sometimes chenier ridge. At the western coast of Bohai Gulf and along the coast of Louisiana in the Gulf of Mexico, such lagoons are typically developed. Instead of being chenier ridge, the form can be a slightly silted zone in the geomorphology. The depression of lagoon is often with saline water. This type of lagoon is limited in scope in the Shandong Peninsula. They can be subdivided into two sorts:

(1) *Depression lagoon at nearshore plain.* There is no ridge morphology between the lagoon and sea, nor there exists any tidal inlet linking the sea. Tidal water overflows into the lagoon only when exceptionally spring tide occurs and has a certain sedimentation. Lagoons ①–③ at the southern coast of the Laizhou Bay belong to this sort.

(2) *Lagoon on the clay beach behind chenier ridge.* This sort of lagoon is different from the predecessors in having small ridge or chenier ridge. The sea water can still flow into the lagoon in spring tide. At present, its sedimentary rate is slow and the composition is with mainly silt-clay. The lagoons ㊺ – ㊽ at the northern coast of the Laizhou Bay belong to this sort.

Type III, Estuarine lagoon

The coastal small estuaries along the Shandong Peninsula were valley depressions in low sea level. When the sea level rises to the present position, they become estuarine bays. The shapes of these lagoons have close connection with the original valley topography. Some are long and narrow, some are oval. As the sediments in these lagoons are mainly terrestrial sand, the formation time of estuarine spits and barriers varies. Sometimes there are multiplex forms. The sedimentary profile of estuarine lagoon shows that it is fine in the lower part, coarse in the upper part, reflecting physical geographical conditions since the Holocene Epoch. First of all, the climate is evidently changed. Although the salinity of the water in estuarine lagoon is low and greatly varies seasonally.

The origin of lagoon falls in the the following three sorts:

(1) Lagoons behind barriers are mainly formed with transverse clay-sand sources. Most of them were formed at the early period of the Holocene Epoch when sea level rose and coastal barriers were pushed shorewards. There are three evidences ascertaining this origin: a) Underlying the barrier and transgression sand layer, there appears a layer of discontinuous lagoonal sediment; b) There is no obvious marine trace between lagoon and land; c) Under the modern lagoonal layer, there is an terrestrial alluvial layer or weathering crust. Most lagoons in the Shandong Peninsula, Fujian and Guangdong belong to this formation.

(2) Lagoons behind the spits and barriers are formed by longshore sediments. After transgression, the longshore sediments along the coast are increased and spits and barriers formed at the shallow. At the same time, lagoons are formed. The evidences for ascertaining the origin of lagoons of this type are as follows: a) Underlying the modern lagoonal layer is beach sandy layer, containing a

certain amount of shell remains. It is completely different from the upper new lagoonal phase; b) eroded or eroded-accumulated often forms remains in interface zone between the lagoon and land. Spits or pennate spits are formed by longshore sediments. The spits are cut by waves and become a barrier. The lagoons south of the Luanhe Estuary and at the western coast of Taiwan are distinct examples of this type.

(3) Coastal barriers or spits jointly formed by longshore and transverse sediment supply. The sands of the spit of lagoon ⑨ at the eastern coast of the Laizhou Bay is partially from the Laizhou Bay shallow.

Now, some typical lagoons at the Shandong Peninsula are discussed as follows:

(1) Lagoon at the eastern coast of the Laizhou Bay: Lagoons are extensively developed from Western Penlai to Hutouya of Yiexian. They were formed after the transgression of the Holocene Epoch. The nearshore spits and barriers developed on the shallow by longshore sediment supply, because terrigeneous materials increased in amount. There is an old eroded marine cliff approximately the same height of 5 m isohypse between the lagoon inside the barrier and the land. It is the maximum limit of transgression, formed before the development of the lagoon. The width of the lagoon is usually within 2 km, the narrow part is only 200–300 m.

The multiple pennate spits at Diaolongzui are typical in coastal geomorphology. In the middle of lagoon ⑦, there are two old spits cut by tidal channels. As the spit is not high enough and is remolded afterwards, it is not abvious in geomorphology. According to sedimentary characteristics and analysis of the old aerophotograph, it has been proved that the old spits are extended into two branches in the middle of the lagoon, thus dividing lagoon ⑦ into an inner one and an outer one (Fig. 2-14).

Most part of lagoon ⑧ is filled by eolian sands, with only a wet depression of less than 200 × 200 m in area. The sediments filling this lagoon are from two ways. One is the wash over fan before lagoon ⑨ is formed during big storm, particles are coarse, being mainly coarse and medium sands, mixed with a small quantity of fine sands and silt, M_z=1.6ϕ, σ_1=2.6. The other way is the action of eolian sands. Although large part of this lagoon has become land, it has not been utilized because the particles of sands are coarse.

The chief sediments of lagoon ⑨ are coarse and medium sands, together with fine sand to clay, i.e. Type (C) in Fig. 2-31 and Table 2-3. The medium, coarse and fine sands flushing over from the northern foreshore of the spit are mixed and silted into the loose and soft lagoon.

There have been many shell organisms in lagoons ⑦ – ⑨ both in past and at present. We collected the shell sample under one meter of surface in the lagoon 150 m west of Danshan. The radiocarbon dating of the sample shows its age is 3780 ± 210 yrs BP. The lower laminar of this shell-containing lagoonal layer is a clay silt of bay facies.

(2) Lagoons at the eastern coast of the Shandong Peninsula. From lagoons ㉖ – ㊲, the coastline is zigzag, with many bays. There are here typical chain-shaped lagoons and abondaned lagoons buried by eolian sands in China. In lagoon ㉘, sediments are evidently divided into upper layer and lower layer. The lower layer is green grey clay and silt clay, M_z=7.1ϕ, that is Type (A) sediment in

Fig. 2-31. This layer contains many species of shells, underlay by weathering crust. The upper layer is clay fine sand, chiefly the outcome of eolian sands and sedimentary type is (B), M_z=4.8ϕ, σ_I=0.92.

Owing to the action of eolian sands, the growing of oyster in the lagoon is directly threatened. Large amount of oysters in the lagoon die, only other orgnisms, such as *Mya* sp. Lucinidae, *Neverita* didyma survive.

Lagoon ㉙ is a tombolo type lagoon along the northern coast at the easternmost terminal in this area. It is an abondaned small lagoon, completely buried by eolian sands. The surface of the lagoon is around one meter, composed of fine eolian sands.

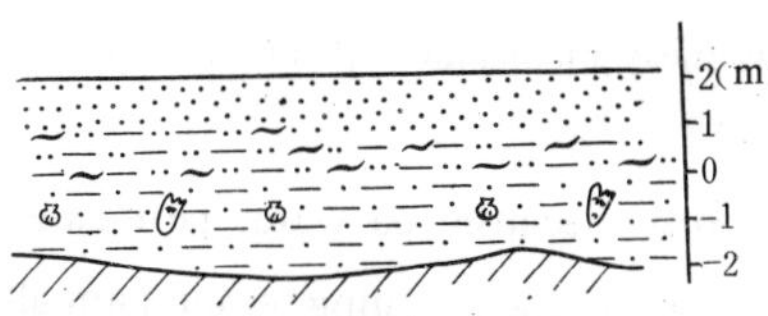

Fig. 2-32. Sedimentary profile of lagoon ㉙, Shandong Peninsula. (for location, see Fig. 2-30)

The sedimentary type is (B). The lower part is a layer of peat containing much black humus. The peat is contained in clay silt and fine sand layer. Its size is finer than that of the surface layer. The degree of dispersion is not so good as that of the surface layer either. This layer downwards is an original lagoon layer not acted by eolian sands, containing much more shells. The sediment is green grey type (A), (Fig. 2-32).

The chain lagoons ㉙ – ㉛ are composed of nine lagoons of different shapes and different dimensions. Lagoon ㉚ includes the present five closed small lagoons. No river flows into these five lagoons, nor are they influenced by eolian sands, therefore a better lagoonal state remains. Again, because there is no inlets linking the lagoon to sea, so the water here is not salty. The plant remains are abundantly concentrated here, forming a layer of black peat, not more than one meter thick. There have not been any marine shells growing in the past or at present. Under the lagoonal layer, there exists no transgression marine facies. The lagoonal peat layer bassets at the foreslope of the barrier, indicating the barrier is retreating. At present, it is extremely necessary to prevent the barrier sands from being carried away and to protect these typical chain-shaped lagoons.

Yuehu ㉛ is the last one of the chain lagoons. Its northeastern part has been silted up by eolian sands. At present, water quality in the lagoon is about the same as sea water. In this lagoon, there are many shell organisms. The water in it is still 1.7 meter deep even in low tide. In the middle part of the lagoon is silt-clay, becoming coarse towards shallow area in a concentric circles. The fine sands are at intertidal flat.

Lagoon ㉟ is between the twin tomblols. In the center of the floor, it is silt-clay, not easily permeated. Yet at the barrier, there is still a thick layer of fine sand. Its permeability is good. Not long ago, the bay mouth was blocked up artificially to keep the water fresh, but it has not succeeded

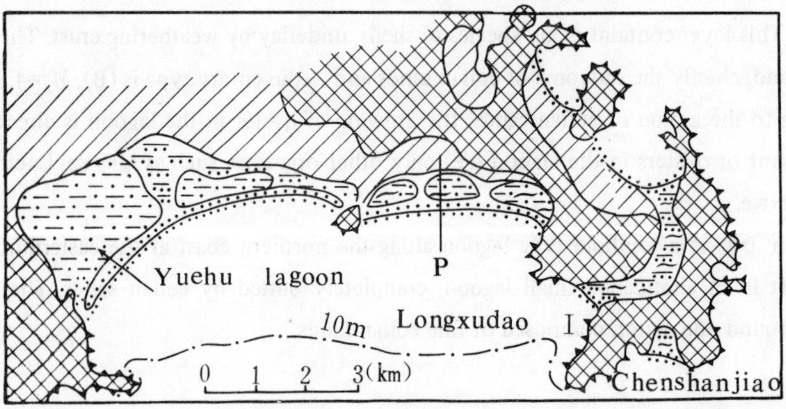

Fig. 2-33. Chain lagoons of eastern Shandong Peninsula. Line P is the profile of Fig. 2-29.

in achieving the expected low salinity because sea water permeates into the lagoon through the barrier. Water salinity can only be close to 5‰ in raining season annually. In long arid season, the salinity increases to 8–11‰.

(3) Lagoons at the southern coast of the Shandong Peninsula: Lagoons ㊶ – ㊸ are located at small estuaries that are being speedly filled up by alluvial sands. The lagoonal water area is small and shallow. Sand accounts for three-fourths of the area. The composition of sediment is mainly fine sand, with some medium sand and silt. The vertical sequence is coarse in the upper layer and fine in the lower layer. The upper layer is type (A)-1 in Fig. 2-31, the lower layer is type (A)-2. Because the underlying lagoonal layer of the bay is high in salinity, so it is hardly to be reclaimed. Small alkali-tolerant bushes are growing on the sandy land. The remains of shell organisms of the lower layer are well kept. (see Table 2-4)

The decadance of sessile organisms in the sedimentary layer of this sort of lagoon is an important mark of great change in lagoonal environments, showing that the lagoonal water area is filled up by the increasing river sand. After the lagoonal area is greatly reduced, the tidal prism is also greatly decreased. As a result, the inlet is nearly closed. The lagoonal area is full of fresh water. The water quality is changed, making the original environment that is suitable for many sessile organisms change into unsuitabe for shells. From Fig. 2-34, one can see a basic process of the lagoonal sedimentary changes.

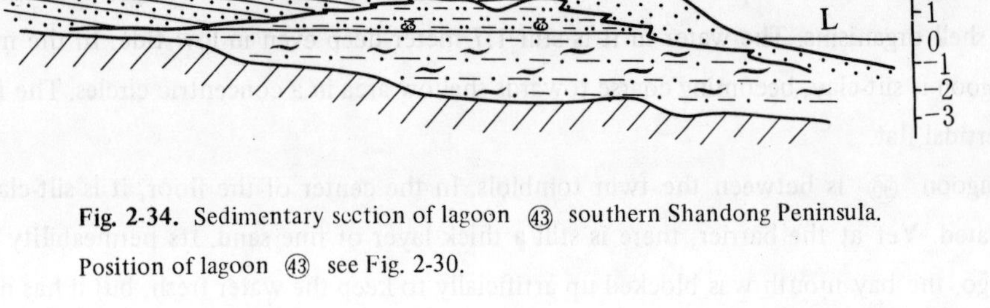

Fig. 2-34. Sedimentary section of lagoon ㊸ southern Shandong Peninsula.
Position of lagoon ㊸ see Fig. 2-30.

VI. Coastal Dunes

The coastal sand dunes are extensively distributed along China's sandy coast. From north to south, virtually all the sandy coasts are being remolded by wind to different extents. At some sections, wind sands are active and large scale sand dunes are built. At some sections unfavourable to wind action, only low sand layers are formed. According to preliminary surveys and statistics, the amount of coastal eolian dunes formed by wind in China is given in the following table:

Province	Liaoning	Hebei	Shandong	Fujian	Taiwan	Guangdong	Total
Dune coast (km)	65	75	95	90	50	210	585

Some typical eolian sand dunes and layers are chosen for discussion as follows:

1. Eolian coast in western Liaoning

Along the 50 km long sandy coast from Yianghe Estuary to Luanhe Estuary in Changli County, there is the largest coastal eolian in China. The plane distribution is like a sheep-horn, or a long and slim triangle with small curvature. The area of the dune is 35 km^2. Fig. 2-35 shows the width of dune is gradually widened from north to south. Its height is also increased gradually southwards. The northern part of the sand dune is 5 – 7 m high in general, but is more than 30 m high in the southern part.

The sediments, of which the coastal dunes are composed, are fine sands of good dispersion and small quantity of medium sands. The content of shell fragments is less than 1%. We analysed 12 samples on the dune and found that various parameters are quite similar to each other. As can be seen from the cumulative frequency curve (Fig. 2-36) that jumping particles are absolutely dominant, accounting for over 95% of the total sand samples, with individual one reaching 98%. M_z=2.4 –2.6ϕ, σ_1=0.3 – 0.5, being in positive skewness and single peak, and the kurtosis is big.

Where do the sand sources of such huge coastal eolian dunes in Jidong come from? How are the dunes formed? Why are these dunes concentrated at the northern coast of the Luanhe Estuary? All these questions should be clearly answered.

We hold that sands are obviousely transported in small quantity southwestwards from the Yianghe River. Yianghe is only a small intermittent stream, with a drainage area of a bit over 100 km^2. It has no large quantity of sands flowing into the sea, so it is unable to supply the large sand dunes. The two capes at Beidaihe and Qinhuandao protruding into the sea interrupt the longshore sand sources from north.

The average amount of sediments flowing from Luanhe into the sea is 18 million tonnes, 75% of which is concentrated in May to August when southeast wind prevails. Under the action of waves

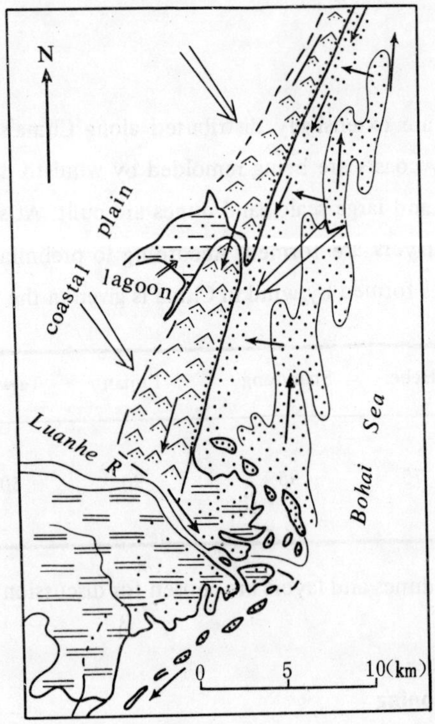

Fig. 2-35. Coastal dunes of Changli, Hebai (Abbreviated as Jidong dunes).

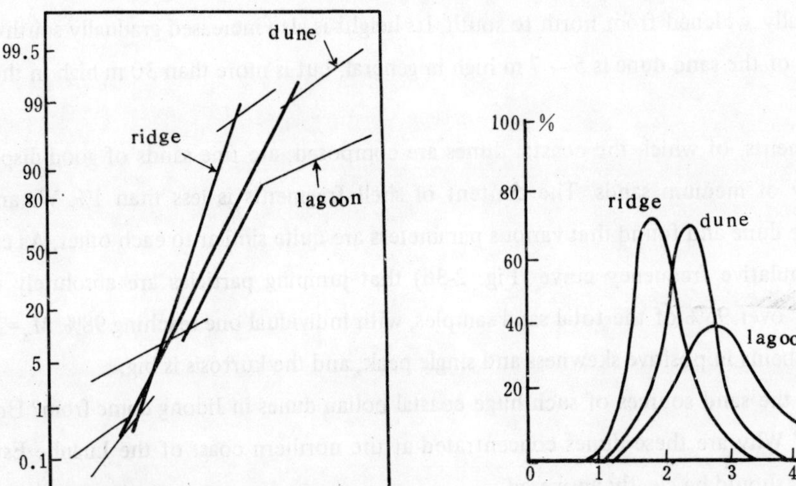

Fig. 2-36. Grain-size curves of dunes, lagoons and nearshore ridges of Changli, Hebei (Data from authers and Lu Haoxue).

from southeast, considerable amount of fine sands are transported northwards. The subaqueous sand ridge north of the Luanhe Estuary is ceaselessly supplied with sands from the Luanhe Estuary. The sediments on the subaqueous sand ridge are similar with those of the coastal eolian dune, both being composed of fine sands, (Fig. 2-36). In summer, the coastal sand ridges in Jidong are gradually enlarged. Under 11–14 m/s in velocity of northeast wind, fine sands on the coastal ridge move towards the sand dunes at 30 cm high and along the smooth beach with few grass. If one stands for several minutes, one centimeter of fine sands will accumulate at the back wind side of one's feet.

Sands on beach are transported towards dunes chiefly in autumn and early winter. At present, the huge Jidong Dune is slowly moving westwards. The causes are two: a) Although wind ENE is dominant, it is not an unique strong wind; b) Although east wind is relatively strong and can transport sands westwards, the northwest dry air current from Yanshan is often 14–20 m/s in velocity, and can transport a part of sands by east wind towards southeast. At present, the southern section of Jidong Dune is moving westwards in a low speed. On average, it moves ten meters westwards annually. Some trees at the southwestern edge of the dune are being buried by eolian sands.

The southern edge of Jidong Dune reaches the river bed of Luanhe. The north of the river bed is entirely different from the south. At the southern side of the river, there is no noticiable dune found at the beach and coast. It shows that wind sand is the most active in autumn, not in winter. In autumn, the river bed of Luanhe has still a certain amount of discharge flowing into the sea. In such case, all the sands are moving southwards from the dune under the action of northwest wind, and after reaching the river bed the sands are carried to the sea by river water. If wind sand were mainly active in winter, then the frozen river bed would become an ice bridge, favourable for wind sands to transport southwards. At the southern bank of Luanhe dune of certain size should have been formed. In fact, there is not any size of dune at the southern bank of Luanhe. This fact is enough to show that wind sand is not active in winter.

Besides, the retreating of eroded coast at Jidong Dune also supplies a certain amount of sand for developing dunes. Since the coastal sand ridge is transferring station for the fine sands on subaqueous sand ridge to be transported to dunes, the coast supplies the dune with sand sources, because it is eroded backwards. At most part of Jidong Dune, lagoonal layer bassets on the beach. This is a strong eveidence (Plate 2-9). The organic content of this kind of grey-black sand-clay layer is between 3 –7%.

Apparently, the sand sources of Jidong Dune are as fallows: a) The coastal barrier of Beidaihe and Yianghe in the northeastern part is eroded retreating, and blown to dune area by strong northeast wind; b) This section of coastal sand ridge is eroded shorewards, the sands are transversally moving under the action of east wind and entering into the dune; c) The sediments form Luanhe are transported northwards when it is mainly south wind in summer. A series of sand ridges and shallows are formed offshore. Then the sands are transferred from these ridges and shallows, entering the dune by transversal transportation. This is the largest sand sources; d) The sands of Jidong Dune are blown to the river bed of Luanhe by north wind and flown into the estuarine area along with the river water, Mixing with the sands transported northwards from the estuary. At present, the dune is being enlarged rapidly. The northern old dune is eroded, and the southern new dune is being enlarged. Ripples of the dune are quite clear (Plate 2-8). Why the sand-protecting forest zone on the dune is hardly to be built up is that the dune is growing and very active.

The area of dune and nearshore sand is being transported in an counter-clockwise cycle for a long period of time, that is an extremely interesting action. Under the action of northeast and northwest wind, large quantity of sands is transported southwards. This particles of fine sand are carried to estuary by river water as soon as they drop into the Luanhe River bed. Although northeast wind prevails in this season, the fine sands that flow into the sea along with river water are not to be directly transported northwards. They are either kept in the delta area or transported southwards, entering the chain sand isles. The fine sands that are kept at the delta are sand sourçes transported northwards by wind waves from southeast.

2. Eolian shell dune

Wind power has played a direct role in forming and disappearing of the shell sand ridge at the western coast of the Bohai Gulf. We discussed three main forming stages of the shell (chenier) ridge in the littoral Bohai Gulf, and have fully considered the action of wind on the transportation of shell fragments. Wind actions are positive and effective, and at the same time it is also variable. It exerts two influences on chenier ridges; a) Under the action of shoreward wind, shell fragments on the chenier ridges are transported to the depression behind the slope or to the top of the ridge, making the depression behind the ridge become a layer of laminar structure of shell and clay-silt. At last, the shell layer can be 7–8 m higher than the lagoonal flat or the top of the ridge; b) Under the action of offshore wind, the shell fragments on the ridge are blown towards beach. The shell fragments settle down on the slope of the ridge, and soon are brought back to the ridge by waves. If the clay at the tidal flat is not thick, waves can still stir the clay and push the shell fragments towards the ridge by storm. At the clay flat where clay layer is thick, shell fragments may not be restored on the ridge, because the action of wind waves is weak.

In November 1964, we visited the Qikou chenier ridge when east wind happened to be 11 –17 m/s in velocity. The strong northeast wind carried shell fragments, blowing against our faces as soon as we were near the top of the chenier ridge. The diameter of mest shell fragments blowing aganist us is 2 –3 mm, sometimes 5 mm in size.

In the growing stage of the shell ridge, the accumulation of shell is larger than consumption, the action of wind is still important. This can be illustrated by the following equation:

$$A > B + C$$

In the equation, A represents shell quantity supplied for ridges from the tidal flat; B represents the quantity of shells transported longshore from the ridge to the estuary; C represents the quantity of shells blown to the depression behind the ridge under wind action.

When the shell ridge is developed to the fullest and begins to stop growing, the equation is:

$$A \simeq B + C$$

From then on, the quantity of beach sediments is all the more increased. The gradient of tidal flat becomes gentle. After the shell organism of the tidal flat dies off, the quantity of shells transported shorewards decreases. Because the slope is gentle, waves is weak and shells are buried by clay-silt layer. Therefore input of shells is smaller than output. The equation is: $A < B + C$

This is the mark that the chenier ridge is entering its decaying stage.

The above quantitative relations and Fig. 2-37 clearly shows that wind power ceaselessly takes part in the whole process of the development of chenier ridge.

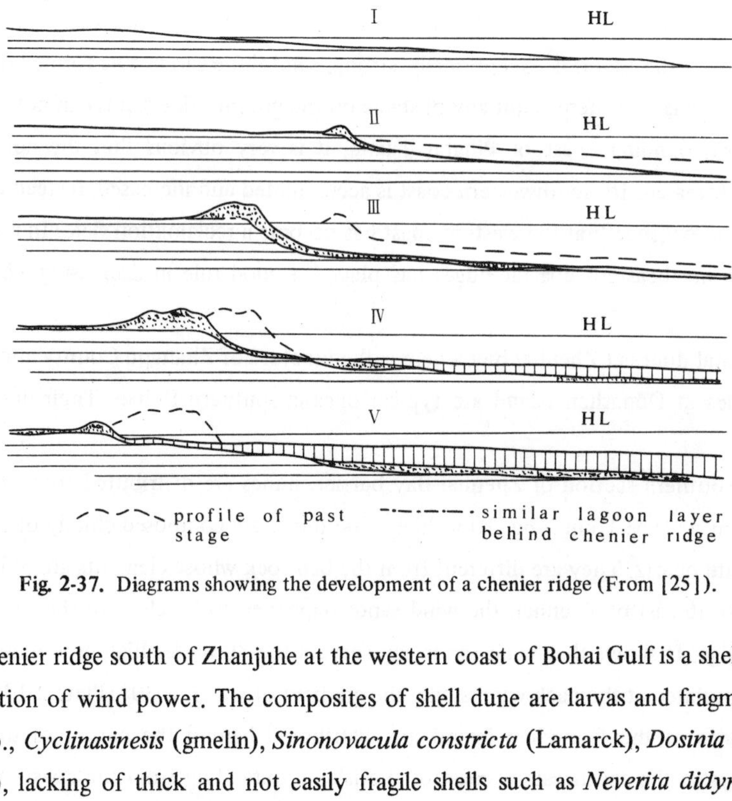

Fig. 2-37. Diagrams showing the development of a chenier ridge (From [25]).

The large chenier ridge south of Zhanjuhe at the western coast of Bohai Gulf is a shell dune growing under the action of wind power. The composites of shell dune are larvas and fragments of thin-shlled *Aloidis* sp., *Cyclinasinesis* (gmelin), *Sinonovacula constricta* (Lamarck), *Dosinia (phacosoma) japoinca* (Reeve), lacking of thick and not easily fragile shells such as *Neverita didyma* (Roding), *Meretrix meretrix* (Linnaeus), and *Rapana thomasiana* Crosse, etc. Shell dune has apparent and irregular slant laminated layers. The lamination tilting landwards is 5°–25°. The incline of shell dune top is little. The back slope of shell dune is about 20° steep. The ridge is 8 m high, 300 m wide. Behind the shell dune, there is a 300 – 400 m wide accumulation area mixed with shell fragments and clay-silt, with unapparently laminated layer. Because the topography is high and the capillary action is cut by the shell fragments, the underlying salty water is not easy to rise. Soil is not easy to be salified. Therefore crops grow well, being an "oasis" on the littoral clay zone.

In the shell-concentrated small estuarine areas at Qikou and Zhaujiabao, shell fragments are often found pushed by wind to the surroundings of wild Jujube grove, making small dunes there 2–3 m higher than the ridge.

3. Dunes at Fujian coast

In Pintan Island, Jinjiang County and Dongshan Island, coastal dunes are very well developed.

Pintan Island is known for its strong wind. Wind sand is much active. The wind sand deposits prevail the flat ground and depressions in the whole island. The coastal ridges (beach) facing northeast and north at the island are at present seriously eroded by wind power. Most fine sands blown away by wind, are filling the vast depression and lagoonal area. We observed on the spot that the fine sands on the ridges of the northeastern bays are constantly blown to southwestern beach during north wind over 9 m/s in velocity, and further transported southwestwards over the filled depression. It has been proved by drilling data that a ten meters deep channel has been filled by wind sands. Because wind on the island is exceptionally strong, sand dune can not be formed on the flat ground. The other reason is that there is not any obstacle on the ground, fine sands can not be retained either.

In the surrounding coast of Pingtan Island, it is very obvious that the northeastern coast is eroded and retreated. Its southwestern coast is accumulated and increased, instead of developing into dunes. It is an evidence that the western coast has become a reclamation field. In recent years, forestation on the northeastern coastal ridges has played a good role in decreasing wind power and fixing sands.

The coastal dunes at Zhenhai Bay where is the juncture of Zhangpu County and Longhai County, and the dunes at Dongshan Island are typical ones in southern Fujian. Their dimentions are fairly large.

At the southern section of Zhenhai Bay barrier, dunes are distributed from the coast edge in a triangle, 1 km wide, 2.1 km long, 10 m high. The dunes are composed chiefly of fine sands of white and grey white quartz. They are different from the bed rock whose elements are of basic igneous rock.

At the southeast of Zhenhai, the wind sands happened to be close to the southern slope of the mountain ridge. It shows that wind sands can surpass the 30 meter high mountain ridge of bed rock and be accumulated at the back wind slope, the gradient of which is 30° (Plate 2-10).

The dunes at the Dongshan Island are distributed behind the barrier of the Dongsheng and Huidong Bays in the southeastern coast of the island, extending from the coast towards land. It is 3 km wide. The development of wind sands has made the ground covered with a layer of pure white fine sands of 5–10 m thick, or even much thicker.

The dunes in the Dongsheng Bay are distributed on the hill slope. The highest dune is 70 m above the sea level. The thickness of sedimentary layer of wind sands is different, increasing gradually from shore to the land. At present, the wind sand of the Dongsheng Bay tends to be stable. There are two reasons. a) At present, sands on the beach and foreshore are mainly coarse sands, small gravels, with little medium sands. Fine sands are even much less, lacking of fine sand sources that are necessary for building up dunes; b) On nearly all the sand area are planted *casuarina* trees, growing thickly.

The dunes are little remolded by wind, with small-size dunes near the coastal bushes being accumulated. Their heights are 2–3 m and they are not large-size. Moreover, they are seldom moved (Plate 2-11).

An 8 km long and 1.5 –4 km wide tombolo, called Gongdou Tombolo protrudes from the southern part of the Dongshan Island. The Huidong Bay is situated to the east of the tombolo, and the Xipu Bay to the west. The Huidong faces the open sea. It is a coast under direct action of east wind. The dynamic conditions of the two bays are obviously different. The wave height in the Huidong Bay can reach 4.5 m when maximum wave occurs, but the wave height in the Xipu Bay is less than 2 m, under the same conditions.

Based on data of 28 drillings on this tombolo, its surface layer is around 10 m thick (different in various places), composed of pure grayshite fine sands. Its underlying layer is green-grey sandy mud of many oysters and shells. The layer sequençe shows that where the tombolo is located today, it was just the bay sediment in the past, or at least, it is a shallow bay behind the barrier or a lagoonal environment. (Fig. 2-38).

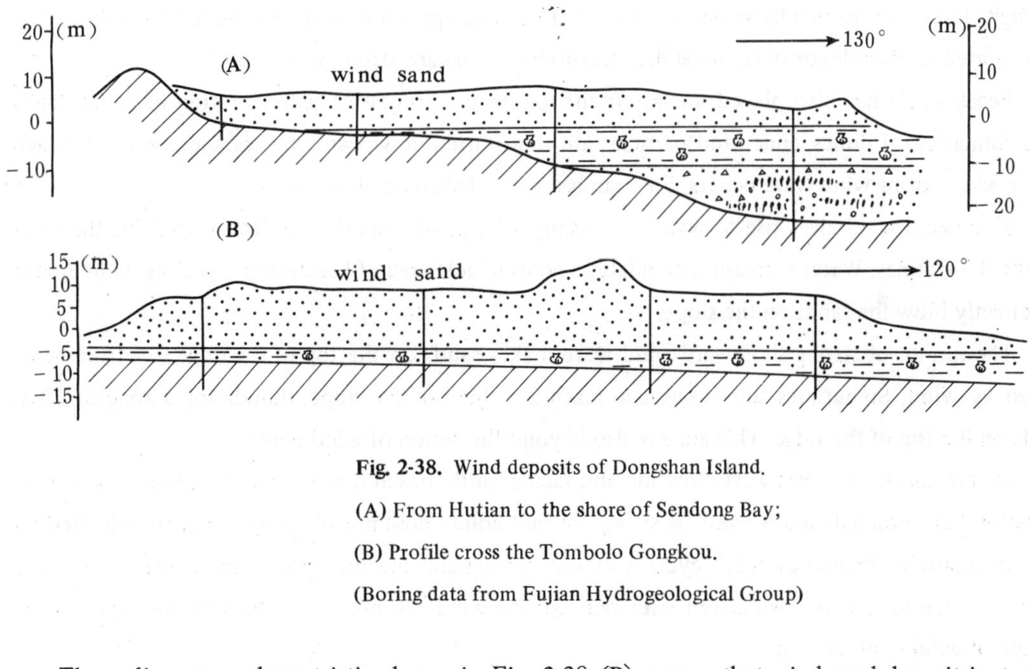

Fig. 2-38. Wind deposits of Dongshan Island.

(A) From Hutian to the shore of Sendong Bay;

(B) Profile cross the Tombolo Gongkou.

(Boring data from Fujian Hydrogeological Group)

The sedimentary charactristic shown in Fig. 2-38 (B) proves that wind sand deposit is strengthened later on, blowing the sands on the barrier to the shallow behind the barrier. In East Bay and West Bay of the Gongkou Tombolo, the content of beach sand is quite different. Wave is strong in East Bay and sand ridge is high and large. It is composed of coarse, medium and coarse sands and small gravels. While in West Bay, wave is weak, sand ridge is small and sands are fine. Undoubtfully, sands of the Gongkou Tombolo are blown from East Bay. Eolian sands blow from east to west without any topographical obstacles before trees are planted to protect sands.

After analysing four sand samples on the traversal profile of the tombolo, we found out that only sands on the barrier in East Bay are exceptionally coarse, being 0.5 –1ϕ. While the rest of the sample sands are 2.5 –2.8ϕ. The degree of dispersion of the tombolo in East Bay is the poorest.

5. On the formation of dunes

The study of forming process of coastal dunes has given rise to general attention. Detailed expositions on the coastal sediment have been given in many treatises [69,74]. Pneumatic machines and ultraspeed photography are used to survey the starting of sand movement and moving orbit under the action of different wind power, which helps to understand the movement of sand particles and sedimentary process.

Strong, dominant wind–northeast wind occuring in autumn and early winter is the main dynamics for constructing China's coastal dunes. The time when northeast wind prevails is exactly the arid season in China. The air current blowing from north to south remains dry and is favourable for the activity of wind sands. Northeast wind happens after small wave caused by west wind (offshore wind) in early autumn. Most fine sands on the ridge remain on the top of it and has not yet been brought to nearshore area by strong wind wave. This is an opportunity favourable for fine sands to be transported landwards (or to lagoonal depression) by shoreward strong wind.

Beach cycle has also played an important role in the process of sedimentation of eolian sands and formation of dunes. Fine sand particles are transported downwards to nearshore zone by beach cycle when strong wind wave occurs. Fine sands are avoided to be blown away.

Shoreward wind gives rise to water-increasing, a frequent and effective factor checking the development of dunes. Water-increasing inundates a considerable area of beach slope, making wind unable to directly blow the sands on the slope.

The run-up of strong shoreward wind wave makes sands on the top of the ridge free from being acted by wind. Sometimes the run-up can reach 2 m high of the slope, inundating a zone of 20 m wide on the top of the ridge. This zone is also beyond the action of wind sands.

On the coastline in east-west direction and facing south, beach dunes are seldom found. The main cause is that, although south wind is strong, in fact eolian dose not play any or a little role. Besides the unfavourable factors as beach cycle, water-increasing and the run-up of wave, most strong wind occurs in summer, being associated with rain. Strong wind can not move the wet fine sand, so the action of wind sands is weak.

At present, the coasts where large-scale dunes developed are in north-south direction of close to north-shouth direction, and facing east, whether they are in northern China (such as Jidong Dune) or in southern China (such as Fujian, Guandong) (Fig. 2-35), slanting to the coastline. Wave crest line is not verticle to the coast. Wave action must be weakened through diffraction. There is no any big water uplifting and water run-up. So, a considerable area of beach slope is still within the scope of wind field. It is conducive to build dunes by transportation of the fine sands by wind.

Under the present climate environment, the action of wind sands leads to the extensive development of dunes and to the destruction of lagoons at many coastal sections or to the state of the lagoons being filled up. Meantime, the action of wind sands make China's many coastline retreat. The speed of such retreat is considerable fast in coast of north-south direction. Many facts have made us realize that, at present, the quantity of sand supply to the coast from nearshore floor is very limited. It is the coastal sand bodies that supply dunes with sands. This is the chief cause for the retreat of coastline.

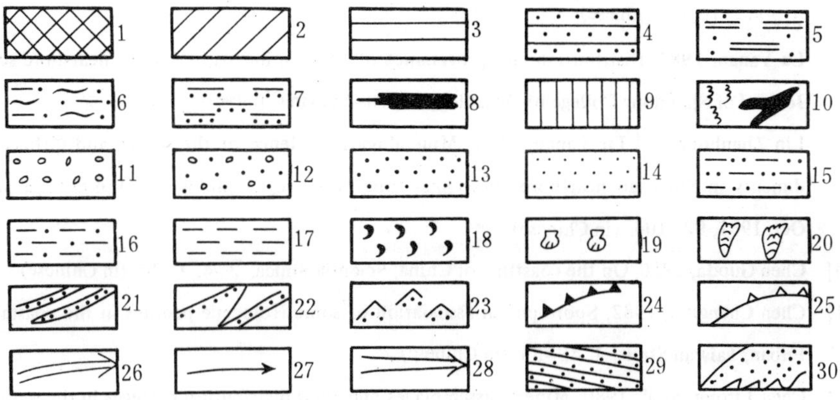

Fig. 2-39.

Fig. 2-39. Explanation of illustrations in this chapter.
1. Bed rock or hill; 2. Continental facies; 3. Clay or silt of marine facies; 4. Marine-alluvial sand; 5. Alluvial plain or delta deposit; 6. Lagoon deposit, ~ with organic matters; 7. Dry lagoon; 8. Swamp or peat; 9. New marine layer; 10. Bench and bed rock in bottom; 11. Pebble-cobble; 12. Mixture of sand-gravel; 13. Midcoarse sand; 14. Fine sand; 15. Silt or clay-silt; 16. Sandy clay or deposit of lagoon and bay; 17. Clayey mud; 18. Shell and its detritus; 19. Mollusc intact or in original position; 20. Oyster; 21. Coastal spit formed by sand and gravel; 22. a) Beach ridge, b) barrier; 23. Dune; 24. Marine cliff; 25. Old Marine cliff; 26. Suspended sediment transport; 27. Rolling transport; 28. Direction of wind action; 29. Alluvial; 30. Sand fan.

References

[1] Wang Xing, 1980, Landscape of Taiwan, Dujia Press, pp58–62, 158–159. (in Chinese)

[2] Jen Meingo, 1965, Quaternary changes in sea-level and their effects on coastal morphology, Oceanologia et Limnologia Sinica, 3, 295–305. (in Chinese)

[3] Ren Meie (Jen Meingo) and Zeng Chenkai, 1980, Late Quaternary continental shelf of East China. Acta Oceanologica Sinica, 2, 94–111. (in Chinese)

[4] Tian Zhengping, 1977, A study of the heavy minerals and its sorting of Wanyiagan Shallow in western Taiwan, Oceanographic Acta, No. 15, 96–101. (in Chinese)

[5] Li Yuanfang, Gao Shanming and An Fengting, 1982, A preliminary study of the Quaternary marine strate and its paleogeographic significance in the Luanhe delta region, Oceannologia et Limnologia Sinica, 5,433–439. (in Chinese)

[6] Li Qinyuan, 1935, On the emergence-submergence of coastline in China, Acta Geographica Sinica, 2(2), 2, 129–166. (in Chinese)

[7] Jen Mingta, Liang Shaolin, 1965, A genesis of longshore single ridges of Qinhuandao area, Geological Review, 3, 200–210. (in Chinese)

[8] Lu Yanan, 1982, Distribution and provenance of heavy minerals in Rizhao nearshore sediments, Shandong, Jouranl of Shandong College of Oceanlogy, 1, 78–86. (in Chinese)

[9] Lin Zhenhong and Lu Yanan, 1981, Mineralogical evidence of the source and diffusion of the surface sediments in Shijiusuo nearshore zone, Shandong Province, Oceanlogia et Limnologia Sinica (Supplement), Oct. 1981, 92–104. (in Chinese)

[10] Chen Guoda, 1950, On the coastline of China, Scientia Sinica, 2–4, 2–28. (in Chinese)

[11] Chen Chenghui, 1982, Sporo-pollen association of some Holocene profiles in the coastal region of southern Fujian, Taiwan Strait, 1, 45–53. (in Chinese)

[12] Chen Lirong, et al, 1980, Mineral assemblages and their distribution patterns in the sediments of the Gulf of Bohai Sea. Oceannologic et Limnologia Sinica, 3, 204–210. (in Chinese)

[13] Feng Wenke, et al, 1982, Preliminary study on submarine relief of the northern South China Sea, Acta Oceanologica Sinica, 4, 472. (in Chinese)

[14] Zhao Huanting, 1980, Geomorphologic feature of embayed coast from Yamen to Moyang River, Oceanologia et Limnologia Sinica, 2, 121–133. (in Chinese)

[15] Zhan Jinwen and Zhao Xitao, 1979, Chronological study on the stratum and new tectonics of Late Quaternary, coastal of souther Fujian and eastern Guandong, Seismologia et geologica Sinica, 3, 27–36. (in Chinese)

[16] Zhao Xitao, et al, 1979, Sea level changes of the eastern China during the past 20,000 years, Acta Oceanologia Sinica, 2, 269–281. (in Chinese)

[17] Xia Dongxing and Liu Zenxia, 1981, Technique of remote sensing used to the coastal investigation, J. of Marine Techique, 5, 54–62. (in Chinese)

[18] Gao Mingdao, et al, 1980, Study on the source of mud-sand in the port of Tanggu, Studia Marina Sinica, No 16, 33–57.

[19] Qin Yunshang and Zhen Tiemin, 1978, A preliminary study of the peat bed on the floor of the north-western Bohai Sea, Studia Marine Sinica, 3, 4–8. (in Chinese)

[20] Han Yousong, 1980, The ostracean reefs and the Xinhe paleo-coastline, *ibid,* 16, 59–66. (in Chinese)

[21] Chai Yuee and Chai Aizhi, 1980, Development of Geomorphology on the eastern coast of Laizhou Bay, Shandong, Transaction of Oceanology and Limnology, 1, 28–34. (in Chinese)

[22] Chai Aizhi and Li Xinyuan, 1964, Some features of coral reefs in southern coast of Hainan Island, Oceanoloia et Limnologia Sinica, 2, 205–218, (in Chinese)

[23] Chai Aizhi, 1978, On the formation of the Zhifu Tombolo. *ibid,* 1, 1–14. (in Chinese)

[24] ————, 1980, Development of the accumulational geomorphology of the spits Diaolongzui, *ibid,* 11, 204–210. (in Chinese)

[25] ————, 1981, A preliminary study of the origin of the chenier ridges of Bohai Gulf, Studia Marine Sinica, 18, 117–132. (in Chinese)

[26] ————, 1982, Diffusion of sediments of the Changjiang River discharging into the sea. Acta Oceanologica Sinica, 1, 78–88. (in Chinese)

[27] ————, and Chai Yuee, 1983, A study of depositional causation and characteristics of coast eolian sand in coast area in China, J. of Desert Research, 3, 1–10, (in Chinese)

[28] Yang Yaolin, The Summary of tidal flats of Taiwan, Oceanographic Acta, No. 15, 133–136. (in Chinese)

[29] Exploitation and utilization of tidal flats in Taiwan, Oceanographic Acta (Taiwan), 16, 59–62.

[30] Barrell, J., 1925, Marine and terrestrial congromerates, Geol. Soc. Amer. Bull., V. 36, pp 279–341.

[31] Bird, E.C.F., 1973, Australian coastal barriers, in Barrier Islands, Dowden Hutchinson and Ross Inc. Publishers, pp 410–426.

[32] Bloom, A.L., 1970, Paludal Stratigraphy of Truk Ponape and Kusiae eastern Carolina Islands. Geol. Soc. Am. Bull., V. 81, pp 1895–1904.

[33] Boothroyd, J.C., 1978, Coastal sedimentary Environments, Edited by R. A. Davis Jr. Springer Verlag New York pp 287–360.

[34] Bruun, P., 1966, Tidal inlets and littoral drift stability of coastal inlets, V. 2, pp 139.

[35] Bruun, P., 1978, Stability of Tidal Inlets Elsevier Amsterdam, pp 13–37.

[36] Byrhe, J. V., D. O. Leroy and C. M. Riley, 1959, The Chenier plain and its stratigraphy southwestern, Louisiana, Trans. Gulf. Coast. Assn. Geol. Soc., V. 9, pp 237–259.

[37] Curray, J. R., F. P. Shepard and H. H. Veeh, 1970, Late Quaternary sea-level studies in Micronesia: CARMEARSEL Expedition, Geol. Soc. Amer. Bull., V. 81, pp. 1865–1880.

[38] Beaumont, Elie, 1845, Lecons de geologie pratique, Paris, pp 223–252.

[39] Dietz, R. S,. 1963, Wave-built, marine profile of equilibrium, and wave-built terraces, a critical appraisal, Geol. Soc. Amer. Bull., V. 74, pp 971–990.

[40] Dolan, Robert, 1965, Seasonal variations in beach porfiles along the Outer Banks of North Carolina, Shore and Beach, Oct.

[41] Emery, K. O., and Hiroshi Niino, 1967, Straitgraphy and petroleum prospects of Korea Strait and the East China Sea, Geophysical Exploration, V. 1 No. 1, Geol. Survey Korea, pp 1–19.

[42] Ewing, Maurice, W. J. Ludwing, and John Ewing, 1963, Geological investigations in the submerged Argentine coastal plain, Geol. Soc. Amer. Bull., V. 74, pp 275–292.

[43] Fairbridge, R. W., 1961, Eustatic changes in sea level, In physics and Chemistry of the Earth, Vol. 4, Pergamon, New York, pp. 99–185.

[44] Fisher, J. J., 1967, Origin of barrier island chain shorelines Middle Atlantic states. Ged. Soc. Am. Sp. paper 115 pp 66–67.

[45] Fisher, J. J., 1968, Barrier island formation, Discussion Geol. Soc. Am. Bull., V. 79, pp 1421–1426.

[46] Gilbert, G. K., 1885, The topographic features of lake shores, U. S. Geol. Survey 5th Ann. Rept. pp 69–123.

[47] Gould, H. R., and E. McFarlan, Jr., 1959, Geologic history of the chenier plain, southwestern Louisiana,

Trans. Gulf Coast Assn. Geol. Soc., V. 9. pp 261–272.

[48] Guilcher, Andre, 1959, Coastal sand ridges and marshes and their environment near Grand Popo and Ouidah, Dahomey. In 2nd Coastal Geography Conference, R. J. Russell, Chm. Coastal Studies Inst. Louisiana State Univ., Baton Rouge, pp. 189–212.

[49] Hayes, M. O., 1979, Barrier island morphology as a function of tidal and wave regime, in Barrier Island Edited by S. P. Leatherman Academic Press, New York, pp 1–28.

[50] Halsey, S. D., 1979, Nexus: New modal of Barrier island development, in Barrier island Edited by S. P. Leatheman, Academic Fress, New York, pp 185–210.

[51] Hoyt, J. H., 1967, Barrier island formation. Geol. Soc. Amer Bull, V. 78, pp 1125–1136.

[52] Hoyt, J. H., 1969, Chenier versus fassier, genetic and stratigraphic distinction, Amer. Asso. Petroleum Geol. Bull., V. 35, pp 298–306.

[53] Inman, D. L., 1953, Areal and seasonal variations in beach and nearshore sediments at La Jolla, California, Beach Erosion Board, Corps Engrs. Tech. Memo., No. 39, pp 134.

[54] Inman, D. L. and R. A. Bognold, 1963, Littoral processes. In The Sea, vol. 3, N. M. Hill, ed., Interscience New York, pp. 529–553.

[55] Johnson, D. W., 1919, Shore Processes and Shoreline Development, Wiley, New York, pp 684.

[56] Joseph R. C., 1973, Transgression and regression, in Marine Geology, Macmillan Company, New York, pp 175–203.

[57] LeBlanc, R. J.,, and W. D. Hodgson, 1959, Origin and development of the Texas shorelines, Gulf Coast Assn. Geol. Soc, Trans., V. 9, pp. 197–220.

[58] Kuenen, Ph., 1950, Marine Geology: New York John Wiley and Sons Inc, pp 568.

[59] Leatherman, S. P. Edited, 1979, Barrier Islands from the Gulf of St. Lawrence to the Gulf of Mexico Academic Press, New York, pp 325.

[60] Milliman, J. C. and Emery, K. O., 1968, Sea levels during the past 35,000 years, Science 162, pp 1121–1123.

[61] King, C. A. M., 1972, Beaches and Coasts, 2nd ed., Edward Arnold London, pp 321–353, 670.

[62] McCann, S. B., 1979, Barrier island in the Southern Gulf of St. Lawrence, Canada, its Barrier Islands, Edited by S. P. Leatherman Academic Press, pp 29–64.

[63] Moore, D. G. and J. R. Curray, 1964 B, Wave-base, marine profile of equilibrium, and wave-built terraces: discussion. Geol. Soc. Amer. Bull., V 75, pp 1267–1274.

[64] Otvos. E. G., Jr., 1970, Development and migration of barrier islands, northern Gulf of Mexico. Geol. Soc. Amer. Bull., V. 81, No. 1, pp 241–246.

[65] Nüno, Hiroshi, and K. O. Emery, 1961, Sediments of shallow portions of the East China Sea and South China Sea, Ged. Soc. Amer. Bull., V. 72, pp 731–762.

[66] Pierce, J. W. and Colquhoun, D. J., 1970, Holocene evolution of a portion of the North Carolina coast. Geol. Soc. Amer. Bull., V. 81, pp 3697–3714.

[67] Phleger, F. B., 1969 B, Some general features of coastal lagoons, In Coastal Lagoons, A. A. Castanares and F. B. Phleger, eds., Univ. Nacional Autonoma Mexico, pp. 5–25.

[68] Phleger, F. B. and G. C. Ewing, 1962, Sedimentology and oceanography of coastal lagoons in Baja California, Mexico, Geol. Soc. Amer. Bull., V. 73, No. 2, pp 145–182.

[69] Richard, A. D. Jr. Edited, 1977, Coastal Sedimentary Environments Springer-Verlag, New York, 420 pp 361–383.

[70] Russall, R. J. and H. V. Howe, 1935, Chenier of Southwestern Louisiana Geogs. Rev., V. 25, pp 449–461.

[71] Ruzyla, K. V., 1973, Effets of erosion on barrier island morphology at Fire Island, in Coastal Geomorphology, New York, pp 219–339.

[72] Shepard, F. P., 1973, Submarine Geology, Third ed., Harper & Row, New York, pp 517.

[73] Sorensen R., Basic Coastal Engineering, 1978, John Wiley & Sons, New York, pp 227.

[74] Rosen, P. S., 1979, Aeolian dinamics of a barrier island system, in Barrier Islands, Edited by Leatherman S. P. Academic Press, New York, pp 81–98.

[75] Postma, H., 1969, Chemistry of coastal lagoons, In Coastal Lagoons, A. A. Castanares and F. B. Pheleger, eds., Univ. Nacional Autonoma Mexico, pp 421–429.

[76] Swift, D. J. P., 1075, Tidal ridges and shoal retreat massifs, Marine Geology, 18 pp 105–134.

[77] Swift, D. J. P., 1975, Barrier island genesis: evidence from the central Atlantic shelf, eastern U. S. A. Sedimentary Geology, 14, pp 1–43.

[78] Wrigth, L. D., 1977, River Delta, in Coastal Sedimental Environments Edited by R. A. Davis, Jr. Springer-Verlag, New York, pp 1–68.

[79] Van Straaten, L. M. J. U., 1965A, Coastal barrier deposits in South and North-Holl and, in particular in the areas around Scheveningen and Ijmuiden, Mededeel, Geol. Soc., Nieuwe ser., No. 17 pp 41–75.

[80] Thorn, R. B., 1960, The design of sea defence works, Butterworths Publications Limited, pp 29–30.

[81] Wright, L. D. and Coleman, J. M., 1973, Variations in Morphology of major river deltas as functions of ocean wave and River discharge regimes, Am. Assoc. Petral. Geol. Bull., V. 57, pp 370–398.

[82] Leontyev, O. K., 1969, Feandrean transgression and the genesis of barrier bars. Quaternary Geology and Climate, Edited by H. E. Wright, pp 146–149.

[83] Zenkovitch, V. P., 1959, On the genesis of cuspate spits along lagoon shores, Jour. Geol., V. 67, pp 267–277.

By Cai Aizhi and Cai Yue-e
Xiamen University, Xiamen

Chapter Three
Tidal Mud Flat

I. General Background

In China, coast with tidal mud flat is more than 4000 km long, or about 25% of total lenght of mainland coast-line of China (Fig. 3-1). It is Mainly distributed along the coastal plain of Jiangsu Province, deltas of the Changjiang River, Huanghe River and Zhujiang River, Bohai Gulf, Laizhou Bay and bays along the coast of Liaoning, Zhejiang, Fujian and Guangdong Provinces.

Sedimentation of tidal mud flats has received increasing attention from marine geologists. But up to the present, detailed sedimentological research is largely focussed on tidal flats flanking barrier islands, such as those of the North Sea coast of the northern Netherlands and northwestern Germany. But as only about 5600 km of the world's coastlines are barrier island coastlines, their value as analogs for the study of sedimentary rocks must be open to question. As pointed out by Klein (1975), perhaps open coast tidal flats were more characteristic of tidal flat sedimentation in the past. Recent valuable works by Boersma et al. (1981) and Nio et al. (1980) concerning sedimentation of intertidal shoal of the Scheldt are not the representatives of coastal tidal flats, as their depositional environments are considerably different. Therefore, a highly important topic for future research is the nature of depositional processes and sediments occurring on tidal mud flats flanking open coasts.

Tidal mud flat of China, with great width (>10 km) and a variety of sedimentary environments, may furnish some models for the study of sedimentation of mud flat in the world.

1. Classification of tidal mud flat

Tidal mud flats of China may be grouped into different categories. According to tidal range, macrotidal flat, mesotidal flat and microtidal flat (Davies, 1964) may be distinguished. Tidal flats of Zhejiang and Fujian Provinces belong to macrotidal, tidal flat along Jiangsu coast is mainly mesotidal, whereas tidal flats of Bohai Gulf and Guangdong Province have the smallest tidal range, i.e. microtidal.

The geographical location of tidal flat affects its hydrological condition; hence, it is closely related to characteristics of sedimentation of mud flat .

According to geographical location, tidal mud flat of China may be classified into:

(1) *Embayment type.* The coast of Zhejiang, Fujian and Guangdong Provinces is largely ria coast with rocky promontories and bays. Tidal mud flat is chiefly located at the head of bays,

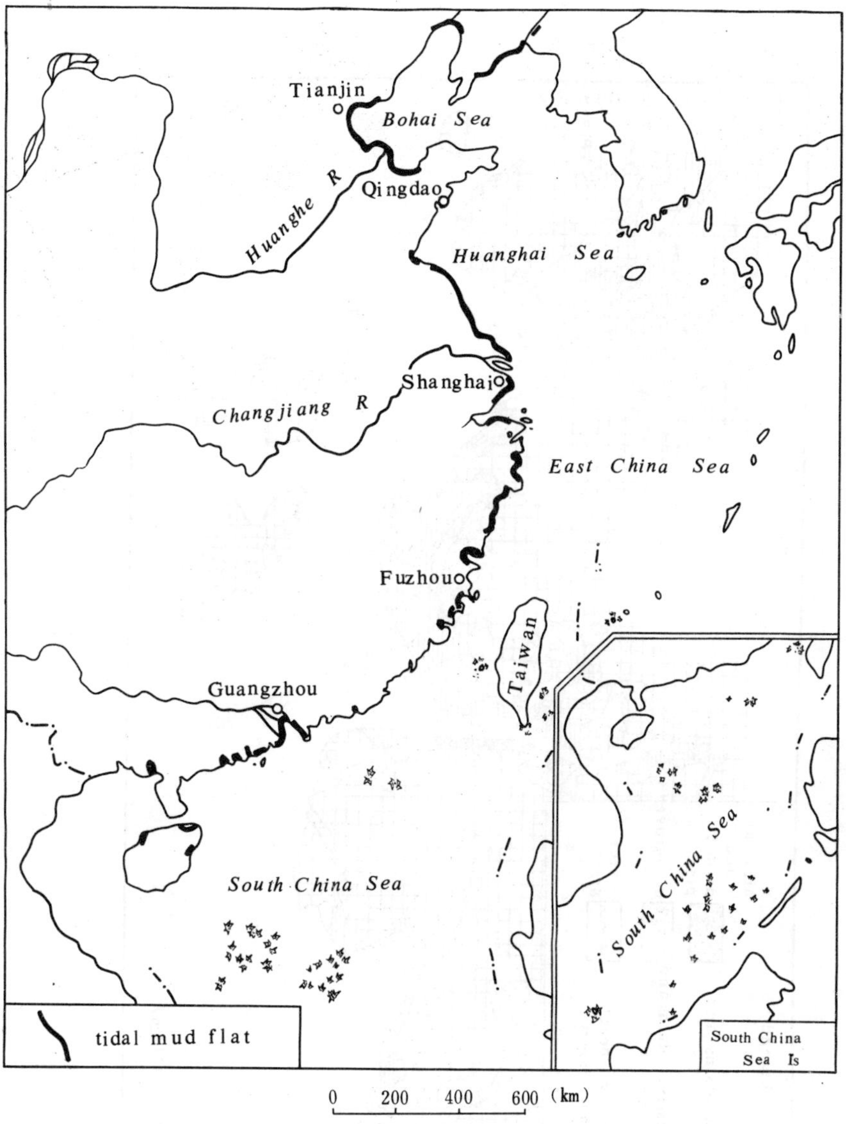

Fig. 3-1. Sketch map showing distribution of tidal mud flat in China.

which are often semi-enclosed or with islands at mouths of the bays, as Taizhou Bay and Yueqing Bay, Zhejiang Province and Guanghai Bay and Zhenhai Bay, Guangdong Province. This type of mud flat is characterized by its discontinuous distribution, short stretches of mud flat being separated by rocky coast and sand beaches (Fig. 3-2).

(2) *Open coast type.* As mud flats along the Bohai Gulf and Laizhou Bay and in Shanghai Municipality. Although mud flat along the coast of Jiangsu Province belongs essentially to open

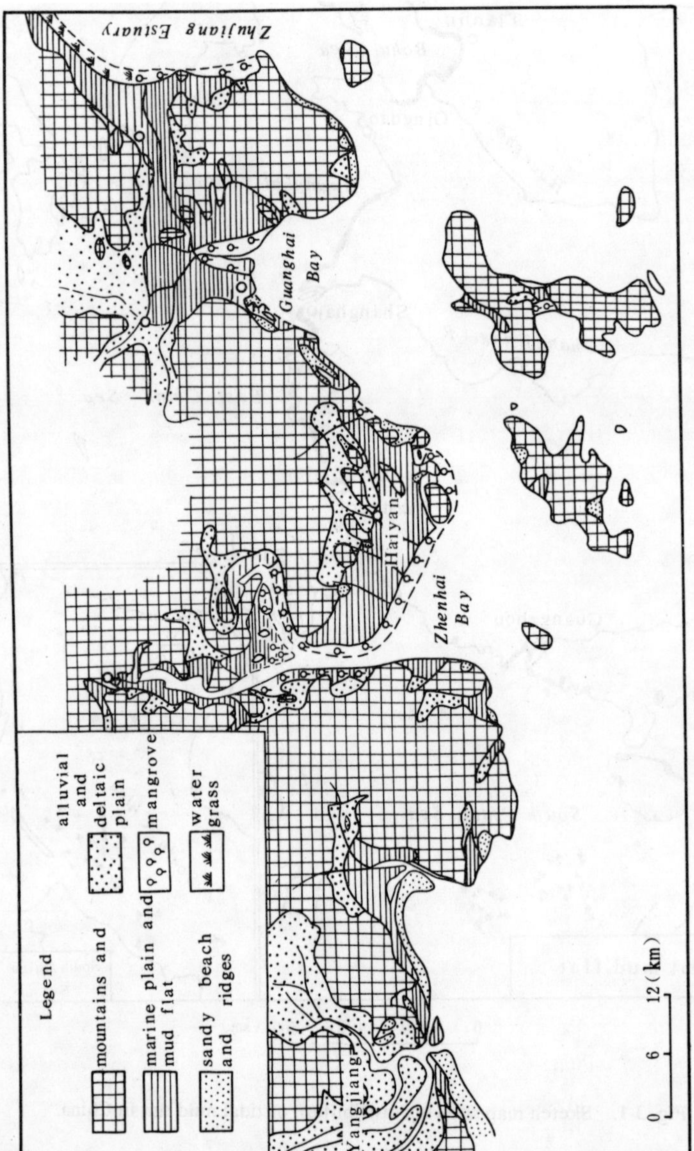

Fig. 3-2. Tidal mud flat of embayment type along the coast of western Guangdong (After Zhao Huanting et al., 1980).

coast type, but here, as a greater part of the coast is sheltered by large submarine sand ridges, its hydrological conditions are somewhat different from open coast without offshore submarine ridges, and it is regarded as a subtype of open coast mud flat.

(3) *Estuarine type.* This type of mud flat is chiefly found along the estuary of the Zhujiang River, Qiantang River and Changjiang River. Affected by fresh water discharge of rivers, water along the mud flat has low salinity. Estuarine mud flat along the south bank of the Changjiang River and Qiantang River Estuarys is scoured by strong currents and is easily slumped. Consequently, in general, it is only suitable for reclamation by construction of solid sea dikes, but not for aquaculture, The mouth of the Qiantang River is a wide estuary; owing to plentiful supply of suspended sediment from the Changjiang River mud flat along its south bank is prograding rapidly and is very wide with an area about 2 million *mu*[1)]. It is the largest estuarine mud flat in China. Since the establishment of the People's Republic of China, more than 900,000 *mus* of mud flat have been reclaimed.

Tidal mud flat of China may be further grouped into prograding and receding, depending on the balance between amount of sediment supply and strength of wave and tidal current in the littoral zone. For example, in Jiangsu Province, owing to sudden decrease of sediment supply after shifting of the outlet of the Huanghe River into the Bohai Gulf in 1855 (Fig. 3-3), the coast near the mouth of the abandoned Huanghe River has changed from prograding into receding and the surface of mud flat has been lowered by the scouring of wave and current. It is estimated that since 1855, the coast around the abandoned Huanghe Estuary has been retreated about 20 km. There are some localities, where although sediment supply has not changed, but owing to stronger wave and tidal current, mud flat and coast is receding, as coast near Lusi, Jiangsu Province. Receding mud flat is, as a rule, much narrower than prograding mud flat and has steeper gradient and coarser sediments. For example, mud flat near the mouth of the abandoned Huanghe River is often only 0.5 km wide, with a gradient about 4%.

It may be noted that the original subaqueous slope, on which marine mud was deposited, has a considerable influence on the gradient of mud flat. For example, south of the Qiantang Estuary, mud flat has developed along rocky ria coast, the original subaqueous slope of which was quite steep. For this reason, along the coast of central and southern Zhejiang Province, the average gradient of prograding mud flats (>1 ‰) is much steeper than that of Jiangsu Province, although sediment of the former is finer than that of the latter. The slope of tidal mud flat along the Bohai Gulf, formed only during the last one hundred years (i.e. after 1855), is gentler, because it has been built on the subaqueous slope of the North China Plain.

2. Source of the Sediment

A greater part of sediment of tidal mud flat in China comes from three major rivers, i.e. the Huanghe River, Changjiang River and Zhujiang River, either directly by littoral transport, or indiectly

1) 15 *mus* equal to 1 hectare.

from re-suspension and re-transport of fine sediments of inner shelf which were previously deposited by these rivers.

The Huanghe River is well known in the world for its extremely heavy load. Although its annual discharge (Lijin Station, average of 1950–1977) is only 0.8% that of the Amazon, but its suspended sediment discharge is 2.2 times as much (Table 3-1) (Pang Jiazhen et al. 1980). Moreover, its outlet to the sea had frequently shifted its position during the historical time. The latest great shift occurred in A.D. 1128–1855 when its outlet changed from Shandong Province to northern Jiangsu, i.e. a distance of 420 km from north to south (Fig. 3-3). Consequently, the Huanghe River has played an important role in building tidal mud flat of China, In fact, two longest stretches of mud flat–mud flat along Jiangsu coast and that along the Bohai Gulf and Laizhou Bay owe their origin to the materials from

Table 3-1 Annual suspended sediment discharge

River	Station	Sediment discharge (10^8 ton)	Year	Md (micron)*
Huanghe R.	Lijin	11.18	1950–1977	7
Changjiang R.	Datung	4.72	1951–1976	8.6
Zhujiang R.		0.83		6.5

*Md of suspended sediment was analysed by Hangzhou University using a Coulter counter.

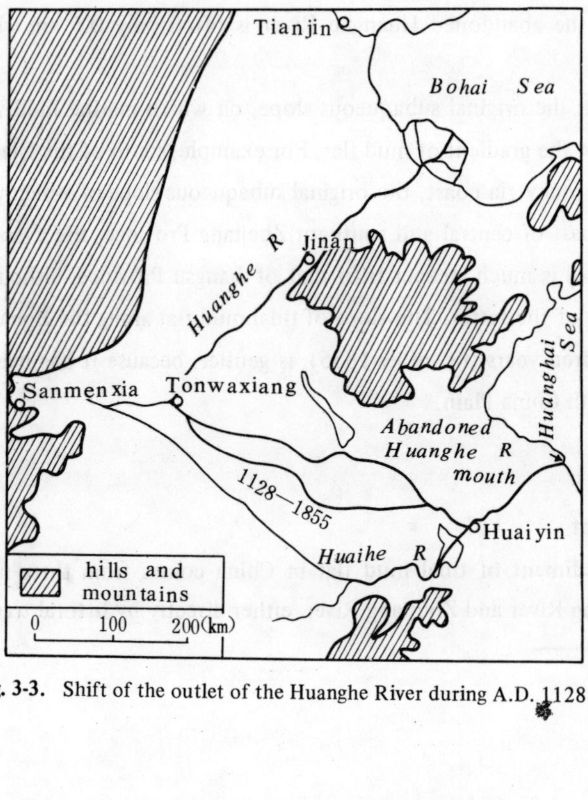

Fig. 3-3. Shift of the outlet of the Huanghe River during A.D. 1128–1855.

the great river.

Considering complicated details of shift of the outlet of the Huanghe River, we took 10×10^8 tons as average annual suspended sediment discharge of the river. It is estimated that more than 4000×10^8 m^3 of sediment were discharged into the coastal zone of Jiangsu during the last southward shift of the outlet of the Huanghe River. Therefore, it is fair to say that the Huanghe River is the important builder of tidal mud flat of China. A major proportion of suspended sediment discharge of the Huanghe River as well as of the Changjiang River is fine and very fine silt. As the coarse population of sand-size particles is very small and clay-size particulates are mostly settled down near the river mouth due to flocculation and biochemical processes, the suspended sediment in the littoral drift consists mainly of silts. Consequently, tidal mud flat of China is chiefly made of all grades of silts. This is one of the distinct characteristics of sedimentation of tidal mud flat of China, which is greatly different from tidal flat of the Netherlands. Thus, in mud flat of China, the difference between "coarse" and "fine" layers is actually that between coarse or medium silt and fine or very fine silt. For example, in interlayered beddings of intertidal flat of the Taizhou Bay, Zhejiang, fine-grained layers consist of very fine silt, whereas coarse-grained layers are formed by medium silt. For this reason, Wentworth's grain-size scale is used in this chapter, because it divides silt into four grades, useful for distinguishing differentiation of grain-size of sediments of mud flat of China.

Suspended sediment of the Changjiang River and Zhujiang River plays an important role in the formation of embayment mud flat along the coast of Zhejiang Fujian and Guangdong. According to oceanographical data, grain-size parameter of surficial sediments, detrital minerals, clay minerals and satellite images, the diffusion of suspended sediment from the Changjiang River is chiefly southward along the coastal zone and may reach the mouth of the Minjiang River, Fujian (Yun Caixing et al., 1981, Cai Aizhi, 1982[1]). In other words, there is a notable "mud stream" more than 800 km long carrying the Changjiang River mud from its mouth along the coast toward south. This perhaps is only surpassed by the great Amazon "mud stream" along the coast of NE South America described by Eisma (Nedeco, 1962; Eisma et al., 1971). This mud stream ensures a continuous supply of material for the development of embayment mud flat along the coast of Zhejiang and Fujian.

Suspended sediment from the Zhujiang River drifts along the coast ot the west, reaching the Guangzhou Bay near Zhanjiang Harbour. Grain-size and chemical elements of sediments of coastal plain and mud flat in western Guangdong are similar to those of the Zhujiang River mouth, indicating that mud of the Zhujiang River plays an important part in the formation of mud flat in this part of China (Zhao Huanting, 1980).

II. Tidal Mud Flat of Jiangsu Province

The coast of Jiangsu Province, situated on the north of the Changjiang River mouth, with more than 90% of the coastline fringed with mud flat, has the longest, continuous stretch of mud flat in

1) See also *Practical Oceanography*, 1980, No. 4, pp 1–6, 30–35; 1981, No. 1, pp 31–37; 1981, No.4 pp 20–25.

China. It is an open coast, facing the Huanghai Sea on the east. But in the middle part of the coast, there is a large offshore submarine ridge field, about 200 km long from north to south and 100 km wide from east to west. A greater part of the submarine ridge field is less than 15 m deep and during low tide, some parts are even exposed. Wave from the open sea, in passing through the shallow water of submarine ridge field, spends a large part of its energy, and consequently, only plays a minor role in the shaping of mud flat of Jiangsu Province. On the other hand, the propagation of tidal wave is less impeded by submarine ridges. Therefore, among hydrodynamic factors, tidal current assumes a highly important role in sedimentation of mud flat.

Tide in shallow sea of Jiangsu Province is of regular semi-diurnal type. Its southern part is dominated by advancing tidal wave from the Pacific and its northern part by anticlockwise rotary tidal wave from the Huanghai Sea; tidal current of the former is from SE to NW, while that of the latter is from NE to SW. These two tidal currents of different directions converge near Jianggang, resulting in spectacular tidal bore and unusually large tidal range. Suspended sediments in the tidal currents deposit at the converging point, forming a large E-W trending offshore shoal of Jianggang–Tiaozini.

The coast of Jiangsu is mesotidal, with mean tidal range 2.5–3.0 m. But the coast between Jianggang and northern Rudong is macrotidal. Here, mean tidal range is over 4 m and the largest tidal range attains more than 7 m. Velocity of tidal current is also great in nearshore zone of northern Jiangsu, with mean velocity of 2.5 knots per hour during spring tide, but in shallow sea south of Doulonggang, the maximum velocity may reach more than 9 knots. Velocity of the flood is greater than the ebb.

Sediments of mud flat of Jiangsu are mainly of silt size (4–8ϕ); evidently, they come chiefly from the Huanghe River. In the last major shift of outlet of the Huanghe River, the river flowed through northern Jiangsu to the sea for more than 700 years. If we take mean annual suspended load of the Huanghe River as 10×10^8 tons, more than 120,000 km^2 of new land may be created by deposition of the Huanghe River silt, enough to build wide coastal plain and mud flat east of Fan's dike in northern Jiangsu and large subaqueous delta off the present mouth of the abandoned Huanghe River. But part of material of mud flat south of Jianggang comes from the Changjiang River. This conclusion is supported by analysis of satellite images, oceanographical data and clay-mineral association of surficial sediments of the inner shelf, nearshore zone and mud flat.

Clay-mineral associations of the abandoned Huanghe River and the Changjiang River are completely different. The former is characterized by high content of montmorillonite and the latter is rich in kaolinite and chlorite, but poor in montmorillonite (Table 3-2).

Clay minerals of surficial sediments of the inner shelf off Jiangsu (122°E) is similar to those of the Changjiang River at 32°N but similar to those of the abandoned Huanghe River at 33°30′N (Table 3-3).

Clay mineral analysis of 17 samples from supratidal and upper intertidal flat of Jiangsu shows decreasing content of montmorillonite and increasing content of chlorite and kaolinite from north to south. It is suggested by Chen[1)] that ratio of montmorillonite to chlorite and kaolinite (M: C+K)

1) Chen Bangben et al., manuscript, 1983.

reflects more completely the change of clay mineral association. When the ratio is >1, montmorillonite is superior to chlorite and kaolinite; when the ratio is <1, montmorillonite is not an important clay mineral. In Jiangsu, from mouth of the Guanhe to Jianggang, M : C + K is >1 and samples from mouth of the abandoned Huanghe River have a ratio of 2.47. But M : C + K of samples from south of Jianggang is <1 and a sample from Qidong has a value only 0.36[1)] (Table 3-4).

Table 3-2 Clay-mineral associations of the abandoned Huanghe River and the Changjiang River mouth

	Mont. (%)	Chl. (%)	Kaol. (%)
Abandoned Huanghe R.	23.97	8.10	8.89
Lower Changjiang R.	5.95	13.06	14.29
Bottom sediment of Changjiang R. mouth	2.0–4.2	18.7–20.8 (C + K)	

(After Zhu Fengguan and Zhang Zhizhong)

Table 3-3 Clay-mineral association of surficial sediments of the inner shelf of the Huanghai Sea (122° 30′E)

Locality	mont. (%)	Chl. (%)	Kaol. (%)
33°30′ N	22.90	6.20	5.93
31°45′ N	6.52	10.57	10.57

(After Zhu Fengguan 1981)

Table 3-4 Mont./Chl. +Kaol. in surficial sediments (0–20 cm) of mud flat of Jiangsu

	Locality	Mont./Chl. + Kaol.
S	Qidong	0.36
↓	Rudong	0.81
	Dongtai	1.25
	Dafeng	1.53
	Sheyang	1.17
	Binhai	1.43
N	Xiangshui	1.28

(After Chen[1)], 1983)

Ten samples from mud flat near Wanggang and Dafeng, show high content of montmorillonite and low content of chlorite + kaolinite which is similar to alluvium of the abandoned Huanghe River but distinctly different from the Changjiang River (Fig. 3-4).

X-ray spectra of clay minerals of suspended particulate of the Changjiang River mouth and of the

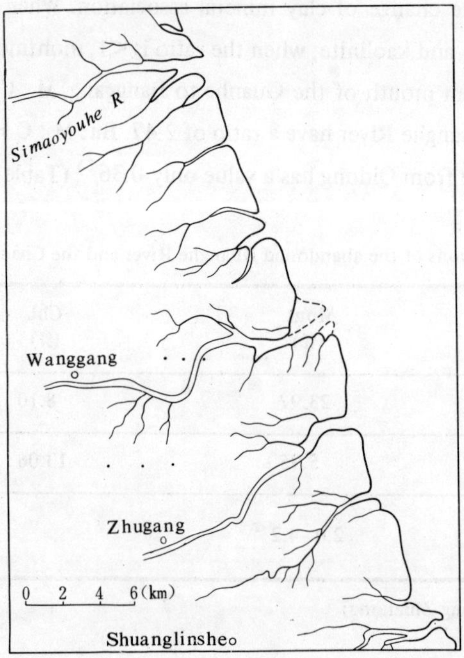

Fig. 3-4 (a). Location of the samples for clay mineral analysis.

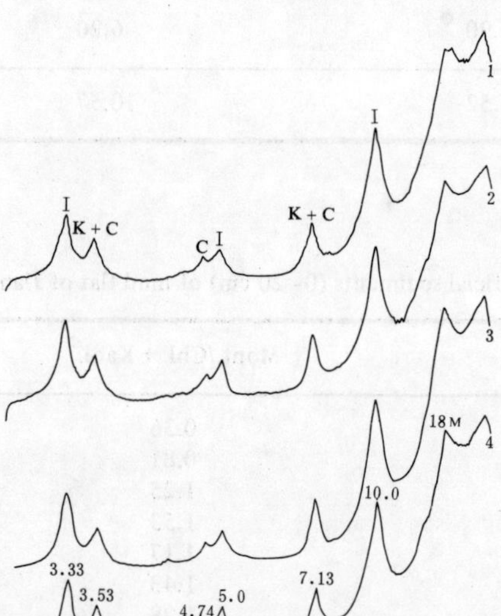

Fig. 3-4 (b). X-ray spectra of separates $<2\mu$ from tidal flat sediment in Wanggang and its adjacent areas, Jiangsu Province.

M – montmorillonite; I – illite; K + C – Kaolinite + Chlorite.

Location of samples:

1. Simaoyou; 2. Wanggang; 3. Zhugang; 4. Shuanlinshe.

Compare this figure with Fig. 4–20 (X-ray spectra of sediments from the Changjiang River mouth).

lower reaches of the Huanghe River yield the same results. Clay mineral association of bottom sediment of the Changjiang River mouth is characterized by Mont. < Chl. < Kaol., while the opposite is true for bottom sediment of the lower reaches of the Huanghe River (Luokou Station) where Mont. >Chl. >Kaol. (Yang Zuosheng et al. 1983).

As suspended sediment of the lower reaches of the Huanghe River comes mainly from loess, its Ca content is far more greater than that of the Changjiang River. This also constitutes one of the important characteristics of the Huanghe River sediments. CaO content in Malan loess of Shanxi Province is 7.7–8.6% and $CaCO_3$ content in loess is generally 8–16%. CaO and $CaCO_3$ contents of mud flat sediments in Wanggang are 5.18% and 8.7% respectively, which is similar to loess and bottom sediment off the present Huanghe River mouth, but remarkably higher than those of the Changjiang River. Infra-red spectra (Perkin Elmer 599 B spectrometer) of 10 samples of $<2\mu$ from mud flat of Wanggang and its adjacent region all show distinct calcite peak which is lack in sediments of the Changjiang River mouth[1)].

Furthermore, drifting bottle test in coastal waters of Jiangsu confirms that there is a consistent littoral drift from the mouth of the abandoned Huanghe River southward to Jianggang[2)].

From the above, it seems clear that material of mud flat north of Jianggang (about 32°40′ N) comes chiefly from the abandoned Huanghe River, whereas south of Jianggang, a considerable amount of suspended particulates from the Changjiang River is added to the mud flat.

At present, most mud flat of Jiangsu is prograding, but some are receding.

(1) Receding mud flat is mainly located near the mouth of the abandoned Huanghe River.

(a) Morphologically, it is narrow and steep. Generally, its width does not exceed 0.5–1 km and its gradient is about 4 ‰.

(b) The surface of mud flat is subjected to erosion of waves and tidal currents and is being continuously lowered. For example, the surface of mud flat near Xiaodingang (south of Lianyungang City, a large deep water port in China) was lowered at the rate of 25.6 cm per year between 1954 and 1966. Here, width of mud flat is now less than 300 m and the original mud flat is covered with a thin sheet of fine sand and coarse silt. Near Xiaowagang (also south of Lianyungang City), according to regular surveys, the surface of mud flat is lowered at the rate of 16 cm per year since 1971 and recently, expensive coastal protection works have been built to stop serious erosion. Owing to strong scouring action of waves and tidal currents, the surface of mud flat is cut by numerous deep tidal creeks (often more than 1 m deep) and rendered uneven, especially in the seaward part of the flat.

(c) The coast has retreated rapidly. It is estimated that since 1855, the coast near the mouth of the abandoned Huanghe River has retreated nearly 20 km. According to a regular survey in recent 13 years, the coast north of the mouth of the Sheyanghe still retreats 77 m per year.

(d) Along the landward side of the mud flat, a discontinuous low chenier about 0.5–1 m high

---

1) Spectrometer analysis by Yang Hofu, 2nd Institute of Oceanography, N.B.O.

2) Drifting bottle test made by lst Institute of Oceanography, N.B.O. in 1980–1981. More than 1500 bottles were released.

can be seen. This is a new chenier in making (see Plate 3-1). But it must be pointed out that near the mouth of the abandoned Huanghe River, there is often no chenier. Probably this is due to the fact that the accumulation of chenier needs a continuous supply of a large amount of shells from tidal flat and consequently, tidal flat outside chenier must be suitable for the flourishing growth of shellfishes. If mud flat is subjected to strong wave and current and surficial sediment about 10–15 cm thick is frequently scoured and stirred, very little benthic molluscs will live in sediments of the flat (Cai, 1980). Near mouth of the abandoned Huanghe River, the coast retreats at a rate of 100–110 m per year and mud flat fringing the coast is seriously scoured. A little farther from mouth of the abandoned Huanghe River, mean rate of coastal recession is only about 30 m per year and wave and current on mud flat are comparatively weak. This explains why new cheniers are found on mud flat a little farther from mouth of the abandoned Huanghe River, but not on mud flat near its mouth. It seems that rapid receding coast is not favourable for the development of cheniers.

A section of coast about 30 km long near Lusi, Qidong County is also retreating. This is because the coastline faces NE, which is the direction of strong wind and lacks the shelter of submarine sand ridges. Strong wave and swift tidal current result in recession of the coast which is fringed by narrow mud flat about 3 km wide.

As receding mud flat is narrow and zonation of sedimentary facies is not clear, it cannot be taken as a typical locality for the study of sedimentation of mud flat of China.

(2) Prograding mud flat Prograding mud flat is an ideal locality for the study of sedimentation of mud flat of China, because it is very wide, often more than 10 km, with distinct zonation of sedimentary facies. Except the above two sections of receding mud flat, all mud flats of Jiangsu belong to the prograding type.

by Ren Mei-e,
Nanjing University, Nanjing

III. Tidal Mud Flat of Wanggang, Jiangsu Province – A Case Study

Tidal flat in Wanggang area, more than 10 km wide, is one of the widest mud flats in China. It progrades seaward at a rate of about 50 m per year. Detailed study has been made on sedimentation of this area to provide a model of sedimentation on a prograding mud flat of open coast in China (Fig. 3-5) (Ren, 1982).

Although there is an extensive submarine sand ridge field off Wanggang, but it is separated from the nearest submarine ridge by an open sea about 20 km wide and open to N-NE which is the prevailing and strongest wing direction of the area. Therefore, wave energy on Wanggang mud flat is little affected by the submarine ridge field and its environment may be considered as similar to the open coast. Tidal range is about 3–4 m. In nearshore water, tidal current flows essentially in a N-S direction. Mean velocity of the flood exceeds that of the ebb, and the difference may reach 0.7 m/s.

1. Hydrographical Conditions

A knowledge of hydrographical parameters is necessary for understanding the sedimentary processes and sedimentary facies on the tidal flat. In Wanggang, a fixed profile line was chosen for survey by leveling instruments. Regularly spaced cement rods were placed along that line at an interval of every 500 m, altogether 24 rods being placed. Along the line, on the surface of the mud flat, a 25 hr whole tide hydrographical survey was carried out on 13 and 14 September, 1980, including the measurement of water depth, velocity of tidal current, suspended matter concentration and salinity.

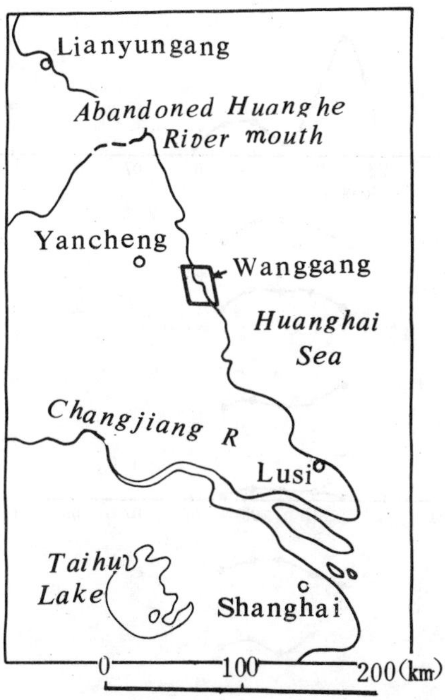

Fig. 3-5. Location map of Wanggang area, Jiangsu Province.

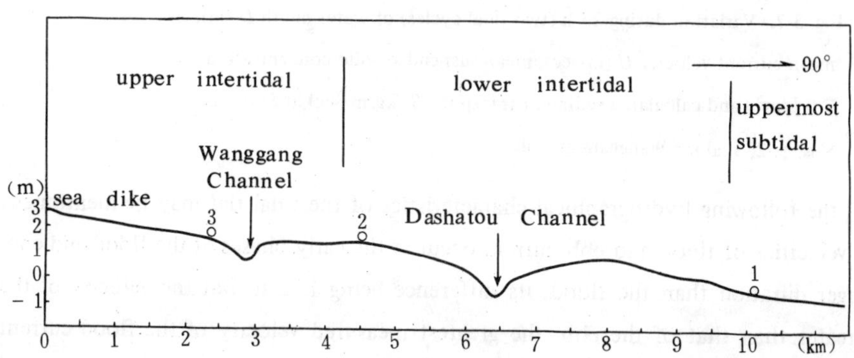

Fig. 3-6. Profile of Wanggang mud flat.

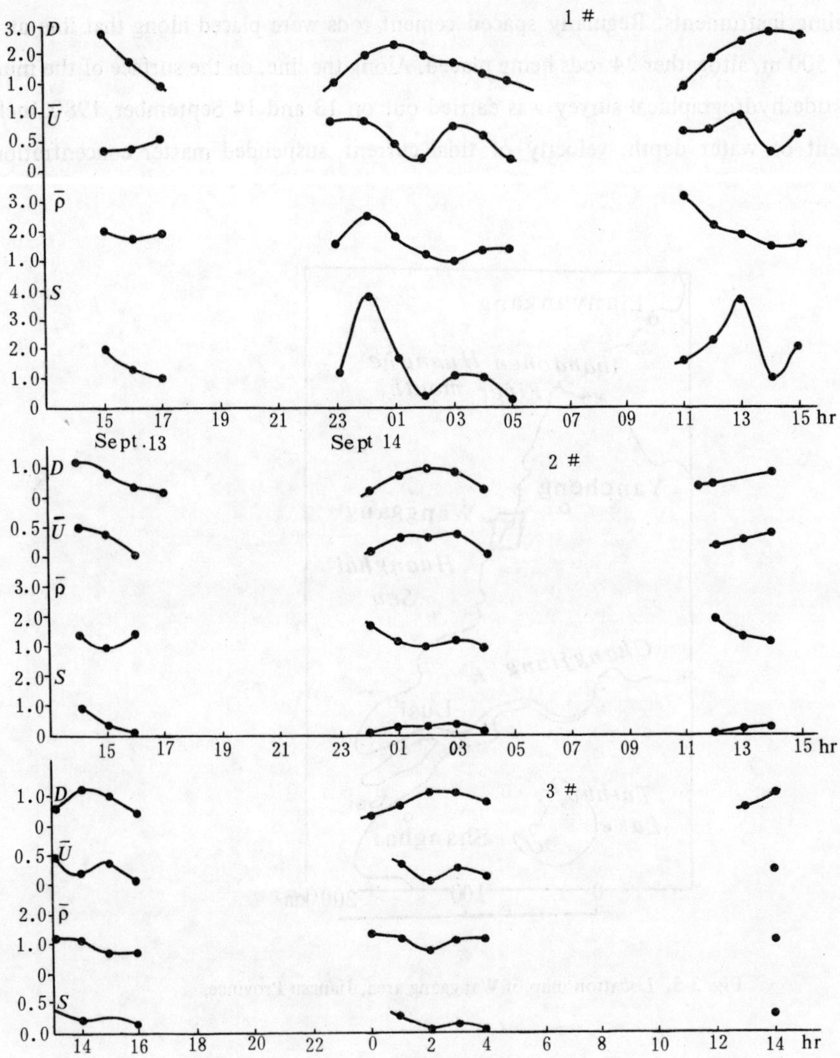

Fig. 3-7. Variation during 25 h (two tidal cycles) of water depth D (m), mean current velocity U (m/sec), mean suspended solid concentration $\bar{\rho}$ (kg/m^3) and calculated sediment transport S (kg/m/sec) at Stations Nos. 1, 2, 3, along Wanggang profile.

From Fig. 3-7, the following hydrographical characteristics of the tidal flat may be mentioned:

(1) Maximum velocities of flood and ebb current occur at the early phases of the flood and ebb. The ebb has a longer duration than the flood, its difference being 1–2 h. But the velocity of the flood current is greater than that of the ebb, the greatest measured velocity of the flood current being 1.36 m/sec.

(2) Suspended matter concentration is larger during the flood than during the ebb. During the ebb, maximum suspended matter concentration lags behind maximum current velocity ca. 1–1.5 h.

But during the flood, maximum suspended solid concentration roughly coincides with maximum current velocity. This is due to the fact that as the velocity of current increases rapidly during the early phase of the flood, more sediments on the flat are set in motion, thus increasing the suspended matter concentration.

(3) Calculated sediment transport during the flood is 1–2 times than that during the ebb.

(4) The gradual decrease of current velocity, suspended matter concentration and calculated sediment transport toward the coast (Table 3-5) reflects the waning of turbulent action of tidal current on the mud flat with diminishing water depth.

Table 3-5 Maximum value of U, ρ and S during the flood and ebb along Wanggang profile

Station	1			2			3		
max. value	$\bar{U}$ (m/sec)	$\bar{\rho}$ (kg/m)	S (kg/m/sec)	$\bar{U}$	$\bar{\rho}$	S	$\bar{U}$	$\bar{\rho}$	S
flood	0.92	3.07	3.81	0.50	1.84	0.83	0.52	1.36	0.36
ebb	0.76	2.00	2.01	0.44	1.44	0.36	0.38	1.06	0.31

(5) Slack tide period was about 1--1.5 h in our survey, but current velocity <0.05 m/sec was not measured. During one flood-ebb cycle on 22:00 September 13, to 07:00 September 14, duration of current velocity <0.20 m/sec in high water was 40 min at Station 1 and gradually increased to 1.5 h at Station 3.

It may be mentioned that during the ebb period, after maximum ebb current velocity, duration of current velocity <0.20 m/sec is considerably longer than during the flood, This period coincides with maximum suspended matter concentration during the ebb when water depth remains fairly high.

Evidently, the time-velocity asymmetry of the flood and ebb and longer duration of current velocity <0.20 m/sec during the ebb are two striking features of hydrographical conditions in Wanggang area, which probably exerts considerable influence on the sedimentation of mud flat.

2. Sedimentary Characteristics

Profile of Wanggang flat is 11.5 km long and is extremely flat, with a gradient only 0.29‰. From the coast to 8 km seaward, the flat has a gradient as little as 0.16‰. On such a broad, and smooth mud flat, sedimentation is mainly controlled by tidal action, because wave has spent nearly all its energy in passing through broad flat. But as wave supplies suspended sediment to tidal current by stirring up bottom sediment from subtidal zone, it is also an important factor in sedimentation of mud flat. Moreover, during spring tide or extremely severe storm, wave may reach upper intertidal zone or even supratidal zone, thus exerting considerable influence on the shaping of the mud flat.

Sedimentation on tidal flat is chiefly a response to hydrodynamic conditions which vary greatly in time and space. Tidal cycle, spring-neap tidal cycle, seasonal variation in water level, rise and fall of water level induced by wind, etc all have influence on sedimentation. Consequently, surface sediments show a distinct zonation of sedimentary facies that corresponds approximately to the variation of hydrodynamic conditions and duration of exposure.

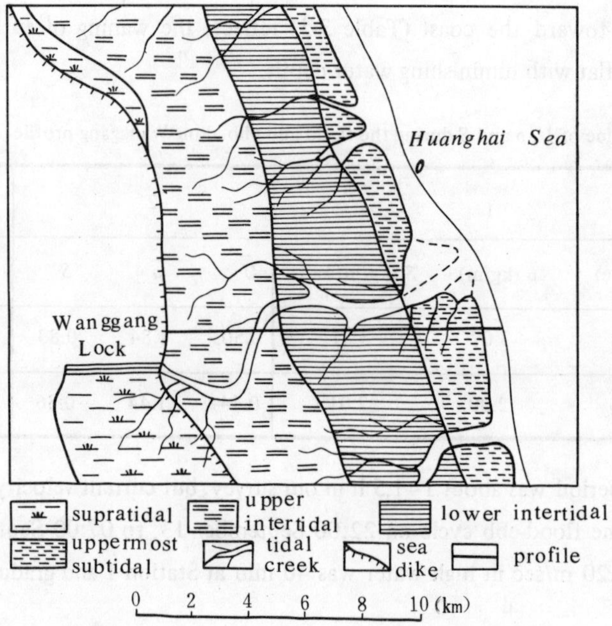

Fig. 3-8. Zonation of sedimentary facies of Wanggang tidal flat.

(1) ***Supratidal zone***

Although there is no supratidal zone in the Wanggang profile, a broad supratidal zone occurs just on its south side. On the lower part of supratidal zone, duration of exposure is about 94–96% and it is under water during spring high water (SHW). But the upper part of supratidal zone is essentially a non-tidal environment as its duration of exposure is 99%, being only under water during exceptionally high water in spring tide, extraordinary large astronomical tide and typhoon therefore, supratidal zone may be further divided into two subenvironments:

(a) *Upper supratidal*

On account of its long duration of exposure, the flat is densely covered with vegetation, chiefly community of *Imperata cylindrics* var. *major*. Sediments are very fine, with Md 6.25 ϕ, the finest on Wanggang flat, but the content of clay is less than 15%. Burrowing organisms *Uca* sp. form distinct burrows (U-shaped, about 40–60 cm deep) and root mottlings are abundant (Plate 3-2). Sediment often shows parallel laminations but discontinuous wavy beddings are also common; the latter may be attributed to dessication cracking or bioturbation (Plate 3-3).

(b) *Lower supratidal*

With shorter period of exposure, only *Suaeda maritima Dumort* and *Suaeda glauca* Bunge sparsely grow on the flat. Grain size of sediments is coarser than that in upper supratidal zone, with Md 5.9ϕ. This is probably due to the fact that during strong typhoon, large waves often break in lower supratidal zone and deposit coarse sediments which mix with fine elements deposited in ordinary weather, thus increasing Md of the sediments. Mud cracks are well developed and *Suaeda maritima* often grow in fissures of cracks, forming beautiful patterns (Plate 3-4).

(2) ***Intertidal zone***

Intertidal zone, about 10 km wide, is the main part of the tidal flat in this area. It may be further divided into two subenvironments, taking average high water level of neap tide in January as a boundary.

(a) *Upper intertidal*

It is about 4 km wide. Sediments are light brown fine silt, with Md 6.18 ϕ, chiefly deposited from suspended matter during high water. Its clay content is less than 15%. Interlayered beddings are common. The upper part of upper intertidal zone is characterized by coarsely interlayered beddings with thick "mud" layers (2–4 cm thick) separated by thin "sand" layers. In the lower part of upper intertidal zone, there are thinly interlayered beddings comprising alternative thin layers of sand and mud. Thickness of individual lamina is usually 1 mm–1 cm. "Sand" layers are mostly yellowish brown, consisting mainly of medium–fine silt: "mud" layers are light brown, consisting largely of silty clay (Plate 3-5). In the lowermost part of upper intertidal zone, small ripple beddings are siometimes well preserved, indicating that this part of tidal flat is subjected to strong wave action during storms.

There is no natural vegetation on upper intertidal flat. But in its upper part, *Spartina anglica* has been grown in recent years which accelerates the rate of accretion.

Burrowing organisms are abundant in upper intertidal zone. Crabs (*Ilyoplax dentimerosa* Shen) are most numerous, forming 10–50 burrows per m^2 of surface of the flat. *Bullacta exorate* Philippi and *Aloiodis laevis* Hinds are found in large numbers in the lower part of upper intertidal zone. Therefore, bioturbation is intense.

(b) *Lower intertidal*

Along the Wanggang profile, the width of lower intertidal zone is more than 5 km. Coarser sediments with Md 4.38ϕ often form coarsely interlayered beddings with thicker sand layers (2–5 cm thick) separated by thin mud layers (<2 cm thick). Tidal creeks are well developed. Here, many small tidal creeks flow into the main creeks, which are generally 30–12 m wide and 1–1.2 m deep. Meanders and point bars are well developed in tidal creeks. Sediment of tidal creeks is a little coarser than that of the adjacent flat. In point bars of tidal creeks, sediments are rather similar to those of streams, with the same fining upward sequence and cross beddings.

In the lower part of lower intertidal zone, wave ripples are well developed due to relatively strong wave action and coarser material (Plate 3–6). In tidal creeks, tidal currents are strongly reversing which has resulted in herringbone cross beddings in tidal creek sediments. As position of tidal creeks frequently shift, former tidal creek sediments are covered by sediments of tidal flat after the abandonment of tidal creeks and thus incorporated into tidal flat sequence. Therefore, herringbone cross beddings are also often found in sediments of lower intertidal flat (Plate 3-7).

(3) ***Subtidal zone***

Only the uppermost part of subtidal zone (above SLW) has been studied. Grain size of sediments is slightly finer than that of lower intertidal zone, with Md 4.92 ϕ, in which 29.7% is fine sand, 63.2% silt and 7.1% clay. Dominant color of sediments is blue-gray and consequently, they are called "blue sand" by local people. The prevailing sedimentary structure is more uniformly horizontal thin bedding (Plate 3-8). Small cross-beddings and herringbone cross-beddings can also be seen in its upper part. Various kinds of ripples are formed by wave action on the surface of the flat. In the profile, wave ripple beddings are often interlayered with horizontal sand layers. The latter are probably formed in a high energy environment. According to Inman, when velocity of wave propagation exceeds 90 cm, sediments move along a plane bed, forming horizontal beddings (Inman, 1957).

From the above facts, it is evident that on Wanggang mud flat, several progressive changes in sediments can be seen from sea to land.

(a) Grain size becomes gradually finer, with the exception of lower supratidal and lower intertidal zones where wave action is relatively strong during storms and ordinary weather. But sorting becomes gradually poor from sea to land, indicating that wave action on sediments is progressively smaller and proportion of suspended matters greater (Table 3-6). Evidently, this is due to the fact

Table 3-6 Grain size of sediments of Wanggang flat in autumn

Zone	Supratidal	Upper intertidal	Lower intertidal	Subtidal
Number of samples analysed	10	18	18	5
Type of sediments	upper part: fine silt lower part: medium silt	fine silt	coarse silt	coarse silt
Mdϕ	6.02	6.18	4.38	4.92
QDϕ	0.62	0.70	0.95	1.16
SKϕ	0.07	0.03	0.29	0.12

that sediment transport and deposition on Wanggang mud flat is chiefly determined by tidal current in particular, large amount of suspended sediment carried by flood current is an important factor in sedimentation of mud flat.

In Wanggang area, velocity of flood current is greater than that of ebb and it decreases progressively from the sea toward the coast, bringing finer sediments to the upper part of the tidal flat. Furthermore, in a tidal cycle, as maximum velocity of flood current occurs near low water, during the beginning of the flood when water is still shallow on the flat, velocity of the flood is already great, and fine sediments deposited in the previous tide cycle are set in motion and carried by the flood current toward the coast. This is also an important factor causing the coarsening of sediments in lower intertidal zone.

The relative small percentage of clay and sand particles in the sediments is attributed to their source from the abandoned Huanghe River.

(b) Color changes from blue-gray, typical of reducing conditions in upper subtidal zone, to light-grayish yellow in lower intertidal zone, yellowish brown in upper intertidal zone and brown in supratidal zone, indicating progressively stronger oxidizing conditions.

(c) Content of heavy minerals decreases from the sea toward the coast, roughly corresponding to progressive decrease of grain size. Content of SiO_2 increases with the increases of grain size, but content of other chemical compounds all increases with the decrease of grain size, Content of FeO and Fe_2O_3 is chiefly influenced by a higher content of dark-colored minerals in subtidal zone (Table 3-7).

Table 3-7. Chemical composition (%) of sediments of Wanggang tidal flat

Chemical composition (%)	SiO_2	Al_2O_3	FeO	Fe_2O_3	TiO_2	CaO	MgO	$CaCO_3$
Supratidal	64.74	11.29	1.38	2.45	0.59	5.58	2.08	9.15
Upper intertidal	63.80	11.53	1.39	2 58	0.64	5.57	2.17	10.03
Lower intertidal	70.53	9.53	1.28	1.98	0.65	4.63	1.54	8.94
Subtidal	67.15	10.13	1.70	2.73	0.85	4.96	1.87	6.69

(d) Sedimentary structures (Table 3-8)

All these progressive changes relate chiefly to waning wave and tidal current and more prolonged subaerial exposure on the mud flat from the sea toward land.

It may be noted that on prograding mud flat of China, there is a distinct steepening of gradient and corresponding coarsening of sediment in the lower part of intertidal zone. For example, in the

Table 3-8 Sedimentary structures in sediments of Wanggang flat

Zone	Sedimentary structure
Supratidal	fine laminations, often disrupted by root mottling and burrowing; wavy bedding common
Upper intertidal	mainly thinly and coarsely interlayered bedding, intense bioturbation
Lower intertidal	mainly coarsely interlayered bedding, sometimes with small cross bedding and herringbone cross bedding
Upper subtidal	more uniform thin bedding, wave ripple bedding

Wanggang profile, from sea dike to 8 km seaward, average gradient is only 0.16‰, but from 8–10 km, it suddenly steepens to 0.7‰, an increase of 4.4 times. In Xinyanggang profile, north of Wanggang, Jiangsu Province, tidal flat is 11 km wide, also one of the widest prograding mud flats in China. Here, from sea dike to 10 km seaward, average gradient is only 0.12‰, but it increases abruptly to 1‰ between 10–11 km, an increase of 8.3 times and Md of sediments in the lower, steeper part is 60 microns, in contrast to 17.4 microns in the upper, gentler part. The same is true for prograding mud flat of the Taizhou Bay, Zhejiang Province. This is probably due to the fact that lower part of intertidal zone is near the breaking zone of waves and consequently, the wave action in this part of mud flat is relatively stronger than in other parts. The relatively strong wave action in the lower intertidal zone and upper subtidal zone of Wanggang is clearly shown by the presence in SEM photographs of many V-shaped and crescent impact marks on the surface of quartz grains (>63 microns) (Plate 3-9).

It is interesting to mention that flaser and lenticular beddings are rarely seen in sediments of intertidal and upper subtidal zone of Wanggang area. Probably this is attributed to difference of sedimentary environments between this area and the North Sea tidal flat of Europe. As sea floor sediments of the inner shelf of the North Sea is largely fine and medium sand (Schuttenhelm, 1980), a copious supply of sand is contributed to the tidal flat under the combined influence of littoral drift and tidal currents. Therefore, a greater part of sediments of the North Sea tidal flat is sandy which is a necessary condition for the formation of lenticular and flaser beddings. On the other hand, sediments of tidal flat of Wanggang area, derived from the abandoned Huanghe River, are mainly grades of silt. This probably explains the predominance of horizontal interlayered bedding rather than flaser or lenticular beddings on intertidal flats of Wanggang area.

Moreover, tidal creeks are numerous on tidal flat of the North Sea. Strong tidal currents, flowing in and out through tidal creeks, dominate sedimentation of tidal flat of the Netherlands, forming conspicuous wave and current ripples which are the basis for the development of flaser and lenticular beddings. As tidal creeks are shallow and less numerous in Wanggang, tidal currents mainly flow on mud flat as a uniform water sheet. Therefore, although there are some wave and current ripples on mud flat of Wanggang, they are of small magnitude and can not easily be preserved. As pointed out by Reineck, the formation of flaser bedding is closely related to grain size of sediments and in the North Sea coast, flaser bedding is chiefly found in sediments of fine sand grade (Reineck, 1967, 1972).

Therefore, owing to difference of depositional environment, interlayered beddings rather than flaser and lenticular beddings are typical in Wanggang. This is similar to tidal flat of the Colorado River delta of America (Thompson, 1975).

3. Effect of storms

The great effect of storms on coastal morphology and sedimentation has received increasing attention from geologists. The effect of a single severe storm on coastal zone often surpasses the results of normal processes in several years. Moreover, storms often leave their characteristic deposits on tidal flat preserved in geological strata.

Storm wave along the coast of China is chiefly due to two reasons: Firstly, typhoon. The record making high water level often occurs when the time of strong typhoon coincides with spring tide. For example, during Typhoon No. 7, 1980, the amount of rise of water level along the coast of Haikang, Guangdong Province reached 5.94 m, the largest in China. The coast of Jiangsu was visited by 76 typhoons during 1951–1972, in which there were 15 typhoons that caused strong wind above Beaufort scale 6, averaging once per 1.5 years. Secondly, winter storms. Strong wind over Beaufort scale 6 is usually due to cold air mass from the north and lasts from late September to late April, next year. The influence of one cold front on coast of Jiangsu lasts 5–7 day, but sometimes exceeds 10 days, when strong northerly wind of Beaufort scale 6–8, even 9 is blowing. During 1951–1972, 84 strong cold air masses passed Jiangsu coast, causing strong wind above Beaufort scale 6. In other words, there were almost 4 strong cold air masses affecting Jiangsu coast every year.

In addition to direct action of strong waves, typhoon and winter storm cause a great rise of water level along the coast and an increase of velocity of flood current which has great influence on sedimentation and morphology of tidal flat. Owing to extraordinary high water level, sedimentary zones of mud flat might shift landward for a considerable distance. For example, during No. 14 typhoon, 1981, position of intertidal zone in Wanggang shifted nearly 3 km landward and supratidal zone was flooded by seawater. Hence a catastrophic event like typhoon will leave its mark in tidal flat sedimentation.

Typhoon 8114 originated on August 27, 1981. At 8 a.m. September 1, its center gradually moved northeast toward Jiangsu coast. The maximum wind velocity in its center was more than 45 m/s. From September 1 to 3, wind velocity on the coast of Jiangsu increased to Beaufort scale 8–9 and occasionally to 10 and the direction of wind was N-NE. At some coastal protruding section, strong wind of Beaufort scale 11 was recoreded (Fig. 3-9).

It was in the season of high sea level (early August of the lunar calender) that Typhoon 8114

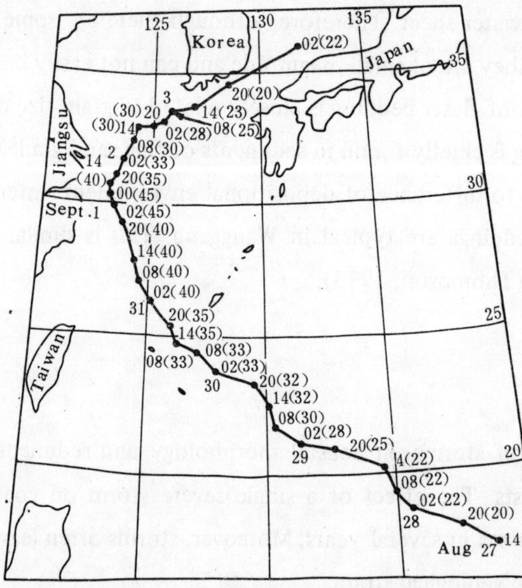

Fig. 3-9. Track of center of No. 14 typhoon, 1981.

( ) in the figure indicating greatest wind velocity (m/s).

broke out. Therefore, on September 1, extraordinary high water levels were recorded at many stations on the coast of Jiangsu, exceeding the historical records. At some coastal stations, water level raised nearly 3 m in one hour, which was more than twice faster than in ordinary spring tide and consequently, tidal current was very swift during the typhoon. At the same time, wave energy was greatly increased. On September 2, maximum wave height at Lusi oceanographical station was almost 3 m higher than in ordinary weather. At oceanographical station of Lianyun Port wave heihgt of 3.6 m was recorded during Typhoon 8114.

(1) ***Effect of Typhoon 8114 on receding mud flat***

As receding mud flat of Jiangsu is rather narrow, the surface of the flat is strongly scoured by storm waves. On the coast north and south of the mouth of the abandoned Huanghe River, originally there was a low, discontinuous chenier ridge or shell bank along the high water level. After typhoon, the chenier ridge was eroded and retreated. The chenier near Dalabakou was originally 1 m high and 20 m wide but during the typhoon, it was nearly wiped out. Shell bank from Shuangyangkou to Biandangang essentially disappeared. These facts show severe erosion of supratidal zone by storm waves during typhoon and indicate that coast subjected to strong wave action is unfavourable for the formation and preservation of cheniers.

Results of regular survey of three profiles of mud flat, north of the Sheyanghe showed that in winter and spring (January 6 – June 2, 1981), intertidal flat was accreting; but during summer, it suffered from scouring and its surface was lowered 0.34–0.5 cm per week. During Typhoon 8114, from August 29 to September 4, 1981, intertidal flat was lowered 5–9 cm in one week, 10–20 times higher than in ordinary weather. In these three profiles, the upper part was lowered while the lower part was aggraded, the equilibrium point between scouring and aggradation being near mean water level. Profiles above equilibrium point are concave upward and the area of scouring is larger than that of aggradation. It may be noted that upper part of the profile is concave upward, similar to profile of sandy beach shaped by waves. This proves that during the typhoon, wave becomes the chief agent on tidal flat and action of tidal current is no longer important (Fig. 3-10).

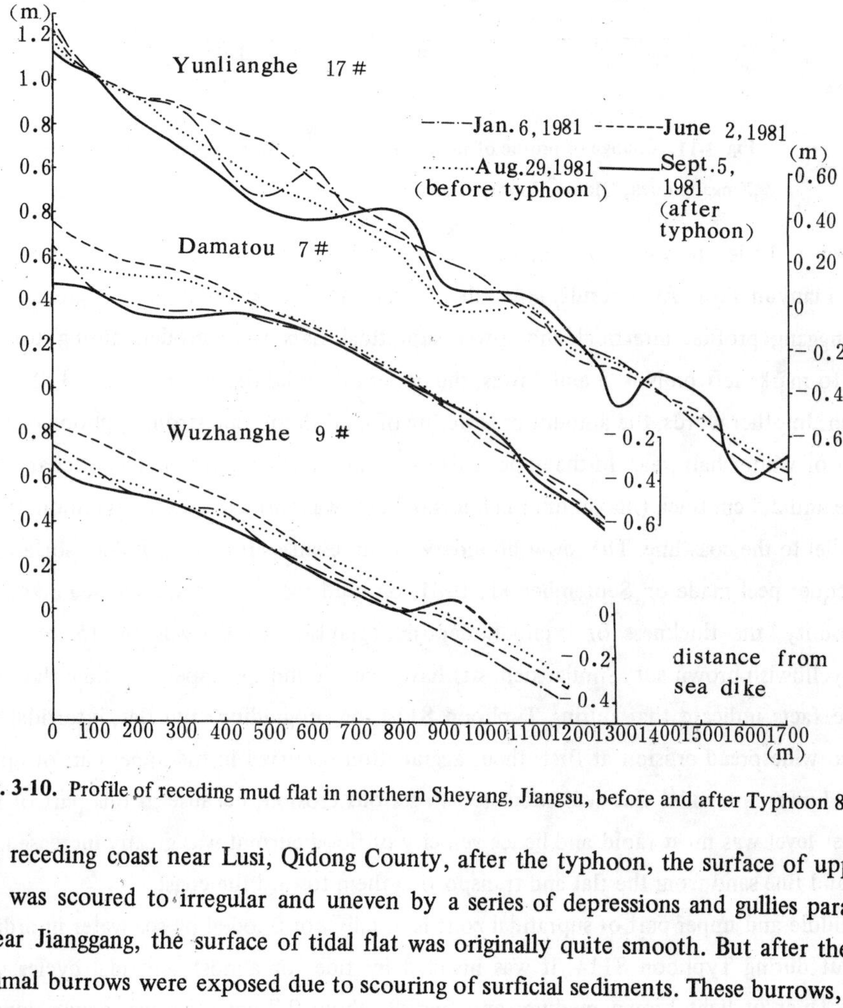

Fig. 3-10. Profile of receding mud flat in northern Sheyang, Jiangsu, before and after Typhoon 8114.

On receding coast near Lusi, Qidong County, after the typhoon, the surface of upper part of mud flat was scoured to irregular and uneven by a series of depressions and gullies parallel to the coast. Near Jianggang, the surface of tidal flat was originally quite smooth. But after the typhoon, many animal burrows were exposed due to scouring of surficial sediments. These burrows, filled with iron compounds, are more resistant to erosion and consequently stand out as small projections (2–8 cm high) above the surface or tidal flat, known as "mud nails" by local people. The height of "mud nails" roughly represents the minimum amount of the lowering of tidal flat by wave and current during the typhoon (Plate 3-10).

On mud flat near Xingang Lock, Qidong, the profile was lowered from sea dike to 3600 m seaward after the typhoon and the largest scouring occurred at about 500 m from the sea dike (Fig. 3-11)

(2) ***Effect of typhoon on prograding mud flat***

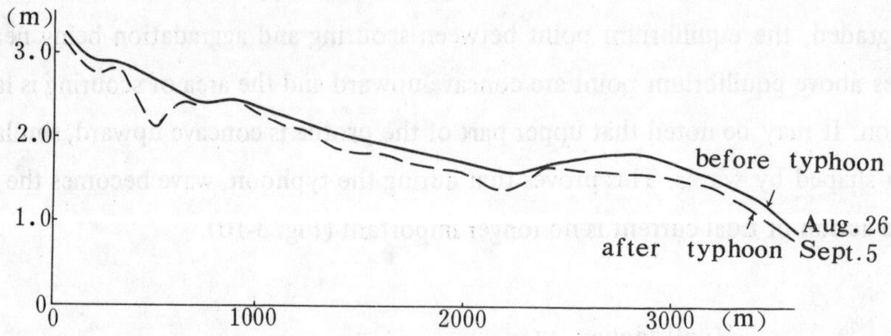

Fig. 3-11. Change of profile of tidal flat before and after Typhoon 8114, Xingang Lock, Qidong County, Jiangsu.

Typhoon 8114 led to a great rise of water level along the coast of Jiangsu, 2.43 m near Lusi and 1.32 m at Lianyun Port. As a result, strength of tidal current on mud flat was greatly increased.

In Wanggang profile, intertidal and lower supratidal flats were eroded during the typhoon. According to marks left on Nos. 5 and 7 rods, the upper intertidal flat was lowered 13–15 cm during the typhoon. In other words, the amount of lowering of mud flat in one strong typhoon exceeds total aggradation of winter half year. In the upper part of upper intertidal zone, a zone of coarse silt about 300 m wide and 2.5 cm thick (maximum thickness 6.7 cm) was formed after the typhoon. It stretched nearly parallel to the coastline. The lower boundary of this coarse silt layer is a scour surface. According to a lacquer peel made on September 11, 1981, on mud flat 1300 m outside sea dike, Yangkou, Rudong County, the thickness of typhoon deposits (grayish "sand") was 14–15 cm on a scour surface of yellowish-brown silt. Similar deposits have been found on upper intertidal flat near Jianggang. These facts indicate that during Typhoon 8114, on prograding mud flat, intertidal zone was subjected to widespread erosion at first; then, aggradation occurred in the upper part of upper intertidal zone. Lower intertidal flat was subjected to serious erosion, because in this part of mud flat, rise of water level was most rapid and hence velocity of flood current was greatly increased, scouring coarse silt and fine sand from the flat and transporting them toward the coast.

The middle and upper part of supratidal zone is usually not flooded by sea water in ordinary spring tide, but during Typhoon 8114, it was invaded by tide for almost ten tidal cycles. After the typhoon, a layer of light brown medium and fine silt about 0.7 cm thick was newly deposited on upper supratidal flat (covered with *Imperata cylindrica*) near Wanggang. Similar deposits were also found on upper supratidal flat near Jianggang.

In Wanggang, Jianggang and Yangkou, Rudong, eroded depressions, roughly parallel to the coast, were found on the seaward side of supratidal zone after the typhoon. They were probably eroded by storm waves in slack tide period.

It is interesting to mention that after the typhoon, a miniature chenier about 50–100 m wide and 3–8 cm high was formed just below sea dike near Yangkou, where tidal flat was slowly prograding in the ordinary weather. The fact that most shells in the chenier have not been broken or abraded shows that they were newly upturned from lower part of the flat and transported and deposited there during the typhoon.

Thus, for prograding mud flat of Jiangsu, lower intertidal zone was scoured during the typhoon. But the other parts of mud flat was scoured at first and then followed by accretion, or only accretion. Therefore, the gradient of mud flat was steepened and its sediment coarsened. It seems that for prograding mud flat, the chief agent on the flat is still tidal current during the typhoon on account of its great width and very gentle gradient, except near high tide time when wave action becomes relatively strong. This is quite different from receding mud flat where wave assumes a leading role in the typhoon.

In offshore submarine ridge field, during the typhoon, strong waves often eroded the bottom of shallow sea and materials of sea bottom were thrown upon the seaward side of shoals which may be exposed in low tide, forming ramparts of fine sand or coarse silt. For example, on the seaward (east) side of a shoal in Jiangjiasha, a rampart about 2–3 m high was seen after the typhoon. Evidently, it was deposited by storm waves like ramparts of coral blocks and cobbles on the coast of coral islands in the ocean (Siddiquie, 1980; Stoddart, 1971).

(3) ***Typhoon Deposits***

During Typhoon 8114, prevailing wind in the vicinity of Wanggang was NE, perpendicular to the trend of coastline. The wind reached Beaufort scale 8–9 and occasionally 10–11. The highest water level at Wanggang station was 5.02 m, breaking the historical record. The rapid rise of water level in the flood reduced time-difference of seawater flooding between tidal creeks and adjacent flat and changed the difference of water depth between tidal creeks and adjacent flat. Under ordinary conditions, ratio of water depth is 2 : 1, but during high water in Typhoon 8114, it was only 1.3 : 1.

Owing to these unusual hydrographical conditions, typhoon deposits were distinctly different from sediments of the ordinary weather. Typhoon deposits were mainly coarse silt (Md 4.93 ϕ) with good sorting, while the underlying deposits were mainly medium and fine silt (Md 6ϕ) with poor sorting. There were very little fine population (very fine silt only 1.0%) in the former, while there was a considerable amount of fine population in the latter (very fine silt 15.7%). Therefore, in grain size frequency curve, the former shows a very sharp peak, while the latter shows a broad peak (Fig. 3-12).

Typhoon deposits were separated from the underlying sediments by a clear scour surface (Plate 3-11), which can also be seen in three profiles across mud flat north of the Sheyanghe mouth (Fig. 3-10).

The characteristic sedimentary structures associated with typhoon deposits are:

(a) *Cross beddings* Both trough and planar cross beddings are found. A set of cross beddings is usually 1–2 cm thick, Foreset laminae dip toward the coast, but sometimes show bipolar directions formed by tidal currents.

(b) *Wave ripples* Perfect wave ripples sometimes may be found in foreset laminae, showing wave action in the deposition of typhoon deposits.

(c) *Horizontal beddings* The lower part of typhoon deposits usually has thin horizontal bed-

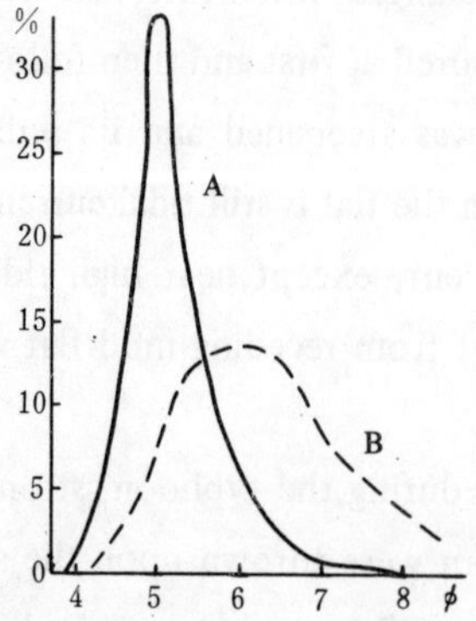

Fig. 3-12. Grain size frequency curve of typhoon deposits and the underlying sediments on upper intertidal zone of Wanggang.

A–Sediments of Typhoon 8114;

B–Sediments of ordinary weather.

dings, as at No.9 rod in Wanggang profile (lower part of upper intertidal zone). But small-scale cross beddings are common in their upper part. This is probably due to the fact that after scouring phase, with gradual waning of wave and current, coarse sediments are deposited. But in the early phase of deposition, current velocity is still great, resulting in horizontal beddings which represent plane bed forms of the upper flow regime. With further decreases of the current strength, cross-beddings are formed.

In the upper part of lower intertidal zone, NE of Zhugang Lock, deposit formed by 8114 typhoon is 6.14 cm thick, and it is 14–15 cm thick in upper intertidal zone, east of Yangkou.

4. Beach cycles

On the subject of beach cycles much research has been made on sandy beach, but little on tidal mud flat. It has long been realized by many Chinese scientists that sedimentation on mud flat also undergoes evident changes due to seasonal changes of hydrographical regime. For example, on mud flat of the Taizhou Bay, there is a coarsening of sediments on the surface of the flat during autumn, but it will return to fine-grained next soring. Based on observations and repeated surveys along the Wanggang profile, a preliminary study was made on the effects of cycles of flood and abb during a single tide cycle, of cycles of spring and neap tide and of annual cycles of variation of sea level along the coast.

(1) *Effect of a single tide cycle*

The genesis of horizontal interlayered bedding is a subject of some dispute. According to Reineck, the bedding is produced from tidal changes during a single tide cycle, "sand" layers deposited during periods of current activity and "mud" layer during slack tide period. But other geologists argue that slack tide period is often too short to deposit a conceivable mud layer (McCave, 1976) Tide in Wanggang area is semi-diurnal; duration of a single tide cycle is only about 12 h and slack tide period is very short (only about 1 h) which is scarcely enough to deposit a mud layer on the tidal flat. In the profile, at No. 4 rod, continuous survey was made during September 27 – October 4, 1981 (spring tide period) every day after ebb tide in day time. It has been found that in one day, no visible interlayered bedding was formed. Here, the surface of intertidal flat is without water during slack tide period after the ebb; only in slack tide period after the flood, the surface of intertidal flat is under water which will affect mud deposition. But slack period after the flood lasts only 40 min to 1.5 h, too short to deposit a notable mud layer.

Settling and deposition of fine-grained sediments is a very complicated problem. In sea water, fine silt and clay are often agglomerated by biological processes into larger composite particles which may have a settling rate equivalent to coarse silt and fine sand grains, i.e. an increase of settling rate of several orders of magnitude. The combined effect of scouring lag and settling lag (Postma, 1967) causes deposition of finer particles toward land and affects zonation of sediments of mud flat from the sea to coast. But a large number of organisms live in the upper layer of mud flat. Their burrowing and other activities frequently stir surficial sediments, causing resuspension of sediments or impeding their compaction. Therefore, detailed study on interactions between organisms and fine-grained sediments is necessary for a correct understanding of mechanism of mud deposition on tidal flat (Drake, 1976).

(2) ***Effect of spring-neap tide cycle***

According to data of Wanggang station (1961–1964), the difference of mean high water level in spring and neap tide is 0.59 m and annual mean HHWL in spring tide is 1.0 m higher than LHWL in neap tide. But in August, the greatest difference amounts to 1.37 m (average 1969–1978). It is estimated that current velocity in spring tide is roughly twice that in neap tide. Different hydrodynamic conditions in spring and neap must result in some discrepancies between deposits of spring and neap tide.

According to our observation at Nos. 4 and 5 rod (upper intertidal) on Wanggang profile, sediment deposited in spring tide is chiefly yellowish brown "sand" and that deposited in neap tide is chiefly light brown "mud".

As the average gradient of tidal flat is 0.29% , changes of water level in spring-neap tide involve a lateral shift of position of HWL on the flat. The difference of mean HWL between spring-neap tide implies a lateral shift of HWL for a distance of 2030 m and the difference of annual mean HHWL and LHWL between spring-neap tide implies a lateral shift of HWL for a distance of 4850 m. During half a month period, when spring tide gradually changes to neap tide and consequently HWL gradually retreats seaward, area of pelitic sediment deposition also shifts toward the sea. Therefore, a broad zone of tidal flat is covered by fine sediments. With the change of neap tide to spring tide, these fine sediments have already been deposited on the surface of flat for 7–10 days and possess considerable cohesiveness. Therefore, they would not be entirely stirred up again by strong current of the next spring tide but will be covered by coarse sediments brought in by flood current of spring tide. Thus, a spring-neap deposition cycle is established with rhythmic alternation of sand deposition in spring tide and mud deposition in neap tide and thinly interlayered beddings are formed which, for this reason, are chiefly located in intertidal zone.

It may be noted that grain size frequency curves of "sand layer and "mud" layer are rather similar, but the former consists mainly of medium silt, with Md 5.35ϕ, while the latter is chiefly fine and very fine silt, with Md 6.10 ϕ. Moreover, the frequency curve of spring tide deposit shows a weak but distinct fine tail. This is probably due to the fact that during the spring tide, part of "mud" deposited in the previous neap tide is resuspended and mixed into coarse sediment brought by spring tide. Furthermore, there is also a small amount of fine sediment brought in by tidal current in spring tide. During slack tide, this "mud" will be settled down, thus forming a fine tail. But as current in neap tide is not strong enough to resuspend or transport coarse sediment, there is no distinct coarse tail in frequency curve of neap tide deposit. Nevertheless, some coarse population may be brought in during strong wind in neap tide and therefore, a small amount of medium and coarse silt is also included in neap tide deposit (Fig. 3-13).

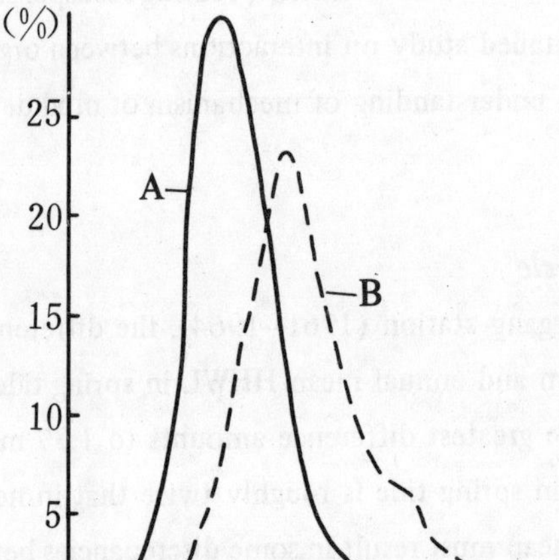

Fig. 3-13. Grain size frequency curve of sediments of the upper part of upper intertidal zone, Wanggang.

A–Sediments of spring tide;

B–Sediments of neap tide.

(3) ***Effect of annual cycle of variation of monthly mean sea level***

There is a marked seasonal variation of sea level along the coast of China. Generally speaking, sea level is highest in August and lowest in January, the difference being 35–50 cm on the Huanghai Sea and 50–70 cm on the Bohai Sea (Li Kunping, 1982). At Wanggang station, monthly mean HWL in September (highest in the year) is 0.68 m higher than that in January (lowest in the year) (Fig. 3-14).

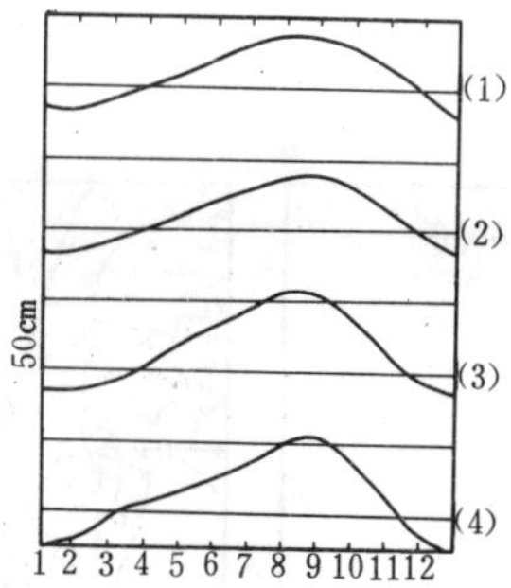

Fig. 3-14. Variation in monthly sea level at some tide-gauge stations in central Jiangsu.

(1) Xinyanggang Lock, average 1957–1967;

(2) Doulonggang Lock, average 1959–1968;

(3) Wanggang Lock, average 1960–1978;

(4) Dongtaihe Lock, average 1960–1964.

This has caused lateral shift of 2720 m of mean high water line on mud flat in a year which brings about seasonal changes of micromorphology and sediments of mud flat.

In the middle and lower part of lower intertidal zone of the Wanggang profile, there is a zone of alternative "sand" (coarse silt) and "mud" (fine silt) deposition about 3 km wide. In spring, 1980, the seaward edge of mud flat (i.e. flat covered with mud) lay at No. 14 rod. In mid and late July, 1980, it retreated to No. 11 rod. It retreated further landward to Nos. 8–9 rod in September and to No. 8 rod in mid-October. On the whole, the seaward edge of mud flat retreated landward for about 3 km from spring to mid-October in 1980.

The seasonal change of width and position of different zones of sedimentary facies can be clearly seen from a comparison of satellite images. For example, in Landsat, image of September 10, 1979 (high HWL), the boundary between lower intertidal and upper intertidal zones and that between upper intertidal and supratidal zones are shifted more landward than in Landsat, image of April 20, 1977 (low HWL), with the result that the width of lower supratidal zone is much reduced (Fig. 3-15) (Tang, 1982).

It was also in mid-October that serious erosion of mud flat occurred, forming conspicuous micromorphological forms–flat-topped, elongated "mud bastions", dissected by an irregular patterns of depressions (Plate 3-12) and the surface of this part of tidal flat was scattered with many "mud boulders" (Plate 3-13), indicating erosive action of wave and currents in autumn. But these "mud bastions" are ephemeral features, because with landward retreat of mud flat, coarse sediments gradually cover dissected surface of the mud flat. In spring 1981, we revisited No. 8 rod and found that tidal flat was remarkably smooth and mud bastions seen in late autumn had totally disappeared (Plate 3-14).

Therefore, it is evident that there is an erosive phase between the alternation of sand and mud deposition. In the central part of lower intertidal zone, thin mud layers deposited are ofter eroded away. But in the upper part of lower intertidal zone where current is weaker, thin mud layers are preserved, forming coarsely interlayered beddings with thick sand and thin mud layers. In the upper

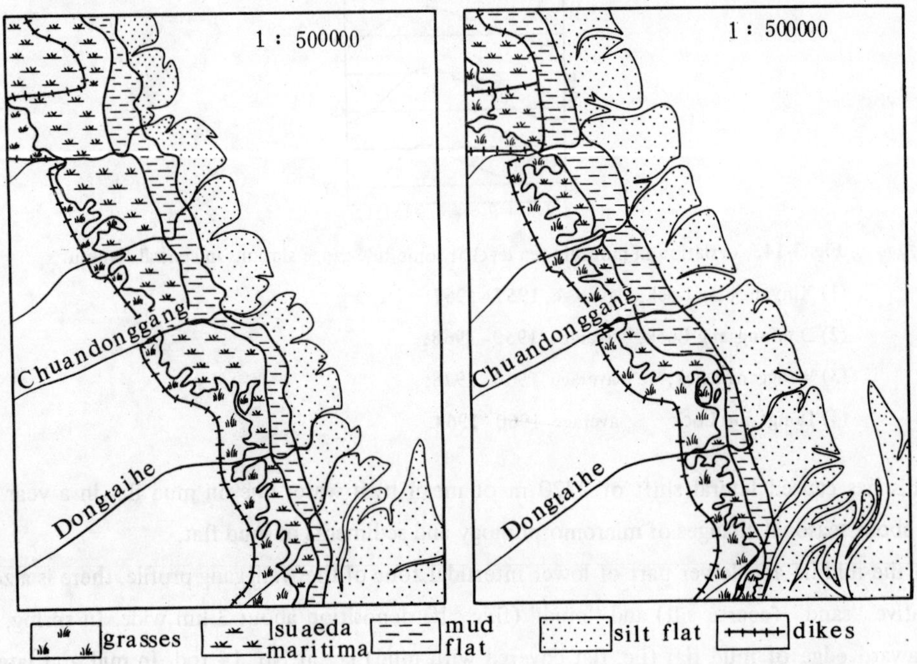

Fig 3-15. Seasonal changes of sedimentary zones on mud flat near Wanggang, Jiangsu.

part of upper intertidal zone, mud deposition is the rule and only during spring tide in higher HWL period, thin sand layer is deposited, forming coarsely interlayered bedding with thick mud layers separated by thin sand layers. Thus, seasonal fluctuation of HWL has a profound effect on sedimentation and on micromorphological features of the tidal flat.

5. Vertical sequences

With seaward growth of the mud flats, supratidal zone of sedimentary facies will gradually overlap intertidal zone of sedimentary facies, which, in turn, will overlap subtidal zone. Therefore, in vertical sequences, zones of sedimentary facies from bottom to top correspond to successive zones of sedimentary facies on the present tidal flat from sea to the coast. In Wanggang area, the rate of seaward growth of the mud flats is about 50 m per year. Thus, the coastal plain 10 km inland from the present sea dike was a tidal flat about 200 years ago. This depositional regression by mud flat accretion allows us to reconstruct tidal flat sequences from vertical profiles in the coastal plain exposed by engineering works.

In shanghai Farm inside the present sea dike, north of Wanggang, a profile more than 2 m high is exposed. The locality was a supratidal zone covered with *Suaeda maritina* before the construction of sea dike in 1978. In the exposure, the following three layers may be distinguished:

(1) The uppermost part (2.2 m. above sea level) is supratidal deposit, which contains abundant roots and is intensely bioturbed.

(2) Upper intertidal deposits (between 1.12 – 2.20 m) *Md* 6.03 – 6.21 ϕ. Between 1.82 – 2.20 m, bioturbation is strong. A layer enriched by Fe_2O_3 can be seen between 1.20 – 1.82 m, and its lower limit roughly corresponds to the original water table or mean sea level. Owing to long exposure of sediments above 1.20 m, iron compounds seep down along plant roots, burrows and cracks, forming many dark brown patches and aggregates which are largely distributed above the water table.

(3) Lower intertidal deposits, *Md* 5.66 – 5.80 ϕ. Thinly and coarsely interlayered beddings are common.

In an excavated profile at Dongbatou, west of Wanggang, about 10 km inland from the present sea dike, similar vertical sequence can be seen (Fig. 3-16).

A significant feature in the vertical sequence is the presence of many layers of storm deposits. In sedimentary layer of upper intertidal zone, Dongbatou, at elevation 1.15 – 1.85 m, there are 7 layers of storm deposits, totalling 25 cm thick, which is 36% of the total thickness of this sedimentary section (Plate 3-15). This may be compared to 4.5 cm storm deposited sand in 20 cm thick sedimentary section on the average in the mud area south of Helgoland, Germany (Reineck, 1967). In profile of Shanghai Farm, two layers of typhoon deposits can also be distinguished in a section of upper intertidal deposit, 23 cm thick. The lower typhoon layer (layer 1 in Plate 3-16) is 4 cm thick, in which clear small cross beddings can be seen. Together with the upper typhoon layer, 3.5 cm thick, typhoon deposits constitute 32% of the total thickness of this sedimentary section. SEM photographs of quartz grains (>63 microns) from both No. 14 typhoon deposits on upper intertidal flat of Wanggang and storm deposit layer 2 in Dongbatou section show clear impact scars on the surface of quartz grain and the surface of scars is clean, almost free from any precipitates. These surface textures clearly show strong wave action and high velocity flow during the deposition of typhoon deposits (Plate 3-17). Thus, hazard events must be considered as of considerable importance in geological history.

by Ren Mei-e, Zhang Renshun and Yang Juhai
Nanjing University, Nanjing

IV. Mud Flat Near Jianggang, Tongtai County, Jiangsu Province

Mud flat near Jianggang, Jiangsu Province, is a special type of mud flat. Within a short distance, prograding flat and receding flat alternately appear (Fig. 3-18). Hydrodynamic condition near this area is rather unique. Here, tidal currents from north and south converge, resulting in a spectacular tidal bore about 1 m high. The converging point is known as Erfenshui[1)] by the local pople. After convergence, velocity of tidal currents rapidly decreases and large quantity of sediment is deposited, forming a large submarine shoal – Tiaozini, about 40 km wide. In this area, mean tide range is 4.08 m and the largest tide range reaches 6.78 m. The surface of Tiaozini is dissected and cut through by strong tidal current (maximum velocity about 2 m/s), and a series of tidal channels are formed. The parallel arrangement of shoals and tidal channels is unique along mud plain coast of China and, perhaps, is also a special type of mud plain coast in the world. Owing to migration and lateral erosion

1) In Chinese, Erfenshui means a water divide between two opposite currents.

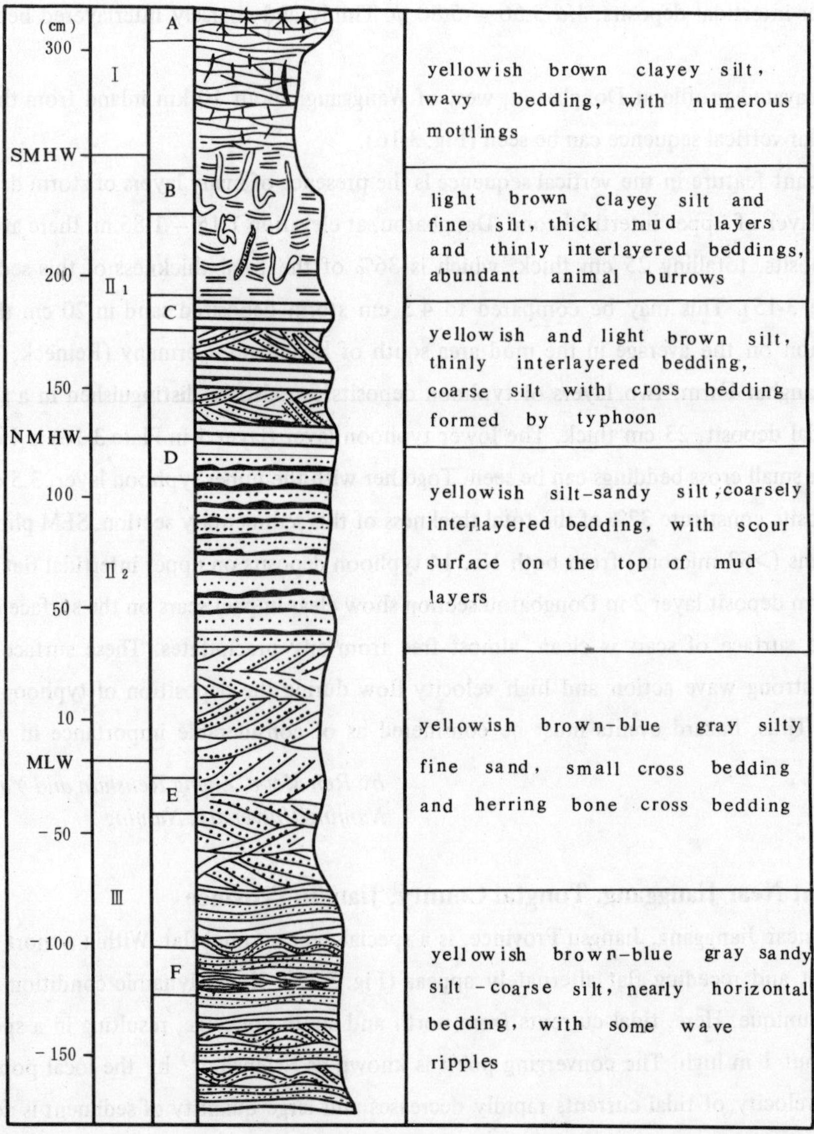

Fig. 3-16. Reconstruction of a sequential model of a mesotidal mud flat from an exposure at Dongbatou, west of Wanggang.

I–supratidal; II_1 –upper intertidal; II_2 –lower intertidal; III–subtidal

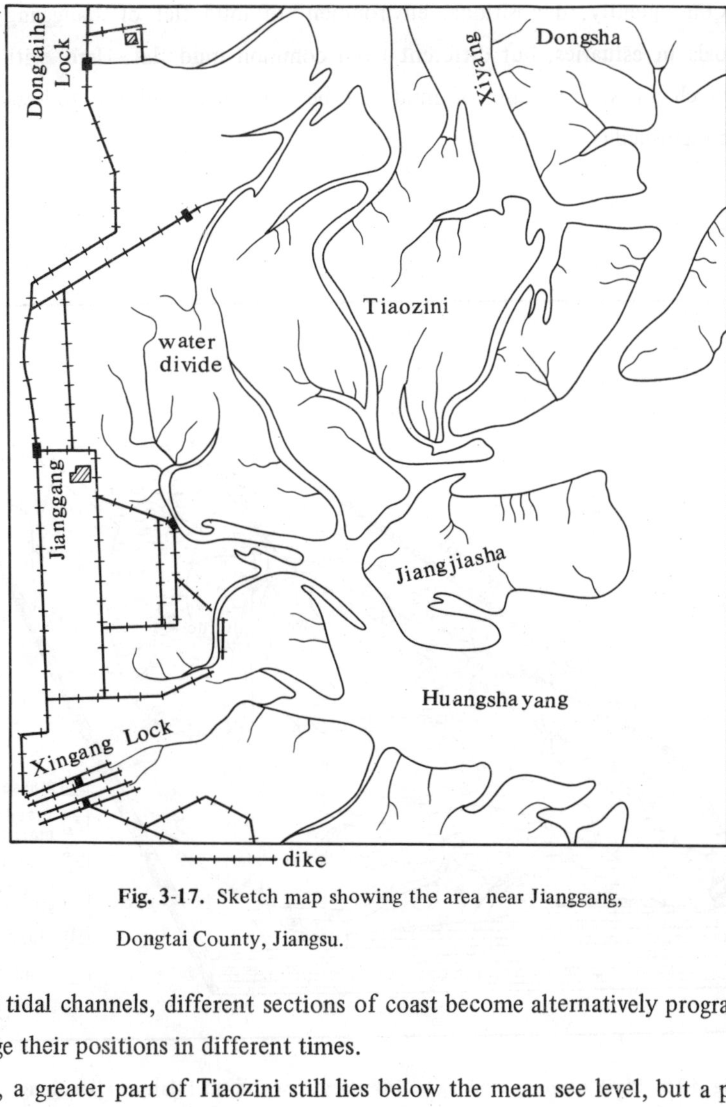

Fig. 3-17. Sketch map showing the area near Jianggang, Dongtai County, Jiangsu.

of meanders of tidal channels, different sections of coast become alternatively prograding and receding which change their positions in different times.

At present, a greater part of Tiaozini still lies below the mean see level, but a part of which is already exposed during low tide and a part even stands above mean high water level, becoming supratidal. Tidal channels in the shoal are often rather deep. In deep channels, maximum velocity of the ebb is greater than that of the flood, but on the tidal flat beside the channel, the opposite is true. This indicates that flood current spreads broadly over tidal flat but ebb current is concentrated in channels. Here, owing to the existence of submarine sand ridges offshore, wave is weak in coastal waters and under strong wind of Beaufort scale 6–7, wave height is less than 1 m. Therefore, the dominant hydrological factor that shapes tidal channels and controls sedimentation on tidal flat is still tidal current.

Owing to the existence of tidal channels running approximately parallel to the coast, the direction of tidal current on mud flat is chiefly parallel or oblique to the coastline, quite different from mud flat of other loaclities where tidal current on mud flat is generally perpendicular to the coastline. Moreover, in tidal channels, tidal current is strongly reversing and flows with great velocity. Consequently, depositional environment of mud flat of Jianggang area is similar to intertidal shoals in estuaries, but different from common mud flat. Here, with increasing distance ırom tidal channels, sediments of mud flat become finer and the following sedimentary zones may be distinguished (Fig. 3-18).

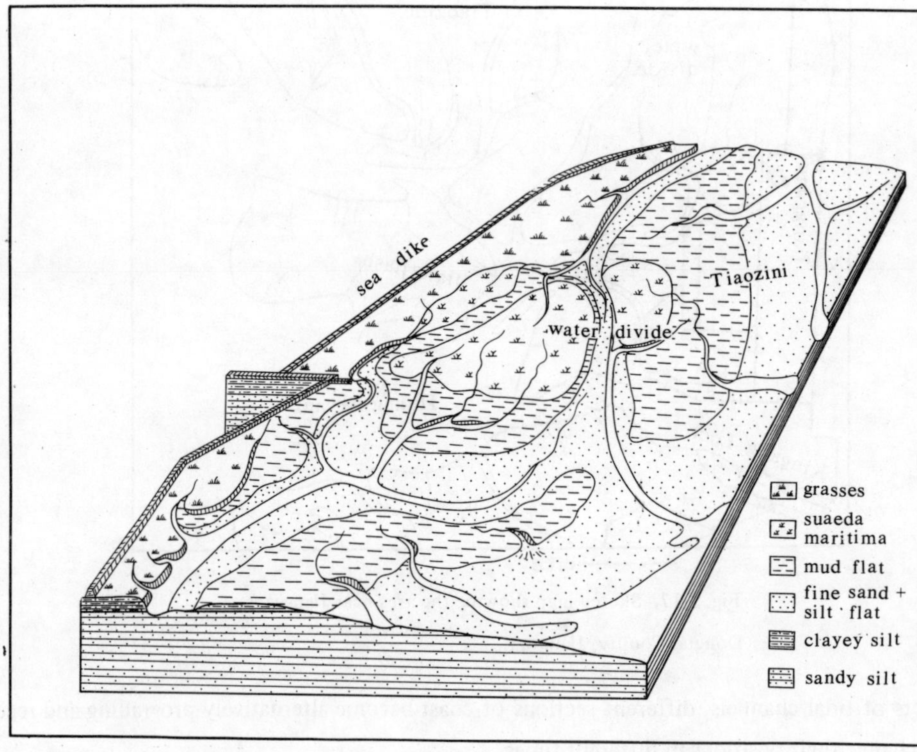

Fig. 3-18. Block diagram showing sedimentary zones on tidal flat near jianggang.

grasses = upper supratidal; *Suaeda maritima* = lower supratidal;

mud flat = upper intertidal; fine sand + silt flat = lower intertidal

(After Zhang Dachu, 1981).

(1) Upper supratidal locally known as grass flat, chiefly covered wtih *Artemisia capillaris* community and *Imperate cylindrica* community. Cover degree is 60–90%. Md of sediments 5.8–6.1 ϕ.

(2) Lower supratidal also known as *Suaeda maritima* flat. Md of sediments 5.5–5.8 ϕ.

(3) Upper intertidal also known as "mud" flat. Md of sediments 5.4–5.6 ϕ.

(4) Lower intertidal also known as fine sand and silt flat. Md of sediments 3.1–5.3 ϕ. Many kinds of current and wave ripples can be seen on the surface of the flat.

From the above facts, it can be seen that sediment of tidal flat of Jianggang area is coarser than that of Wanggang due to greater velocity of tidal current (maximum velocity of flood current on intertidal flat is 1.8 m/s). In sediments of intertidal flat, cross beddings are well developed and reactivation surfaces are found, indicating stronger tidal current in this area.

On a bluff on the concave side of a large tidal channel near Jianggang, a section about 7 m high is exposed. On this exposure, vertical sequence corresponding to sedimentary zones of the present tidal flat may be observed:

(1) Upper supratidal altitude above 4.5 m. Yellowish brown clayey silt.

(2) Lower supratidal altitude 3.5–4.5 m. Chiefly fine laminations but wavy or curled beddings can be occasionally seen (Plate 3-18).

(3) Upper intertidal altitude 2.5–3.5 m. Browish gray clayey silt, with cross-beddings. In the lower part, small penecontemporaneous foldings can be seen (Plate 3-19).

(4) Lower intertidal altitude <2.5 m. Grayish-white silt and fine sand.

by Ren Mei-e and Zhang Renshun
Nanjing University, Nanjing

V. Mud Flat Along the Coast of the Bohai Sea

Mud flat along the coast of the Bohai Sea is one of the longest mud flats in China. It is more than 500 km long, second only to Jiangsu coast.

The Bohai Sea is a microtidal sea. M_2 tidal wave system enters the Bohai Sea through the Bohai Straits, forming two left rotary tide wave systems, the no-tide points of which are located outside Qinhuangdao and the old Huanghe River mouth respectively. Therefore the Huanghe River mouth is microtidal. Mean tide range is only 0.6 m in the mouth of the Shunxiangou, but it increases to 1.2 m at the head of the Bohai Gulf. The coast along the Laizhou Bay is also microtidal (Dou Zhenxing, 1981).

Material of mud flat along the coast of the Bohai Sea also comes from the Huanghe River. The transport and diffusion of the Huanghe River sediment in the sea are greatly influenced by oceanographical conditions. First, tidal current.Outside the mouth of the Shunxiangou is a no-tide point where tide range is very small but tidal current is strong. From this point, tidal current flows and carries the Huanghe River sediment NW toward the Bohai Gulf and S and SE toward the Laizhou Bay. Therefore, before 1976 when the Huanghe River entered the sea near Diaokou, a greater part of the Huanghe River silt was transported toward the Bohai Gulf. But after the shifting of the Huanghe River mouth southward from Diaokou (along the Bohai Gulf) to Qingshuigou (along the Laizhou Bay), a marked change of sediment transport pattern has occurred and now, a major part of the Huanghe River sediment is transported by tidal current southward and southeastward to the Laizhou Bay instead of NW to the Bohai Gulf. Therefore, suspended solid concentration of seawater in Tianjin New Harbour has decreased almost 100% and the subaqueous delta of the Haihe River has suffered from serious erosion[1)].

Secondly, residual current. In nearshore zone of the Huanghe River Delta, under the southerly wind, the direction of surficial residual current is from the Laizhou Bay to NW, i.e. to the head of the

---

1) The decrease of suspended solid concentration in Tianjin New Harbour is partly due to new engineering measures.

Bohai Gulf. Under the northerly wind, the surficial residual current is from NW to SE. But in the bottom layer, the residual current is from mouth of the Laizhou Bay to W or NW both under southerly and northerly winds. As more than 80% of the Huanghe River silt discharages into the sea during flood season of July to October when southerly wind prevails, more silt is transported by residual current to the Bohai Gulf.

Thirdly, wave. According to Pang, under wind-driven wave, sediment in nearshore zone of northwestern Huanghe River delta is chiefly transported to the west but in nearshore zone of eastern Huanghe River delta (i.e. south of the Shunxiangou) the direction of sediment transport is mainly toward south.

From the above facts, it can be seen that because of the shifting of the Huanghe River mouth to the south in 1976, at present a major part of the Huanghe River silt is transported by tidal current and wave toward south, although a lesser part of the Huanghe River silt is still moved northwestward to the Bohai Gulf by residual current.

Sediment discharge from the Huanghe River has a marked influence on the distribution of grain size and mineral assemblage of surficial sediment of the Bohai Sea and the distribution of suspended sediment concentration in surficial sea water of the Bohai Sea. The mean annual sediment discharge of the Huanghe River is 11.18×10^8 tons in which 36% is finer than 6 ϕ (fine silt). A greater part of sediment coarser than 6 ϕ is deposited in subaerial delta and nearshore zone; only fine particles finer than 6 ϕ is diffused to the sea, entering the Bohai Gulf and the Laizhou Bay. Therefore, in the map of distribution of surficial sediments of the Bohai Gulf, there is a distinct zone of silty mud in the western part of the Bohai Gulf, stretching northward to Nanpu. Surficial sediment of the Laizhou Bay and southern part of the Bohai Gulf has high content of muscovite, carbonates, plagioclase and hornblende and low content of magnetite. Sediment of the lower Huanghe River has the same characteristic mineral assemblage. But surficial sediment of those part of the Bohai Sea influenced by sediment discharge from the Liaohe River and Luanhe River has very low content of muscovite, carbonates and hornblende (Chen Lirong, 1980). This shows that the Huanghe River sediment mainly diffuses to the Laizhou Bay and southern part of the Bohai Sea.

Influenced by sediment discharge from the Huanghe River, suspended solid concentration of the surface sea water in the Huanghe River mouth, Bohai Gulf and Laizhou Bay is very high; especially outside the old Huanghe River mouth, suspended solid concentration in July, 1958 and 1962 amounted to 100–150 mg/1 (Qin Yunshan, 1982) (Fig. 3-19).

Mud flat along the coast of the Bohai Sea is distributed in the following three regions: (1) Along the Bohai Gulf. It is about 275 km long from the old Huanghe River mouth to Nanpu. (2) Along the Huanghe River Delta, chiefly mud flat along inter-distributary bays between spits of river mouth of different periods. (3) Along the Laizhou Bay. Up to the present, only mud flat along the Bohai Gulf has been studied in some detail.

1. Laizhou Bay

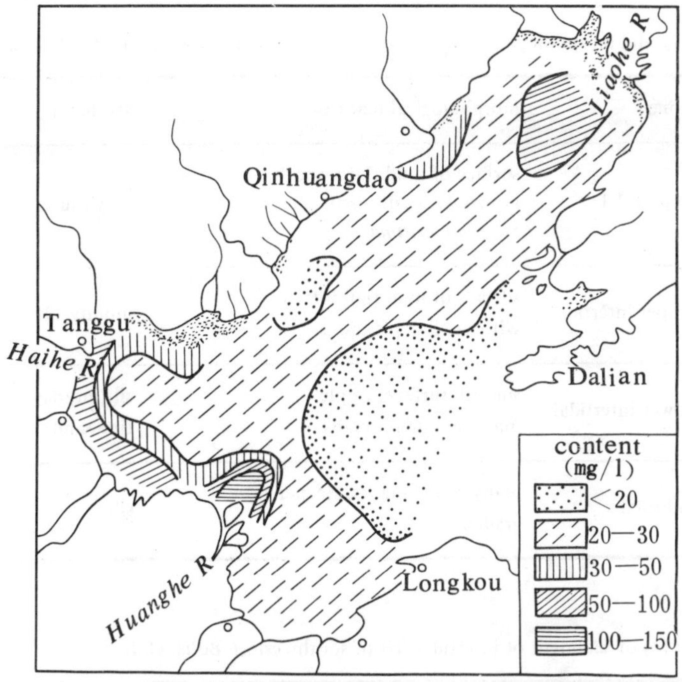

Fig. 3-19. Suspended matter in surface sea water of the Bohai Sea (July 1958 and 1962, after Qin, 1982).

Generally speaking, mud stream from the present Huanghe River mouth can only reach north of the Xiaoqinghe mouth. Only under strong northerly wind, can the Huanghe River silt reach the mouth of the Xiaoqinghe. The coast of the Huanghe River delta north of the mouth of the Xiaoqinghe is now fringed with wide mud flat. For example, near the mouth of the Guanglihe, north of the Xiaoqinghe, mud flat is 12 km wide outside the present sea dike. But along the Laizhou Bay, east of the Laomihe, sediment of tidal flat is a little coarser. At the head of the Laizhou Bay, oceanographical energy is low and consequently, mud flat is stable or a little prograding although sediment supply is not plentiful. Sedimentary and morphological zonation of a mud flat along the coast of Shouguang County, Shandong Province, is as indicated Table 3-9.

2. Bohai Gulf

Along the Bohai Gulf, mud flat is well developed in its southwestern part (i.e. south of the mouth of the Haihe River). Average width of mud flat is about 3 km and its gradient is 0.5–1% . Similar to other regions, sediment of mud flat along the Bohai Gulf generally becomes finer from the sea toward land (Table 3-10). It consists of all grades of silt, indicating its origin from the Huanghe River.

In the Bohai Gulf, flood velocity is commonly greater than ebb velocity. Consequently, amount of sediment transported to tidal flat by the flood exceeds that carried back by the ebb, and mud flat is continuously prograding, like mud flat along the coast from Tianjin New Harbour to Qikou (42 km long). But between the Dakouhe and Jiajiapu, as the coastline is facing NE, i.e. facing the direction of

Table 3-9 Sedimentary and morphological zonation of a mud flat in Shouguang (modified after Fan Yisun)

Range (km)	zone	morphological features	sediment
Sea dike–1 8	supratidal	surface of mud flat smooth, grown with salt plants like *Suaeda maritima*	silty mud
1.8–3.9	upper intertidal	surface of mud flat dotted with small depressions	muddy silt
3.9–7.0	lower intertidal	smooth surface, ripple marks abundant crabs	silt interlayered with fine sand
blow 7.0	subtidal	many ripple marks, steeper gradient	silt

Table 3-10 Grain size (ϕ) of sediment of intertidal flat of southwestern Bohai Gulf

Locality	upper part	middle part	lower part
Tanggu	7.5	7.1	3.7
Qikou	6.9	6.4	5.7
Zhaojiapu	8.1	6.7	6.1

(after Li Congxian et al., 1982)

the strongest wind of the region and is under strong wave action, the coastline is eroded and receding, mud flat is narrower (about 1.5 km wide) and gradient is steeper.

In Lichuhe profile, south of Tianjin New Harbour, the following sedimentary zones may be distinguished (Fig. 3-20).

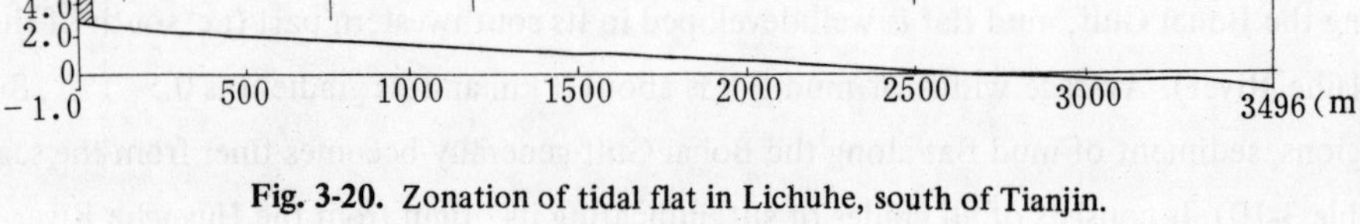

Fig. 3-20. Zonation of tidal flat in Lichuhe, south of Tianjin.

I–Supratidal; II_1 –Upper intertidal;

II_2 – Middle intertidal; II_3–Lower intertidal;

III–Subtidal.

(revised after Wang Yin et al., 1965)

(a) Supratidal zone Chiefly clayey mud (Md <5 μ), mud cracks are well developed. Small *phragmites communis* and *Suaeda glauce* are found near cheniers.

(b) Upper intertidal zone Chiefly mud (Md 5–10 μ).

(c) Middle intertidal zone Coarse silt (Md > 50 μ) interlayered with mud, usually mud in upper layer and silt with shell fragments in lower layer. On the mud flat, elongated mud bastions 3–18 m long and several to tens cm high can often be seen. After strong storm, these bastions are buried under mud.

(d) Lower intertidal zone Here, as flood current is still strong, coarser material, chiefly coarse silt, is deposited. Owing to stronger wave and tidal current, surface of mud flat is scattered with many small ripple marks.

(e) Subtidal zone From low water line to sea, sediment becomes gradually finer again (Wang Ying, 1965).

The above zonation is roughly similar to sedimentary zonation of Wanggang, Jiangsu Province. But owing to difference in hydrodynamic conditions and sediment supply, mud flat in southwestern Bohai Gulf is far narrower than that of Wanggang, but its gradient is steeper.

Seasonal variation of sedimentary zones of mud flat along southwestern Bohai Gulf is similar to that of Wanggang. The zonation is most distinct in spring. In summer, when wind and wave are stronger, with coarsening of sediments, mud bastions are eroded and subtidal zone extends landward. After autumn, with weakening of oceanographical energy, zonation is gradually reestablished.

During the sixties , tidal flat of the Haihe River subaqueous delta was about 5 km wide. Because the delta was rapidly prograding seaward from the middle of the ninteenth century to the beginning of the twentieth century, the gradient of mud flat was rather smooth, only 0.5% . Here, sedimentary zonation of intertidal flat is similar to that of Lichuhe. Burrowing animals in upper intertidal zone are mainly crabs and those in lower intertidal zone are mainly mollusks. Before the construction of tide gate at the Haihe River mouth in 1958, delta of the Haihe River had been prograding seaward. But after 1958, as sediment discharge from the Haihe River to the sea reduced more than 5/6, the coastline of the Haihe River delta receded, forming low erosional scarps with small new cheniers on the scarps and mud boulders at the foot of the scarps. In a few years after the construction of the tide gate, the coastline of the delta retreated at a rate of 20–30 m per year. During recent years, owing to diminishing supply of sediment from the Huanghe River, the subaqueous delta of the Haihe River has further suffered from serious erosion (Coastal Studies Group, Nanjing University, 1974) (Fig. 3-21).

An important characteristic of coastal morphology of the Bohai Gulf is a series of four cbenier ridges on the coastal plain. Cheniers can also be found in coastal plain of other parts of China, but their magnitude and degree of preservation cannot be compared with those on the west coast of the Bohai Gulf. Indeed, cheniers on the west coast of the Bohai Gulf may be taken as typical in China and perhaps are one of the best developed cheniers in the world.

The positions and ages of these four chenier ridges are as follows: (Fig. 3-22) (Table 3-11).

In this region, the second chenier ridge is the largest which now remains as discontinuous hummock, 4–5 m high and more than 200 m wide at its base. This may be compared with cheniers on the

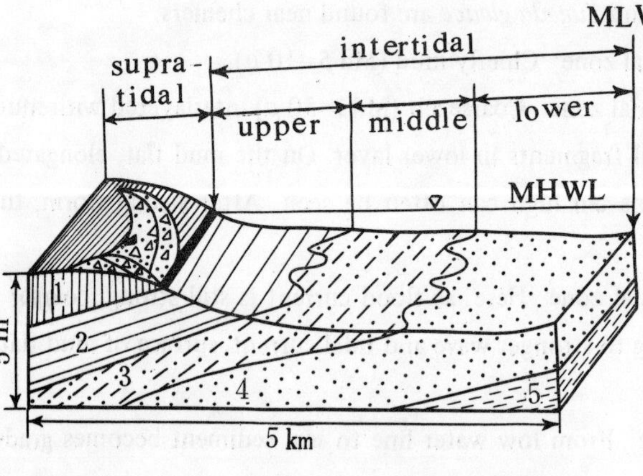

Fig. 3-21. The subaqueous delta of the Haihe River.
1. Mud and clay; 2. Mud; 3. Silt and fine sand;
5. Silty mud of subtidal zone; 6. Shell debris.

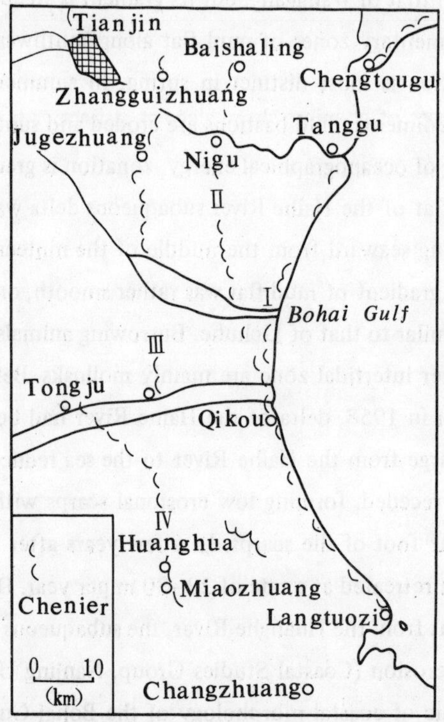

Fig. 3-22. Cheniers on the west coast of the Bohai Gulf.
(after Zhao, 1980)

Mississippi delta, where their width is generally a little more than 100 m and their height does not exceed 4.5 m. On the west coast of the Bohai Gulf, the original ridge-like form of cheniers has often been greatly obliterated by human activities. For example, the second chenier ridge at Baishaling was 5 m high in 1957, but now it has been levelled in many localities. The third and first chenier ridges are only 1–2 m high and are scattered on the coastal plain as discontinuous fragments.

Table 3-11 Chenier ridges on the west coast of the Bohai Gulf and their age

	Location	Age (yr BP) (^{14}C dating)
4th chenier	Tongju-Miaozhuang	4700–4000
3rd chenier	Zhangguizhuang–Jugezhuang–Changzhuang	3800–3000
2nd chenier	Baishaling–Nigu–Qikou	2500–1100
Lst chenier	Chengtougu–Tanggu–Qikou	probably 600–500

(after Zhao Xitao et al., 1980)

Cheniers are chiefly made up of shells and shell fragments with interlayers of coarse silt and fine sand. Main species of shells are *Area subcrenate, Mactva quadrangularis, Meretrix meretrix, Cyclinna sinensis, Rapana rhomasiana, Umbonium thomosi, Terebra aussumieri, Polynices diayma, Sinonovacala constricta, Solon* sp., *Ostrea gigas* and *O. rivularis.* All these species are at present living on intertidal flat or in shallow waters. Deposits of cheniers show clear stratification, usually horizontal or gently dipping. In the seaward slope of cheniers, beds dip to the sea and dip angle is generally 5°–6°. In their landward slope, beds dip to the land but dip angle is a little steeper, about 6°–8° (Plates 3-20, 3-21).

Similar to deltas of other major rivers of the world, the origin of chenier ridges on the west coast of the Bohai Gulf is mainly due to shift of the Huanghe River mouth. During historical time, the river had not only flowed through northern Jiangsu Province into the Huanghai Sea, but also had frequently changed its outlet when it entered the Bohai Sea. When the Huanghe River flowed into the Huanghai Sea, sediment supply to the west coast of the Bohai Gulf was greatly reduced and the coast retreated. At that time, with steepening of profile of tidal flat, the role of wave on tidal flat was increased. As fine sediment (mud) was carried away by wave, the material of tidal flat was coarsened to medium-coarse silt which is most suitable to the growth of most shells that build cheniers. Abundant mollusks in tidal flat ensure a plentiful supply of shells to cheniers. During the formation of cheniers, profile of coastal zone was probably similar to that of sandy coast. Shells and shell fragments were continually washed to the coast by breakers and accumulated near the high water line, gradually forming ridges. Thus, the mechanism of formation of cheniers is much like that of sand bars of sandy coast. When the Huanghe River entered the Bohai Sea again, the west coast of the Bohai Gulf became a prograding, mud coast once more, and chenier ridges ceased to develop. They are gradually fronted by tidal mud flat and lie a considerable distance from the present coastline, standing out as notable marks of ancient shorelines on the coastal plain. Similarly, since the outlet of the Huanghe River into the Bohai Sea also changed frequently in the historical time, certain sections of the coast along the Bohai Sea received abundant sediment in certain periods, but were inadequately supplied with sediments during other periods and consequently, small cheniers were formed, Since the Huanghe River has changed its outlet to the sea many times, a series of chenier ridges were built on the west coastal plain of the Bohai Gulf. The west coastal plain of the Bohai Gulf constitutes the most remarkable chenier plain of China where chenier ridges and interridges depressions appear in regular successions, the latter now being occupied by mud plain, marshes or lakes.

From the above facts, it may be visualized that the following conditions are necessary for the formation of chenier ridges:

(a) Moderate steep gradient of intertidal flat According to coring data, the gradient of the foot of cheniers and of the old beach beneath cheniers is 1–1.9% and the gradient of the original intertidal flat is 0.33–0.2%. Under these conditions, the coast can be frequently affected by waves which transport coarser sediment and shells to the coast, forming cheniers. Near the mouth of the abandoned Huanghe River, the most favourable beach gradient for the formation of new cheniers is also 0.2–0.8%. On the other hand, the gradient of tidal mud flat is usually less than 1‰; hence, wave action is very weak and cheniers cannot be formed, Cores at Dakouhe, Qikou and other localities on the coast show that vertical sedimentary sequence consists of mud in the upper and lower layers between which is an interlayer of shells and silt about several to tens cm thick. The upper mud layer, about 3 m thick has been deposited since 1855 when the Huanghe River entered the Bohai Sea again, while the shell-silt interlayer was formed during A.D. 1185-1855 when the outlet of the Huanghe River shifted to Jiangsu and sediments of tidal flat along the Bohai Gulf became coarsened.

(b) The formation of cheniers needs a long time. According to ^{14}C dating of shells from chenier ridge at Qikou, in a sedimentary column less than 3 m thick, the difference of age between shells of the upper layer and the lower layer amounts to 940 years. Hence, it takes about 600–900 years to build a chenier ridge 2–3 m high. It is evident that the coast must be stable for considerably long time for the formation of a chenier ridge of reasonable dimension. For this reason, rapid receding coast is unfavourable for the formation of cheniers. Coast near the mouth of the abandoned Huanghe River in Jiangsu furnishes a good example. Here, since the Huanghe River shifted northward into the Bohai Sea, low cheniers have been formed near the high water line in many localities, except where the coast retreats rapidly.

Owing to huge quantity of sediment discharge of the Huanghe River, changes of its outlet has exerted profound influence on sediment supply of the adjacent coastal zone. Since China is famous for very long historical record and rich archaeological findings, chenier ridges on the west coast of the Bohai Gulf are indeed one of the most typical cheniers in the world, The ages of different chenier ridges made with reference to historical and archaeological evidences generally conform to ^{14}C datings. For example, cultural remains of the late period of Warring States to the Western Han Dynasty have been discovered in the second chenier ridge and large sites and tombs belonging to the period of Warring States discovered in the third chenier ridge. Their ages correspond to ^{14}C datings. According to historical record, before the Western Zhou Dynasty, i.e. about 3400 years BP, the Huanghe River entered the sea near Tianjin for the first time and its mouth was situated somewhere north of the present Luanhe River mouth. Probably, tne third and fourth chenier ridges were formed in that time. Between 602 B.C. and 132 B.C., the Huanghe River shifted southward, entering the Bohai Sea through Shandong Province. As a result, the second chenier ridge was gradually built up. Finally, the first chenier ridge was formed between A.D. 1194 and A.D. 1855, when the Huanghe River flowed through Jiangsu Province into the sea, but it gradually ceased to develop after the river had shifted its course northward in 1855. Between 1938 and 1945, the Huanghe River overflowed at Huayuankou, near Zhengzhou (Henan Province) and entered the sea through Jiangsu again. But on the west coast of the Bohai Gulf, only a shell layer about 10 cm high was deposited near the high water line, as eight years are too short to form a conceivable chenier ridge.

On the coastal plain of central Jiangsu and Shanghai, a number of chenier ridges can also be found, In addition to shells, the material of these ridges is mainly sand–fine sand with good to very good sorting. They are transported by saltation. Sediment of cheniers contains high content of heavy minerals, chiefly amphibole, epidote, garnet, etc., all with a great specific gravity. But micas are almost absent. These characteristics indicate that during the formation of cheniers, wave has considerable energy and sediment has been repeatedly transported and winnowed by waves. This shows that hydrodynamic conditions during chenier formation is entirely different from those for the present mud flat (Gu Jiayu, 1983).

3. The modern Huanghe Delta

The coastline of modern Huanghe River delta, formed since 1855, is about 160 km long, extending from the mouth of the Taoerhe in the west to the mouth of the Xiaoqinghe in the south. In the modern delta, the Huanghe River has repeatedly overflowed and changed its course. From 1855 to the present, 12 major changes of river course have occurred (Fig. 4-1), in which shift of the outlet to Diaokou in 1964 and to Qingshuigou in 1976 are two most important changes that have great influence on morphology and sedimentation of coastal zone of modern Huanghe River delta. Sediment of the Huanghe River bed is coarse, 64% being very fine sand and coarse silt at Qianzuo station near the river mouth. When the river enters the sea, coarse bed load rapidly deposits near the river mouth, forming spits along its subaqueous channel. These spits project into the sea as bird's-foot elongated sand bodies.

From February, 1964 to April, 1965 when the Huanghe River began to enter sea near Diaokou, river mouth spit extended into the sea nearly 10 km. Between 1964 and 1976, with river mouth spits as framework, a small bird's-foot delta was formed. As river mouth spits extended seaward, elongated interdistributary bays were formed between them. These interdistributary bays, only 2–3 m deep and with quiet environment, facilitate deposition of mud and therefore, are called "mud bays" by local fishermen. The bottom of "mud bays" is covered with a layer of fluid mud about 1–5 m thick and coast of bays is fringed with wide mud flat.

In modern Huanghe River delta, the coast near the abandoned river mouths is subjected to vigorous action of wave and current and the old spits and coastline are rapidly eroded and retreated. Study of satellite images shows that from 1976 to 1981, the coast near the old river mouth near Diaokou retreated nearly 6 km and the old subaqueous spits almost belevelled and disappeared.

Whereas at the present river mouth (Qingshuigou), the coast has been rapidly prograding since 1976 and in some localities, it extended as much as 12 km seaward from May, 1976 to May, 1977.

From the above facts, it can be seen that on account of frequent shifts of outlet of the Huanghe River, different parts of the coast of modern Huanghe River delta have undergone progradation or recession, and prograding and receding coasts often change their position through times. Consequently, sediment of different sectors of tidal flat and subtidal zone also varies. When the river enters the sea near interdistributary bays, coarse silt may cover old mud layer, as at the present river mouth.

Because the Huanghe River flowed through the Xiaoqinghe into the sea in 1895 and 1937 and the southward diffusion of the Huanghe River silt since the river changed its outlet to the Qingshuigou in 1976, the present tidal mud flat along the mouth of the Xiaoqinghe is quite broad. Between the high and low water lines, mud flat along its north bank is 7–9 km wide and that along the south bank 5–6 km wide, and its gradient is about 0.45%. Moreover, inside the high water line, there is an extensive swamp covered with *Phragmites communis, Suaeda maritima*, etc., which is only flooded by seawater in exceptionally high tide. Hence, including supratidal swamp, tidal mud flat along the mouth of the Xiaoqinghe amounts to some 10 km wide, constituting one of the widest mud flats along the Bohai Sea (Chen Bingan, 1982).

VI. Tidal Mud Flat of Embayment Type

Tidal mud flat of embayment type, which is mostly prograding seaward, is mainly distributed along the coast of Zhejiang, Fujian, Guangdong and Liaoning Provinces. Although it consists of short stretches, but as bays are often deeply indented, penetrating far into inland, its total length is significant. It is estimated that coast with mud flat of embayment type is about 1000 km long in Zhejiang Province and 500 km long in Fujian Province. At present, a detailed study on sedimentation of mud flat of embayment type has been made is Taizhou Bay, Zhejiang which will be treated in Section 7 of this Chapter as a typical example of mud flat of this type in China.

(1) Mud flat of Zhejiang Province is mainly found along the coast of the Hangzhou Bay, Taizhou Bay and Wenzhou Bay. In general, its width is about 4–6 km, with average gradient about 1‰. Estuarine mud flat along south coast of the Hangzhou Bay, with abundant supply of sediment from the Changjiang River due to its proximity to the mouth of this great river, has a width nearly 10 km.

Zonation of Zhejiang mud flat is roughly similar to that of Jiangsu. It often has a steeper gradient (>1‰) in the middle part of the flat due to stronger wave action there. Sediments of Zhejiang mud flat consists mainly of fine–very fine silt (Md 6–8ϕ), with thinly interlayered bedding.

Erosion and accretion of mud flat is closely related to wind and wave. In Yongqiang area, Wenzhou, during No. 4 (July 3–5), No. 6 (July 14–15) and No. 16 (August 17–20) typhoons of 1964, huge waves lowered the surface of mud flat and brought about a coarsening of sediments. But after the typhoon, the mud flat was quickly aggraded with a mud layer 10–15 cm thick (Li Congxian, 1982).

Sediments of embayment flat of Zhejiang come mainly from the Changjiang River and inner shelf. Owing to growth of intertidal shoal – Chongming Island in the river mouth, the north branch of the Changjiang River has been quickly silted up in recent years and more than 95% of river discharge and sediment load passes through the south branch into the sea and drifts southward. This is favourable for sediment transport toward the south. The East China Sea littoral curretn, influenced by fresh water from the Changjiang River, has a lower salinity (<31‰) and a lower density. According to *T-S* diagrams made by Lin Chuanlan from data of monthly oceanographical profiles of 1975–1979, the fresh water plume from the Changjiang River mouth appears as a broad belt in the shallow sea south of 32° N and west of 123° E during winter, spring and autumn. It extends southward to south of 28° N. In summer, because of the influence of southerly wind, the fresh water plume from the Changjiang River extends northeastward to about 127° E, but there is still a narrow belt of diluted water flowing southward along the coastal zone to south of 28° N (Lin Chuanlan, 1982)[1]. It seems that suspended sediment of the Changjiang River is transported southward along the coastal zone in the whole year. The First Institute of Oceanography, National Bureau of Oceanography has carried out drifting bottle test for three consecutive years in the inner shelf of southern Huanghai Sea and northern part of the East China Sea. The result of the test proves that south of the mouth of the Changjiang River, in the surficial layer of the sea, there is a littoral current drifting southward in the whole year as far as northern Fujian (Fig. 3-23).

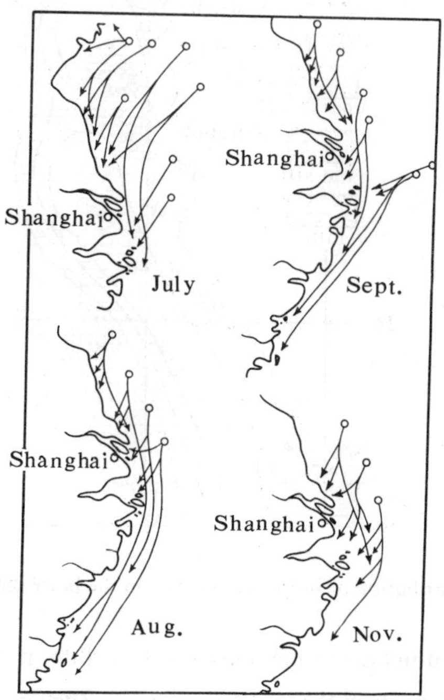

Fig. 3-23. Tracks of drifting bottles in the coastal water of East China (after lst Institute of Oceanography NBO, China).

---

1) Lin Chuanlan et al., 1982, A preliminary study on water movement and its seasonal variations in the East China Sea, Practical Oceanography 1982(3), 23–30.

From salinity data of the nearshore water of the East China Sea, it can be seen that a low-salinity plume about 30–50 km wide extends from the mouth of the Changjiang River to the coastal water of northern Fujian. This further indicates that southward drift of the diluted water from the Changjiang River is rather stable. It may be visualized that finegrained sediment of the Changjiang River is continuously drifted southward by the littoral current and supplies material for the building of mud flat along the coast of Zhejiang. Part of fine sediment carried by the littoral current is deposited on sea floor of the inner shelf, forming a distinct mud belt (Md 8–9 ϕ) in the inner shelf from the mouth of the Changjiang River to northern Fujian, (Qin Yunshan, 1963; Cai Aizhi, 1982) (Fig. 3-24).

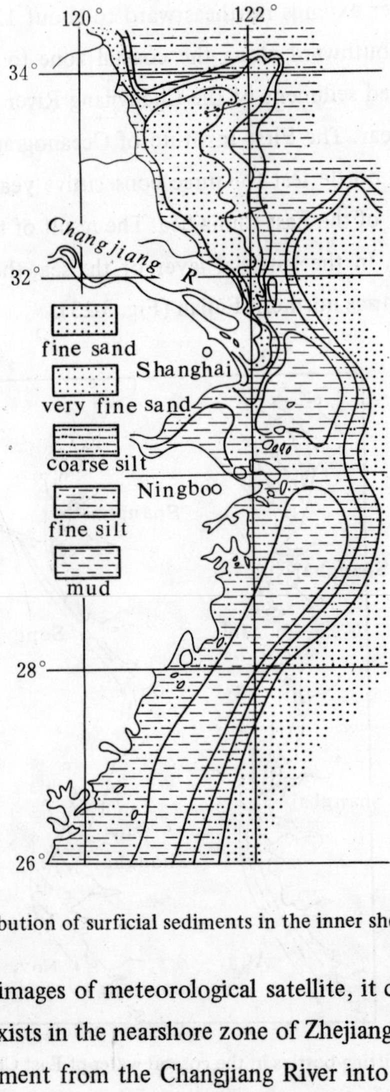

Fig. 3-24. Distribution of surficial sediments in the inner shelf of East China.

From Landsat images and images of meteorological satellite, it can be seen that a zone of turbid water about 30–50 km wide exists in the nearshore zone of Zhejiang and northern Fujian, indicating the extent of diffusion of sediment from the Changjiang River into the sea. During summer, under the southeast monsoon, the diffusion of the Changjiang River sediment reaches the mouth of the Oujiang River, Zhejiang; but in winter and spring, when the northerly wind is predominant, the sediment of the Changjiang River can be carried as far south as to the mouth of the Minjiang River, Fujian (Yun Caixing, 1981).

It may be noted that the amount of accretion of mud flat and shallow sea near Zhejiang is remarkably out of proportion to sediment load of rivers of this province. For example, the annual volume of accretion in shallow sea of Wenzhou area less than 10 m deep is 30 million tons, but the annual sediment load from the Oujiang River, Feiyunjiang and Aojiang is only 5 million tons. This shows the great role of the Changjiang River as source of sediment for this region.

In clay-mineral association, illite is the main claymineral in sediments of mud flat of Zhejiang, with a small amount of chlorite and kaolinite. This is similar to claymineral association of mud of the inner shelf and sediments of the mouth of the Changjiang River. Therefore, the material of mud flat of Zhejiang comes directly and indirectly from the Changjiang River, the former from the mud stream of the Changjiang River and the latter from stirring up of sea floor sediment of the inner shelf by waves. Owing to great tide range of Wenzhou region (maximum range is 7.21 m), tidal current is swift and its sediment concentration is also very high. In the shallow sea 8 km off the mouth of the Feiyunjiang, the maximum mean sediment concentration reached 3.48 kg/m^3 in July, 1976. Sediments stirred up by waves from the inner shelf is carried by strong tidal current to the nearshore zone and deposited there as wide mud flat of embayment type along ria coast of Zhejiang (Ji Zhongchun, 1981; Liu Changyu, 1982)[1)].

(2) Mud flat along the coast of western Guangdong the Zhujiang River, with annual load 83.36 million tons, is one of the great rivers of China. Owing to sediment discharge from the Zhujiang River, mud supply in littoral zone of Guangdong is plentiful. It is estimated that about 1/5 of sediment discharge of the Zhujiang River is deposited in river mouth, but a major part of fine suspended sediment (<0.01 mm) is carried away by littoral current. The littoral current along Guangdong chiefly flows toward SW, carries Zhujiang River sediment westward to the coast of western Guangdong and Leizhou Peninsula and deposits it in bays, forming tidal mud flat. This littoral current, known as the South China Sea littoral current, is deeply influenced by fresh water from the Zhujiang River and consequently has low salinity, which in summer, sometimes may be less than 10‰, or less than 1/2 of salinity of South China Sea warm current in the open sea. Under the northeasterly wind in winter, its width reaches 30 knots.

In the inner shelf of western Guangdong, sea-floor sediment largely consists of mud. Clayey organic mud can be found from the mouth of the Zhujiang River to Nanpeng Island and sediment containing >2.0‰ organic matter spreads westward to the Guangzhou Bay. The characteristic pattern of distribution of surficial sediment of shallow sea shows that fine sediment from the Zhujiang River is drifted westward in the nearshore zone about 10–30 knots wide. From the mouth of the Zhujiang River to the west, the limit of "mud line" shifts consistently toward the sea until to the east of Leizhou Peninsula, and the distribution of mud reaches –40 to –50 m isopleth (He Yueqiang, 1982) (Fig. 3-25).

Head of bays in western Guangdong is fringed with mud plain and mud flat. Its gradient is gentle,

---

1) Liu Changyu et al., 1982, Study on source of mud in coastal region of southern Zhejiang, Symposium on Marine Sedimentation, abstracts.

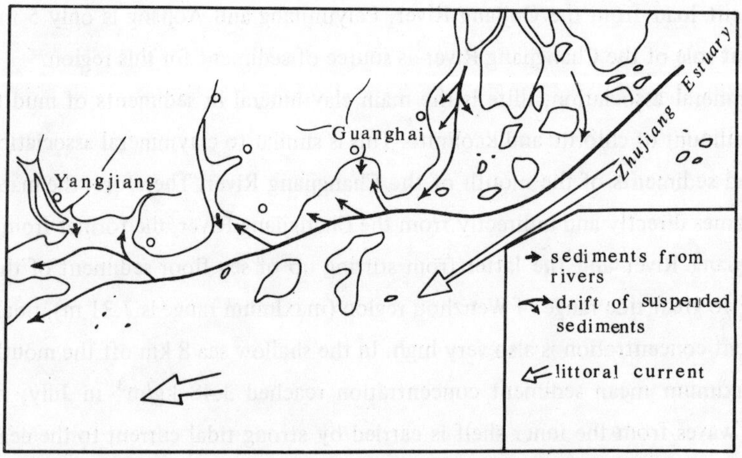

Fig. 3-25. Littoral drift along the coast of South China.

often less than 1‰. Because its material comes chiefly from the Zhujiang River, grain size and chemical composition of the sediment are similar to those of mud flat at the mouth of the Zhujiang River, 80–90% of the sediments has Md 0.05 mm to < 0.005 mm. Chemical elements in the sediments are mainly Si, Al Fe, Ca, Mg, B, Mn, Ti and Va. Here dead bodies of benthic mollusks also form main constituents of sediments of mud plain and mud flat.

The formation and evolution of mud flat in western Guangdong are closely related to tidal current and wave. Here, velocity of the flood is greater than that of the ebb. Near Haiyan, velocity of flood current is 0.53 m/s while velocity of the ebb is only 0.4 m/s. Sediment concentration of the flood is also generally higher. As a result, sediment carried to and deposited on mud flat by flood current can be only partly carried back by ebb current to the sea. Consequently, mud is gradually accumulated at bay head, forming mud flat and mud coastal plain. Mud stirred up by storm waves from sea floor of shallow sea is added to littoral mud stream, increasing its concentration and consequently, accretion by tidal current is accelerated. For example, near Zhenhai Bay and Haiyan, after typhoon, surface of mud flat is often accreted 10–30 cm. Furthermore, mud flat at bay head of western Guangdong is usually covered with mangrove, which also accelarates accretion of mud flat. It is estimated that coastal mud plain at bay head of western Guangdong was prograding seaward at a rate about 10 m per year during the last 200 years (Zhao Huanting, 1980).

The growth of mangrove on mud flat of South China (Guangdong, Guangxi, Fujian and Taiwan) causes a distinct change in chemical composition of sediments and in sedimentation processes of mud flat. Consequently, mangrove mud flat forms a special type of mud flat in China. Although mangrove can grow in many kinds of sediments of tidal flat, it flourishes best on mud flat. For example, the most luxuriant mangrove in northern Hainan Island is found at the head of the Puqian Bay, where sediment of tidal flat is very fine, Md 0.01–0.05 mm, and more than 30% of the sediment is smaller than 0.01 mm. Owing to gentle gradient and greater width of tidal flat formed by mud, the extent of mangrove is also very wide.

From the sea to the coast, mangrove mud flat may be divided into four sedimentary facies zones:

(1) Subtidal zone. It is situated below the mean low water and sediments are slightly coarser. For example, subtidal zone of mud flat in the Puqian Bay, Hainan Island is formed chiefly of muddy coarse silt which, subjected to frequent wave action, has good sorting.

(2) Mud flat without mangrove in lower intertidal zone. It is situated seaward of mangrove flat and is under water during ordinary high tide. In high water, wave can carry coarser sediments to and deposited in this zone which, therefore, also has coarser sediments with good sorting.

(3) Intertidal mangrove flat. Situated between mean high water level and mean water level, it is well grown with mangrove vegetation. As *Rhizophera mucronata* and other plants have large quantity of supporting branches, vegetation on mud flat is very dense and surface of the flat is covered with many fallen branches and leaves. Owing to dense vegetation, strength of tidal current and wave is much reduced and mangrove flat becomes a relatively calm environment where fine-grained mud is deposited. Because rates of biological and physical sedimentary processes are increased by mangrove, mud flat is rapidly accreted and its surface is often covered with fluid mud, dozens of centimeters thick, and becomes very swampy. Here, organic content of mud is very high; for example, it is 3.06% in the Puqian Bay, i.e. about three times higher than that of mud flat without mangrove in lower intertidal zone and supratidal zone. This mud often has bad odor coming from NH_4 and H_2S gases produced by decay of mangrove plants.

(4) Supratidal zone. Situated above mean high water level, the influence of tide on mud flat is weak. Here, with lowering of ground water table, sediments of the flat becomes well aerated and consequently, its organic matter gradually decomposes and mangroves are replaced by semimangrove plants and salt grasses and shrubs (Mao Shuzhen, 1980)[1)].

Table 3-12 Grain size of sediments of mangrove flat in Puqian Bay, Hainan Island

	Subtidal Zone	Intertidal zone without mangrove	Intertidal zone with mangrove
Md (mm)	0.058	0.057	0.025–0.035

(From Mao Shuzhen, 1980)

With progradation of mangrove flat, these sedimentary zones overlap successively on the next sedimentary zone from coast toward the sea, forming characteristic vertical sequence of mangrove coastal plains.

by Ren Mei-e
Nanjing University, Nanjing

1) Mao Shuzhen et al., 1980, Mangrove swamps of northern Hainan Island.

VII. Tidal Flat Deposits of Taizhou Bay, Zhejiang Province

1. General conditions

The coast of Taizhou Bay is situated in the middle part of the coast of Zhejiang facing the East China Sea. This bay is being filled with sediments continuously. As a result, the coast, like most of Zhejiang coast, has been progressing. The present Taizhou Bay is the unfilled remnant of an old bay (Fig. 3-30).

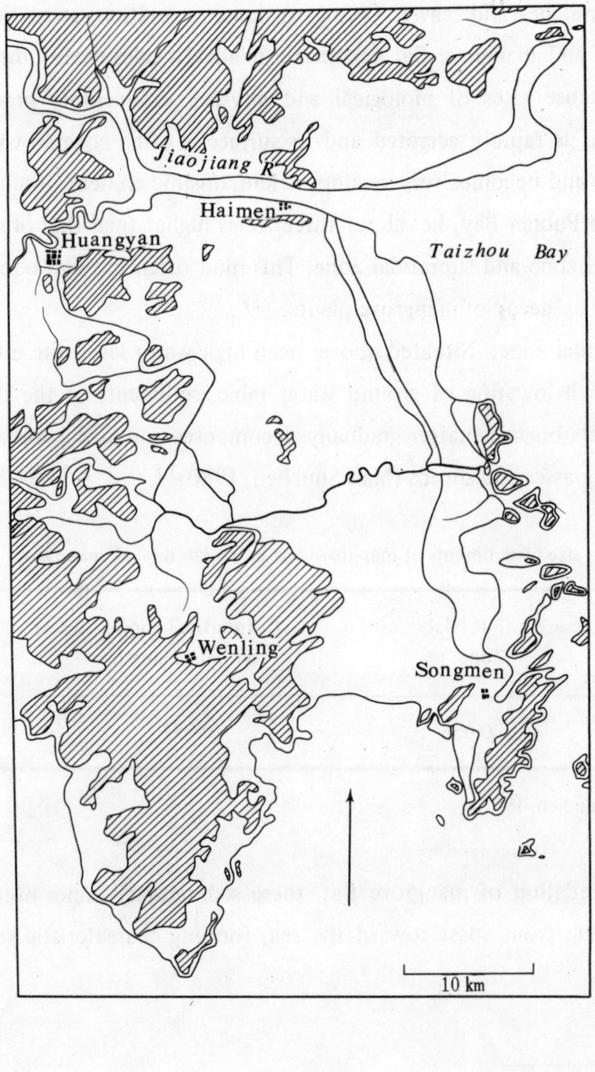

Fig. 3-30. General map showing location of the coastal area along the Taizhou Bay, Zhejiang Province.

The bays along the coast of Zhejiang Province, including Taizhou Bay, were formed in the Late Mesozoic. At that time violent tectonic movements brought about two series of gigantic faults, one trending NNE–SSW and the other NNW–SSE. The base rock of primary Taizhou Bay is a faulting block of subsidence, its outline reflects the above-mentioned faulting trends. Since Late or Middle Pleistocene, a great deal of loose sediments, were deposited on the base rock of faulting subsidence the maximum thickness being over 180 m near the present coastline. The lowermost layer of the whole vertical sequence is continental cemented sandy gravel, overlain by thin marine bluish grey fine sediment layer with shells. This sedimentation process repeated four times, the uppermost layer (surface layer) reveals the Holocene marine sediments (possibly may be lacustrine deposits at the bottom of this layer), with a maximum thickness of 73–82 m, which gradually becomes thin landward and disappears finally at the margin of old bay or terminates at a place about 1–3 km above the present tidal reach of river valleys. At present, at the margin of the primary Taizhou Bay and at the bottom of river valleys the Holocene sediment layer is in turn overlain by alluvium, diluvium, etc.

In the Taizhou Bay with a gentle plain of sediments, the surface elevation of which is slighly higher than the local mean spring high water, and the sediments primarily contain fine silt and very fine silt (Md 0.015–0.006 mm). The plain is surrounded by hilly mountains consisting of the Mesozoic volcanics.

There is a chenier in the middle part of the Wenling-Huangyan plain lying on the south of the Jiaojiang Estuary. The chenier extends from Haimen on the bank of the Jiaojiang Estuary toward the south, and has a shape of slight arc, with its length about 15.5 km. In general, its height, width, sediment thickmness and grain size decrease from north to south. According to field surveys the maximum height of the chenier is 3.4 m above the local surface and has a maximum sediment thickness of 5.8 m. The sediments in the north of chenier are yellowish brown medium and fine sands (Md 0.20–0.39 mm) and become gradually coarser to south (Md 0.011– 0.053 mm), and contain a large number of shells. These characteristics of the chenier show that the chenier is the product of the southward transportation of sediment from the Jiaojiang Estuary and was formed under the condition of intensifying wave. The chenier also indicates that progradation of the Taizhou Bay coastline temporarily slowed down or even stopped.

The cheniers lie not only on the south of Jiaojiang but also on the south of many estuaries of Zhejiang. All of them have made a major geomorphological difference between the south and the north of an estuary. On the basis of historical records we realize that a series of sea walls have been built to the east of the chenier on the Wenling-Huangyan plain since the 15th Century, and at least, this chenier had been formed before 12th Centary. After formation Wenling-Huangyan plain has advanced about 10 km toward sea. The newest seawall was built in 1976. The broad tidal flats are distributed on the east of this seawall; a narrow belt of grassy salt marsh is on its land side and the intertidal flat is on the sea side.[1)]

1) Hangzhou University, A report of local coastal observatory of Taizhou Bay, Zhejiang (1964–1966), 1975, 1.

The grassy salt marsh outside the seawall with a width of 300–500 m only is located above mean spring high water. The intertidal flat with a gentle slope (0.9%) lies between mean water and mean spring low water with a width about 5000–5500 m, decreasing toward the Jiaojiang Estuary, and stretches in a belt parallel roughly to coastline. Different positions of this tidal flat are characterized by distinct drainage, landforms, sediments and organisms. According to these characteristics the intertidal flat can be divided into three main zones (from land to sea): the high tidal flat with fine sediments, the middle flat with channels, ridges and coarse sediments, and the low tidal flat. Outside the tidal flat is the subtidal zone, a wide and shallow Taizhou Bay water area less than 10 m in. Its sediments are finer than the tidal flat sediments.

2. Hydrological conditions

The Jiaojiang River is the main river discharging into the Taizhou Bay, whose runoff, transportation of sediments and their variations influence directly the developmental processes and sedimentary characteristics of the Taizhou Bay. The Jiaojiang River belongs to the mountain stream of monsoonal rain type. Average annual water discharge of the Jiaojiang River is 125 m^3/sec, and its water discharge at the flood season (from May to June and September) is 74% of the annual output. The annual sediment discharge is 1.1 million tons, and over 2/3 of which concentrate in flood season. In accordance with salinity determination, the fresh water of the Jiaojiang River can extend onto isobath of 10 m of sea area. At the same time no flood current occurs in the Haimen section. The sediments of the Jiaojiang River transporting into the Taizhou Bay, occur also in the flood season. According to the results of granular analysis of suspended sediments obtained from tidal limit of Jiaojiang, their Md 0.01 mm accounts for 2% only. Since most of the coarse sediments deposit above the Niutoujiang section (east of Haimen) where the river bed of Jiaojiang becomes narrow suddenly most of the coarse components of suspended sediments have very little effect on the sedimentation of the coast of the Taizhou Bay. On the contrary, the total suspended sediments (about 0.5 million tons) are transported mainly into the Taizhou Bay and have a greater effect on it.

The type of tidal wave in the area of the Taizhou Bay is standing wave and there is a maximum velocity of the tidal current near the middle tide level. The tide of the Taizhou Bay is regular semi-diurnal one. The ratio of amplitude of main constituents $(\frac{HK_1+HO_1}{HM_2})$ is equal to 0.25. The average tidal range is more than 4 m and the maximum is over 6 m. For this reason, the coast of the Taizhou Bay belongs to the macrotidal type. The tidal current in this bay is rotary current with the trend of the major axis of WNW–ESE (the direction of food strength is about 300° and that of the ebb strength is about 120°). According to the field observation during the middle tide, the maximum velocity of tidal current reached 1.5 knot (July 17, 1980), and there is few difference between flood and ebb tidal currents. When the tidal current comes to the tidal flat, it will gradually become rectilinear current, and its direction is basically perpendicular to the coastline. But when the water reaches middle tidal water on the low tidal flat, it still has the characteristics of rotary current.

On the basis of the tidal water statistics of the Taizhou Bay for 20 years (1959–1979) we may see that the average tidal range of this area exhibits a regular seasonal variation, with a minimum in January and a maximum (above 4.1 m) in July through October. Within a year the smallest and largest monthly values of average tidal range might differ by 0.36 m.

The above-mentioned tidal characteristics are bound to be reflected in geomorpholgical and sedimentary processes of the coast of the Taizhou Bay. In a work, the tide is a primary hydrographic factor which contributes to coastal sedimentary processes of the Taizhou Bay.

As for the wind condition of the Taizhou Bay, the northerly is the dominant followed by southwesterly and the prevalent is the north-northeast. For the seasonal changes the main wind direction in June and July is southwesterly, and that of the other mouths is north-north easterly (Fig. 3-31).

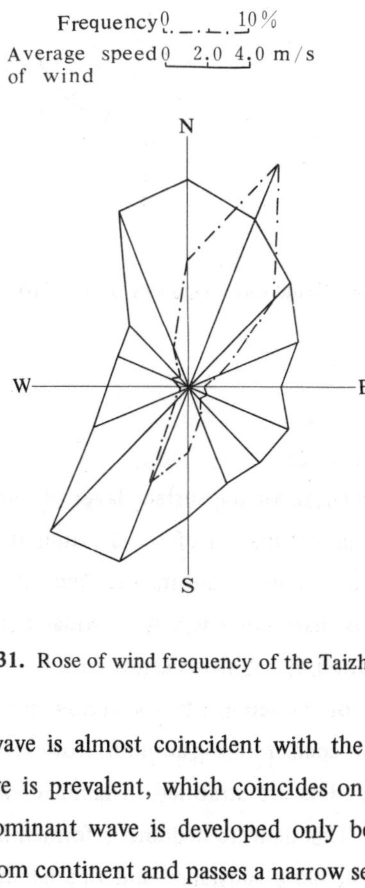

Fig. 3-31. Rose of wind frequency of the Taizhou Bay.

The direction of wind wave is almost coincident with the wind direction in the bay within a year the north-northerly wave is prevalent, which coincides on the whole with the wind frequeny. However, the direction of dominant wave is developed only between northerly and northeasterly because the westerly comes from continent and passes a narrow sea area (Fig. 3-32).

Within a year the largest waves occur usually from August to November. Swell comes from east and east-northeast for entire year, and it is clear that swell is restricted by the fetch. Wave has a great influence on coast development and sediment transport of the Taizhou Bay. Sea current is another hydrographic factor affecting the coast of the Taizhou Bay. From the sedimentation viewpoint, the longshore current of the East China Sea has great influence on the finer sediment supply of the Taizhou Bay, because it flows southward near the coast of Zhejiang during the winter half year. In addition, the Taiwan warm current also has some influences on the sedimentation of the Taizhou Bay.

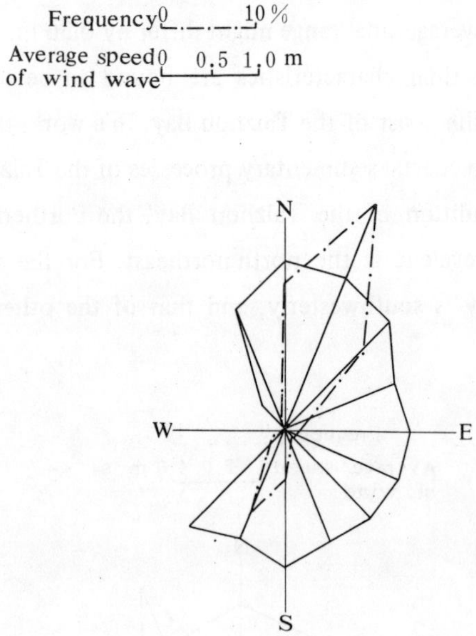

Fig. 3-32. Rose of wind wave frequency of the Taizhou Bay.

3. Characteristics of the sediments

(1) ***Grain size of the sediments*** (Table 3-12)

On the basis of the grain size analysis of the surface layer of sediments taken from the upper tidal flat (from middle tidal flat to grassy marsh) of the Taizhou Bay coast on the south of the Jiaojiang Estuary , we know that the sediments are mainly fine silt and very fine silt with a little medium silt, of which the fine silt is distributed widely on tidal flats and accounts for the main sedimentary type. The grain size variation of sediments is not great.

Table 3-12 shows that grain size of the sediments has a seasonal change, and the sediments tend to become finer during spring time but coarser during autumn. However, this difference is slight.

The deposits on the upper tidal flats are stratified in laminae, which are easily recognized in the field work when the mud is dry. This delicate bedding is formed by the interlamination of thin layers of coarse material in the finer basic substance and this lamination is formed by the tidal currents. However, on the basis of the grain size analysis of the two layers that the median diameter varies between very fine silts and medium silts (0.006–0.020 mm) and at times even in the fine silts only, we know that the difference of grain size is slight between the two layers.

The characteristics of the grain size in surface sediments concentrated in fine silt grading are manifested not only on the upper tidal flat but on the east plain of chenier. In addition, the deposits in high tidal flat on the north of the Jiaojiang River are also fine silts (Table 3-13).

Table 3-12 Characteristics of grain size of the tidal flat sediments

Time period	Type	Weight percentage of grain (mm)				Md	Sorting coefficient	Asymmetric coefficient	Sample number
		>0.1	0.1 –0.05	0.05 –0.01	<0.01	(mm)	(So)	(Sk)	
Spring, 1964	fine silt	0.74	5.77	51.22	42.27	0.0143	2.73	0.48	1
	silty clay	0.38	2.85	42.88	53.94	0.0088	2.63	0.65	2
Autumn, 1964	fine silt	0.43	4.87	56.18	38.91	0.0146	2.08	0.64	7
	silty clay	0.56	3.90	42.13	53.41	0.0089	2.51	0.72	3
	clay	0.47	0.15	32.50	66.89	0.0057	2.61	0.70	1
Spring, 1965	fine silt	0.31	4.63	51.98	43.08	0.0126	2.45	0.56	16
	silty clay	0.20	2.23	42.69	54.88	0.0083	2.83	0.54	15
	clay	0.53	2.00	32.44	65.03	0.0065	1.73	0.52	2
Autumn, 1965	fine silt	0.32	6.34	55.32	38.02	0.0169	2.20	0.62	16
	silty clay	0.20	3.36	40.50	55.94	0.0082	2.66	0.72	9
	clay	0.13	4.01	31.29	64.47	0.0059	2.85	0.82	2

Table 3-13 Comparison of sedimentary type between tidal flat on the north of the Jiaojiang River and the Wenling-Huangyan Plain

Region	Type	Weight of grain (mm)				Md (mm)	(So)	(Sk)	Sample number
		70.1	0.1–0.05	0.05–0.01	<0.01				
Tidal flat	fine silt	0.42	5.81	52.99	40.78	0.0140	2.18	0.75	6
Plan	fine silt	4.64	0.96	51.26	43.14	0.0109			7

In comparison of the sediments on the upper tidal flat of the Taizhou Bay with those in its adjacent waters (Taizhou Bay and Jiaojiang Estuary) we may see the marked difference between the two (Table 3-14).

Table 3-14 and 3-12 show that the fine sand is the main sediment type of the Jiaojiang Estuary, and the medium and fine silts and a small amount of coarse silt come next, exhibiting an uncontinuous sedimentary type. The fine components of sediments are extremely similar to the sediments on the upper tidal flat of the Taizhou Bay, but the sediments on the upper tidal flat are slightly finer, and its sorting is poor. Moreover, the fine silt and coarse silt components of sediments of the Taizhou Bay, the tidal flat and the river bed of the Jiaojiang Estuary are very few. These characterisitics and regional difference of the sediments are closely related to its source and transport processes.

Table 3-14 Comparison of the sediments between the tidal flat of the Taizhou Bay and its associated waters

Region	Type	Weight percentage of Grain (mm)					Md	So	Sk	Sample number
		0.25	0.25–0.10	0.10–0.05	0.05–0.01	<0.01	(mm)			
Jiaojiang Estuary	medium sand	66.33	32.93	0.27			0.2840	1.18	1.02	3
	fine sand	31.45	58.81	1.26			0.2148	1.23	1.03	12
	fine site			7.30	62.41	30.29	0.0204	2.33	0.66	5
	silty clay			5.90	38.63	55.47	0.0084	2.97	0.95	3
	clay			0.90	33.44	65.67	0.0055	3.32	0.71	2
Taizhou Bay	fine silt			6.60	65.62	28.08	0.0230	2.19	0.61	3
	silty clay			6.51	41.49	52.00	0.0092	3.87	0.57	1
	clay			1.77	28.87	69.27	0.0045	3.52	0.78	15
Tidal flat	fine silt			4.63	51.98	43.08	0.0216	2.45	0.56	16
	silty clay			2.23	42.69	54.88	0.0083	2.83	0.54	15
	clay			2.00	32.44	65.03	0.0065	1.73	0.52	2

(2) *Mineral composition*

The samples for mineral determination were taken only from high and middle tidal flats. Its results are shown in Table 3-15.

Twenty-one sets of main minerals have been determined (Table 3-16), in which, eighteen sets are heavy minerals and three sets are light minerals. Biotite makes up the majority of the heavy minerals, then comes chlorite (10% and more) followed by a small amount of others. On every position of the

Table 3-15 Mineral composition of the tidal flats

Station	Weight percentage of grain																			
	Heavy mineral																	Light mineral		
	Biotite	Muscovite	Phlogophite	Chlorite	Chloritoid	Epidate	Zoisite	Hernblende	Tremolite	Actinolite	Limonite	Magnetite	Henatitie	Ilmenite	Columbite -Tantalit	Zircon	Apatite	Rutile	Feldspar	Quartz	calcite
DA1-1	54	5	2	12	0.8	3	1	5	0.5	NO	5	1.1	0.3	1	5	4	0.3	NO	84.1	15.9	af
DA2-1	53	4	0.5	13	1.6	4	0.9	6	2	0.2	5	2	1.1	0.2	2	5	0.5	f	78.1	21.9	af
DA5-1	49	3	1	12	0.3	10	1	6	2	1	5	2	2	0.5	2	3	0.5	NO	70.6	29.4	af
DA6-1	50	2	1	10	0.5	9	1.7	7	2.7	1	6	2.8	1.2	0.4	2	3	0.2	f	71	27	2
DA8-1	52.9	3	2	12.4	0.7	4.5	2	6	1	0.5	6	1.5	0.8	0.2	3	3	0.5	NO	71.2	28.8	af

* f=a few; NO=never present.

upper tidal flat, the amounts of heavy minerals are the same, but biotite, muscovite and chlorite show an increasing tendency from the middle part of middle tidal flat to the high tidal flat and the lower part of middle tidal flat. On the contrary most of the others increase toward the middle part of the middle tidal flat. Among the light minerals the amount of feldspar is the highest, then quartz and a little calcite. The feldspar content of light minerals increases clearly but that of quartz is contrary to the case. It is obvious that distribution variation of minerals on tidal flat is an inevitable outcome of the changes of hydrodynamic structure of the tidal flat.

The minerals in deposits on the tidal flat of the Jiaojiang Estuary and the Taizhou Bay have the following characteristics:

(1) There is no difference in mineral sets in the three regions, except in mineral amounts. For example, the chlorite amount ranks the first in the Taizhou Bay, and then come the Jiaojiang Estuary and the tidal flat. The amount of zircon, a characteristic mineral in the river basin of Jiaojiang is. much more in the Jiaojiang Estuary than in other two regions.

(2) The amounts of heavy minerals are different from light minerals in the deposits in the aforementioned three regions (Table 3-16). The mean weight percentage for the light minerals is 95.6% in the Taizhou Bay, 87.1% on the tidal flat and 78.3% in the Jiaojiang Estuary.

The minerals in tidal flat deposits of the Taizhou Bay are in agreement with those of the Mesozoic volcanics and intrusion within river basin of the Jiaojiang River. However, each set of the minerals has a change in amount in the transport processes.

Table 3-16 Comparison of minerals in the deposits between tidal flat and its adjacent waters (in weight percentage)

	Jiaojiang Estuary					Tidal flat					Taizhou Bay				
	No.1	No.2	No.3	No.4	Mean	No.1	No.2	No.3	No.4	Mean	No.1	No.2	No.3	No.4	Mean
Heavy mineral	22.9	17.8	21.6	24.4	21.7	13.9	7.3	25.0	5.8	12.9	4.5	3.5	4.9	4.9	4.4
Light mineral	77.1	82.2	78.4	75.6	78.3	86.1	92.7	75.0	94.2	87.1	95.5	96.5	95.1	95.1	95.6

(3) *Chemical characteristics*

It is difficult to examine the distributional difference of chemical components in tidal flat deposits, because only two samples have been taken from upper tidal flat for chemical analysis. This section only gives an outline of the similarity and difference in chemical characteristics of deposits between the Jiaojiang Estuary, Taizhou Bay and tidal flat. As can be seen in Table 3-17, SiO_2 and K_2O in deposits decrease from the Jiaojiang Estuary to the Taizhou Bay. On the contrary, Al_2O_3, Fe_2O_3, MgO, Na_2O, CaO and FeO increase from the Jiaojiang Estuary to the Taizhou Bay. However, the amounts of the most of the chemical components of the tidal flat deposits rank the middle between those of the Jiaojiang Estuary and the Taizhou Bay. All of these changes correspond to the differences of the environmental conditions and grain size of deposits among the three regions. It may be seen that amounts of SiO_2 and K_2O agree with the distribution of coarse deposits. On the contrary, the amounts of Al_2O_3 are in agreement with the distribution of finer deposits. The contents of Na_2O and MgO would be related to salinity of deposits and the variation in amounts of CaO in the above-mentioned different regions may result from the numbers of marine shell fragments. It is obvious that salinity and marine biodetritus in the Taizhou Bay deposits are higher and more than those in the other two regions, in which the contents of Na_2O, MgO and CaO of tidal flat are higher than that of the Jiaojiang Estuary.

Table 3-17 Comparison of chemical component in deposits between tidal flat and its adjacent waters (in content percentage)

	SiO_2	Al_2O_3	K_2O	Fe_2O_3	Na_2O	MgO	FeO	CaO	TiO	MnO	P_2O_5	H_2O^+	H_2O^-
Jiaojiang	79.78	9.70	4.31	1.48	1.71	0.53	0.74	0.29	0.18	0.12	0.02	0.89	0.38
Taizhou Bay	57.15	15.63	3.19	4.38	2.25	2.94	1.66	3.73	0.67	0.12	0.14	3.05	2.88
Tidal flat	71.84	12.79	3.69	2.74	1.69	1.64	1.21	0.79	0.49	0.14	0.13	2.08	1.47

(4) *Biogenic deposit*

In a survey of the upper layer deposits on tidal flat, we took some samples to determine microfossils. From the results of determination, it can be seen that foraminifera numbers, individual size and species have some differences among the various sections of tidal flat.

In the upper part of high tidal flat, the foraminifera numbers are smaller, about dozens of grains per 50g dried sample. Most of these are euryhaline species, such as *Ammonia beccarii* var, *Nonion* sp., and a small numbers of these are stenohaline. such as *Florilus decorus, Epistomonella naraensis, Bolivina* sp., etc. The simple differentiation of foraminifera is lower (<9) and its individual diameter is small (mean diameter is about 0.23 mm) (Table 3-18).

Table 3-18 Individual size of foraminifera on various position of tidal flats,Taizhou Bay

Section	High tidal flat	Middle part of middle tidal flat	Lower part of middle tidal flat
Size (mm)	0.23	0.29	0.34

The foraminifera numbers greatly increase in the middle part of the middle tidal flat, and the maximum reach 590 grains per 50g dried sample. Among them, *Ammonia beccarii* var. is still predominant and the others are *Ammonia convexidorsa, Nonion* sp. *Epistominella naraensis, Spiroloculina* sp., *Triloculina* sp., *Nonionella basiloba, Elphidium advenum* etc. In comparison with foraminifera on the upper high tidal flats, the foraminifera on this section have a larger individual diameter (0.29 mm) and a simple differentiation (10–15).

In the lower part of the middle tidal flats the foraminifera numbers decrease clearly (about 50 grains per 50g dried sample), and most of these are also euryhaline species, such as *Ammonia beccarii* var. and *Nonion* sp., etc. Besides there are a certain example of stenohaline species such as *Cuttulina* sp., *Spiroloculina laevigata* etc. The smiple differentiation of foraminifera on this section is 10, less than those deposits in the upper and middle parts of middle tidal flats but greater than those deposits on the high tidal flats. Besides, the mean individual diameter increases further to 0.34 mm (the maximum can reach 0.5 mm). From the above-mentioned characteristics of foraminifera in the lower part of middle tidal flats it can be seen that they also belong to transitional type between marine and continental microbiota, but their marine characteristics are more clear than those deposits on the tidal flats above them.

In addition, in tidal flat deposits there are a number of ostracoda, which generally vary in number from several to dozens of grains per 50g dried samples. The variations of ostracoda and foraminifera on the tidal flats are similar but ostracoda are less than foraminfera. The proportion of ostracoda to foraminifera in numbers is 1:5–1:7. The ostracoda species are also less in number. The euryhaline *Sinocytheridea latiovata* is the most important constituent of ostracoda. Besides,

Hemicytheropteron sinensis, Loxoconcha viva, Tanella opima Chen, *Leguminocytherels,* etc were found.

In the sediment analysis we found that all the sediment samples of the Taizhou Bay and a part of sediment samples in the Jiaojiang Estuary have some stenohaline foraminifera, such as *Epistominella naraensis* and *Bolivina* sp.

In summary, a small number of foraminifera, which lived at the water depth of 30–60 m, were found in the tidal flat of the Taizhou Bay and its adjacent waters. This fact demonstrates that the sediments of tidal flats of the Taizhou Bay may come from inner continental shelf.

4. Geomorphological and sedimentary zonation

As has been mentioned at the beginning of this section that the tidal flat can be divided into three main geomorphological and sedimentary zones: the grassy mash zone (supratidal zone), the intertidal zone and the subtidal zone. The intertidal zone can be further divided into three subzones from land toward sea. In fact, each zone or subzone is a physiographic unit. Therefore most of the scholars have the same views on the zonation of tidal flat, but are different in denomination of the zones or subzones due to the differences of the recognition and the object studied. The geomorphological and sedimentary characteristics of various zones or subzones of the tidal flats will be explained below (Fig. 3-33).

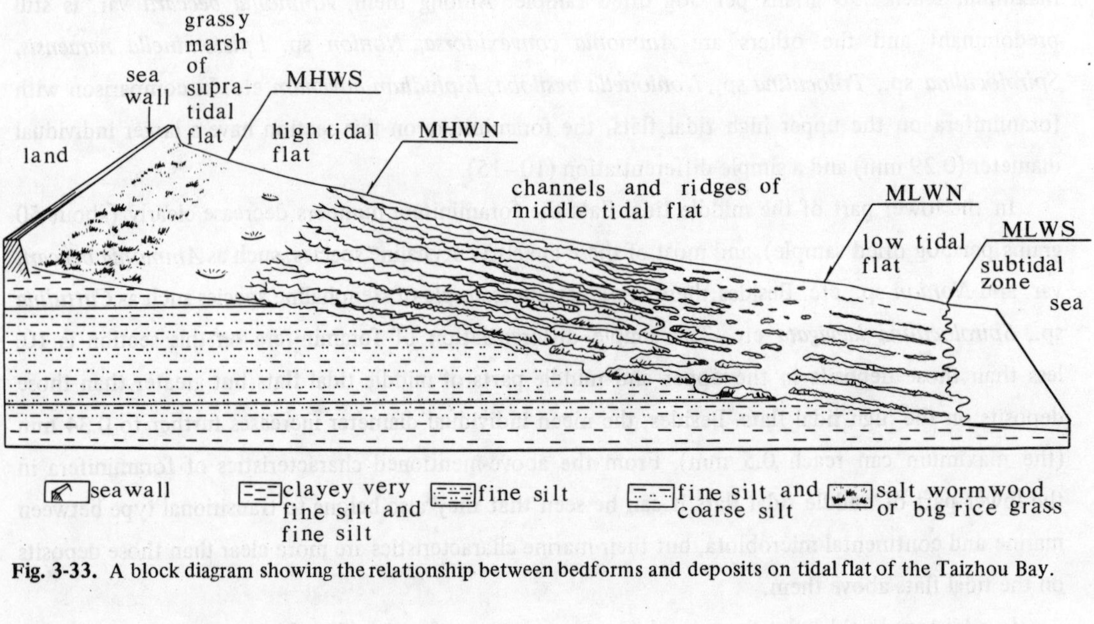

Fig. 3-33. A block diagram showing the relationship between bedforms and deposits on tidal flat of the Taizhou Bay.

(1) *The grassy tidal marsh zone (Supratidal zone)*

This zone lies close against the east of the newest seawall and above mean spring high water, which is submerged by sea water only at the higher spring high water. The width of this zone is only 500 m, because the newest seawall was built no more than 17 years ago. The salt marsh is even with a gradient of 0.8‰ . The vegetation is salt tolerant. The foot of seawall on the land side of salt marsh has been eroded and a 0.5–1 m high cliff is thus formed when higher high tide comes together with dominant waves. The eroded muds are pebblelike to scatter in the upper part of salt marsh thinly laminated with thicker clayey fine silts and very thin silts. The coarser components (0.05 mm) of the deposits are less than 7% and coarsest mainly of terrigenous materials, including feldspar, quartz, mica and a small amount of other heavy minerals. The laminae are easily to be recognized due to the existence of mica in deposits.

(2) *The high tidal flats subzone* (Plate 3-22)

The high tidal flats (between the mean high water springs and mean high water neaps) have a gradient of 1.2‰ –1.3‰, and the gradient tends to increase above 2.5‰ on the tidal flat of the Jiaojiang Estuary. In fact, the intertidal flat as a whole is slope between the salt marsh and subtidal zone, and yet its slope is quite gentle because it is wider. From the geomorphological viewpoint, J. L. Davies (1972) denominated the intertidal flat "intertidal slope", which is suitable for the coastal area of the Taizhou Bay. When the high tidal flats are submerged in tidal waters, the annual mean frequency is about 33% for the upper part of high tidal flat and increases gradually to 90% for the lower part of high tidal flats. Because the time during which the upper of high tidal flats is covered by water, is very short, there are sparse grass patches scattered on this place and some times cracks occur in the upper layer of deposits. On the contrary, there are no other plants in the lower part of high tidal flats besides algae.

The width of the high tidal flats is about 1000 m which increases from north to south. The surface is generally smooth, only in the lower part embryonic tidal gullies are developed and ripple marks occur occasionally.

The deposits of high tidal flats are also thinly laminated clayey fine silts and fine silts. The flaser bedding can be seen once in a while in the upper layer deposits.

(3) *The middle tidal flats subzone* (Plates 3-23, 24, 25)

The middle tidal flats (between mean high water neaps and mean low water neaps) is a very complicated region of tidal flats for its deposits and surface features, and has a width about 2400 m and a steeper slope than other zones of tidal flats. Most parts of the middle tidal flats are affected by tidal currents which flow back and forth twice every day. Therefore the middle tidal flat is an active region in an intertidal zone.

The relief on the middle tidal flats is greater than that on other parts of the tidal flats. The downcutting of tidal channels and the interchannel ridges with a great number of ripple marks are the main geomorphological characteristics in this region. The tidal channels cut down clearly from here and form a channel system whose main channels extend parallelly toward east by south.

In the upper part of a middle tidal flat the tidal channels generally are 15–20 cm in depth and 20–50 cm in width, but they increase to 60–70 cm and 180–240 cm respectively in the middle part of a middle tidal flat. The tidal channels have a steeper wall and the capture phenomenon between them in this region. It is also here that the tidal flats are separated by channels in a form of undulating. In the lower part of a middle tidal flat the depth of channel is reduced to 30–40 cm and the channel systems become netlike so the interchannel ridges are usually spindlelike.

The sedimentary characteristics of the middle tidal flat are the great variation in the sediments, the complex sedimentary structures, coarser grain of sediments and the thicker layer of sediments. The above-mentioned characteristics are mainly found in the middle even in the lower part of the middle tidal flat .

Table 3-19 Comparison of grain size of sediments between the channel bottom and the surface of the middle tidal flat (Taizhou Bay, Nov., 1964)

Location	Weight percentage of grain (mm)				Md (mm)	So	Sk
	70.1	0.1–0.05	0.05–0.01	<0.01			
Channel bottom	1.78	20.48	71.13	6.61	0.0405	1.321	0.80
Tidal flat surface	0.86	5.48	44.56	49.10	0.0102	2.227	0.67

From Table 3-19, it can be seen that the sediments on the tidal flat surface are much finer than those at the channel bottom, but coarser than those on the high tidal flats. The grain size of sediments at the channels bottom is the coarsest and is well sorted as compared with those in other parts of tidal flats of this area. This results obviously from the concentration of the tidal currents and the increase of current velocity at the channel bottom.

The deposits generally contain bluish grey coarse silty laminae (2–3 mm thick) and thin brown fine silty laminae. The flaser bedding and the lenticular bedding are also common in the deposits, resulting from waves and stronger tidal currents, and probably the changes of channel course. During the spring tide on July 2, 1965, we observed the bluish grey coarse silts about 2 cm thick on the brown fine silt layer. This demonstrates that the above discussion is reasonable.

(4) ***The lower tidal flats subzone***

The lower tidal flats (between the mean low water neaps and mean low water springs) has a slope of 1.4–1.5‰ and a small relief of flat surface with a number of thin flows and few tidal channels. The deposits are very fine and fine silts (Md 0.01–0.005 mm) with a poor sorting.

The lower tidal flat lying close to the river bed of the Jiaojiang Estuary is very narrow and has a steeper slope. In a word, the slope of intertidal zone is steeper than that in other parts of tidal flat, either the supratidal zone or the subtidal zone. The slope of every subzone in the intertidal zone becomes steeper successively from land to sea. Thes change of slope is especially manifested on the tidal flat near the Jiaojaing Estuary. It is clear that the feature of the intertidal flats in the coastal area of the Taizhou Bay is a convex profile, which corresponds to the hydrodynamic processes of the tidal flats.

5. Sedimentation of tidal flats

The sediments of tidal flats are mainly brought by the flood currents, scattered over the flats from the Taizhou Bay and the deeper waters. There are also sediments provided by local erosion on the tidal flat itself that are redistributed on it, and by a small amount of organic material produced on the tidal flat itself. The velocity of tidal current decreases and the time, when the maximum current velocity occurs, also defers successively along with the increase in height of tidal flats from sea to land. The middle part of high tidal flat, the upper part of the middle tidal flat and the subtidal zone have the maximum flood velocities of 29 cm/sec, 31 cm/sec, and 100 cm/sec respectively, and the mean flood velocities of them are 6.8 cm/sec, 7.1 cm/sec and 49 cm/sec respectively (during the spring tide on 15 June, 1965). During the middle tide period on 17 July, 1980, the lower tidal flat and subtidal zone have the maximum flood velocities of 50 cm/sec and 78 cm/sec respectively. Usually, the suspended sediments of the Taizhou Bay have a median diameter of 0.005–0.020 mm, similar to those on the tidal flats. The tidal currents carry the sediments basically in suspension to the tidal flats. The gradual decreasing velocity of the tidal currents causes a reduction in capacity of water and results in a gradual decrease in amounts and grain size of the load. During the flood tide period, the sedimentation might be expected to produce a series of sediments decreasing in grain size from lower tidal flat to high tidal flat. Therefore the coarser differentiated load will gradually settle down without disturbance or waves during the flood tide period (Fig. 3-34).

As can be seen from Fig. 3-34, around the flood slack, grain size of suspended sediments along the vertical line on the lower tidal flats is slightly coarser than that of subtidal zone, in which the median diameters of suspended sediments at the bottom of lower tidal flats and in subtidal zone are 0.01 mm and 0.007 mm respectively.

Fig. 3-34 also shows that a tendency of the suspended load decreases from subtidal zone toward the tidal flat during the flood slack. However, the changes of suspended load are related to not only tidal velocity but also water depth on the tidal flat. According to the data taken from various field surveys, the suspended load above the high tidal flat is generally higher than that above the middle tidal flat. For example, the mean suspended load during the flood period above the high tidal flat and in the supper part of middle tidal flat are 12.77 kg/m^3 and 7.63kg/m^3 respectively (during the middle tide period on 16 October, 1965). At the same time, the sediment discharges per square meter above the high tidal flat and in the upper part of middle tidal are 1544 m^3 and 7736 m^3 repectively. These showed obviously tendency of sediment discharge gradually decreasing on the tidal flats from sea toward land. It is understood that the above-mentioned tendency is the main factor in the progressive development of the tidal flats.

The suspended loads along the vertical line are different among characteristic times during the flood tide (Fig. 3-34). As the appearance of flood stength, the suspended load in the subtidal zone is higher than that on the lower tidal flat, and both have a uniform distribution along the vertical line due to the great velocity . At this moment the subtidal zone and the lower tidal flat would be eroded. As the occurrence of flood slack, the suspended load increased near the bottom, the increments on the lower tidal flat increased greatly (Fig. 3-34) and formed a layer of high sediment concentration, thus bringing about the sediment accumulation on the lower tidal flat. The sedimentation during the flood tide period is closely related to the settling lag of the sediments. During the ebb tide period the current velocity gradually increases along the tidal flat from land toward sea. For example, the high tidal flat, the upper part of middle tidal flat and the subtidal zone have mean ebb velocities of 5.8 cm/sec, 7.7 cm/sec and 48 cm/sec respectively (during spring tide on 15 June, 1965). They are slightly lower than those during flood tide of the same tide period.

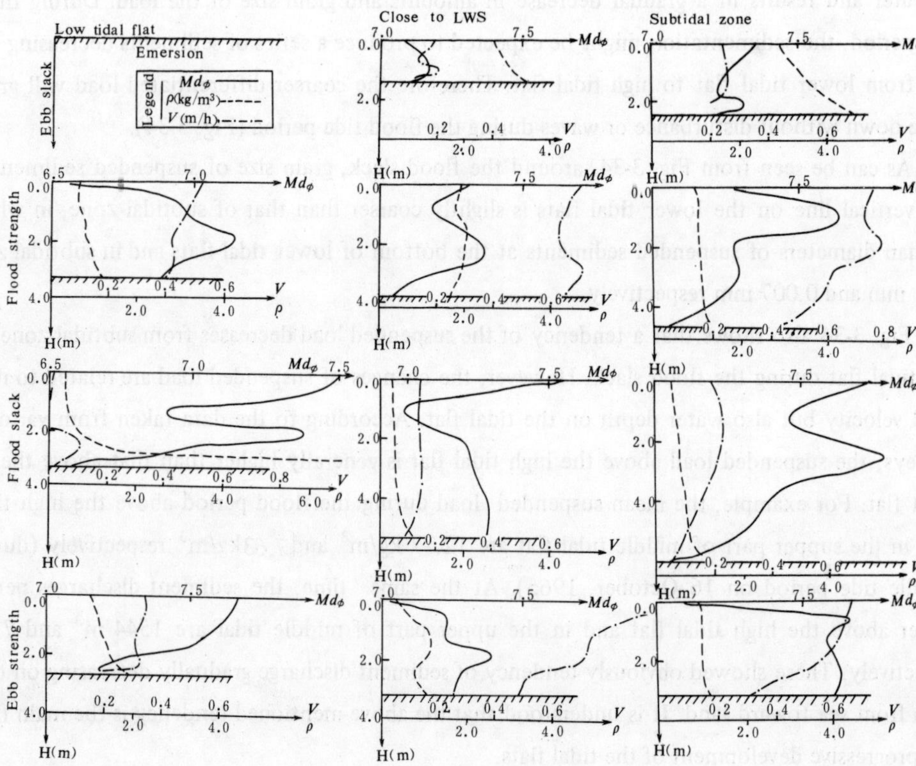

Fig. 3-34. The change in median diameter and content of suspended sediments with tidal velocities at some individual times on the lower tidal flats and in subtidal zone.

In comparison with the ebb tide of the same tide period, the flood tide has higher suspended load. During the ebb tide on high tidal flat and in the upper part of middle tidal flat, the mean suspended loads are 7.63 kg/m^3 and 7.5 kg/sec respectively, and sediment discharge per square meter are 868 m^3 and 5165 m^3 respectively, being 676 m^3 and 2570 m^3 less than those of the flood tide at the same point and time. This shows that the landward side of middle tidal flat has a net income of sediments in a tidal cycle and the scouring has begun on the middle tidal flat, especially in the tide channels, as evidenced by the coarser sediments and the ebb current ripples at its bottom.

Alternation of the scouring and deposition and its amount on the tidal flats vary with the alternation of spring tide, and neap tide. As a change from neap tide to spring tide for monthly tide, the sediment discharges and its depositional effects during the flood tide on the tidal flats gradually increase, respectively. At the same time, the scouring effect is also greater in the ebb tide and the positions of scouring and deposition gradually shift upwards from the lower part of tidal flat. As a change from spring tide to neap tide every month, its scouring and depositional effects on the tidal flats are contrary to those at a change from the neap tide to the spring tide. For the development of tidal flats, the former process caused mainly the high tidal flat to increase in height and the lower tidal flat to increase in width, and the latter caused the high tidal flat to increase in width. As a result, the tidal flats will advance uninterruptedly with the deposits unless the storm waves arise.

Tendency of seasonal scouring and deposition within a year is basically similar to that within a month. The seasonal variation in tidal level (comparison between the highest and the lowest monthly values) is much less than the daily variation in tide level within a month. The ratio of them is 1:6. Therefore the latter is the more effective factor in the change of scouring and deposition on tidal flats.

Wave is another siginificant factor in effect on bedforms and sedimentation of tidal flats. The frequency of occurence of wave heights, which is 3–4 in scale (its mean value is 1.67 m) is 67–87% every month. In accordance with water depth being 1.28 of the wave height at breaking point, it is roughly estimated that wave breaking will occur at the water depth of 2.1 m. In fact, it is no more than 2 meters due to the wave decay, so the breaking point occurs commonly in the middle part of middle tidal flat at the high tide and occurs in the lower part of high tidal flat when the higher high tide occurs. Therefore, the wave processes mainly concentrate on the middle tidal flats. Thus the changeable bedforms and complex textures, coarser granules and well sorted sediments must all occur on the middle tidal flats. It is clear that the wave action on the high tidal flat is not great under ordinary conditions, but the storm wave, which is brough by the typhoon during August through October, can affect all the high tidal flats even the salt marsh outside the seawall, thus bringing violent scouring to this area; however it is not of an yearly occurrence.

The organic sedimentation is manifested in salt marsh and the upper part of high tidal flat. The organism, which is mainly grass, can reduce the wavc erosion and retain the sediments brought by the flood currents. The organic effects on the other part of the tidal flats are not very great.

In a word, the characteristics of sedimentation in each zone of tidal flats from land to sea can be explained as the following.

The salt marsh zone is a region in which vegetation has a significant effect. The effect of wave and tide is rather less and the sedimentation is very slow under ordinary conditions.

The high tidal flat is a stable sedimentary zone but its sedimentary rate and thickness are small due to shorter duration of flood tide on it. In addition, because the duration of flood slack is very short, very fine sediments cannot easily reach onto its surface and the granular sorting is rather well. Wave effects generally are the weakest unless the storm wave, brought by typhoon, occurs.

The middle tidal flat is a zone which is not only affected by tides but also uninterruptedly reformed by waves, therefore this is an unstable and complex sedimentary region.

The lower tidal flat is a strong sedimentary region. Its sediment are the finest (Md 0.0086 mm) and dominantly poor sorting. Wave effects of the lower tidal flat are weaker than those of the middle tidal flat.

6. Conclusions

This study has demonstrated that the plain along the coast of the Taizhou Bay is a chenier plain in substance. The chenier has been formed before the 12th century (after ^{14}C dating, its correct date is about 2800 BP). Now the plain on the east side of chenier is about 10 km in width due to the shoreline progress.

The tidal flat lies to the east of the newest seawall which has a width of 5–6 km with a gentle slope, the mean gradient being 1%. The salt marsh, or the supratida zone is very even, as subtidal zone, but the intertidal flat is a slope which increases gradually in its gradient from land to sea and forms a convex profile. The characteristic of this is suited to the hydrodynamic condition and regional differences of bedforms and sediments arising therefrom.

In general, the sediments of tidal flats are much finer in grain size than those of river bed of the Jiaojiang Estuary but slightly coarser than those of the bottom of the Taizhou Bay. Most of the sediments on tidal flat have a median diameter of 6–7.5 ϕ. The coarse silt component of the sediments is very few. The changes of horizontal distribution of sediments are not great except those of middle tidal flat. The changes of vertical distribution of sediments on tidal flats and coastal plain which were formed by progression of tidal flats toward the sea, are not too clear to be distinguished easily as well as the horizontal distribution of them.

The Taizhou Bay is a macrotidal area with a mean tide range over 4 m. The tide is a primary factor in the development of tidal flat along the coast of the Taizhou Bay. The sediments are gradually differentiated and settled as the flood currents move from the lower part to the upper part on the tidal flats. The waves modified this process and turned the sediments of middle tidal flat into a complex one. The storm wave caused by the typhoon disturbed the developmental process of tidal flat. On average, the rate of natural progress of tidal flat along the coast of the Taizhou Bay is about 4 m per year.

by Feng Huaizhen, Hangzhou University, Hangzhou

References

Boersma, J.K. et al., 1981, Sedimentology, V. 28, pp 151–170.

Davis, J.L., 1964, Zeit. Geomorphologic, V. 8, pp 27–42.

Davis, Jr. R.A. (ed.), 1978, Coastal Sedimentary Environments, Springer-Verlag.

Dionne, J.C., 1973, Jour. Sed. Petrol. V. 43, pp 848–856.

Drake, D.E., 1976, Suspended sediment transport and mud deposition on continental shelves, in D.J. Stanly et al. ed. Marine Sediment Transport and Environmental Management, 127–158, John Wiley and Sons.

De Jong, J.D., 1977, Sed. Geol., pp 13–23.

Eisma, D. et al., 1971, Contr. Mineral. and Petrol., V. 31, pp 321–334.

Hawley, N., 1981, Sed. Geol., pp 699–712.

Inman, D.L., 1957, Wave generated ripples in nearshore Sands, Beach Erosion Board, Corps Engrs. Tech. Mem., No.79.

Klein, G. dev., (ed.), 1976, Holocene Tidal Sedimentation, Dowden, Hutchingson and Ross, Inc.

Klein, G. dev., 1975, Epilogue, Tidal sedimentation some remaining problems, in Tidal Deposits (ed. by R.N. Ginsburg), 407–410, Springer Verlag.

McCave, I.N., 1970, Jour. Geophysical Research, pp 4151–4159.

Morton. R.A., 1979, Sed. Geol., pp 105–122.

Nedeco, 1962, Demarara Coastal Investigation, Delft.

Nio, S.D. et al., 1980, Sed. Geol., pp 263–279.

Postma, H., 1967, Sediment Transport and Sedimentation in Estuarine Environment, in Lauff (ed), Estuaries, 158–179, Horn-Shafer Co.

Reineck, H.E., et al., 1967, Zeitmessungen an Gezeitenschichten, Natur. Mus. Frankfurt, pp 97, 193–197.

Reineck, H.E. et al., 1967, Mar. Geol. V. 5, 227–235.

Reineck, H.E. et al., 1968, Sed., Geol., pp 99–104.

Reineck, H.E., 1972, In Recognition of Ancient Sedimentary Environments (Rigley ed.).

Reineck, H.E. et al., 1973, Depositional Sedimentary Environments, 97–113, 2nd ed., 1980, Springer-Verlag.

Ren, M.E. et al., 1983, Sedimentation on tidal mud flat of Wanggang area, Jiangsu Province, China, in Proceedings of International Symposium on Sedimentation on the Continental Shelf, With Special Reference to the East China Sea, Vol. 1, pp 1–19, 1983, China Ocean Press.

Schuttenhelm, R.T.E., 1980, Mar. Geol., 34, M27–M37.

Siddiquie, H.N., 1980, Mar. Geol., 34, M11–M20.

Sioddart, D.R., 1971, Coral-reefs and Islands and Catastrophic Storms, in Applied Coastal Geomorphology (J.A. Steers, ed.).

Terwindt, J.H.J. et al., 1972, Sed., Geol., pp 85–98.

Thompson, R.W., 1975, Tidal flat sediments of the Colorado River Delta, Northwestern Gulf of California, in Tidal Deposits (ed. by R.N. Ginsburg), pp 57–66.

Yang Zuosheng and J.D. Milliman, 1983, Fine-grained sediments of Changjiang River and Huanghe River and Sediment Sources of East China Sea, Symposium on Sedimentation on the Continental Shelf, With Special Reference to the East China Sea, Vol. 1, pp. 436–446, China Ocean Press, 1983.

Cai Aizhi, 1982, Acta Oceanol. Sinica, V. 4, pp 78–88. (in Chinese)

Chen Bingan, 1982, A Dynamic Analysis of mud-sand off the Xiaoqinghe river mouth, Marine Science Bulletin, V. 1, pp 57–70. (in Chinese)

Chen Lirong et al., 1980, Oceanol. et Limnol. Sinica, V. 11, pp 46–64. (in Chinese)

Coastal studies Group, Nanjing University, 1974, The Subaqueous delta of the Haihe and Silting Problem of Tianjin New Harbour, Jour. Nanjing University (Natural Sciences Ed.), 1974 (1), 80–89. (in Chinese)

Dou Zhenxing et al., 1981, Acta Oceanol. Sinica, V. 3, pp 355–369. (in Chinese)

Geng Xiushan, 1981, Acta Geographica Sinica, pp 423–433. (in Chinese)

Gu Jiayu et al., 1983, The Cheniers of the Northern Coastal Plain of Jiangsu Province, Acta Sedimentologica Sinica, V. 1, pp 47–59. (in Chinese)

He Yueqiang et al., 1982, Tropic Oceanology, V. 1, pp 58–71. (in Chinese)

Ji Zhongchun, 1981, Scientia Geographica Sinica, V. 1, pp 77–84. (in Chinese)

Li Congxian et al., 1982, Oceanol. et Limnol. Sinica, V. 13, pp 48–59. (in Chinese)

Li Kenping et al., 1982, Acta Oceanol. Sinica, V. 4, pp 529–536. (in chinese)

Pang Jiazhen et al., 1979, Oceanol. et Limnol, Sinica, V. 10, pp 136–141, V. 11, pp 295–305. (in Chinese)

Qin Yunshan et al., 1982, Acta Oceanol. Sinica, V. 4, pp 191–200. (in Chinese)

Ren, M.E. et al., 1983, Marine Geol. and Quaternary Geol., V. 3, No. 4, pp 1–24. (in Chinese)

Tang Wenzhou et al., 1982, Classification of coast zone in Jiangsu Province using landsat image and digital data, Jour, Nanjing Univ., Natural Sci, Ed., 1982, No. 4, pp 939–948. (in Chinese)

Yuan Yaochu et al., 1982, Acta Oceanol. Sinica, V. 4, pp 1–11. (in Chinese)

Yun Caixing et al., 1981, Oceanol. et Limnol, Sinica, V. 12, pp 391–401. (in Chinese)

Zhao Huanting, 1980, Oceanol, et Limnol, Sinica, V. 11, pp 121–133. (in Chinese)

Zhao Xitao et al., 1980, Chenier Ridges on the West Coast of the Bohai Gulf, Kexue Tongbao, 25, pp 243-247.

Ren Mei-e et al., 1979, Marine Practice, No. 3, pp 1–7. (in Chinese)

Chin Yunshan et al., 1982, Geology of Huanghai Sea and East China Sea, pp 39–51, Science Press. (in Chinese)

Feng Huaizhen et al., 1964, Collected Oceanic Works, No. 2 pp 27–43, China Ocean Press. (in Chinese)

Feng Huaizhen, 1964, Chinese Society of Oceanology and Limnology Treatise Abstract, pp 54–55, Science Press. (in Chinese)

Wang Zongtao, 1982, Marine Geology Research, No. 2, pp 79–88. (in Chinese)

Bi Aohong et al., 1964, Collected Oceanic Works, No. 2, pp 47–55, China Ocean Press. (in Chinese)

Davies, J. L., 1980, Geographical variation in coastal development, 2nd ed., pp, 169–177. Longman, London.

Davis, R. A. Jr., 1978, Costal sedimentary environments, Springer-Verlag, New York, pp 69–159.

Fairbridge, R. W., and Bourgeois, J., 1978, The encyclopedia of sedimentology, Dowden, Hutchinson & Ross, Inc., Pennsylvania, pp 795–805.

Friedman, G. M., Sander, J. E., 1978, Principles of Sedimentology, John Wiley & Sons, Inc., pp 537–553.

Ginshurg, R. N., 1975, Tidal deposits, Springer–Verlag, New York, pp 5–91.

King, G. A. M., 1972, Beaches and coasts, 2nd ed., St. Martin's Press, New York, pp 137–162.

Klein, G. dev., 1976, Holocene tidal sedimentation, Dowden, Hutchinson & Ross, Inc., Pennsylvania, pp 6–243.

Komar, P. D., 1976, Beach Processes and sedimentation, Prentice-Hall, Inc., New Jersey, pp 295–299.

Reading, H. G., 1978, Sedimentary environments and facies, Blackwell Scientific Pub., Oxford, pp 169–177.

Reineck, H.E., and Singh, I. B., 1973, Depositional sedimentary environments, Springer-Verlag, New York.

Chapter Four
Modern Estuary Sedimentation of Some Large Rivers in China

I. Basic Characteristics of Modern Estuary Sedimentation

Fistly, the Huanghe River, the Chaogjiang River and the Zhajiang River flow from west to east.

The material composition of modern estuary sediment in China is both restricted by the matter properties of the river basins and submerged by climate in various regions. The detrital minerals in sediments of the Huanghe River are characterized by the high contents of common bornblends and carbonate. The content of CaO varies in range of 13.13% to 9.63% in the chemical composition of bottom sediments in the Huanghe River and the near seas. It has not be found in other estuarine sediments in China. These characteristics of sediments of the Huanghe Estuary are inherent in the loess of the Huanghe River valley, and are the results of the influence of dry climate and mid-dry climate in the region of North-West China. Surface-layer bottom sediments in the Zhujiang (Pearl) Estuary have the characteristics of being rich in iron minerals[1)]. In addition, the content of organic matter in the bottom sediments is higher than that of other rivers in China, which is closely relative to the Zhujiang River flowing through the tropics and subtropics regions. As far as the clay minerals of modern estuary sediments in the Huanghe River, the Changjiang River and the Zhujiang River are concerned, except illite content being the highest in the three estuaries the montmorillonite is higher in the Huanghe River and the kaolinite is higher in the Zhujiang River. These are all reflection of incoming sediments and climate influences of the river basins. The Huanghe River is world famous in its heavy concentration. In average, over 1.1 billion tons of sediments are transported into the estuarine region each year. The rate of land forming was calculated as about 23 km^2 per year, and the rate of extension of coastline was caloulated to very from 0.15 km/yr to the 0.42 km/yr. The Changjiang Estuary was situated in the place of the Tanyang or more western region in the time of the greatest transgression of the sea in the Holocence epoch. Since it has about 0.5 billion bons of sediments transported into the estuary region each year, so it now enters into the sea in the place of east to Shanghai. The Changjiang Delta is developed by the Changjiang River in the extension rate of 20 m each year. The Zhujiang River is the third largest river in China. The total amount of suspended sediments discharge and bottom loads discharge are more than 0.1 billion tons each year, which have been accumulating in the mouth region to form the rich and net-viened the

1) Zhu Sulin et al., Study of neavy minerals of surface-layer sediments in the Zhujiang Estuary and its near coast, 6, 1982.

Zhujiang Delta.

The estuaries are the place of trasition from continental deposit to marine deposit. In general speaking, the estuaries are the complex and changable sedimentary environment formed by two sources of materials, the sea source and land source; two dynamics, the runoff and tidal current; two media, the fresh water and salt water, which form the modern sedimentary system being composed of suspended sediments, fluid mud and bottom materials. The tide ranges of the Huanghe Estuary and the Zhujiang Estuary are both less than 2 m. The tide limit of the Huanghe Estuary is only some km, and is about fifty to one hundred sixty km in the Zhujiang Estuary. Both estuaries fall into the microtidal ones. The tidal range of the Changjiang Estuary is 3.2 m . It falls into the meso tidal estuary, where its tidal current limit is situated in the place of 300 to 540 km up the mouth. But, the tidal current limit will move a certain distance downtream when the flood peaks of the river flows downstream in flood seasons. In above estuarine regions, the runoff and the tidal current have variou complex combination types. The sediments transport in the form of suspended sediments and bed load obey the following dynamical sorting rule of estuarine sedimentation: One is that it is mainly controlled by runoff, and the sedimentation property has still the characteristics of intermittent sedimentation of river; the other is that it is mainly controlled by both of runoff and tidal flow . The former is characterized by intermittent deposit and the later by fine grainn sediments, redoposit and well sorting. According to the dynamic storing of sedimentation, we may consider the sandy place of bed load as the down limit of the runoff action. So, we say runoff transporting sand and tidal current carries mud into tidal estuaries. There are phase differences between tidal level and tidal current velocity in the Changjiang River and Zhujiang River. Maximum velocity of flood and ebb tides appears near the middle tidal level where the sediments become coarser. Most of the materials of fine sediments deposited in the place between middle tidal level and high tidal level.

The grain sizes of suspended sediments in some large rivers in China are all finer. The flocculation takes place in the region where fresh water mixes with saline water. Thus, one of basic characteristics in modern estuary sediment is the flocculent deposit. The results analysed by Coulter Conter Model TA II have showed that the average median diamater of suspended sediments in the Huanghe Estuary is 7 microns, and that of suspended sediments of the Changjiang Estuary and Zhujiang Estuary are 8.6 microns and 6.5 microns respectively. These sediments belong to the category of fine sediments, and they fall into the range of opeimum flocculating diamaters, owing to flocculation, the settling velocity of fine sediments will increase 3–7 times, which speed up the deposition of the mouth bar and in some estuaries certain scale of fluid mud will be formed, such as that in the South Channel of the Changjiang Estuary.

In respect to the longitudinal distribution, the space distributing rule of modern estuary sediments obeys the abovementioned rule of the dynamic sorting of sedimentation. They become finer as they are being brought downstream. However, the sediments in sand bar are more coarser than that in other places. This fact can be summarized as a sedimentation model of coarsefine-coarse-fine. In the river bed, where ebb tidal current including runoff predominants, the sediments are usually coarser; and in the channel of predominated by flood tidal current the sediments are finer. Thus the sediments in the Changjiang Estuary are coarser in its northern region and smaller in its southern region. The grain size of sediments in Lingdingyang in the Zhujiang Estuary is coarser in the eastern region and finer in the weastern region.

It should be pointed out that the dispersion of sediments carried by the Huanghe River into the sea influnces not only the Bohai Gulf and the Laizhou Bay but also the much larger regions, for the circulation and coastal currents in the Bohai Sea. It may pass through the strait of southern portion of the Bohai Sea to the Huanghai Sea and becomes a part of suspensions of the nearshore current.

II. Modern Sedimentation of the Huanghe Estuary

1. Introduction

The Huanghe River is world famous in its aptness of deposition, avulsion and displacement. The lower reaches of the Huanghe River have breached more than 1500 times, and major shifts river course was 26 times in the last 3000 yr. The last shift of its course took place in 1885, which led to its estuary moving from the northern region of Jiangsu Province to Shandong Province and the river flows into the Bohai Sea up to now. The depositing plain formed before this time is called the old delta, and the depositing plain developed after this time is called the recent delta, as shown in Fig. 4-1.

The river segment of 130 km from the Beizhen to the river mouth is celled the estuary, which consists of three regions, the mouth region, the delta and the offshore region; among them Ninghai is the top of swing small delta. The Huanghe Estuary had transversely swinged frequently in a period of twelve years. The southwestern region is more higher than the northeastern region in the small delta where there are numerous old streams, between which is low-lying land. The estuary spit remained in the swinging prossece of the Huanghe Estuary have joined each other to form the radial coastline. From the coastline to the region 15 m deep is called the offshore region. The tidal range in the Huanghe Estuary is less than 2 m. Its tidal limit is only 30–50 km; in addition, it only inflances the river segment near Ninghai in the time of storm tide. The tidal current limit is much smaller, being less than 2 km[1)].

2. The Characteristics of inputting water and sediments

The average annual sediment load in the Huanghe River is 1.1 billion tons. The rate of land forming and extension of coastline as mentioned in preceding section is the biggest in some larger rivers in China, which is closely related to the characteristics of inputting water and sediments and

1) Wang Kaichen, Estuarine information and law of fluvial processes in the Huanghe River, 1980.

offshore dynamic.

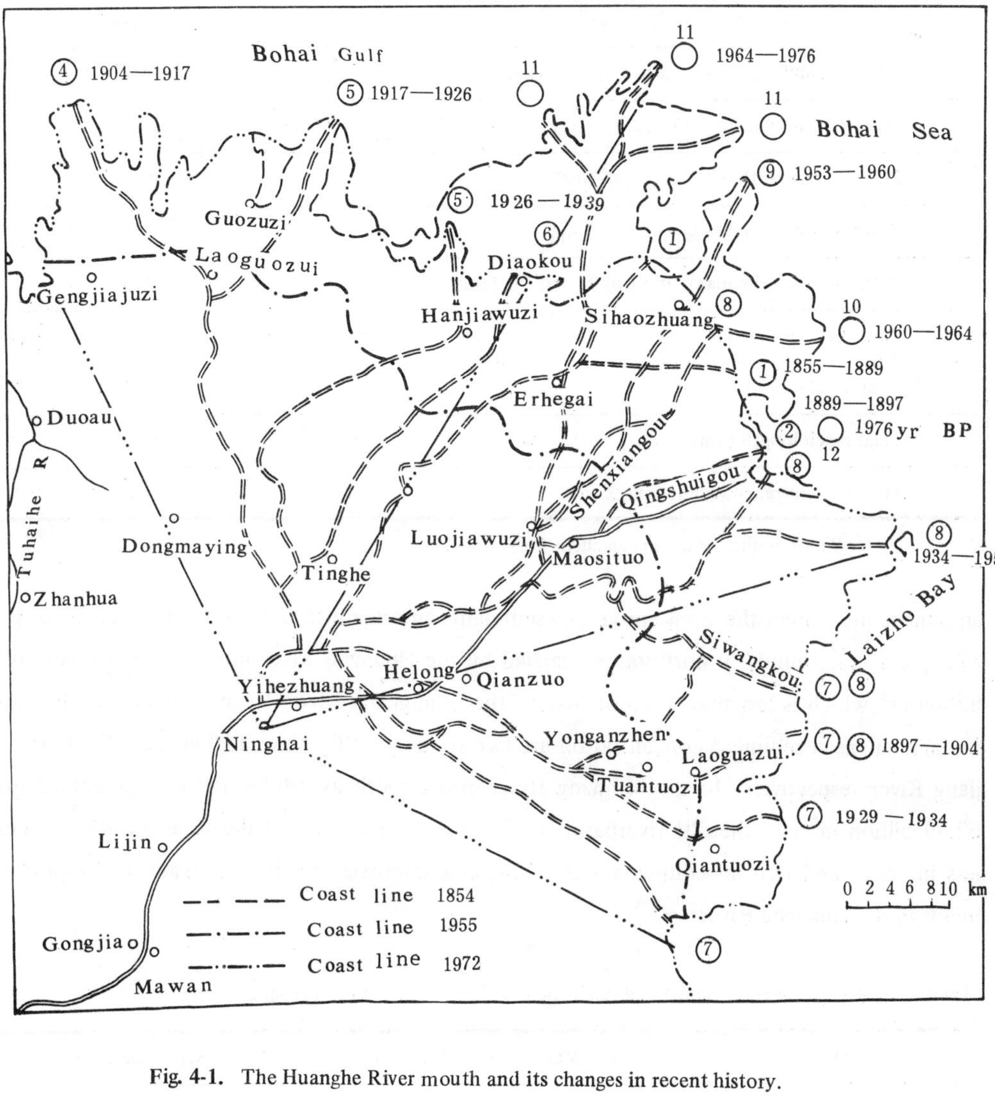

Fig. 4-1. The Huanghe River mouth and its changes in recent history.

There are no tributaries converged into the river in the estuary of the Huanghe River. Its condition of inputting water and sediments is similar to that in lower reaches of the Huanghe River.

We will make a further discussion on the data of the Lijin station, as can be seen in Table 4-1.

This Table has shown the statistical results of the data in Lijin station from 1950 to 1977. It follows that the total amount of sediments carried by the Huanghe River into the estuary is huge. The sediment distribution is that 24% of the total amount was deposited on the top of the delta, 40% on the coastal area and the rest 36% disperses into the deep sea by various dynamic factors. There fore, about 76% of the suspended sediments in the Huanghe River enters the Bohai Sea. Its total

Table 4-1 Basic characteristics of hydrology in the Huanghe River

Average annual runoff ($10^8 m^3$)	442.8
Average discharge (m^3/sec)	140.0
Average annual suspended load (10^8t)	11.18
Average annual sediment concentration (kg/m^3)	25.3
Maximum stream flow (Q_{max}) (m^3/sec)	10400
Minimal stream flow (Q_{min}) (m^3/sec)	0
Ratio between the Q_{max} and the Q_{min}	. . .
Ratio between the maximum runoff and the minimal runoff	10.6
Ratio between the maximum suspended lood and the minimal suspended load	8.68
Mean tidal range (m)	0.61–1.13
Tidal reach length (km)	15–30
Tidal current reach longht (km)	0–2

(After Wan Kauchen and Pang Jiazhen 1980, 1982)

amount is more than the total amount of sediments entering into the seas from other rivers in China. Yet, the average annual runoff volume carried by the Huanghe River into the estuary is only 44.28 billion m^3 which is less than 1/20 of that of the Changjiang River. Its average annual discharge and the annual mean sediment concentration are two times and fifty times higher than that in the Changjiang River respectively. In the Minjiang River in the southeast China, the average annual runoff is 62.49 billion m^3 and then its riverbasin area is only 1/13 of that of the Huanghe River. Therefore, less in water and rich in sediments is the basic characteristic of inputting water and inputting sediments in the Huanghe River.

Table 4-2 Monthly average runoff and suspended load at the lijin station from 1950 to 1977

Month	Jan.	Feb.	Mar.	Apr.	May.	June	July	Aug.	Sep.	Oct.	Nov.	Dec.
Runoff ($10^8 m^3$)	13.69	12.26	22.57	24.70	23.43	19.46	50.56	77.61	71.61	65.05	41.77	20.78
Percentage of total annual runoff (%)	3.1	2.8	5.1	5.6	5.3	4.4	11.4	17.5	16.2	14.7	9.3	4.7
Suspended load	0.037	0.048	0.227	0.302	0.271	0.245	1.71	3.49	2.61	1.47	0.635	0.125
Percentage of total annual sediment load (%)	0.3	0.4	2.0	2.8	2.4	2.2	15.3	31.2	23.4	13.2	5.7	1.1

The annual change of inputting water and suspended load of the Huanghe River and the distribution of that in a year is controlled by the climate in the drainage area. (Table 4-2 and Fig. 4-2).

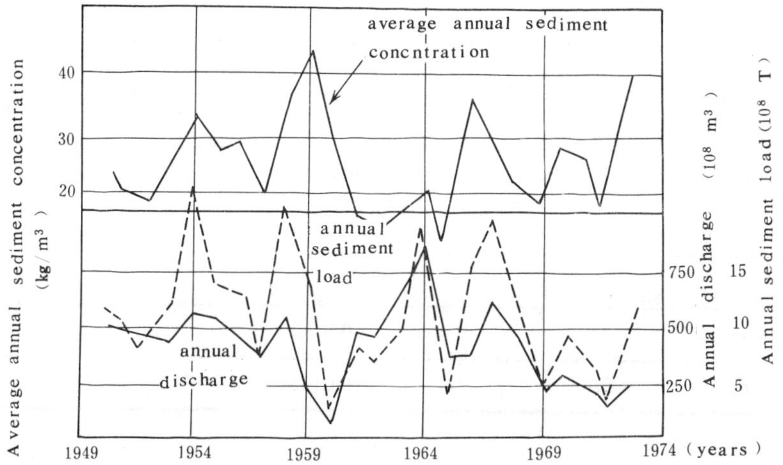

Fig. 4-2. The process of annual runoff and sediment load in the Jijin station.

Fig. 4-2 has shown that in the changes of the annual runoff and the annual sediment load in the Huanghe River there exists relative period of the rich water and rich sediments for three to four years, which has increased sharply the estuarine swinging. It has been shown in Table 4-2 that the monthly runoff and the monthly sediment load are not fairly well-distributed in a year. The runoff of the freshet (from July to October) is 59.8% of the total annual runoff, 40.2% of which belongs to other months. The monthly runoff of February and August is the maximum and the minimum one in a whole year. The non-homogeneous distribution of the total annual suspended sediment load is greater than that of annual runoff in the Lijin station. The maximum monthly suspended sediment load of the Lijin station is in August, and the minimum one is in January. When storm floods come down from watershed, the sediment load in July to October can be as much as 83.1% of the total annual sediment load, and that in other months account for only 16.9% of that.

3. Suspended sediments and bottom sediments of the Huanghe River

The sediments transportation of lower Huanghe River is mainly the suspended load, and the bed-load is less than 1% of the total annual sediment load. The annual mean value of grain sizes of the suspended sediments in each station along the middle-lower Huanghe River are shown in Table 4-3. The table show that there exists local change of median diameter because the branch flowing into the major stream and sediments depositing along the river, but, they become finer as they are transported downstream.

Table 4-3 Change of median diameter of suspended sediment along the middle-lower Huanghe River

Station	Median diameter of suspended sediments
Lanzhou	0.026
Baotou	0.0172
Wupu	0.0338
Longmen	0.329
Shaanxian	0.0310
Qinchang	0.0282
Aishan	0.0191
Lokou	0.0175
Lim	0.0185

(After Qiannin, et al, 1965)

The grain size of the suspended sediments in the Lijin station of the Huanghe Estuary analysed with the Coulter Counter Model TA II by Hangzhou University is shown in the following table.

Table 4-4 Content of grain size of suspended sediment in the Lijin station of the Huanghe River

Grading (mm)	0.063 – 0.032	0.032 – 0.016	0.016 – 0.008	0.008 – 0.004	<0.004	Md_ϕ	Note
Content (%)	7.10	12.10	25.10	23.70	32.30	0.0070	July, 1982

The season changes of median diameter in middle-lower Huanghe River is very obvious. It can be clearly seen in Table 4-5 that the sediments are smaller in flood season, and coarser in dry season of the Huanghe River. This is due to the silt of storm floods being mainly the results of surface erosion of direct runoff in the ground surface in flood season, and the sediments of dry season being the retransported matters from river bed erosion.

One of the important characteristics of the suspended sediments in the Huanghe River is that the average sediment concentration is the highest in China, and the concentration in flood season is extremly high. The highest concentration measured from station between Longmen to Lijin from 1966 to 1977 varied in arange of 933 kg/m^3 to 222 kg/m^3, and the sediment concentration being more than 1500 kg/m^3 has been measured in a storm flood. From Xiaolangdi to Lijin, in July to August 1977, in two short storm floods the total amount of 0.906 billion tons sediments has deposited which is equivalent to the total depositing amount of the whole flood season in the downstream of the year. It has been known from this that the intensity of erosion and deposition in the lower Huanghe River is extremly great. There are huge amount of suspended sediment exchanging with bottom sediments.

Table 4-5 Season change of median diameter in the middle-lower Huanghe River

Station	Jan.–Mar.	Apr.–Jun.	Jul.–Sep.	Oct.–Dec.
Lanzhou	0.0114	0.0235	0.0258	0.0453
Baotou	0.0302	0.0088	0.0203	0.0311
Wupu	0.0522	0.0264	0.0275	0.269
Longmen	0.0709	0.0225	0.0306	0.0504
Shaanxian	0.0462	0.0362	0.0272	0.0375
Qinchang	0.0375	0.0292	0.0197	0.0320
Gaochun	0.0307	0.0209	0.0148	0.0252
Aishau	0.0191	0.0176	0.0142	0.0247
Lokou	0.0192	0.0225	0.0136	0.0186
Lijin	0.0224	0.0222	0.0155	0.0217

(After Qian Nin, et al, 1956)

The distribution of grain size indicate that the mean size of the grains becomes smaller as they are being brought downstream. The sediments of the major channel bed is coarser than that of the river dright. The bottom sediments of the reach upward from Qinchang station are very fine sand. The bottom sediments downward from this section all fall in the category of coarse silt, and their particle composition and median diameter has changed slightly along the river. The bottom sediments of the wide and shallow section has changed slightly on the cross section; the particle size of the major channel bed is generally coarser than that of the river drift in the narrow and deep section. The matters of the major channel bed in the flood season are coarser, which will become finer because the fine sediments coming from riverbasin are deposited along the river in dry season.

According to the data for mineralogical research by Chen Lirong and other's (Institute of Ocenology Academa Sinica), the results of their study showed that the detrital minerals are made up of 30 minerals components in the Huanghe Estuary, as shown in Table 4-6.

The above table shows that the feldspar types of light minerals are 60.2% of the total, among them the playioclase and the orthoclase contents are 53.7% and 6.5% of the total amount respectively. It is worthwhile to point out that the high contents of the calcite and others carbonate minerals in the light minerals is a basic characteristic of light mineral composition in the Huanghe Estuary. It is quite evident that this basic characteristic is greatly in accordance with the high content of carbonate minerals in the wide distributing loess in the Huanghe River Basin. In heavy minerals the hornblende content is 36% of the total amount, the epidote ranks the second in content, and iron, garnet etc the third. If the heavy mineral component of detrital minerals in the above table is com-

Table 4-6 Detrital minerals of bottom sediments and their contents in the Huanghe Estuary

Heavy minerals (%)		Light minerals (%)	
Magnetite	0.1	Plagioclase	53.7
Ilmenite	8.1	Orthoclase	6.5
Garnet	7.7	Carbonate minerals	7.8
Zircon	1.1	Others	32
Hornblende	36.0		
Muscovite	1.5		
Epidote	27.4		
Pyroxene	2.6		
Titanite	4.0		
Others	11.5		

(After Chen Lirong, Luan Zuofeng and Zhen Tiemin, 1980)

pared with the heavy mineral component in the loess analyzed by Liu Donsheng, it can be seen that the contents of the hornblende and epidote in the estuarine bottom sediments is higher than that in the loess, especially for hornblende, its content is three times higher than that in loess, therefore it is the main characteristic mineral of the sediments in the Huanghe River.

The suspended sediments and bottom sediments in the Huanghe Estuary are all finer. Besides the detrital mineral component, the clay minerals occupy an important place in the bottom sediments. Fig. 4-3 shows that the result of the X-ray diffraction spectrum of the bottom sediments ($<2\mu$) in the Huanghe Estuary. The X-ray of diffraction spectrum of the samples untreated, and treated by physical and chemical methods shows that the peak values of 9.98Å, 4.99Å and 3.33Å are all the characteristic peaks of the illite. The peak value of 14.25Å in the X-ray diffraction graph is the characteristic value of the montmorillonite. Its peak value moves into the peak value of 17.12Å when the samples are treated with glycerine acid, which causes its layers exponding. According to this peak value it is differ entiated from chlorite. The appearance of basic surface reflection peak value of 7.0Å and 3.58Å in Fig. 4-3, and the disappearance of the peak value of 7.09Å in 550°C verify the existance of kaolinite. The peak value of 14.25Å and 7.09Å in chlorite is overlapping with that in kaolinite. In addition to the above differences, when the samples is treated with hot hydrochloric acid (6*N*), the peak value of 14.25Å disappears, which indicates the chlorite has been dissolved, accordingly, the existance of the trioctahedron chlorite has been identified. The above analysis shows that the major clay minerals of bottom sediments are illite, montomorillonite, kaolinite and chlorite in the Huanghe Estuary. Their relative contents are shown in Table 4-7.

The chemical compositions of the bottom sediments in the lower Huanghe River, the bottom sediments in offshore region and that in the Huanghe River Basin are shown in Table 4-8.

From Table 4-8, it can be clearly seen that the common characteristic of chemical composition

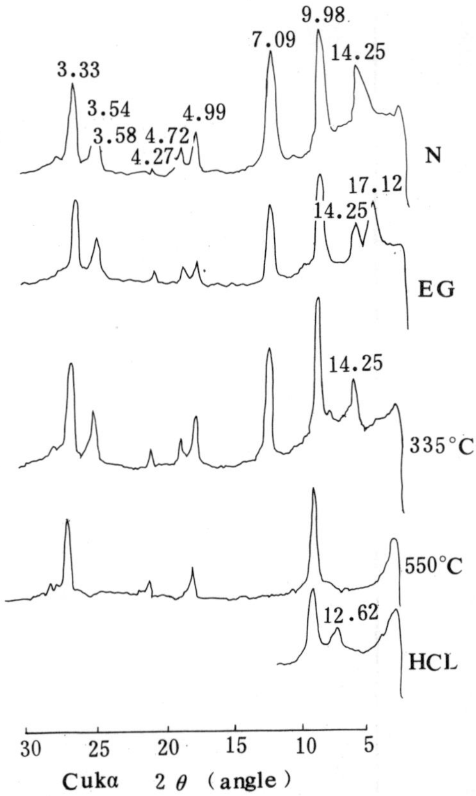

Fig. 4-3. X-ray diffraction graph of bottom sediments (< 2μ) in the Huanghe Estuary.

Table 4-7 Major clay minerals and their relative contents (%) in the bottom sediments of the Huanghe Estuary

Treating method	Untreated	Glycerine	Hydrochloric	335°C	550°C	Relative content(%)*
Illite	9.98 4.99	X	X	X	X	67
Montmorillonite	14.25	17.12	X	O	O	13
Kaolinite	7.09 3.58	X	X	O	Δ	8
Chlorite	14.25 7.09 4.72 3.54	X	Δ	X	X	12

Note: X– no reaction; O– reacted; Δ– disappearing.
Heating time is two hours in each case, the inter distance is Å.
*– The caltulation method after that of Chen Peiyuan.

Table 4-8 Comparison between the chemical composition of bottom sediments in the lower Huanghe River, the offshore region and the loesses

Chemical Composition	SiO_2	TiO_2	Al_2O_3	Fe_2O_3	FeO	MnO	MgO	CaO	Na_2O	K_2O	P_2O_5	H_2O^+	H_2O^-	S	Volatile matter	Total
Malan loess	58.83	0.68	9.64	3.02	2.14	0.18	2.33	8.14	1.78	2.55	0.17	3.85	1.07	0.50	4.99	99.89
Lishih loess	59.10	0.64	10.35	3.50	1.57	0.15	2.28	7.64	1.69	2.56	0.16	3.88	1.44	0.50	4.49	99.92
Wucheng loess	50.27	0.55	9.40	3.47	1.26	0.05	2.00	13.57	0.98	2.11	0.14	4.87	1.75	0.51	8.96	99.76
Bottom sediments in the lower Huanghe River	50.28	—	17.44	7.43	—	—	1.39	13.13	—	—	—	—	—	—	—	—
Bottom sediments in the offshore of the Huanghe Estuary	48.85	0.51	17.96	8.83	—	0.16	3.98	9.63	0.95	3.19	0.19	—	—	—	13.82	100.01

After Liu Dongshen (1962), Qian Nin (1965), Institute of Oceanology, Academia Sinica (1978).

for three kinds of loesses and bottom sediments of lower the Huanghe River and offshore is in the trend of rich in aluminium, particularly in the loesses being transported and moved. As the particle size becomes finer, the Al_2O_3 content becomes higher and higher from land to sea. This trend of rich in aluminium is precisely a concrete reflection for the clayey of bottom sediments. It is because that the content of $CaCO_3$ is higher the organic matter in the loesses is poorer. So, the $CaCO_3$ has become the cohsive agent of the loesses, but, when it meets with water, $CaCO_3$ is easy to be dissolved to form $Ca(HCO_3)_2$ and then be carried out, which leads to high CaO content in the lower Huanghe River and offshore region. This is an important characteristic of sediments in the Huanghe Estuary, and is seldom seen in some other large river estuaries in China.

4. Diffusion of sediments transported into the seas

Enormous amount of sediments are transported into the estuarine region by the Huanghe River, among them 76% enters the seas. In the 0.9 billion tons of sediments entering the seas 0.43 billion tons of sediments continuously diffuse into the deep sea by various dynamic factors and the rest accumulates in the front of the estuary to develop the underwater delta.

The offshore region of the Huanghe Estuary is, in fact, formed by the junction of the submerged deltas in the course of each cycle of swing of the Huanghe Estuary.

The colour of bottom sediments in the offshore region is drab, a counter part of sediment colour in the Huanghe River, their size composition is shown in Table 4-9.

Table 4-9 The particle size composition of bottom sediments in offshore region in the Huanghe Estuary

Grading (mm)	>0.25	0.25–0.10	0.1–0.05	0.05–0.01	0.01–0.005	0.005–0.001	<0.001
Mean content	0.50	1.20	2.80	21.60	17.30	30.60	30.20
Sample number	1	61	61	61	54	54	54

(After Institute of Oceanology Academia Sinica)

Above data show that the fine sand content of the sediments entering into the seas from the Huanghe River is very low, but the content of clay grading is about 60% of the total amount. It follows that the sediments are transported into the seas in the Huanghe River in the form of suspension.

According to Qin Yunshan et al, the high concentration zone (>100 mg/1) diffuses only in an area over ten kilometers in front of the Huanghe Estuary; in farther region, it decrease below 50 mg/1. They are not only influenced by the unevenness of the annual suspended sediment load in the Huanghe River, but also closely related to oceanic hydrology and climate factors in the Bohai Sea (Fig 4-4).

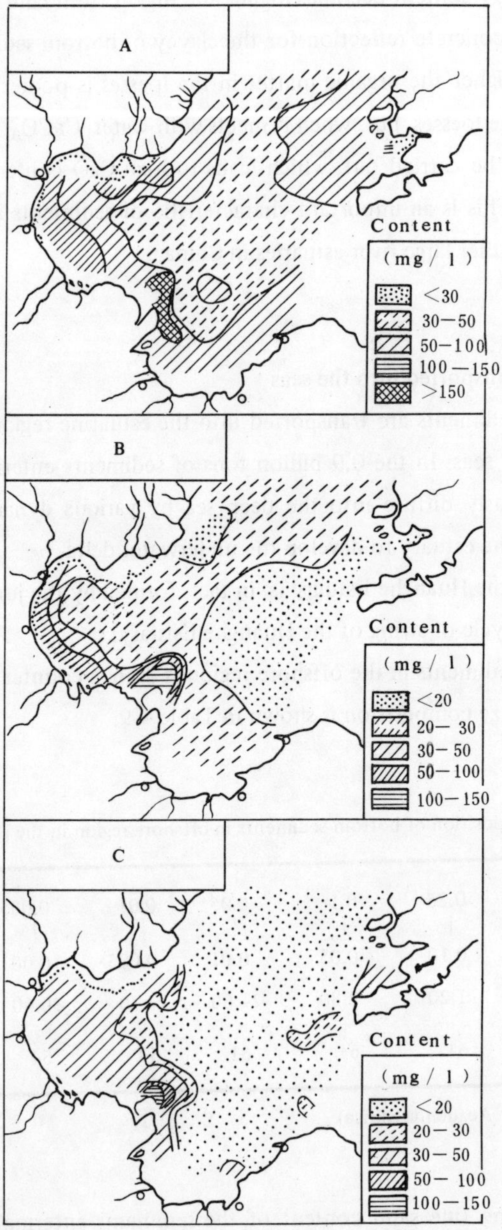

Fig. 4-4. Content of suspended matter in the nearshore area of the Huanghe Estuary.

A–Distribution of suspended matter in surface sea water of the Bohai Sea (April);

B–Distribution of suspended matter in surface sea water of the Bohai Sea (July);

C–Distribution of suspended matter in surface sea water of the Bohai Sea (October).

Fig. 4-4B clearly showns that the interrelationship between the heavy suspended sediment concentration region and surface load of storm flood in autumn in the Huanghe River. There prevails northeast wind in April in the Bohai Sea, which is an annual common wind and is the strongest wind direction, as well in addition, the strong vertical vortex; leads to the ocurring of the highest suspension zone in the region out of the river mouth (Fig. 4-4A) The vertical distribution of sediments carried by the Huanghai River into the Seas is shown in Fig. 4-5.

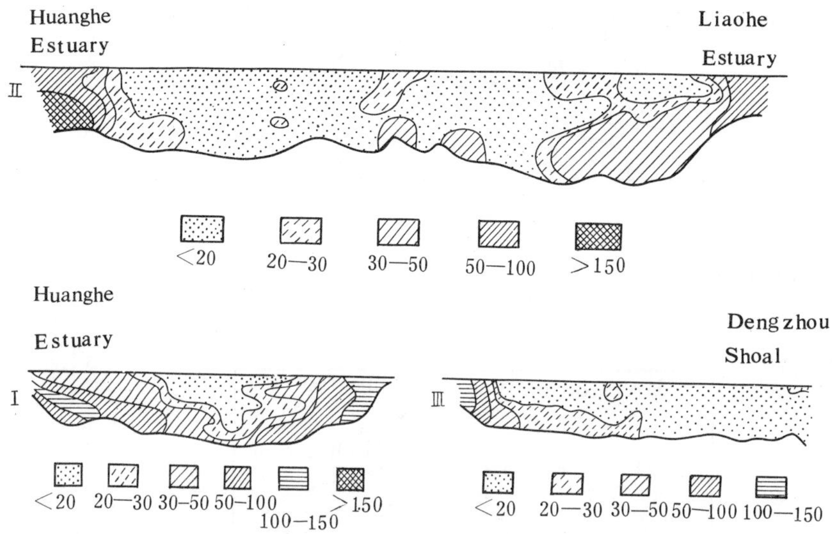

Fig. 4-5. Vertical distribution of suspended matter of the Huanghe Estuary and its nearshore areas.

I– Distribution of suspended matter in section I;

II– Distribution of suspended matter in section II;

III– Distribution of suspended matter in section III.

In the cross section from the Huanghe Estuary to Caofeidian, because the great vertical gradient of sediments suspension disperses in the form of denisty current; and in another cross section, homogeneous vertical mixing of sediments exists obviously. The sediments dispersed northward to the west are more than that to the southeast.

We have mentioned that in chemical composition of the bottom sediments in the Huanghe River, the high content of $CaCO_3$ is its basic character, which is certainly reflected in the dispersion range and dispersion direction of the sediments carried by the Huanghe River into the seas, as shown in Fig. 4-6.

It has been known from Fig. 4-6 that as the distance to the coast increases, the content of $CaCO_3$ will decrease.

Fig. 4-7 shows that the maximum deposition isopachyte is situated in the north by west region out of the Huanghe Estuary, which is in accordance with the direction of resdual flow, tidal current and the direction of monsoon in flood season of this region.

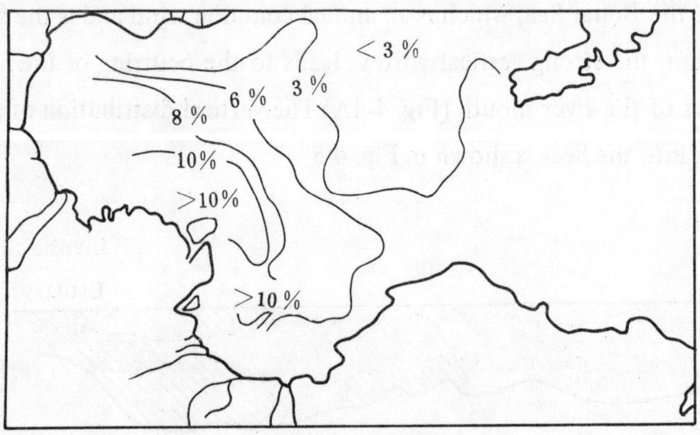

Fig. 4-6. Distribution of $CaCO_3$ content in sediments of shallow sea outside the Huanghe Estuary.

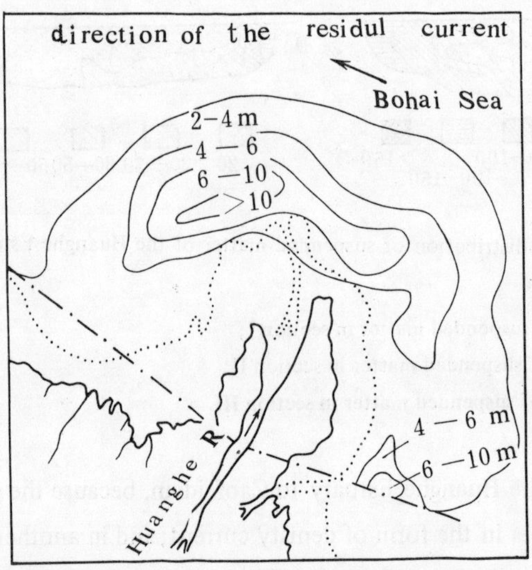

Fig. 4-7. Isopaches of the sediments of the Huanghe Estuary.

III. Modern Sedimentation of the Changjiang Estuary

1. Modern sedimentary environment

The Changjiang River, which has the total length of 6308 km, is the biggest river in China and the third biggest in the world. Tatung in Anhui Province is generally considered as the upper limit of lower tidal reaches of the Changjiang River. The total length of tidal limit is about 300 to 540 km. The river is widened gradually from Chiangin up to the estuary. The river section from Hsiliuchien to

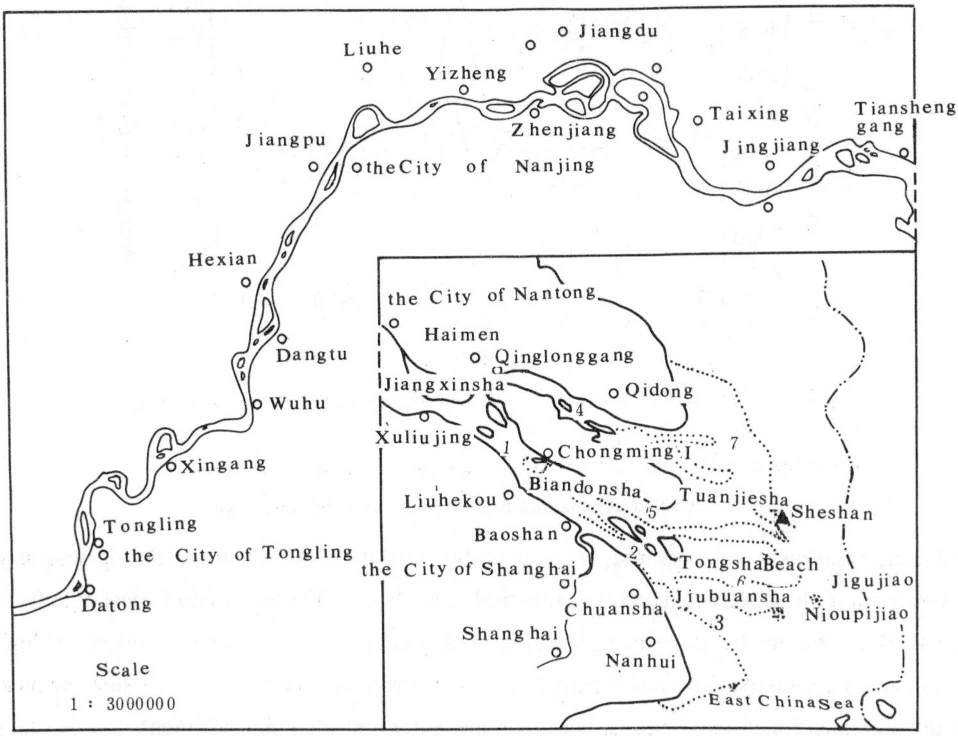

Fig. 4-8. Changjiang Estuary.
1. South Branch; 2. South Passage; 3. South Channel;
4. North Branch; 5. North Passage; 6. North Channel;
7. Qidong Beach.

the mouth of total length about 170 km is widened from 5 km to 90 km. The rate of widening amounts to nine. In this reach, the river is separated by the Chungming Island, the Hengsha Island and the Jiuduansha Shoal, where the North Branch, the South Branch, the North Passage, the South Passage, the North Channel and the South Channel are formed. The estuary has a shape of bell as shown in Fig. 4-8.

(1) *General description of water and sedimentary condition*

The incoming water and incoming sediments of the Changjiang Estuary is governed by the hydrological condition of the whole estuary, which is influenced by both the runoff discharge, the sediment discharge of river basin and the tidal influx, the sediment load of sea area. Both of them vary in periods of several years and of a year. The combination of these variations leads the compex hydrological cor ''tion. Based and changable upon the hydrological data of Tatung station from 1951 to 1976, we calculaied and obtained the following result the annual mean runoff discharge is about 925 billion m^3; the annual mean sediment discharge is about 0.472 billion tons. The quantities of incoming water and incoming sediments are shown in Fig. 4-9. Since only in recent years both runoff

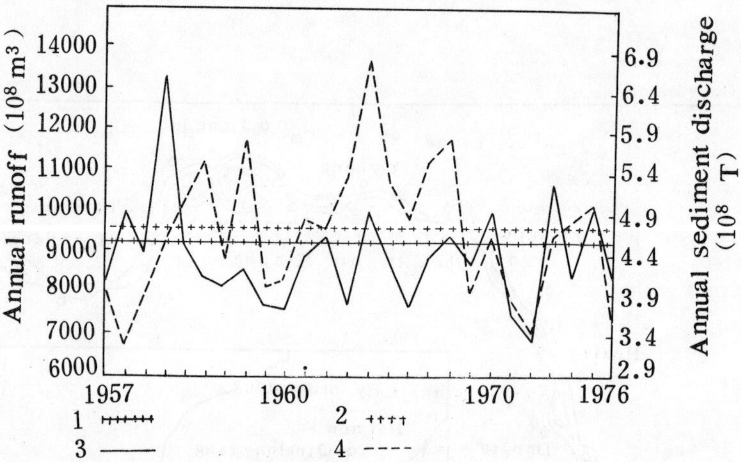

Fig. 4-9. The process of annual runoff and annual sediment discharge in the Tatung tation of the Changjiang River.
1. Average annual runoff; 2. Average annual sediment discharge (no determination, Jan.–May, 1952); 3. Annual runoff; 4 Annual sediment discharge.

and concentration were measured, their periodical variation in river basin during several years is somewhat not clear. According to the statistical analysis on 100 years' hydrological data from Yican and Hankou stations by Changjiang Valley Planning Office, we learn that a continuous high water period and a continuous low water period appears alternatively every 7–8 yr. Using the mean value of the aforementioned two periods, the incoming water and incoming sediments can be summarized into six cases of combination as shown in Table 4-10. It is known from this table that the number of years with high water and high concentration, or low water and low concentration, is about 42% of the total number of years under statistics, and that the number of years with other four combinations

Table 4-10 The combination types of water and sediments of Tatung station in the Changjiang River

Combination type	%	Combination type	%
Low water and low concentration	30.77	high water and low concentration	19.24
Low water and high concentration	30.77	middle water and low concentration	3.85
High water and high concentration	11.54	middle water and high concentration	3.85

is about 58%. In these, the number of years with low water but high concentration amounts to 30.77% of the total number. So the condition of incoming water and incoming sediments of thc Changjiang River Basin in recent years are somewhat complex. The distribution of water and sediments in the Changjiang River during a year is shown in Fig. 4-10. The amounts of incoming water and incoming sediments in flood season are 71% and 85% respectively of their amount available in a whole year. The highest amounts appear in July. In recent years, monthly mean runoff during this month approaches 50,000 m^3/s, which is one sixth of annual mean value; and the monthly mean sediment discharge during this month is over 100 million tons, which is one fifth of the annual mean value. Therefore, the high concentration of incoming water and incoming sediments in flood season is one of the important character of the distribution of water and sediments in the Changjiang River during a year, which is also one of the reasons causing the deposition of sediments during flood season in the mouth region of the river. The tide in the Changjiang Estuary is of the type of regular

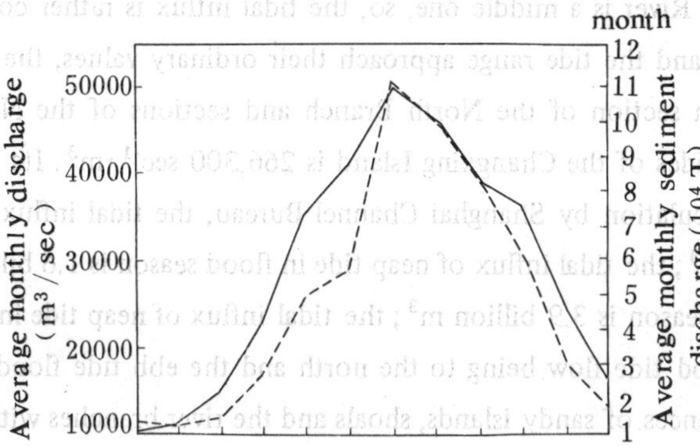

Fig. 4-10. The process of average monthly sediment discharge and average monthly runoff in Tatung station of the Changjiang River.

half-day tide. When the tidal wave of the Pacific Ocean is transmitted to the estuary, it is frictionated and reflected by shallow sea topography, estuarine topography and runoff, and deforms gradually, causing the largest tide range to appear in the region of mouth bar. Up the mouth bar, the tide range decreases and the time of ebb tide increases as shown in Table 4-11.

Table 4-11 The statistical data of synchronous flood tide duration, ebb tide duration and tide range

Station	Spring tide of flood season			neap tide of flood season			spring tide of day season			neap tide of day season		
	T_f	T_e	ΔH	T_f	T_e	ΔH	T_f	T_e	ΔH	T_f	T_e	ΔH
Liuhe	3:54	8:24	2.99	4:11	8:03	2.36	–	–	–	–	–	–
Wnsang	4:00	8:15	2.97	4:25	7:54	2.36	3:59	8:26	3.00	4:58	7:15	1.64
Changshu	4:38	7:37	3.26	5:00	7:20	2.62	4:26	8:00	3.46	5:20	6:44	2.07
East to Jiuduansha	5:33	6:40	3.40	5:50	6:34	2.69	5:36	6:50	3.88	5:51	6:38	1.81
Liuhuashan	5:40	6:35	3.14	5:51	6:33	2.37	5:51	6:29	3.85	5:54	6:28	1.81

(by Shen Jinshan et al. Shanghai Channel Bureau Dec. 1981)
T_f– flood tide duration; T_e– ebb tide duration; ΔH– tide range.

The tidal flow in the Changjiang Estuary, which is more complex than that of the tide, is basically of the type of regular half-day tidal flow. There is reversing tidal flow in the region to the west of the connecting line between the Qidong Shoal and the Hensha East Bank; and, in the region to the east of this line, the tidal current changes to rotation current. This rotation current is clockwise in the sea area outside the isobath of – 10 m. The aforementioned link line is also the boundary of the predominant current of flood and ebb tides, that is, in upper half portion of the North Channel and the whole North Passage, the ebb tide flow including runoff is the predominant current. But, in the lower reach of the North Channel and the whole South Channel, the flood tidal current is predominant. The tide range in the Changjiang River is a middle one, so, the tidal influx is rather considerable. When runoff of the upper reaches and the tide range approach their ordinary values, the total tidal influx passing through the Miaogan section of the North Branch and sections of the North Passage and South Passage on the two sides of the Changxing Island is 266,300 $sec^{-1} \cdot m^3$. Its ratio to runoff is 8.8:1. According to the calculation by Shanghai Channel Bureau, the tidal influx of spring tide in flood season is 4.5 billion m^3; the tidal influx of neap tide in flood season is 1.6 billion m^3; the tidal influx of spring tide in dry season is 3.9 billion m^3; the tidal influx of neap tide in dry season is 1.3 billion m^3. Because the flood tide flow being to the north and the ebb tide flood to the south, as well as becouse of the influences of sandy islands, shoals and the river branches with multiple stages, the space distribution of tidal influx does not coincide with that of the amount of sediment load, as shown in Table 4-12. From this table, we see that in the whole tidal influx of flood tide and ebb tide, the fraction of the North Branch is smaller, besides, sea water and sediments there nove reversely, causing the North Branch becoming narrower and narrower. The values of tidal influxes and the

Table 4-12 The mean current dividing ratio (%) and sediment dividing ratio (%) between channels in the Changjiang Estuary

	Flood tide		Ebb tide		Runoff discharge	Sediment discharge
	Tidal influx	The amount of sediment load	Tidal influx	The amount of sediment load		
The South Branch	78	55	89	76	100	139
The North Branch	22	45	11	24	0	–39
The South Passage	59	55	51	47	40	40
The North Passage	41	45	49	53	60	60
The South Channel	69	66	60	65	12	61
The North Channel	31	34	40	35	88	39

sediments in the South and North Passages differ slightly, but the runoff discharge and the sediment discharge in the North Passage is obviously larger than that in the South Passage. As to the South Channel and the North Channel, the situations are entirely different. The tidal influx of flood tide and ebb tide and the amount of sediment load of tidal flow in the South Channel amount to more than 60% of the total tidal influx and sediment load in the estuary. The runoff of the South Channel is only 12% of the total value, but its sediment discharge is 61% of the total value. In the North Channel, the runoff amounts to 88% of the total value, but the sediment discharge is only 39% of the total value. Therefore, between the South Channel and the North Channel, the condition of water transport and sediment transport are quite different. Besides, the water and sediments in the Changjiang Estuary is also influenced by the nearshore current, including the near shore current of North Jiaugsu and the warm current of Taiwan, wave formed sediments and salty water wedge; all these factors cause the Changjiang Estuary to be a complex dynamical sedimentary environment.

(2) *Mixture of fresh water and salty water in the Changjiang Estuary*

In the region of tidal current limits, especially in the mouth region with a length of more than one hundred kilometers, mixture of fresh warer and salty water is one of the important environmental conditions, which has a great influence on the estuarine sedimentation and the deposition intensity.

The periodic variation of runoff and tidal current of the Changjiang River and the development of the river bed topography of the mouth both influence the distance of reverse flow of salty water and the extent of the mixture of salty water and fresh water. In recent twenty years, the upper mouth of the North Branch has been becoming shallower and narrower, and has lost the function of discharging the runoff. The current velocity of flood tide is higher than that of ebb tide in the Chinglungkang Section, so is the variation of salinity, leading to the salty water with salinity greater than 5‰ flows reversely towards the South Branch. Investigating the reverse flow of salty water in each channel of the Changjiang Estuary, we get the conclusion that the salty water reaches the farthest in the North Branch. In other places, it reaches farther in the South Passage than in the North Passage and in the South Channel than in the North Channel.

There are three fundamental types of mixture of salty water and fresh water in the Changjiang Estuary: the weak mixture type, the gentle mixture type and the strong mixture type. Generally speaking, weak mixture type appears in neap tide of flood season, where water is separated into layers. Strong mixture type appears in spring tide in dry season, where difference in salinity of water is small. In most cases, the mixture of water is mainly of the gentle type where salty water expands wedgewise.

In the Changjiang Estuary, where the mixture of water is mainly of the gentle type, the distributions of current velocity, concentration of flood tide and ebb tide vary on account of longitudinal and vertical gradients of density. In the course of variation of near bottom current in flood and ebb tide from downstream flow to up stream flow, there are low water slack and high water slack of null net current velocity. The range in which it moves is the so-called slack water zone. According to the analysis of data taken from the South Channel, we have concluded that its upper limit reaches Xiaojiuduan and its lower limit is near the beacon light No. 54. The total length of slack water zone is more than 30 kilometers. The position of slack water zone in the Chingjiang Estuary is roughly that of the mouth bar. Near the slack water zone, fresh water mixes with salty water; there after, large amount of fine sediments carried by fresh water and large amount of electrolyte supplied by salty water flocculat near their boundaries, to form water layers with high concentration of sediments. Flocks fall in high speed to provide the available condition for the formation of fluid mud. Therefore, the mixture of fresh water with salty water and flocculation is one of the important reasons of the deposition of the mouth bar in the Changjiang Estuary.

(3) ***Medium condition***

The sedimentary condition of medium in the Changjiang Estuary has some influences on the transportation and deposition of fine sediment particles.

The pH value of near bottom water in dry season in the Changjiang Estuary varies from 7.43 to 8.48 and its mean value is 7.98; in flood season, the pH value of near bottom water there is in range 7.43–8.48, and its mean value is 7.71. In flood season, the pH value of river water is in range 6.98–7.46, and its mean value is 7.22. The aforementioned data show that the liquid phase and the solid phase of bottom materials are both weakly alkaline sedimentary environment. The pH value of medium has great influence on the development of clay minerals. When the pH value of medium is 7.0–8.5, illite is the most stable among these minerals. Therefore, besides the influence of sediment sources, the abundant content of illite in bottom materials in the Changjiang Estuary is associated with the particular pH value available for the existance of this mineral. We choose the ratio of Fe_2O_3 to FeO as the criterion of oxidative or reductive environments. If this ratio is greater than one, then the environment is an oxidative one, or else, it is a reductive environment. In the case of the Changjiang Estuary the ratio is greater than one, so the bottom there is oxidative. Besides, the hydration coefficient of H_2O/MgO is between 12.7 to 1.89 and the bottom in the Changjiang Estuary is only a weakly oxidative environment. In such an environment, although illite in the bottom materials of the Changjiang Estuary acts somewhat as action of dekaliation, it is still in the primary stage. So illite is relatively stable.

2. Suspended sediments in the Changjiang Estuary

The particle size data of suspended sediments in the Changjiang River and the estuary are fewer than the concentration data taken there. In the past, the granumeter method was usually used to analyze the particle size distribution of sediments. Since 1976 the Counter Model TAH has been used in the Hangzhon University, and it takes the advantage of higher accuracy: the results obtained by the former is generally coarser than other methods.

(1) ***Concentration of suspended sediments***

The concentration data of the middle and lower reaches of the Changjiang Estuary are shown in

Table 4-13 The monthly mean concentration in years of main survey station in middle and lower reaches of the Changjiang Estuary (kg/m^3)

Station	Years	Month												Whole year
		Jan.	Feb.	Mar.	Apr.	May.	Jun.	July.	Aug.	Sep.	Oct.	Dec.	Nor.	
Yichung	1950–1972	0.073	0.045	0.122	0.318	0.928	1.130	1.910	1.770	1.380	0.745	0.474	0.200	1.180
Hankou	1954–1972	0.200	0.168	0.241	0.412	0.459	0.552	0.830	0.865	0.775	0.629	0.470	0.316	0.613
Tatung	1954–1972	0.125	0.117	0.164	0.316	0.469	0.505	0.809	0.785	0.740	0.576	0.367	0.232	0.547

(After the Changjiang Valley Planning Office, 1980)

Estuary, where the concentration is higher than the upper reaches and the outside of estuary, as the tidal waves in the sea become standing ones when they are transmitted into the estuary. Because of distortion of tidal waves, there is a phase difference of one fourth of a period between tidal level and tidal current velocity, so that the maximum current velocity of flood tide and ebb tide appears near the middle tidal level. During high and low water slacks, the settling time of sediments in flood tide is longer than that in ebb tide. Besides, a net displacement of sediments takes place towards river source, when the current velocity of flood tide is greater than that of ebb tide and transported distance of sediment particles in flood tide is longer than in ebb tide; and still more sediments are transported towards river source in spring tide. Flocculation and delay effects of fine sediments intensify this process. This process repeats again and again; as a result, a high concentration zone is formed.

(2) ***Particle size of suspended sediments***

The results of statistical analysis of particle size data of suspended sediments down Hsiliuchien are shown in Table 4-15. In the table, shows that the mean value of median diametaers of suspended sediments is 0.0086 mm, and its minimum and maximum values are 0.0047 mm and 0.0172 mm respectively. The very fine silt grading with particle diameters a range of 0.008 mm –0.004 mm is the predominant grading of suspended sediments in the Changjiang Estuary. The mean content of this grading exceeds 45%. The sum of this grading content and the clay grading content is greater than

Table 4-15 The characteristic value of the suspended sediments in the Changjiang Estuary

Median diameter (mm)		Predominant grading range and its content (%)			Content of clay grading (%)	
Mean value	Extreme value	Grading range	Mean content	Extreme content	Mean value	Extreme value
0.0086	0.0047 – 0.0172	0.003 – 0.004	45.53	29.3 – 73.7	26.53	6.50 – 43.80

70%. Therefore, the suspended sediments in the Changjiang Estuary belongs to the type of fine sediments.

(3) ***The character of vertical distribution of suspended sediments***

The vertical distribution of particle size of suspended sediments is shown in Fig. 4-11. From the above table, we see that the veriation of predominant grading content and clay grading content of surface water layers in different characteristic tidal durations illustrates clearly the strong dynamical sorting effect of suspended sediments in this layer. Much more complex situations occur in the intermedium layer, but in near bottom layer, the clay grading content and major grading content are relatively high.

Table 4-13.

Table 4-14, shows that sediment concentration decreases downstreams. The concentration data taken monthly by each station in a year is quite different. For example, the ratios of the maimum value of monthly mean concentration to the minimum, measured by Yichung and Tatong stations in recent years, are 42 and 7 respectively. Obviously, the range of variation of concentration becomes smaller downstreams, owing to the regulating effect of lakes and the river itself. The concentrations of the Changjiang Estuary down Hsiliuchien are shown in Table 4-14. From this table, we see that the

Table 4-14 The concentration of spring tide in flood season in the Changjiang River (kg/m^3)

River portion	Mean value		Maximum value	
	Flood tide	Ebb tide	Flood tide	Ebb tide
South Branch	1.55	1.16	2.50	2.10
North Branch	0.94	1.05	1.95	3.49
South Passage	1.18	1.17	2.86	2.13
North Channel	1.44	1.22	3.06	2.72
South Channel	1.88	1.64	2.63	4.31

From "The circumstances of the Changjiang Estuary" Shanghai Channel Bureau.

spatial distribution of sediment concentrations of these channels has the character of increasing gradually from the North Passage to the South Passage and from the North Channel to the South Channel. As far as the concentration in the Changjiang Estuary during a period of flood tide and ebb tide is concerned, the maximum concentration during flood tide in the South Channel and the North Passage and the average concentration during flood tide in the North Passage are lower than that of ebb tide; in other place, they are higher during flood tide than during ebb tide. The probable reasons for these phenomena are that the time duration of flood tide and that of ebb tide are unsymetrical, i. e., the duration of high current velocity in flood tide is longer than that in ebb tide, causing sediments to deposit when water overflows shoals in high water; or that dilution occurs due to the downstream runoff of the Changjiang River; or that the sea water with larger density increases the ability of sediment transport and the influence of wave formed sediments in the mouth zone. All the above reasons cause the sediment concentration in flood tide to become higher than that in ebb tide. Besides, there is an untra-high concentration zone from Jiangying to the mouth in the Changjiang

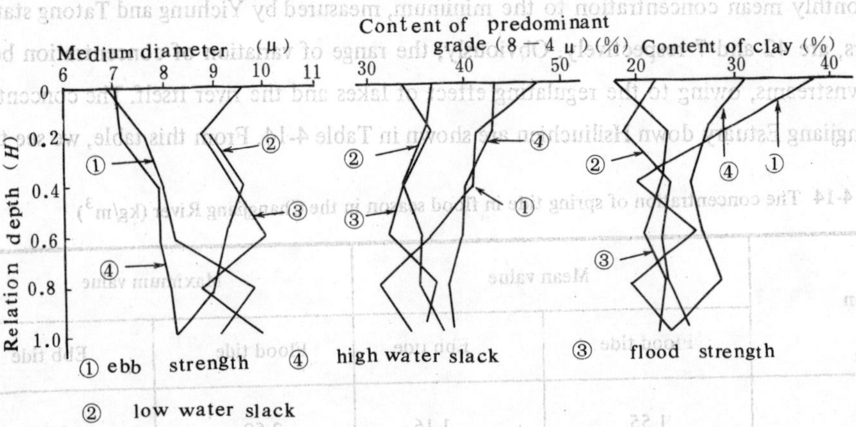

Fig. 4-11. Vertical characteristic of suspended sediment at the Baoshan Iron and Steel Works situated by the South Branch.

The graphs in this figure may be summarized into the low water slack-flood strength type or the high water slack-ebb strength type. This characteristics of vertical distribution of suspended sediments are directly related to both the sediments supplied by flood tide and ebb tide and the sorting action of water flow with different turbulent intensities, i.e. they are influenced by many factors.

(4) *Variation of particle size of suspended sediments with tides*

The tidal waves in the Changjiang Estuary become more and more unsymmetrical when they are transmitted upstream, so that the duration time of ebb tide is longer than that of flood tide, and that the current velocity of ebb tide is greater than that of flood tide. Therefore, there is a net displacement of suspended sediments towards the seas during a tidal period. Fig. 4-12 shows the varistion of particle size of suspended sediments with tide in different water layers at a vertical near the Baoshan Iron and Steel Works situated by the South Branch. Except the surface layer and the 0.2 *H* layer, two peak values of particle size of suspended sediments appear in high water slack and low water slack in the rest water layers. Usually, the particle size of suspended sediments in flood strength is coarser than that in ebb strength, the reverse is true in high water slack and low water slack. The calculated values of median diameters of suspended sediments of four stations in longitudinal synchronous survey testing using tracing sand in the South Channel are shown in Table 4-16. The ratio of values during low water slack to that during flood strength is approximately unity, illustrating that the suspended sediments in the South Channel is of the same character of low-water slack-flood strength type as in the South Branch. It seems that there is a close relation between the particle size of suspended sediments in a tidal period and the preceeding one.

But this ratio of values during ebb strength to that during flood strength is greater than one, and conversely the ratio of values during low water slack to that during ebb strength is less than one. The

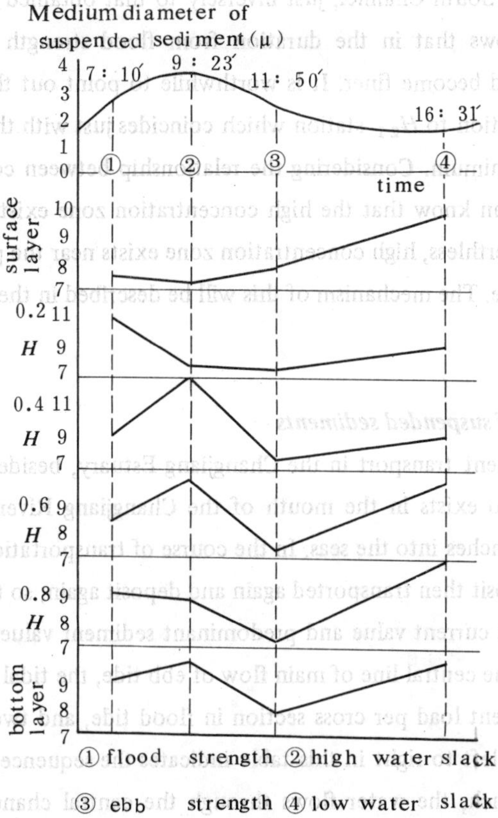

Fig. 4-12. Variation of median diameters (μm) with the tide of the suspended sediments near the Baoshan Iron and Steel Works situated by the South Branch.

Table 4-16 The ratio of median diameters of saspended sediments in the South Channel of the Changjiang Estuary

Ratio	Layer	Station position			
		Chiutuan-denchuan	H_{25}	H_{21}	H_{18}
Ebb strength / Flood strength	0.8H	1.0581	1.7715	1.3205	1.2903
Low water slack / Flood strength	0.8H	0.9628	1.1143	0.9615	1.0968
Ebb strength / Low water slack	0.8H	0.9121	0.6290	0.7281	0.8500

former shows that the particle size of suspended sediments in ebb strength is coarser than that in flood strength in the South Channel, just inversely to that obtained near the Baoshan Iron and Steel Works; the later shows that in the duration from flood strength to low water slack, suspended sediments deposit and become finer. It is worthwhile to point out that the value of this ratio in the section from H_{25} station to H_{21} station which coincides just with the slack water zone, is either the maximum or the minimum. Considering the relationship between concentration and salinity in the South Channel, we can know that the high concentration zone exists mostly near the position with salinity of 10‰, neverthless, high concentration zone exists near the position with a salinity of 20‰ in the time of ebb tide. The mechanism of this will be described in the section of flocculation.

(5) ***Transportation of suspended sediments***

About the sediment transport in the Changjiang Estuary, besides bed load and suspended sediments, fluid mud also exists in the mouth of the Changjiang River. The river down Hsiliuchien is divided into four branches into the seas. In the course of transportation, the suspended sediments are transported then deposit then transported again and deposit again, so the situation is rather complex.

The predominant current value and predominant sediment value in Table 4-17 are based on the calculation value of the central line of main flow of ebb tide, the tidal influx per cross section and the the amount of sediment load per cross section in flood tide, and two index equations in this table. The station No from left to right in this table indicates the sequence from upstream to downstream. From the South Branch, the water flows through the central channel into the North Passage. Although the predominant current and predominant sediments of ebb tide fluctuate along the stream: the general tendency is that they decrease along the stream up to the mouth, where the water and sediments of flood tide are dominant. So the water flow and sediments of ebb tide in the South Branch–the North Passage are both dominant. They also decrease along the stream from the South Passage to the upper mouths of the South Channel and the North Channel. In the South Channel and the North Channel, the situations are obviously different. The predominant current value of ebb tide at longitudinal No 15 station in the upper mouth of the North Channel reaches 0.7226. Up to the middle portion of the North Channel, it changes to flood tide current being dominant. In the South Channel, the water current and sediments of flood tide are both dominant. Only at longitudinal No 30 station in the mouth of the river the predominant sediments appear in ebb tide, which is mainly influenced by the water current of ebb tide overflowing the Jiuduansha Shoal.

Tidal creeks and cross creeks have been developing on river banks and on shoals in the Changjiang Estuary. They are the important channels of sediment exchanging between channels, which is shown in Table 3-8. The different reaches the cross creeks are on, different the dynamics forming the cross creeks is. Generally speaking, the direction of the cross creeks formed by ebb tide flow is mostly by South-West to North-East; therefore, sediments are transported along this direction. The cross creek of Nanmeng is an example; the direction of the cross creeks formed by flood tide flow is by South-East to North-West. The cross creeks of Hengsha and Tongsha are such creeks. At the same time,

Table 4-17 Predominant current and predominant sediments in the Changjiang Estuary

Reach	South Branch					South Passage		North Passage	
Station No.	5	6	9	7′	12′	Δ13	Δ14	14	Δ1
Predominant current	0.7767	0.7171	0.7003	0.6825	0.6911	0.6132	0.5516	0.8406	0.7628
Predominant sediments	0.7384	0.7913	–	–	0.4707	0.6370	0.4965	0.8259	0.7979
Reach	**North Passage**			**South Channel**			**North Channel**		
Station No.	Δ2	Δ5	Δ9	Δ24	Δ27	Δ30	Δ15	Δ18	Δ22
Predominant current	0.6168	0.6278	0.4960	0.4145	0.3522	0.4387	0.7226	0.3937	0.4448
Predominant sediment	0.7072	0.5939	0.4773	0.3030	0.2912	0.7860	0.6581	0.2256	0.4597

* Predominant current = $\frac{Qie}{Qie + Qif}$; Predominant sediments = $\frac{Sae}{Sae + Saf}$.

Qie – the tidal influx per cross section in ebb tide;
Qif – the tidal influx per cross section in flood tide;
Sae – the amount of sediment load per cross section in ebb tide;
Saf – the amount of sediment load per cross section in flood tide.

sediments are exchanged between channels by cross water current overflowing shoals; the input water and output sediments of the aforementioned off mouth station of the South Channel are originated from the sediments, which are carried by the ebb tide current of the North Channel flowing across the Jiuduansha shoal. The incoming water and incoming sediments of several hydrological station in upper portion of the South Channel are influenced by the flood tide in river bank of Nanhui. The biggest change in sediment transportation of the Changjiang Estuary is the appearances of reverse flow of sea water and sediments in the North Branch. When the very big flood occurred in 1954, in the Changjiang River, a large area of accumulation body of flood was formed in upper mouth of the North Branch, which led to the water discharge there being resisted. At the same time, the dynamical axis had moved from the north water way of Peimao to the south water way of Peimao. The aforementioned two changes influenced not only the distributing ratio of water and sediments of the South Branch and the North Banch, but also the sediment transportation of the Channels down the South Branch. According to the hydrological survey data of many years, twenty-five percent of the runoff could flow through the North Branch in 1951. But up to Aug. 1959 the runoff of the North Branch had been less than 1% of the total runoff of the river. The value was –7.5% in 1971, that is, reverse flow of sea water appeared. It is known from this that the flood tide current is strengthened by the influence of accumulation body of flood in its upper mouth. The "tidal bore phenominon" having been appeared since 1959, intensify the reverse movement of sediments in the North Branch. This is that the dynamics change and the accumulation body of flood in the Changjiang River causes the river bed changjiang greatly. The reverse moving sediments and the eroded sediments of the accumulation body in the North Branch cause the Piendan shoal depositing and expanding rapidly, and also influence the expanding of the Chongming Island and the Hengsha East Shoal. Therefore, the sediment transportation direction has been changed only for 20 years, but its influence on the developing of the river bed in the Changjiang Estuary can not be negaleted. As far as the South Branch and the three channels into the seas, which are on its lower reaches, are concerned, suspended sediments is mostly transported towards the seas, especially in the North Passage and the river portion of the South Passage upstream the Hengsha East shoal. The sediment transportations in the South Channel and the North Channel are more complex. Flood tidal flow is the predominant flow in the South Channel, and the water flow is also influenced by the Nanhuan River Bank. Suspended sediments move towards the down mouth of the South Passage. It is known from the above-mentioned analysis that the North Passage and the North Channel are the main channels discharge of water. In all, except the reverse movement of sediments and water flow in the north branch and a part of sediments being transported upstream in the South Channel, all the sediment transports in the Changjiang Estuary are towards the seas under the influences of Coriolis force and the nearshore current. The sediments entering the seas are transported and disperse to the Southeast.

3. The flocculation of sediments in the Changjiang Estuary

In the course of transportation and deposition of fine sediments, chamical changes also take place besides physical changes. Among them, flocculation and antiflocculation, which are the important differences between fine and coarse sediments in the process of deposition, are especially important.

(1) *Flocculation phenomina*

There are two kinds of flocculation in the Changjiang Estuary, which are of similar properties. One is the ion adsorption, which take place when the acid fresh water is mixed with the alkoline sea water. It is usually called coacervation. The other is more complex. In the medium of water, there are water moleculs of strong polarity. The particles of fine sediments with large surface area are covered with a water film consisting of double electrical layers, the adsorption layer and the diffusion layer. It is hydrophobe and sensibive to electrolye. So, when it meets sea water consisting of electrolyte of middle concentration, especially near the boudary layer of fresh water and salty water. The collisions produced by water current and Brownian motion, causing ions to exchange, leading to decrease of potential. When attraction force is greater than repulsion force, single sediment particles will begin to aggragate into colony flocculent coacervative bodies. This is termed flocculation. The flocculation of fine particles has both the physical action and the chemical action, but essentially is a electrochemical action.

Flocculation halo is one of the flocculaiton phenomina, which is observed in the Changjiang Estuary. The settling velocity of suspended sediments with a medium diameter of 0.0086 mm, is 0.0079 cm/s. It consists of piece like clay minerals, and has the character of suspending and floating. Fine sediments and its flocks are continously carried to water surface and appear as milk flocculation halo in strongly tubulent water. The rapid deposition of suspended sediments due to flocculation exists in the Changjiang Estuary. During the time duration near the high water and low water slacks, the lasting time with such a small settling velocity is not long. Therefore, the settling height of fine sediments is limited. But the flocks produced in flocculation will settle in a much greater velocity, which causes the vertical gradient of concentration to increase. At the same time, flocculation will cause the sediments in high concentration layer to settle more rapidly. Fluid mud layer will be formed near the slack water zone of salty water wedge.

(2) *Flocculation condition*

Many factors can influence flocculation. A process of flocculation and antiflocculation in a tidal period is now under discussion (see Fig. 4-13) In the beginning of a flood tide, all the water layers have fresh water with a salinity less than 3‰ . It can be seen that ebb tide current pushes the salty water wedge to lower reaches of No 24 station in the South Channel, although, the accelaration of flood tidal flow is greater. High concentration layers appears continousely in 0.8 H and 0.6 H of water depth in the second and third time point of flood tide. The highest concentration layer in flood tidal period with a concentration of 7.07 kg/m^3 exists in the near bottom layer in the third time point of flood tide. Its corresponding salinity is 10‰ . After the third time point of flood tide, current velocity and salinity will increase continously, but the concentration will decrease. It shows that after vertical gradient of salinity has increased, the vertical circle current will become stronger, which results in the dispersion of sediments towards upper water layers and intermediate water layers. From the fifth time point of flood tide to high water slack, current velocity will changes from 0.2 m/s to zero, at which time, the highest salinity in a flood tide period, which is higher than 15‰ will appear. The high concentration layer will descend to 0.8 H and disappear finally. During the period of ebb tide, the process is similar to that in flood tide. The difference is that the vartical gradient of salinity and current velocity is greater, especially for the water layers above 0.6 H, where, all the layers are filled with fresh water, and concentration is less than 1 kg/m^3 constantly. It is shown from aforementioned fact that the huge amount of fresh water flows downstream into the seas through the upper and intermediate water layers. In near bottom water layers, the salinity is high and stable. At the fourth time point of ebb tide, a high concentration layer with concentration of 2.5 kg/m^3 appears in near bottom water layer. After this time, its value will decrease untill low water slack, when the high concentration layer will disappear. The above analysis have shown that the mixture of saline water and fresh water in the Changjiang Estuary is certainly of gentle mixture type, which may change the vertical distribution of current velocity and concentration. The mixture layers are often at the position of 0.6 H and 0.8 H of water depth. Some layers may reach 0.4 H of water depth. High concentration layers behaviour the same correspondently. This agreement is just the reflection of flocculation of fine sediments in the boundry layer of fresh water and saline water. In addition, it addition, it has been seen from this that middle turbulent water is favourable for the flocculation of fine sediments. Too strong or too weak a turbulence is unfavourable for the flocculation of fine sediments.

The above-mentioned fact is the circumstance of a station in a tidal period. The syncronous data of four hydrosurvey stations in a range of 60 km along the stream in the characteristic tidal time are shown in Fig. 4-14. The highest concentration layer of flood tide exists at the top extreme of saline water wedge, where the salinity is about 10 ‰. The highest concentration layer of ebb tide exists at the down extreme of saline water wedge, where the salinity is about 20‰. We found, in flocculaition test, that illite and kaolinite flocculate at the salinity from 9‰ to 13‰, and montmorillonite flocculates at the salinity from 20‰ to 24‰. Based on both the field data and the laborlatary data, the following important results can be obtained. Illite and kaolinite flocculate in flood tide, which results in the formation of high concentration layer; the flocculation of montmorillonite in ebb tide causes the formation of high concnetration layer. Summarizing the aformentioned fact, we can get the conclusion that except the influence of the dynamics environments of the Changjiang Estuary, the flocculation and antiflocculation of the estuary are both related to the properties of fine sediments and the sedimentary environments. Therefore, the flocculation condition in Changjiang Estuary can be summarized as following: First of all, the Changjiang Estuary is aboundant in fine sediments, which is the materials supply of flocks; the second is that there is electrolyte of middle concentration; the third is that the turburence of water current is suitable. Too strong or too weak a turbulence is unfavourable for the formation of flocks. The time mostly suitable to flocculation in the Changjiang Estuary is the time of neap tide in flood season.

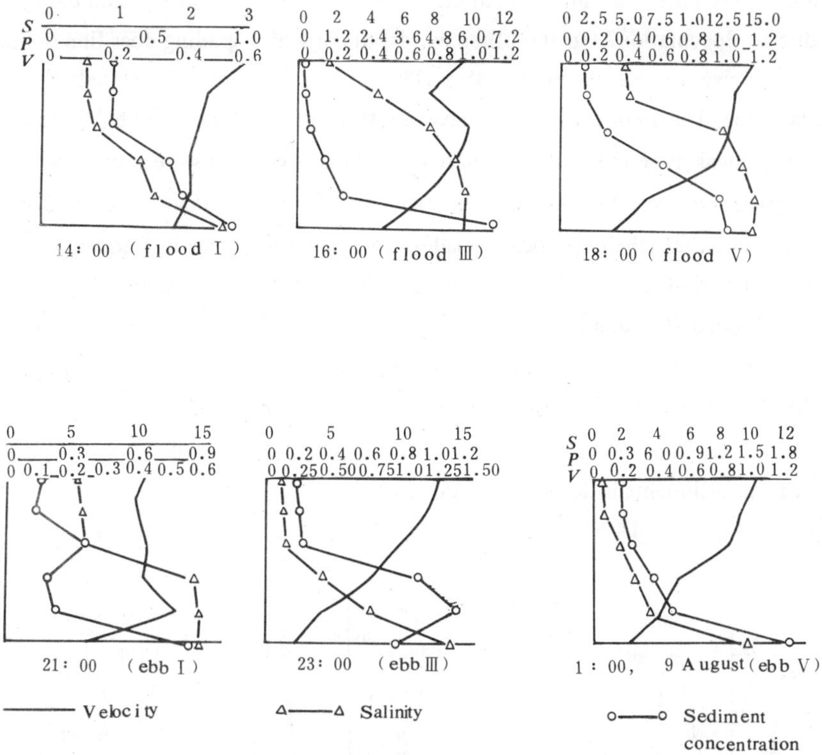

Fig. 4-13. Vertical profile of velocity, sediment concentration and salinity in main tidal time of the longitudinal No. 24 station in the South Channel.

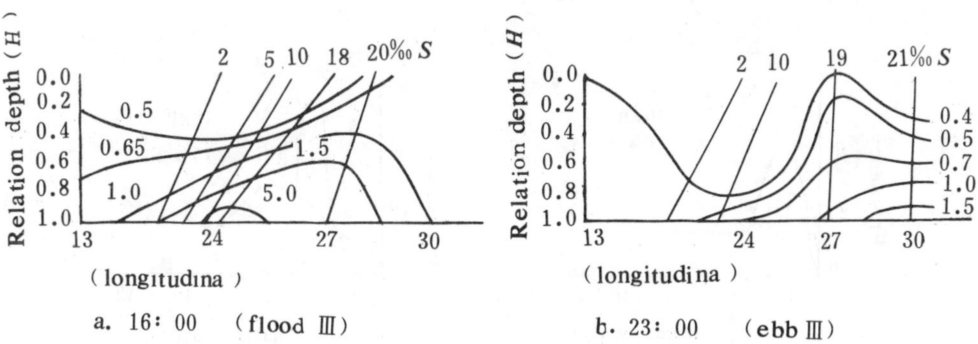

Fig. 4-14. Longtudinal isopleth of salinity and sediment concentration in the South Passage-South Channel.

(3) *The characteristics of flocculation*

The folcks formed in flocculation is a soft, translucent round grains with high water content. Its characteristics will be discussed as follows.

One is the critical diameter of flocculation of fine sediments will not be discussed here. The median diameter is usually used to indicate the sediment particle size. Thus, half the particles in weight is coarser than this, and another half of them is less than this. Sediments, which have different particle grading, may have the same median diameter. Does each grading of sediments particles forms the flocks due to flocculation? On the other words, under which grading does fine sediments flocculate? Fig. 4-15 is given to discuss tiis question. The result of the test of distilled water group is expressed in dash line. The frequency curves have the form of single peak type (a), double peak type (d,c) and multiplepeak type (b); but all the frequency curve of flocks are only of single peak types, and shift left wards. On the other words, all the particles will become coarser due to flocculation. It is worthwhill to notice that all the particles smaller than 8μm, disappear in frequency curve of flocculation, which shows that almost all the particles of the very fine silt grading and clay grading have become flocks through flocculating. So, the particle diameter of 8μm is called the optimum diameter of flocculation. From the frequency curves of flocculation, it has been indicated that most of the flocks has the diameter of 32 to 16μm. Only few of them is larger than 32μm. The test in May 1982 proves that all the particles larger than 32μm will not flocculate. Therefore, the critical diameter of flocculation of fine sediments is defined as 32μm. here.

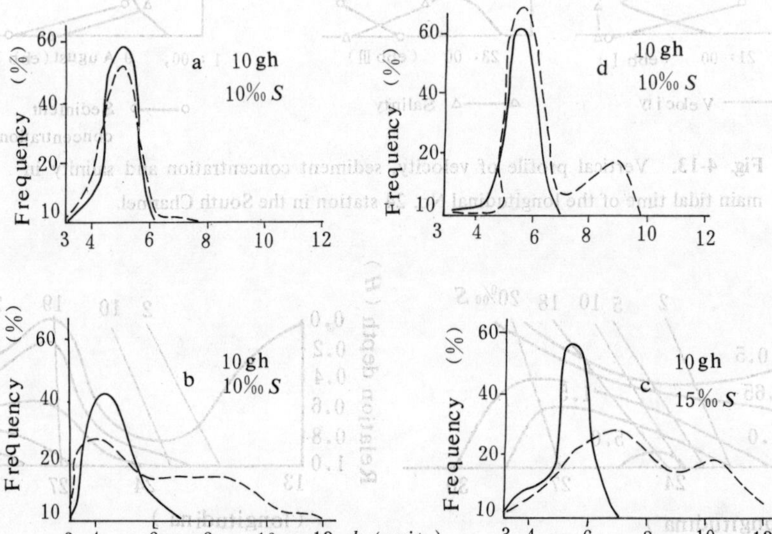

Fig. 4-15. Frequency curve of flocks.

The other is the critical salinity of flocculation of fine sediments. The critical salinity of flocculation of fine sediments discussed here is concerned to the quantity of flocculation. The following characteristics can be seen from Fig. 4-16. First is that the gradient of the cumulative curve of flocks is greater than that of the distilled water group. The curve becomes more gradient as salinity becomes higher. The most gradient curve appears in salinity of 10‰ , which means under this salinity the quantity of flocculation reaches the greatest. This value is defined as the optimum salinity of floccu-

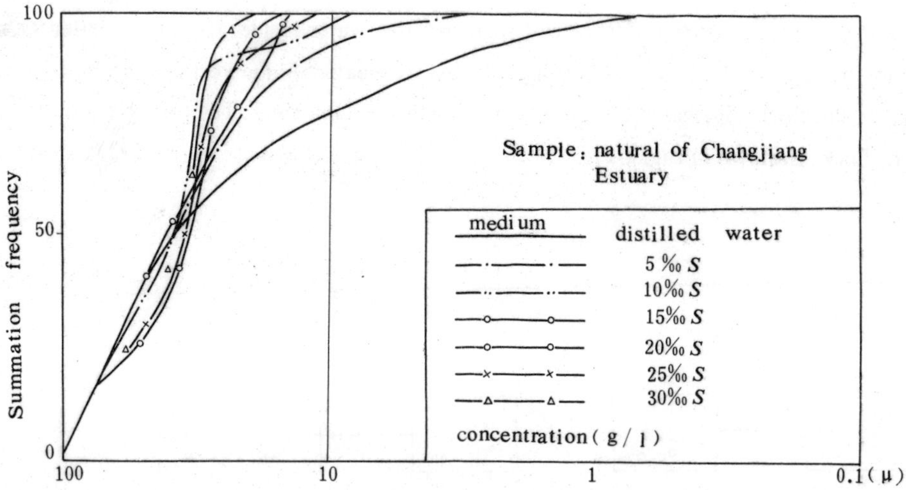

Fig. 4-16. Cumulative curve of flocculation.

lation of fine sediments in the Changjiang Estuary, which will change with the particle size, the clay mineral contents and test apparatus; the second is that the tests points in concentrated near the point of 32μm, and a turning point is formed in the curve. In the upper left of this point, the quantity of flocculation increases, and the particle size will become coarser; in the down right of this point, the particle size will become finer. The cumulative curve have the shape of "S". These phenomina can also be seen in other groups of tests.

The flocks may have different struetures. About this, few research works have been made before. The famous sedimentation scientist, Prof. Qian Ning had suggested to carry out the research on microstructure of flocks[1)]. We found, in the test observation, that there is few flocks with single round grains which usually aggragates into floc aggragates. As shown in Plate 4-1, they form the floc aggragates with different structures:

1) *Catenary structure.* At the beginning of flocculation, the fine flocks dispersed in water aggragates and collides each other continously to form the catenary flock structure, which can be further divided into single catenary structure and double catenary structure, as shown in Plates 4-2 and 4-3.

2) *X-type structure.* It is formed by crisscrossing of flocks with catenary structures, or by settling and aggragating of small flock aggragates as shown in Plate 4-4.

3) *Honeycombed structure.* This kind of flock structure is formed by the developing of above two kinds of structures or by the collision of medium and small flock aggragates. It can be seen from Plate 3-5 that free water exists in this kind of structure, which is the main structure of flocks. It has been observed that stagnant phenomina appear in the settling of flocks. It could be considered that

1) Qian Ning, 1981, Problems on movement of water current with high concentration of sands.

the honeycombed structure has the great influence on the decreasing of settling velocity of flocks. At the same time, the flocks of this kind of structure like sponge, which may act as a "buffer" to the sediment particles with a settling velocity greater than that of it, and pool these particles into the flocks.

4) *Flock aggragate structure.* Its dense and with few free water inside. Its volume is ten to one hundred times of granular floc. So, its settling velocity is greater. Some vortices can be observed.

The size of the flock aggragates can be seen in Plate 4-6 and Plate 4-7. The forming and developing process of flocks is shown in Fig. 4-17.

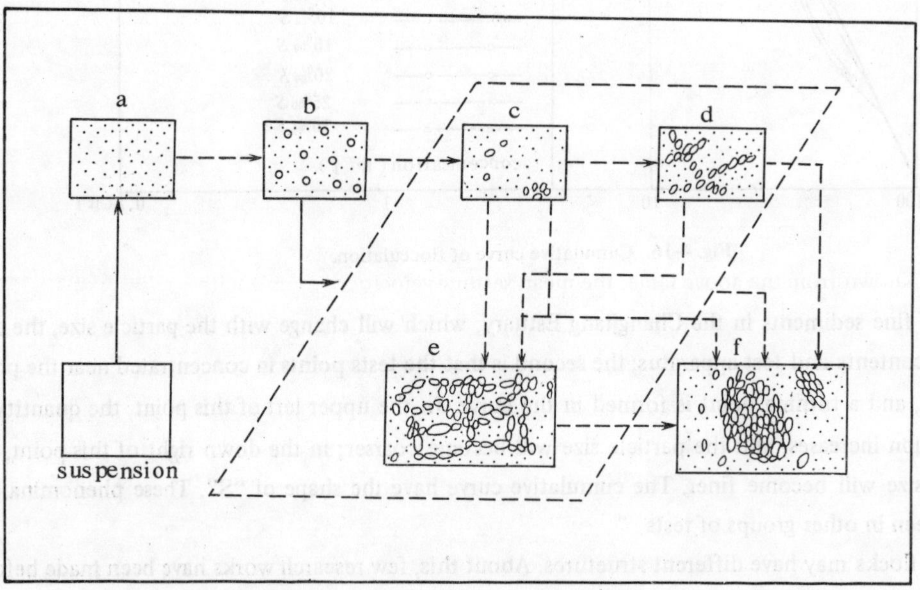

Fig. 4-17. Schematic diagram of flock formation and its developing process.
a–Flocs, b–Floc aggregate; c–Medium aggregate; d–Catenary and X-type structure;
e–Honeycomb structure; f–Large network.

(4) *The influences of flocculation on the sedimentation in the Changjiang Estuary*

Besides dynamical condition, flocculation is one of the important reasons of the rapid settling of sediments in the Changjiang Estuary. The flocks formed by flocculation, which have different size, and different structure, may have great influences on sediment deposition. It can be seen from Table 4-18 that the maximum diameter of flocks is about 7 times as large as the minimum one.

Using microphotograph, we can directly observe the flocks and measure the real size of maximum and minimum flocks most preferably. The results measured by all other methods are smaller than the real value. The size of predominant grading measured by each method agrees to each other, which indicates that the tendency of well-distributed of flocks existes. The particle size of flocks changes with the forming condition of flocculation and settling process. In the results of sedimentation balance measurements, a upper concave line section existing in the curve of the quantity of

Table 4-18 The size of flocks of fine sediments in the Changjiang Estuary*

Observating methods	Maximum diameter of flocks (mm)	Predominant grading of flocks (mm)
Pipette	0.1250	0.0320–0.0160
Sedimentation balance	0.1250	0.0320–0.0160
Microphotography	0.7100	0.0350–0.0070
Coulter Counter	0.1016	0.0250–0.0202

* All the data are the average value of direct measuring data obtained in each group of tests.

flocculation upon time can prove this. The settling velocity of flocks can be seen in Table 4-19. It has been known from the above table, the mean settling velocity of sample with a salinity of 13.26‰ is the greatest. The mean velocity of other samples is smallar than it, but is still greater than that of distilled water. Owing to flocculation, the average settling velocity of fine sediments will increase three to seven times. If we compare the average settling velocity of each grading smaller than optimum particle size of flocculation with the average settling velocity of flocks, then the later is ten to seventy-five times greater than the former. Therefore, for the rapid deposition of sediments in the Changjiang Estuary, although, the ability of water current carrying sediments being weakened is of course the important reason, flocculation causes the settling velocity of fine sediments to increase ten to dozens times which is also the important reason for it. Sediments in the Changjiang Estuary has the tendency of depositing in flood season and eroding in dry season. Neap tide in flood season is most favourable for the flocculation, which causes the depositing speed of sediments in the mouth region of the Changjiang Estuary to increase, especially at slack water zone of saline water wedge in the mouth bar.

The second influence of flocculation on sedimentation of the Changjiang Estuary is that the settled flocks has the stronger ability of resisting erosion. It has been well known that, exchanging ability of positive ions obeys following sequence; $Fe^{3+}>Al^{3+}>H^{+}>Ba^{2+}>Ca^{2+}>Mg^{2+}>K^{+}>Na^{+}>Li^{+}$. In the double electrical layers of sediments in the estuary, which is formed in electrolyte, the ions of Ca^{2+}, Mg^{2+}, which have the stronger ability of displacement, are easily adsorbed on the surface of sediment particles and replace the ions of K^{+}, Na^{+} having lower chemical valence from adsorbtion layer. This influences not only the flocculation quantity, but also the secure degree, then to increase the eroding resistance ability after deposition. From the definition of sodium adsorbed ratio, SAR= $Na^{+}/\sqrt{½(Ca^{2+}+Mg^{2+})}$, it is known that, in two media of sodium adsorbed ratios of 3 and 3.5, the

Table 4-19 Settling velocity of flocks

Grading (mm)	Frequency (fi) 0.125–0.063	0.063–0.032	0.032–0.016	0.016–0.008	0.008–0.004	0.004–0.002	0.002–0.001	Mean radius ($\bar{r}$) (mm)	Deviation of ri (Dri)	Mean settling velocity ($\bar{v}$) (cm/s)
Mean particle diameter of this grading (Di)	0.094	0.047	0.024	0.012	0.006	0.003	0.015			
Mean particle radius of this grading (ri)	0.047	0.024	0.012	0.006	0.003	0.0015	0.0008			
Salinity (‰)										
0	–	–	10.89	14.52	38.12	30.86	5.60	0.0038	1.0423968×10^{-5}	0.0108
3.12	0.93	3.47	38.66	41.44	8.33	7.64	–	0.0087	3.5499382×10^{-5}	0.0478
5.05	–	2.08	30.21	51.46	11.25	5.83	–	0.0076	1.7168781×10^{-5}	0.0320
8.18	–	3.65	54.95	33.59	8.33	–	–	0.0097	1.8813391×10^{-5}	0.0483
13.23	–	16.00	50.40	21.20	12.00	–	–	0.0116	4.0370144×10^{-5}	0.0749
18.87	–	9.23	29.76	36.90	17.26	5.95	–	0.0067	3.514478×10^{-5}	0.0482
26.46	–	10.42	35.16	33.85	7.81	12.50	–	0.0092	3.9566986×10^{-5}	0.0534

number of calcium ions and magnesium ions of the former is more than one hundred times higher than that of the later. But, the eroding rate of the later is greater than that of the former obviously. This tells us that the firmness of flocks formed in the sea water with more calcium ions and magnesium ions will increase obviously and its eroding resistance ability after deposition becomes stronger. In the over saturation displacing tests on flocks formed in the electrolyte of $CaCl_2$, it has been found that neither stirring nor dilluting of the solution can distory the flocks thoroughly after displacement. They can only cause the flocks to become smaller.

The influences of flocculation on the sedimentation of the Changjiang Estuary are also reflected in the corasening of the bed load in flood seasons. Seasonaly, the materials just deposited in flood seasons should be finer. But it has been found that in flood seasons the particles of bed load in the South Channel, where fluid mud exists, are coarser, which, besides the influences of ship navigation and channel dredging, is due to after a quantity of coarser silt materials catched in the process the sorting of particles by water flow of flocculation deposit on the river bed, before the fine particle materials will suspended again, so the coarser materials then remain on the river bed.

4. The fluid mud in the Changjiang Estuary

The existance of fluid mud is one of the important phenomina in modern sedimentary process in the Changjiang Estuary. It constitutes along with suspended sediments and bed load, the sedimentary system of the estuary.

(1) *The characteristics of fluid mud*

The fluid mud exists mostly in the South Channel of the Changjiang Estuary. It exists on the centre of the channel having greater water depth and is different not only to the bed load, but also to the upper layers water. It is a halfliquid lenticle having high concentration, great density, and consisting of fine particles. As the fluid mud has the property of plasticity, so it is termed as Bingham body.

The particle size composition of fluid mud can be seen from Table 4-20. From this table we see that the characters of the particle size composition of suspended sediments, fluid mud and bottom materials do not agree to each other. The mean median diameter of fluid mud is 0.0198 mm, which is between that of bottom materials and suspended sediments, but is obviously finer than that of bottom materials; the predominant grading of fluid mud equals to that of suspended sediments, but is smaller than that of bottom materials; the most obvious difference existing between them is the clay content, which is 11% higher than that of suspended sediments and 8% higher than that of bottom materials. The clay content of suspended sediments approximates that of bottom materials. The high content of clay in fluid mud is not only the important character of particle size composition of fluid mud, but also the reflection of flocculation of suspended sediments in fluid mud. From above analysis, we see that there is not only relationship but also obvious differences between suspended sediments, fluid mud and bottom materials, in the Changjiang Estuary.

Table 4-20 The chracteristic of suspended sediments, fluid mud and bed load of near bottom layer in the South Channel in the Changjiang Estuary

Sample type	Sampling date	Sample quantity	Median diameter (mm)		Priolity grading	Percent content of particles 0.004(mm)	
			Mean value	Changing range		Mean value	Changing range
Suspended sediments	Aug. 1978	16	0.0114	0.0051 | 0.0138	0.008 | 0.016	17.1	4.5–31.0
Fluid mud	Sep. 1978	7	0.0198	0.0037 | 0.0330	0.008 | 0.016	27.8	6.5–47.2
Bed load	Aug. 1978	10	0.0327	0.0034 | 0.0680	0.016 | 0.063	19.9	2.0–32.8

Being analyzed by electromicroscope, thermo-balance, thermo, differential X-ray diffraction and so on, the clay. mineral composition of fluid mud can be described as follow: 1) Analysis by electromicroscope, according to the results of observation and identification on photographs of TEM and SEM, the clay minerals of fluid mud in the Changjiang Estuary is mainly illite, followed by chrolite, montmorillonite, halloysite, vermiculite and so on. There are still quartz, calcite, non crystalline SiO_2, hematite and other nonclay materials, which are shown in Plates 4-8, and 4-11 respectively; 2) Thermo-spectrum analysis, in the dehydration curve obtained with Stanton thermo-balance, the first step of dehydration is the loss of adsorbed water under 110°C. The quantity of loss of weight lies between 0.9%– 4.9%. The second step is the dehydration between 350–650°C, where the dehydration action is mainly done by (OH). The quantity of lost water in this step reaches 4.6%–5.4% as shown in Fig. 4-18. Outside the second dehydration region, there is still some dehydration actions. No absolut boundary exists between dehydration region and non-dehydration region. Generally speaking, the dehydration curve of fluid mud in the Changjiang Estuary still keeps some basic characters of that of illite. At the same time, it is also influenced by the dehydration of other clay minerals. The results obtained by thermo differencial analysis indicates that the first heat absorbing valley exists mostly between 80– 110°C, which is due to the loss of adsorbed water; the second heat-absorbing valley exists between 450–650°C. The shape of this curve section is wider. It is related to the dehydration action of (OH); the heat-absorbing valley in 900°C is not clear; the heat transmitting

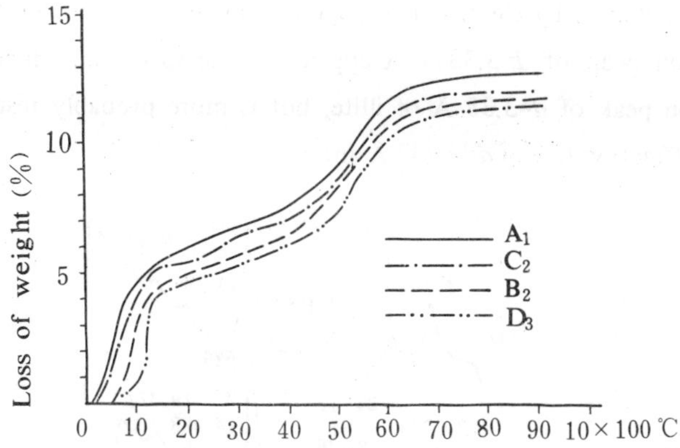

Fig. 4-18. Clay dehydration curve in the South Channel of the Changjiang Estuary.

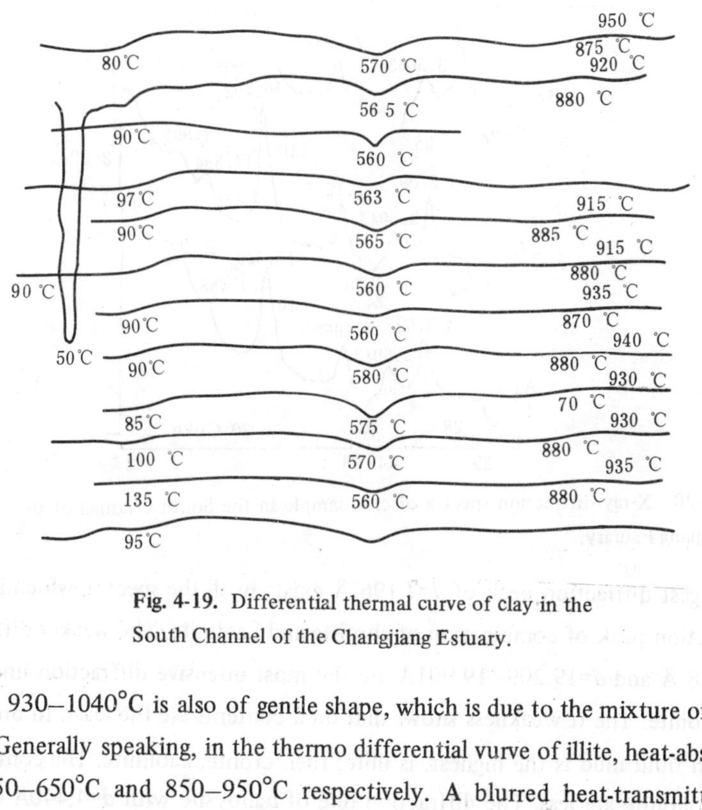

Fig. 4-19. Differential thermal curve of clay in the South Channel of the Changjiang Estuary.

peak between 930–1040°C is also of gentle shape, which is due to the mixture of kaolinite as shown in Fig. 4-19. Generally speaking, in the thermo differential vurve of illite, heat-absorbing valleys exist in 100°C, 550–650°C and 850–950°C respectively. A blurred heat-transmitting peak exists in 930–1000°C. As seen from the results of thermo-spectrum analysis, the peak and valley values of thermo differential curve are on the whole coincident with the steps of dehydration, especially in the second heat effect. Two kinds of curve can be verified and supplied for each other. 3) X-ray structure analysis. The diffraction spectra are shown in Fig. 4-20. In some diffraction spectrums shown in this figure, the wave length used is CuKa=1.5418 Å. A common and obvious characterof these spectra is that the highest diffraction peak of illite appears in each spectrum line. The charateristic value of d=10.329 Å to d=10.526 Å is due to the reflection of X-ray by the crystal surface of (001) of illite, which verifies the results obtained by electromicroscope photographs and thermo-spectrum analysis. Another strong diffraction peak of d=3.3371 Å appears in the spectrum, which is probably the second highest diffraction peak of d=3.35 Å of illite, but is more probably resulted in the combination of the highest diffraction line of d=3.343 Å and it.

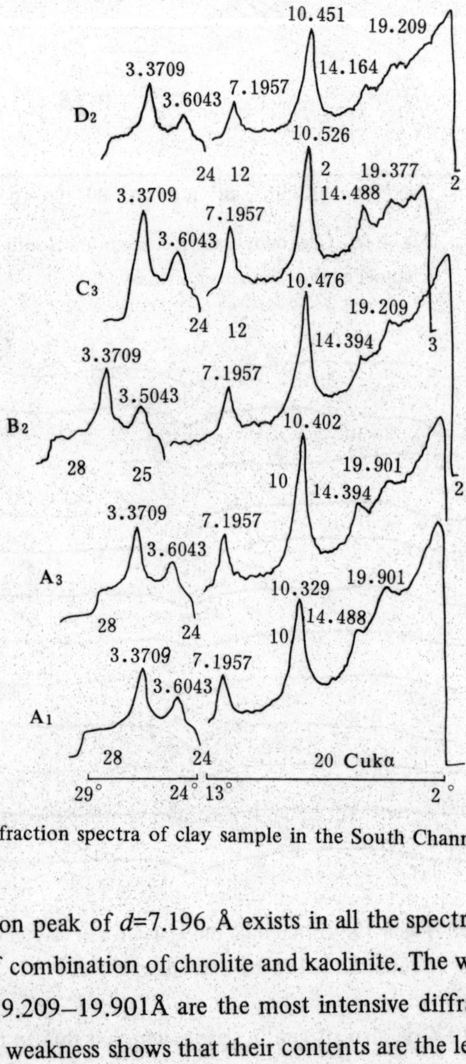

Fig. 4-20. X-ray diffraction spectra of clay sample in the South Channel of the Changjiang Estuary.

The third strongest diffraction peak of d=7.196 Å exists in all the spectra, which is the result of the strongest diffraction peak of combination of chrolite and kaolinite. The weaker diffraction peaks of d=14.164–14.488 Å and d=19.209–19.901Å are the most intensive diffraction lines of vermiculite and montmorillonite. The it weakness shows that their contents are the least. In brief, the mineral, whose content in fluid mud is the highest, is illite, then crolite, kaolinite. The content of vermiculite and montmorillonite are less. The diffraction line of halloysite with d=4.440Å exists in other X-ray diffraction spectra.

According to the data praided by Zhou Rinxi et al, the specific gravity of fluid mud in the surface layer is 1.04g/cm^3. In intermedium layer, it is 1.25g/cm^3. In bottom layer, it reaches 1.38 g/cm^3. There is a vertical gradient in specific gravity, which is the result of layer-built deposition of fluid mud.

(2) *The dimensions and structure of fluid mud*

After the dredging of the South Channel in 1975, serious deposition of channel was found in flood season next year. Almost the whole channel was filled with fluid mud. Based on observatious on fluid mud in the channel, its thickness is generally 0.2–0.6 m. The thickest layer of fluid mud reaches 1.2 m. According to the measurement and calculation done by Shanghai water Channel Bureau, the total amount of fluid mud in a larger dimension of fluid mud in the South Channel reaches 1500 million m^3.

A natural settling test of suspension made by fluid mud in static water had been done in our laboratory. The phenomina of layer-built depositing were observed in all the samples being tested. It can usually be divided into five layers. The miximum number of layer reaches seven. This phenomenon is similar to the inevently vertical distribution of specific gravity of fluid mud, which indicates that layer-built deposition of fluid mud exists both in static water and moving water, so that different structures exist in fluid mud. It is due to the different settling of different clay minerals composing the fluid mud. Table 4-21 is given to illustrate this problem. It has been seen from this table that a great difference exists in the particle size of several major kinds of clay minerals. According to the

Table 4-21 The comparision of properties of three main clay minerals

Clay minerals	The ratio of tetrahedron and octuhedron	Particle size	Displacing ability of positive ions	Substitute of same type	Molecular formula
Kaolinite	1:1	large	8	none or	$Al_4[Si_4O_{10}](OH)_3$
Montmorillonite	2:1	small	100	Aluminium in octahedron is displaced by magnesium	$Al_2[Si_4O_{10}](OH)_2nH_2O$
Illite	2:1	middle	80	Sillicon in tetrahedron is displaced by aluminium ;	$(K,H_2O)Al_2[(Al,Si)Si_3O_{10}](OH)_2$

investigation made by Xu Jiquan, the diameters of kaolinite particle are about 0.1–2 μm; most of the illite particles, have a diameter of 0.1–1μm; the particles smaller than 0.1μm are mainly montmorillonite. It is obvious that the differences in particle size of clay minerals certainly lead to the different settling velocities. The data obtained in settling velocity tests on different clay minerals have shown that when the salinity is 10% , the settling velocity of montmorllonite is 2.10–3 cm/s, which shows that the ratio of settling velocity between these two kinds of clay minerals is ten. It is known from this, the materials composing the fluid mud usually deposit with layer-built style according to the difference of particle size, specific gravity, shape and composition, revealing the reason of structure difference inside the fluid mud.

(3) *The behaviour of fluid mud*

Fluid mud exists in the Changjiang Estuary, but it does not exists in all the channels inputting the seas in the estuary. According to the investigation on the distribution of fluid mud in the Changjiang Estuary, the fluid mud layers have not been found in the North Passage, the South Passage and in the North Channel. Only in the South Channel, the developing of fluid mud has been found. So the behaviour of fluid mud in the Changjiang Estuary is mostly concerned with the South Channel.

It has been pointed out that the tidal flow of the South Channel is at the process of change from bidirectional current to rotation current.

The mixture of saline water with fresh water is mainly of the gentle mixture type. The position of slack water point and slack water zone are relative stable. The fine sediments are easy to deposit and can furtherly develop into fluid mud in the time from spring tide to neap tide. For the time of appearance of fluid mud, most is at the time from June to Augest, when is the flood season of the Changjiang River and the time of the Warm Current of Taiwan with high temperture and high salinity approatching the estuary. The former carries a large amount of fine sediments originating from river basins and to supply aboundant materials for forming of fluid mud. The relations between water dynamics, sediments and medium is certainly complicated. Nevertherless, the time, when the fluid mud moves most frequently, and its quantity is the greatest, is just this time period, and this is not occasional. The above three factors are the necessary conditions.

Fluid mud behaviours in a close relation to the process of forming and disappearing of fluid mud. Obviously seasonal change exists in the fluid mud behavious. There are much fluid mud in flood season during summer half year. It appears mostly in neap tide. In brief, the general tendency of fluid mud behaviours is that it appears in summer-half year frequently and disappears in winter-half year and is formed in the time of neap tide, disappears in the time of spring tide in flood season. Some times, it can exists untill the time of spring tide.

(4) ***The influences of the fluid mud on the sedimentation in the Changjiang Estuary***

The dimentions of the fluid mud in the Changjiang Estuary is smaller, which is smaller than that of Rohr Estuary in French. Although, the Changjiang River is aboundant in sediment discharge, and it is concentrated in flood season, when there are strong river flow and tidal flow. The duration of inititating velocity of fluid mud greater than 0.5 cm/s is longer than the time of fluid mud depositing, which restrains the fluid mud of the Changjiang Estuary to develop in large scale. The distribution of the fluid mud in the estuary is limited in the South Channel. So, the influence of fluid mud depositing in the South Channel is greater than that in other channels. It often enters the centre of the South Channel, in the time of neap tide in flood season. So the water depth of the South Channel in flood season is less than that in dry season by 0.6 m to 1.0 m.

Flood season is the time, when the fluid mud varies and moves frequently, It is usually formed in the changing time from spring tide to neap tide, and disappears in that from neap tide to spring tide. Therefore, the period of forming and disappearing of fluid mud is short, which causes frequen exchanges between suspended sediments, fluid mud and bottom materials. In each process of forming and disappearing of fluid mud, about one fifth of the fluid mud materials can not resuspend, These remaining coarser materials deposit then on the bottom of the channel, denser and denser, and become bottom materials, which lead to the bottom materials to become coarser in flood season. The fine sediments are raised again from fluid mud, and are transported by flood tidal flow, then deposit at the places of middle tidal level and high tidal level of shoals and river banks, or disperse outside the mouth and deposit there again.

Fluid mud is the "transfer station" of sediments downstream. The large quantity of fine sediments are transferred by it to deposit again in the aforementioned mophological positions. It is a half-liquid lenticle. Under the actions of certain bottom gradient and gravity, it moves under water towards the seas with a style of perticular colony movement, and accumulates at the front of the Delta, causing the delta to become gentler, and stretch towards the seas constantly. As the planning deep water channel is inplemented, the above-mentioned tendency will become more obvious.

5. The bottom materials of the Changjiang Estuary

The Changjiang Estuary is a accumulative zone from land sedimentation to ocean sedimentation. The sediments on its bottom are the products by the action of two material resouces, two dynamics and two media in a long period of time. This section mainly describs the particle size, mineral sompositions, chemical composition and sedimentary characteristic of the bottom materials in the reach downstream Nanjing, especially in the mouth region downstream Xiliujiang.

(1) *The particle size composition of bed load*

According to the calculation made by the changjiang Valley Planning Office, the annual amount of bed load transport is 4.35 million tons. The ratio of it and the suspended sediments discharge is 1:11. It has been seen from this that the suspended sediments and bed load transported by the Changjiang River have a great influence on estuarine sedimentation. In order to be easy to discuss the character of the bed load in the Changjiang River, the change of median diameter of the bed load particles of the middle and lower reaches of the Changjiang River is first illustrated briefly here, as shown in Table 4-23.

Table 4-23 The changes of median diameter of the bed load in middle and lower reaches of the Changjiang River

Stream reach	Yichang	Hankou	Tatung	Nanjing	Xi Liujiang
Median diameter (mm)	0.2030	0.1770	0.1660	0.1560	0.0933

(According to the data provided by the Changjiang Valley Planning Office and Hongzhou University)

Yichang is the place where the River enters middle reaches from upper reaches. The annual mean median diameter of years is 0.2030 mm, which belongs to the category of fine sand. Through intermittent transport of long distance to Xi Liujiang, the control section of mouth region in the Changjiang River, the average median diameter reduces to 0.0093 mm. The mean diameter of bed load of each branch in the Changjiang Extuary is shown in˙ following table, which indicates that the median diameters in the mouth region of the river are all smaller than 0.1 mm. Among them, the median diameter of the South Channel is the smallest, and its mean value is 0.0290 mm. So, it is very clear that the median diameter becomes finer downstream. The changing rule of bed load on cross direction in its mouth region can also be seen from Table 4-14. In a word, the particles of bed load in the North Branch are coarser than that in the South Branch. The particles of bed load in the North Passage, and the North channel are coarser than that in the South Channel.

Table 4-24 The average median diameter of each branch in mouth region of the Changjiang River

Branch	North Branch	South Branch	North Passage	North Channel	South Channel
Median diameter (mm)	0.0930	0.0840	0.0607	0.0599	0.0290

Six kinds of sediment, such as fine silt, can be found in the bed load of the Changjiang Estuary as shown in Table 4-25.

Table 4-25 The type of sediments and its percent content in the sediment of the Changjiang Estuary*

Type	Fine sand	Very fine sand	Coarse silt	Midium silt	Fine silt	Very fine silt
Percent content	12.50	32.14	33.94	14.29	1.79	5.36

* The sampes was collected in Aug. 1978.

It has been seen from this table that the content of middle fine sand and fine sand is 45%. The content of silt is about 55%. Among them, very fine sand and coarse silt are two main types of sediments in bed load of the Changjiang Estuary. The contents of which are 66% of the total sample. So the bed load of the Changjiang Estuary is midium by small which reflects the basic character of sediments available in the Changjiang River basin.

The particle size parameters of bed load are the important sedimentary information of the investigation on the developing process of estuary and estuarine mophology. Because of the limited space, the dynamics sorting of sedimentation can not be discussed in detail. So, the particle size and its statistical parameters of each kind of sediment is given in Table 4-26. From this table, it is not difficult to see the following characteristics. As the particle sizes are finer, the sorting of bed load in the Changjiang Estuary becomes poorer and poorer. For example, the average sorting cofficient of fine sand is 0.96, but that of very fine silt, of which the particle size is the finest in the Changjiang

Table 4-26 The mean particle size composition and mean particle size parameters of the sediments in the Changjiang Estuary

Item	Grading (mm)	Fine sand	Very fine sand	Coarse silt	Middle silt	Fine silt	Very fine silt
Particle size composition (%)	0.250 – 0.125	65.86	12.88	1.33	2.83	–	–
	0.125 – 0.063	23.07	57.09	30.52	16.62	8.00	12.24
	0.063 – 0.016	3.51	20.58	43.02	28.99	26.30	13.33
	0.016 – 0.004	5.18	5.14	12.98	30.88	31.20	30.77
	0.004 – 0.001	1.33	1.88	4.54	6.77	12.52	16.39
	0.001	1.04	2.48	7.64	13.91	21.98	27.27
Particle size parameters	$Md\phi$	2.64	3.60	4.43	5.42	6.63	7.61
	$M_Z\phi$	2.92	3.86	5.10	6.25	7.26	7.96
	$\sigma_1\phi$	0.96	1.11	2.03	2.69	3.13	3.37
	$SK_1\phi$	0.40	0.73	0.60	0.50	0.32	0.10
	$KG\phi$	1.51	2.14	1.42	0.93	0.77	0.79

Estuary, rising to 3.37. Besides the different dynamics of sorting, that the fine sediments have the property of flocculation is one of the important reasons causing the above sorting characteristic. Similarly, the particle size of bed load becomes finer towards the seas. The particle composition tends to well distributed gradually. The biased distribution of paeticle size of bottom materials tend to normally distributed. These all reflect the increasing of sea dynamics.

The period change of runoff and tidal flow is the dynamical factor of bed load of the Changjiang Estuary varying in the flood seasons and dry seasons. Taking the South Channel as an example, the section, which crosses the Nanhui river bank, the main channel of the South Channel, the down extrem of the sandy mouth of Shangsha and the flood tidal creek of Shangsha, in Fig. 4-21 is one of the fixed section of bed load in flood season and dry season, which has some representation to the analysis on the change of bed load in flood seasons and dry seasons. As far as the space distribution of sediments is concerned, only very fine sand exists at the Chiutuansha Shoal. In other mophological positions are all silt deposition. It is shown in this figure that the particles of bed load at station B_2 becomes coarser in flood season and vice versa. At both sides of station B_2. The median diameters of bed load particles in flood seasons are finer than that in dry seasons. The predominant grading of

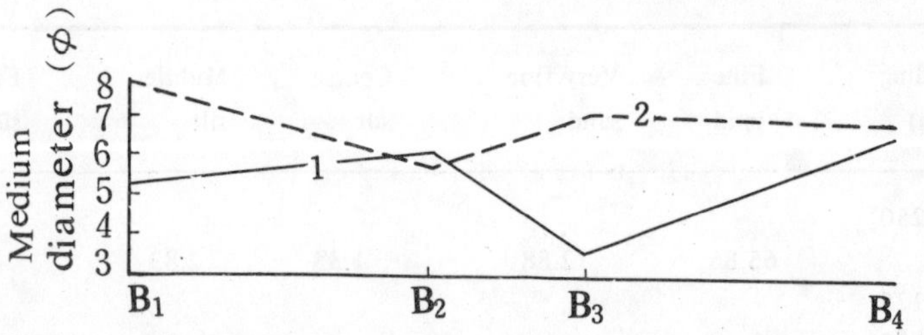

Fig. 4-21. Transeverse change of median size of the bottom sediments in the South Channel Section of the Changjiang Estuary.

bed load in dry seasons is about 0.063–0.016 mm, Its content reaches 48.1 2% ; nevertheless, in flood seasons the predominant grading of bed load is of clay grading, of which the particle size is smaller than 0.004 mm, its content is as high as 49.61%. The sorting cofficient of the former is 2.20, and that of the latter is 1.90, which indicates the sorting degree in dry seasons is better than that in flood seasons. This kind of sedimentary characteristics caused by depositing in flood seasons and eroding in dry seasons also appear at B_3 station. According to the investigation made by Shanghai Water Channel Bureau, it has been found out that the sand mouth stretches downstream by 600 m in flood seasons, and contracts upperstream in dry seasons. From the results of particle size analysis we see that the bed load of A_3 station has the characteristic of being coarser in flood season and finer in dry season, and that of B_3 station has the opposite characteristics. In addition, at the sand mouth of Shangsha, the bed load of the bead of the sand mouth becomes coarser and that of the tail becomes finer in flood seasons and vice versa in dry seasons. The aforementioned fule can be easily seen from Table 4-27. It can also be verified by the particle size parameters. If the particles of bed load are

Table 4-27 The change of sand particle and clay particles at sandy mouth of Shangsha in flood and dry seasons (%)

Station No	A_3		B_3	
Season	Dry season	Flood season	Dry season	Flood season
Sand particles	30.59	70.38	74.85	19.62
Clay particles	25.95	7.56	2.14	40.15

coarser, their sorting begree is better than that of the finer bed load. The deviation distribution of the bed load at A_3 station deviates positively in flood season, and the bed load of B_3 station is negative deviation distribution in dry seasons, which shows obviously that the eroding action of the flood run in the dry seasons leads to the bed load at the tail zone of the sandy mouth having the property of sea sediments. Seeing from above mentioned sedimentary characteristics, the same phenomina of bed load becoming coarser exists in both stations. Nevertherless, as the mophological positions and the time of bed load becoming coarser of these two stations are different, the dynamic conditions causing coarser bed load and the sediments supply are somewhat different. The bed load of A_3 station is eroded by ebb tidal run in flood season and supplied by the incoming sediments of upper reaches, but the bed load at B_3 station is eroded by flood tidal ran in dry season and supplied by the bed load of shoals. The problem of bed load of the main channels in the South Channel being coarser in flood seasons has been discussed before, so is not mentioned here.

(2) *The mineral composition of bed load*

In the Changjiang Estuary, except the materials of sandy island with the shape of wild goose line and shoals having been formed since the Holocene, which consists of mainly the fine sand, all the bed load in other places is finer. Therefore, besides the identification of distrital minerals, the investigation of clay minerals is of great importance in the research on the mineral composition of bed load, which will be discussed as follow.

The results of identification of the light minerals in the Changjiang Estuary has shown that they are mainly composed of feldspar and quartz, and still of some carbonate minerals such as calcite. The ratio of feldspar content and quartz content is 3:7. Although, there are not many kinds of light minerals in the bed load, its weight percent content is usually higher than 90%. In some samples, it is as high as 99%. It can be known from this that light minerals is predominant in the ditrital minerals of bed load in the Changjiang Estuary.

We have not paid great attention to this kind of minerals before. Although feldspar kaolinization is familiar for all of us, only a few research work has been done on it. According to the identification and analysis done by Cheng Lirong et al and us on the heavy minerals of bed load in the Changjiang Estuary, more than 20 kinds of heavy minerals exists in bed load. Their average percent contents are amphibolite, 40%, epidote, 20%, mica, 15%, ilmemite, 8%, magnetite, 3%, some stable minerals such as garnet, zircon, tourmaline, apatite account for in addition to ferrous minerals. It can be known from aforementioned heavy mineral composition of the bed load in the Changjiang Estuary, the combination type of heavy minerals are amphibolite family-epidote falily-mica family.

The clay minerals of bed load in the reach down Chienchou of Nanjing in the Changjiang Estuary have been analyzed with the same instruments are above-mentioned ones. A few spectrum lines of X-ray structure analysis on them are shown in Fig. 4-22. The feature peak value of the clay minerals of the bed load is similar to that of fluid mud, and will not be described here.

The results of observation with electromicroscope are added as follow, by observing the electromicroscope photographs of illite. We found that most of its particles are between 0.1 μm to 2 μm, with a smooth surface and a complete crystal structure, They are shaped like scale and bar (see Plates 4-12 and 4-13), and aggragate together. These characteristics can be distinguished from the factor of hydrothermal alteration. In the observation with electromicroscope, the particles of chrolite are thicker, so they are poor transparent. Although they have the well-developed crystal surface and angle. It can be distingnished from illite by its shape character of straight arris and sharp crystal angle and poor transparent (see Plates 4-13 and 4-14). It is worthwhile to point out that the microphorous

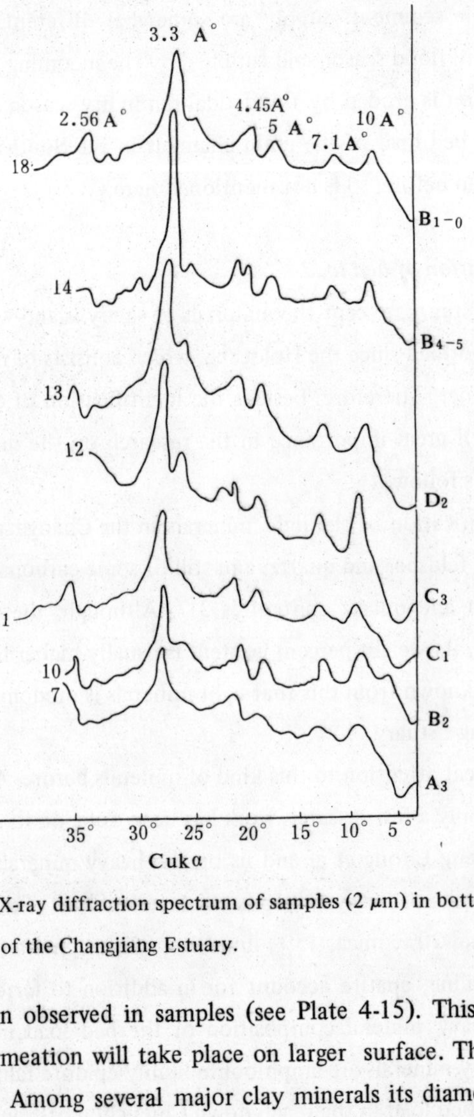

Fig. 4-22. X-ray diffraction spectrum of samples (2 μm) in bottom sediments of the Changjiang Estuary.

structure of chrolite has been observed in samples (see Plate 4-15). This microphorous structure shows that the adsorbion permeation will take place on larger surface. The crystal outline or kaolinite is shown in Plate 4-16. Among several major clay minerals its diameter is the biggest, and is about 4 μm to 1μm. The clear arrises of aforementioned clay minerals indicates their short sedimentation time, and the short transport distance, which are just the basic characteristics of incoming sediments of river basin. The chemical composition of hallosite and kaolinite is the same. But in electromicroscope hallosite has the tube-like crystal outline (see Plate 4-17). Vermiculite is developed by biotite. Its piece-like structure exponds when it is heated, it has crackle, which is its obvious character (see Plate 4-18). The particle size of montmorillonite is fine, and most of its diameter is less than 0.1 mm. It has a large surface area, and the electrical charge carried by it is the most in some major clay minerals. In the observation by microscope. Its appearence of aggragation structure can prove this (see Plate 4-19). In order to compare the difference of clay minerals of bed load in flood and dry seasons, according to the mean content of K_2O, which is 5.5%, in illite, a testing calculation was done. The results of the calculation are shown in Table 4-28. Such a tendency can be

Table 4-28 The illite content (%) in bed load in flood and dry seasons in the Changjiang Estuary

Station	Deep A_1	B_4	C_2	D_2	Survey period
Flood season	62.27	66.36	63.27	58.91	Sep. 1976
Dry season	71.89	70.19	67.82	72.36	Apr. 1976

seen from this table that after a hydrographical period of flood and dry seasons, the illite content increase in range of 2.55% to 13.55%, according to the different distance to the mouth and the different particle size of the bed load. Usually, the nearer to the mouth and the finer the particle size, the clearer this tendency.

Summarizing above-mentioned fact, the contents of major clay minerals in bed load in the Changjiang Estuary are as follows: illite (65.4–78.6%), chrolite and kaolinite (18.7–20.8%), montmorillonite (2.0–4.2%).

(3) ***The chemical composition of bed load***

The chemical composition of 56 samples is shown in Table 4-29. Analyzing on the chemical composition of every kind of sediment in the Changjiang Estuary, we found, except the decreasing contents of some compounds, the content of most chemical compounds increase as the particle size becomes finer, which is closely related to the particle size of the bed load there being middle by fine and the hydro-chemical medium in the estuary. It is well known that the fine sediments not only have a large surface area, but also is transported and settles in the process of continouse mixture of fresh water with saline water. Flocculation therefore intensifiesion adsorption and exchange, which leads to the contents of Al_2O_3, MgO, K_2O, adsorbed water, and the loss of weight by burnning obviously increase. It has been from this table that in the estuarine enviroments, flocculation of clay takes a great rule in the transportation and conceneration of some compounds. The only exception is the content of SiO_2, which in fine sand is 68.09% and in very fine silt is 54.82%. This is a normal changing tendency, because the finer particle size means the increase in clay content and the decrease in sand content (mainly composed of quartz). A peak value of CaO content appears in very fine sand and coarse silt, which is relative to the existance of some ocean biological bodies, such as foraminifera and crushed shell pieces. In addition, it is seen from this table that the content of Al_2O_3 is higher than the total contents of CaO, Na_2O and K_2O and K_2O, Al_2O_3 is therefore oversaturated. This phenominon and the phenominon of the intergrowth of non-crystal SiO_2 and Fe_2O_3 observed nuder electromicroscope both indicate the hydration actions produced on detrital minerals.

6. The mouth bar sedimentation in the Changjiang Estuary

Table 4-29 The chemical composition of each kind of sediments in the Changjiang Estuary

Component	Fine sand	Very fine sand	Coarse silt	Midium silt	Fine silt	Very fine silt
SiO_2	68.09	64.21	62.90	59.83	56.55	54.82
Fe_2O_3	2.92	2.46	3.15	3.69	4.33	4.27
FeO	1.73	1.52	1.67	1.68	1.97	2.20
Al_2O_3	10.97	10.44	12.16	13.62	15.37	15.98
TiO_2	0.89	0.73	0.86	0.92	0.84	0.87
CaO	3.51	4.62	4.28	3.91	3.17	2.93
MgO	2.36	2.26	2.48	2.60	2.99	3.08
K_2O	2.09	2.05	2.25	2.43	2.80	2.89
Na_2O	1.74	1.77	1.64	1.45	1.24	1.37
MnO	0.08	0.06	0.08	0.08	0.08	0.11
H_2O^+	1.55	1.68	2.31	3.00	3.44	3.72
H_2O^-	1.01	0.97	1.57	2.16	3.29	3.22
Burning	4.12	5.53	6.38	6.89	6.74	7.23

The lower reaches of the Changjiang River is a river portion with a lot of sandy islands. The river forms a estuary with aforementioned three stage of branching and four flow divisions. From the deep channels in both sides of the Hengsha Island to the mouth. Inside the isobath of –10 m, there is a accumulating body, usually called mouth bar, which causes this reach becoming the nevigation obstruction portion in the channel into the seas of the Shanghai Harbor.

Mouth bar is the important morphological topography of the river bed in the Changjiang River. Its north limit is the Qidong Shoal in Jiangsu Province, and the Nanhui Shoal is its south limit. The dimensions of above two shoals are rather big. Their composing materials are both finer, that are the sediments from very fine sand to clay; the main composing sediments are midium silt, so, the typical sandy sedimentation of modern estuary has not been formed there. The cross link line of the tops of mouth bar is in the direction of NNE – SSW. The East Chongming Shoal, the Tongsha Shoal and the Jiuduansha Shoal are three topographys in the shape of a saddle. The width of the mouth bar is about 90 km. Its length is 40 – 60 km, and is higher than the isobath of –10 m by 4 – 8 m.

Such a huge mouth bar is formed in a particular estuarine topography and water-sediment condition. After the bed load is transported to the estuarine zone, due to the stream dividing, it forms central bars in different dimensions, such as the Peimaosha Bar and the Liuhosha Bar. Sometimes, water overflows these bars in flood seasons. They also move downstream to form other central bars, and transported in the form of sandy wave. In the reaches where the current velocity of flood strength is greater than that of ebb, spindle-like central bar will develope into hoof-like central bar. The steep slope of this kind of central bar is towards upper reaches and the gentle slope of it towards the seas. This is a special form of central bar developed by bidirectional water flow in a tidal estuary. When numerous central bars move to the mouth bar position, the stream disperses more rapidly and has more divisions; central bars then stablized relatively. After then, the central bars in upper river portion are transported to this position, which causes the central bars to pile up and combine into central bars with larger dimensions, such as the Jiuduansha Central Bar. The above-mentioned fine sediments will flocculate, when fresh water mixes with saline water in the mouth zone, The flocks so produced settle in a speed of ten to dozens times greater than normal ones, which causes the deposition of channels and leads to shallower water and the deposition and expanding of central bars. When the banks of both sides of the mouth bar develop and stretch furtherly, that is, when the river stretches towards the seas naturally and takes the function of protective screen and shield. These central bars often accumulate vertically, to develope into sandy island, and are connected to the north bank finally.

The position of the formation of the mouth bar varies with the different ratio of runoff and tidal run, which is greater than 0.1 in the Changjiang Estuary. So a huge mouth bar accumulating body has formed near the mouth.

According to some drill information in some islands such as the Hengsha Island, bay sedimentation exists under the layer of 28–42 m deep. It is characterized with the greygreen and grey-blue sediments. Most of it is clay and fine silt, which constitutes the horizontal sedimentary structure. Abvoe this depth, there is modern sedimentation formed by river phase – sea phase, which forms the lenticle composing of layers of very fine sand to very fine silt respectively, in which there are horizontal, oblique, undate and other sedimentary structures, which shows the more complex dynamical conditions in its developing.

As the Changjiang Estuary is aboundant in the incoming water and incoming sediments, since the Holocene the wild goose-like sandy islands arranged to the south by east have been formed; but, there was not embryonic form of the Jiuduansha Shoal untill the end of last century, which has been now developed into a large shoal with a total length of 50 km and a width of 13 km.

7. The sedimentary system in the Changjiang Estuary

The Changjiang Estuary is the transition accumulating zone from land to marine sedimentation. In the time of the largest marine transgression in the Holocene the estuary was located near Panyang. It is now situated to the east of Shanghai. It has been stretching towards the seas with a speed of 20 m/y and forming the morden Delta of the Changjiang River.

As a transition sedimentary zone, the Changjiang Estuary is affected and restricted by two environment factors of land and sea. The huge amount of sediments produced in the river basins is transported towards the seas by runoff in the form of suspended sediments and bed load. The latter forms the sandy island and shoal sedimentation between the reaches near and in the mouth. The shoals arranged in wild goose-line and the Dongsha Shoal and the Jiuduansha Shoal are these sedimentations; most of the former flows into the seas, except the downstream deposition. The land source materials having been transported into the seas, combined with the sediments transported by the Northern Jiangsu Coastal Current and the Warm Current of Taiwan are transported by tidal flow into the estuary again and deposit there. So, we usually say that runoff transports sand and tidal flow transports mud. The mechanism of formation of high concentration zone in the Changjiang Estuary is due to the tidal wave there having the property of changing from shallow water tidal wave to standing wave. which causes suspended sediments to produce a net displacement towards the seas. Much sediments are concentrated upstream in spring tide, which along with the flocculation in the mixing of fresh water with saline water and the delay effect of fine sediments, causes the formation of the high concentration zone. Its position moves up and down, according to the different interaction between runoff and tidal flow. A vertical concentration gradient exists in the high concentration zone. Affected by gravity and flocculation, a high concentration layer is formed near the river bottom, which can furtherly develop into fluid mud in neap tide of flood seasons. This is the particular sediment transport form in the Changjiang Estuary. Under the action of gravity, the fluid mud moves along the bottom slope in the colony form and is accumulated at the slope foot of the front of the delta. In the course of disappearing of fluid mud, one fifth of it, the coarser particle materials, remains on the river bed. Besides depositing at the position between high tidal level and middle tidal level of bank and river beach, the other fine sediments disperse towards the seas again. Most of the fine particle sedimentation in the Changjiang Estuary have the characteriltics of flocculation sedimentation. Summerizing the above all, there are two material sources, two dynamics and two media, in the estuarine region of the Changjiang River, which is also influenced by the coastal currents, including the Northern Jiangsu Coastal Current and the Warm Current of Taiwan, and wave, is a complex, changable modern sedimentary environment. In this enviroment, suspended sediments, fluid mud and bed load constitute the modern sedimentary system of the Changjiang Estuary.

(by Zhong Zhizhong)

IV. Modern Sedimentology of the Qiantang Estuary

1. General description of the sedimentary environment

The Qiantang Estuary world-famous for its tidal bore is a typical macrotidal one in China. The total tidal reach of the River is more than 270 km, from the tidal limit-Lucibu down to the linking line between Nanhui and Zhenhai. The Qiantang Estuary is divided traditionally into three reaches, the upper reach is 83 km long, from Lucibu to Wenjiayan, The north bank of the reach consists of Devonian sand rook and the south bank, small moutains consisting of Jurassic and Cretaceous rock systems in various sizes. The bed of this reach is stable. Most of the riverbasinborn materials deposit here, the shoals here consisting of coarse-medium sands present a jagged, interlocking pattern, and there are some islands of varying sizes. This reach is basically not influnced by the sediments derived from sea, and can be classified into an island reach. The middle reach of the Qiantang Estuary is 101 km long, from Wenjiayan to Ganpu, both of its banks are flood-marine plain developed since the Holocene, the terrain of he plain is flat. The reach gradually widens from 1 km at Hangzhou to 20 km at Ganpu. A distinct and a very large sand bar exists in this reach, which is about 130 km long, 10 m high above horizontal linking line between Donglu and Zhapu. It is estimated that the sand bar is of 42.5 billion m^3 in volume. Such a big sand bar consisting mainly of coarse silt is seldom seen in Chinese river sedimentation (see Fig. 4-23). Because the reach is wide and shallow and the interaction

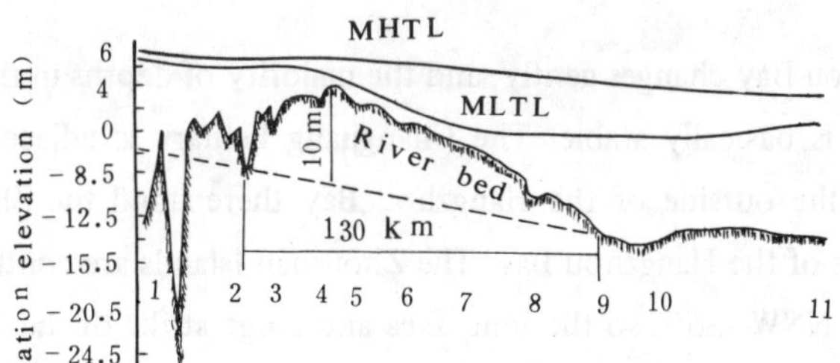

Fig. 4-23. Longitudinal profile of big sand bar of the Qiantang Estuary.

1. Tangjiabu; 2. Wengjiayan; 3. Hangzhou; 4. Qibao; 5. Cangqian; 6. Haining; 7. Jianshan; 8.Zhapu; 9. Ganpu; 10. Jiashanwel; 11. Dayangshan.

between runoff and tidal flow, especially the turbulent tidal bore, in this reach is strong, the trough of this reach waves frequently in cross direction and depth changes strongly in longitudinal direction. The amplitude of depth changing reaches 3–6 m, sometimes even more than 10 m. When the main stream approached that river bank, the lower one of the bank is often drawn up, and the upper one collaps down. The maximum of land retreating in a tidal cycle would reach 245 m. Therefore, the river bank moved and removed repeatedly, the river trough moved here and there, forming a wandering river way: a seldome scene in Chinese estuaries. From Ganpu downward, the estuary comes into its lower reach, namely the Hangzhou Bay, which is about 100 km long and 100 km wide. The widening rate of the Qiantang Estuary reaches 54, if it is calculated from Hangzhou to the linking line between Nanhui and Zhenhai, forming the most typical trumpet estuary in China (see Fig. 4-24). The

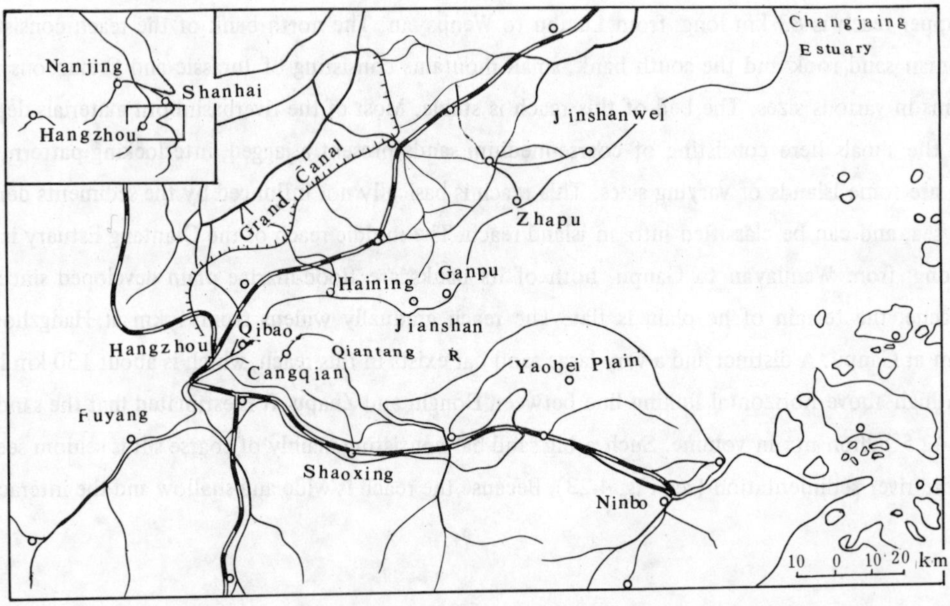

Fig. 4-24. Location of the Qiantang Estuary.

bed form of the Hangzhou Bay changes gently, and the majority of depths in this reach are between 5 m to 10 m. The reach is basically stable. The Changjiang Estuary is adjacent ot the north of the Hangzhou Bay, and in the outside of the Hangzhou Bay there stand the Zhoushan Islands, which form the natural defense of the Hangzhou Bay. The Zhoushan Islands are controlled by two structure lines of NNE–SSW and NNW–SSE, so the long axes and gorge strike of the Islands present NW–SE direction, in consistent with the direction of the flood and ebb tides in he Hangzhou Bay. Moreover, the Zhoushan sea region has the most strong tidal flow in China. Thus, it can be imagined that the discharge (including fluid and solid) of the Shangjiang Estuary and the strong tidal flow of Zhoushan sea region must have profound influence on the morden sedimentation of the Qiantang Estuary.

The Qiantang River is abundant in the water derived from its basin and definient in the sediments derived from its basin. The Qiantang riverbasin is not large, about 490,900 km^2, while its average annual runoff reaches 2.9×10^{10} m^3, and average flow is 921 m^3/sec. The maximum one is 29,000 m^3/sec in flood season, and the minimum one only 15.4 m^3/sec in dry season, with a ratio of 1900. So we say Qiantang Estuary is pronouncedly characterized by mountain-river. The most strinking characteristics in the process of water rinning into the Qiantang Estuary is extreamly abundant in flood tidal flow, reaching about 190,000 m^3/sec. The ratio between bed-building runoff and average flood tidal flow is defined as the ratio of runoff and tidal flow, which is 0.01 in the Qiantang River, indicating that the amount of sea water is much greater than that of incoming water from the river basin, and that the salinewater intrude far to the upper reach of the estuary. The volume of the tidal prism of the Qiantang Estuary is particularly great, the reason of which may be associated with the high tidal range, besides the trumped shape of the estuary. That is, in more detail, the river width shrinked abruptly, the tidal range increase along its intrusion course, The maximum tidal range at Ganpu station reaches 8.93 m and the annual average is 6.45 m, being about twice as high as that outside the Bay. This point can also be seen clearly from the average high and low tidal levels shown in Fig. 4-23. Because the river bed rises abruptly, the tidal waves transform obviously in its propagating process, and form the famous tidal bore near Chienshan. The height of tidal bore near Haining reaches 1–2 m in general. It is measured that the height of a single tidal bore reaches about 3 m, being a very great scenery. When two tidal bores met, the height of the obstructed water column reaches 8–10 m, which had terrible destruction to the bank engineering. The pressure of the tidal bore is measured to be 7 ton/m^2. In such reach with great tidal prism and shallow depth, the tidal current velocity is, of course, great. The mean velocity of verticals at Ganpu section reaches 4–5 m/sec, comparable to swift stream in mountains. Meanwhile, the tidal current takes a great change along its up-course (see Fig. 4-25). The flood tidal velocity and ebb tidal velocity at Ganpu station are about

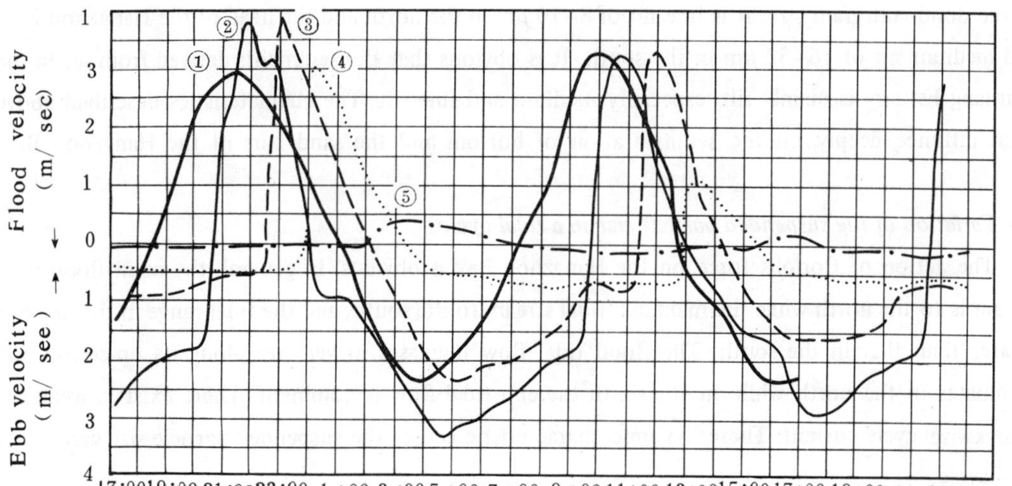

Fig. 4-25. The velocity changes of flood tide and ebb tide along the stream in Qiantang Estuary.

(1)–Ganpu; (2)–Dajianshan; (3)–Haining; (4)–Cangqian; (5)–Zhakou.

the same, while above Haining station, the flood tidal velocity is more than twice of the ebb tidal. The asymmetry of the flood and ebb tidal velocities certainly leads to the asymmetry of the sediment transport. The amount of the river-born sediments of the Qiantang Estuary is low, its average annual sediment discharge is only 6.68×10^6 tons. If we compare it with that of the Huanghe River, the Qiantang River can be regarded as a clean water river. But, the Qiantang estuarine sediments carried by the tidal flow are abundant, the average concentration at Ganpu station attained to 3–4 km/m^3, the maximum is 51 km/m^3. The sediment amount passing the section of Ganpu in a tidal cycle is more than 10 million tons. The amount of sediments derived from sea is much greater than that of riverborn, which will be described in more detail in the sections that follow.

2. Suspended sediments in the Qiantang Estuary

Sea-born sediment is the main source of the Qiantang Estuary sediment. In recent years, the author and his collegues analyzed particle size of suspended sediments using a Coulter Counter Model TAII. The analyzed results of the samples collected in synchronism from three different layers, i.e. surface middle and bottom layer inside and outside the Hangzhou Bay in August 1981 as follows.

(1) ***Basic characteristics of the suspended sediment particle size***

Table 4-30, shows that the Mdϕ of the suspended particles is 12.61 μm, which is the greatest in several Chinese big rivers, the content of the very fine sand (> 63 μm) in each sample is very low, only 1.44% on average, and the clay content is also low (averagely > than 7%), which is the lowest in several Chinese big rivers. Silt contents are more than 91%. The characteristic of the suspended particle size in the Hangzhou Bay is low in very fine sand and clay, and is high in silt. In the spectrum of the dominant grain size, it is fine silt of 8–16 μm in the north and the head of the Hangzhou Bay, and medium silt of 16–32 μm in the south. It is obvious that the sediments derived from sea in the Qiantang Estuary is mainly silt, especially medium and fine silt. The characteristics described above most influnce deeply on the sedimentation of bottom and the sand bars of the Hangzhou Bay.

(2) ***Variation of the suspended particle size in a tidal cycle***

The action of Coriolis force on the Hangzhou Bay is obvious. In general, the main flood tidal stream is to the north while the main ebb tidal stream to the south, and the tidal range in the north is greater than that in the south. The flood tidal flow increases in velocity along its up-course and dominates in the north while in the south the ebb tidal flow is dominant. There exists a weak anticlockwise cycle current. These dynamic characteristic causes the suspended particle size varing in a tidal cycle.

It can be seen from Table 4-31 that in the north the medium size of the suspended particles in flood tide is slightly greater than that in ebb tide, while in the south and in the bay head, it is greater in ebb tide than that in flood tide. As viewed from the space distribution of the suspended particle medium size of the Hangzhou Bay, the particle medium size in the north is somewhat greater than that in the south in flood tide, and *vice versa* in ebb tide. The content of the very fine and coarse silt grading in the south region is two to three times of that in the north region. The predominant grading

Location	<4	Grading (μm)					Average medium grain size (μm)
		4–8	8–16	16–32	32–63	> 63	
North	7.11	30.18	32.03	22.28	6.60	1.68	10.88
Bay head	7.77	27.68	31.08	26.93	6.53	1.02	11.71
South	5.96	22.03	29.00	30.87	10.36	1.62	14.61
Average	6.95	26.63	30.70	26.73	7.50	1.44	12.61

Table 4-31 The grain size components of ther suspended sediment in the Hangzhou Bay within a tidal cycle (%)

Location	Station	Tide	Grading (μm)						Medium size (μm)
			<4	4–8	8–16	16–32	32–63	> 63	
North	H_1	Flood	6.53	28.53	33.31	24.15	6.36	0.88	11.37
		Ebb	7.66	31.22	31.97	20.99	6.95	2.21	10.48
South	H_2	Flood	4.50	20.84	28.92	30.22	12.01	2.53	10.53
		Ebb	5.84	21.92	30.21	31.80	10.42	1.65	14.68
Bay head	H_3	Flood	8.18	27.39	23.63	27.24	5.46	0.97	11.59
		Ebb	7.05	26.94	32.43	27.94	5.90	2.10	12.28

Note: The data are average value over different layers.

of the former and latter are medium silt and fine silt separately. The sediment supply and the exchange of suspended sediments and bottom sediments between inside and outside the Qiantang Estuary is shown in Table 4-32.

Table 4-32 The grainsive components of the suspended sediments in the region out of the Changjiang Estuary and of the Hangzhou Bay (%)

Grading	<4	4–8	8–16	16–32	32–63	> 63	Average medium size (μm)
Out of Changjiang Estuary	12.58	45.50	25.89	10.55	4.59	0.89	7.10
Out of Hangzhou Bay	14.18	28.67	24.80	18.64	10.53	2.63	8.76

Summarlizing the above-mentioned fact, some important acquaintance can be obtained as follows: Firstly, although the sediment discharge of the Changjiang Estuary has great influence on the Qiantang Estuary, it is not quite like the previous result obtained by some scholars that the sediment source of the Chientang Estuary is mainly from the Changjiang Estuary. It is easy to see from the above-mentioned facts that the suspended sediments diffused from the Changjiang Estuary down to the south of the Hangzhou Bay dilute and fine the suspended particles, that is to say, the sediments in the Hangzhou Bay is mainly influenced by the suspended materials finer than fine silt, especially very fine silt, discharged from the Changjiang Estuary. Exactly speaking, the Qiantang Estuary is supplied with the suspended materials finer than fine silt in the Changjiang Estuary, whose influence on the bottom sedimentation of the Hangzhou Bay is greater than that of the sedimentation of the sand bar bodies in the mouth reach. Secondly the pattern of the suspended sediment transport in the Hangzhou Bay is generally that: the sediments derived from Zhoushan sea region or even that from the continental shelf of the East China Sea move toward the bay in flood tide time, while in ebb tide time, the riverborn sediments enter into the bay, the sediments exchange horizontally and vertically in a large range. Finally, due to the characteristics of the suspended sediments in the Hangzhou Bay, 75% of the bottom area of the Hangzhou Bay is occupied by fine silt and the suspended sediments and bottom sediments are the same on the whole the same, which is an important attribution of the sediments derived from sea.

3 Bottom materials in the Qiantang Estuary

By comparison with other estuaries, the Qiantang Estuary is distinctly in abundant sediments derived from sea, great tidal bore action, and a large sand bar seldom found in Chinese river estuaries. Now let us discuss it in view of grain size: minerals and palaemicroorganisms as below:

(1) ***Type and distribution characteristic of bottom sediments***

Most of the bottom sediments in the Qiantang Estuary, except for little pebbles and sand, are finer than coarse silt. The bottom sediments can be classified into 8 kinds (Table 4-33).

Table 4-33 The bottom sediment types and their contents (%)

Type	Medium sand	Find sand	Very fine sand	Coarse silt	Fine silt	Very find silt	Clay
Percent content (%)	0.11	3.16	12.63	23.26	18.95	5.26	1.05

Note: Based on the analyses of the samples taken from October 1960 to May 1981.

Table 4-33 shows that there exist two kinds of sediments, coarse sediments and fine sediments in the Qiantang Estuary. The space distribution of the bottom sediments represents the interaction of two kinds of sediments sources, two kinds of dynamics and two kinds of medium conditions. The pebbles were already deposited at some reaches and the top of the island above the near mouth reach. Such coarse materials as very fine sand were mostly deposited above Wenjiayan. Only in flood season, a few coarse materials may move downstream, scattering in small patches over the mouth reach and no accumulation of sandy center shoal can be formed. As to the distribution of the fine material, whether riverborn or seaborn, they move and deposit in the form of suspended sediments, and are influenced by river bed form, the growth and decline of runoff and tidal flow, and the mixing of fresh and salt waters. The longitudinal distribution of the medium size of sediment particle in the Qiantang Estuary is shown in Fig. 4-26.

From the above figure, the pattern of he longitudinal distribution of sediments is that the grain size of the sediments gets finer from the upreach towards the seas, but it gets coarser obviously at the position of the sand bar. Therefore, the pattern of the longitudinal distribution of sediments can be summed up as coarse-fine-coarse fine. The transverse distribution characteristics of sediments in the mouth reach and the bottom of the bay is, in brief, coarse at the north and fine at the south, resulting mainly from the situations of the flood tidal stream being to the north and the flood tidal flow being dominant below Qibao.

(2) ***Sorting action of dynamic sedimentation***

The Qiantang Estuary, as mentioned abvoe, is a mountain-river-macrotidal estuary. The sorting action of dynamic sedimentation follows the basic rules in a tidal reach. The fine sand, as stated above,

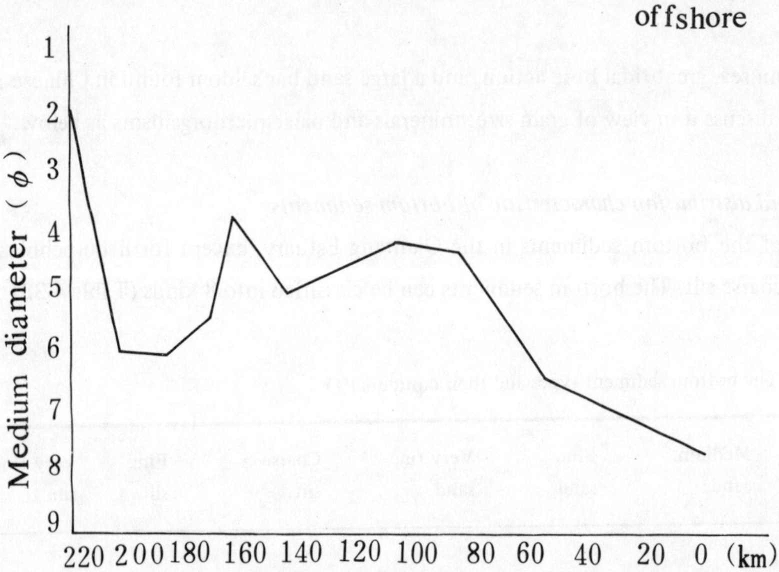

Fig. 4-26. Longitudinal distribution of the medium size of sediment particles in the Qiantang Estuary.

has accumulated in the upward area from Wenjiayan. The fine particles carried by tidal flow have influences on the developing of the Dongjiang Spit, but have no influence on the reach up the Dongjiang Spit. It follows that in mentation is under the control of runoff, and is characterized of seasonally intermitten deposition, same as that of common rivers. The characteristic of bottom sediments in the reach from Wenjiayan to Ganpu is consistent with the landscape of the sand bar and the interaction between runoff and tidal flow. The top of the sand bar where the tidal bore action is the most strong mainly consists of medium silt, coarse silt and very fine sand, the content of sand is higher too, the ratio among sand (0.063 mm), silt (0.063–0.004 mm) and clay (0.004 mm) is 2:6:2 probably. The sorting coefficient is all less than 2.60, except that near Toupen in Xiaoshan where the sorting coefficient is 3.25. On either sides of the sand bar, the materials get finer, the landward slop of the sand bar mainly consists of medium silt and fine silt, and the seaward slop mainly consists of coarse silt and medium silt. The sorting at the seaward slop of the sand bar is poorer than that at the landward slop. By comparing the sediment type distribution map based on data measured in Oct. 1960 and May, 1981, we learn that the distribution range of very fine sand at Ganpu is broader than the latter, and sorting is better too, indicating that the sand bar is scoured at spring tide, and silted slightly at neap tide. Consequently we take Qibao as a demarcation line, above which the sorting action by runoff is dominant and below which the sorting action by tidal flow is dominant, the Hangzhou Bay is a typical trumpet estuary, where, except the top and the north near Ganpu and near shore region of the bay where exist very fine sand and coarse silt, fine sediments finer than middle silt mainly are fine silt, coveres 75% of the area of the bay. By comparing the characteristics of the bottom sediment distribution with the grain size content of suspended sediments shown in Table 4-30, it is known that the bottom sediments are in consistent with the suspended sediments, which is an important attribute of sediments derived from sea. It is worthwhile to point cut that the depth in the Hangzhou Bay does not vary greatly. It is between 5–15 m, and the mean depth is 10 m. Under such a depth, the critical velocity to initiate movements of fine silt exceeds 2 m/sec, so the suspended sediments are difficult hard to initiate once deposited on the bed. Therefore, the suspended sediments in the Hangzhou Bay is easy to suspend and difficult to initiate when it has deposited. The sediments in the mouth reach have undergone many times of transport-deposit cycle. Their sorting coefficient varies between 1.36–2.82, and the average value is 1.94, which, on the contrary, is better than that of the sediments on the sand bar and also better than that of the shallow sea sediments out of the bay. Summarizing above all, for the sorting action of the dynamic sedimentation in a mountain-macrotidal estuary, there exist two processes controlled by both of runoff and tidal flow, the former of which is characterized by intermittent deposit, while the latter by fine particles, many times of deposit-transport cycle and well sorting.

(3) ***Sedimentary structure of the sand bar***

Less riverborn sediments in the Qiantang Estuary has restricted the estuarine delta to develop, which is verified by the very great widening rate of the estuarine plain. When propagating to the Hangzhou Bay from the Zhoushan sea region, the tidal wave is subject to the resistance and reflection of narrowing and arising of the river bed, changing from progressive to standing wave. The asymmetry of the tidal wave gets more and more obvious. Tidal range and current velocity increase upstream, about twice at Ganpu Station of the ones at the link line between Nanhui and Chenhai. Near Naining, tidal range and current velocity reach the maximum. In addition, current velocity and suspended sediment concentration in flood tide are all greater than that in ebb tide, resulting in the amount of sediment transport in flood tide exceeds that in ebb tide and the asymmetry of the ammount of sediment transport, that is, where exists a net sediment transport from sea towards land. Therefore, in the mouth reach of the estuary, the large sand bar accumulation body has developed based on the postglacial shallow sea sedimentary facies. According to the bore data at Qibao, this large sand bar is a lenticle laid in shallow sea deposit layer. The medium diameter of their particles varies between 0.097–0.027 mm and its mean value is 0.0388 mm, therefore, this sand bar consists of very fine sand, coarse silt and medium silt etc, but mainly of coarse silt as the same as its surface layer.

Fig. 4-27 shows that the sand bar is not composed of homogenous particles. There are small lenticle consisting of medium to very fine silt, which appears in horizontal or slant uncontinuous distribution, in the sand bar. Most of these lenticles inside the sand bar have developed in its upper layer while in the lower layer there are few of them. According to the bore data, the thickness of these lenticles is between 0.3–4.0 m. the width is between 10–50 m. The grain size of the materials composing the lenticles are characterized by lower contents of clay and very fine sand. The content of clay is generally between 3% and 30% and the content of very fine sand is less than 10%. Generally speaking, the materials composing the lenticle inside the sand bar is finer than the materials composing the surrounding body. it is worthwhile to point out that these lenticles have developed on the scour troughs which used to be scoured by flood and ebb tidal flows including runoff. Because of strong longitudinal and longtitudinal changes of the sand bar reconstruction of these lenticles occurred. Therefore, a complete lenticle is seldom seen, which accounts for about 2% of the whole volume of the sand bar. It follows that, in the view of a whole, the structure of the sand bar is unitary,

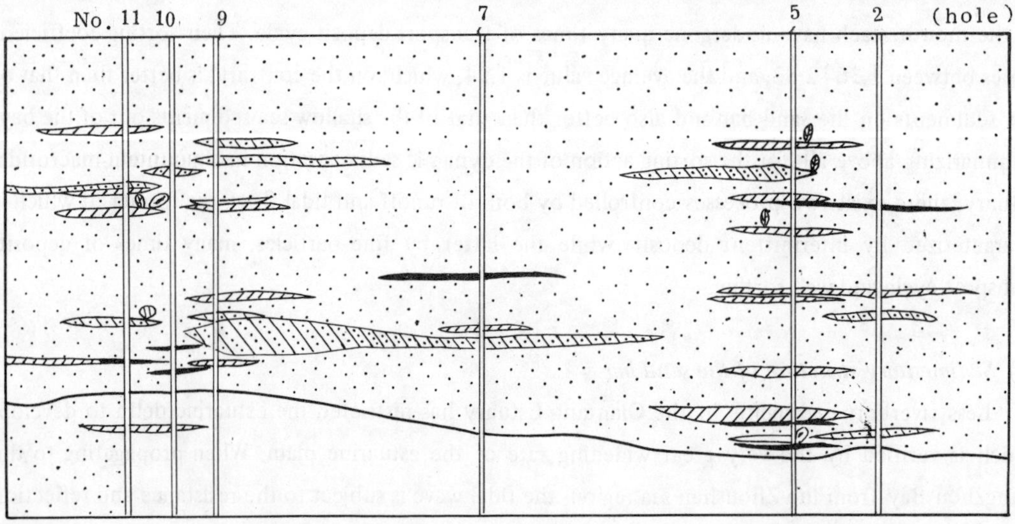

Fig. 4-27. Section of the sedimentary construction of the big sand bar of the Qiantang Estuary.

although there are two layers of sediments, the upper and lower layers, an some small lenticles in it. This precisely reflects the characteristics of sea-born sediments and strong tidal bore in which sediments and runoff are the less important facters. Obviously materials composing the sand bar do not fit with the strong tidal bore dynamics, but why have the large sand bars formed in the estuary? It can be explained by the sediment derivation and the relation between runoff and tide. For the sediment derivation, we will explain it in the next section. According to the concept of the ratio between runoff and tide, presented by Professor Qian Ning, the value of the ratio determines the position of the sediment accumulation. As shown by statistical analyses, when the ratio is greater than 0.10, the sand bar will be formed outside or near estuary mouth, if the ratio is less than 0.02, the sand bar will be formed inside the mouth, if the ratio is between 0.02 and 0.1, the sand bar will be formed at the transitional zone, The ratio between runoff and tide in the Qiantang Estuary is 0.01, so the sand bar appeared inside the estuary. The scale of the sand bar depends on the availability of sediments derived from sea. The particle size of suspended sediments is, as mentioned above, between 8.76–12.61 μm, the dominant grain size is medium and fine silt, the content of coarse silt is mostly less than 10%. Therefore, it certainly undergoes in a long time dynamical sorting to form such a large sand bar consisting of coarse silt. The materials of sand bar are well sorted, having less clay, their resistance against scouring poor. These results in the strong scouring and accumulation in longitudinal and lontitudinal sand bars.

(4). ***Mineral components of the bottom sediments***

The clastic materials in the Qiantang estuary have been sorted regularly by runoff and tidal flow in the light of their grain size, specific gravity and shape. The mineral sorting is accompanied with the grain size sorting resulting in richness in some minerals and poorness and poorness in other minerals. It has been identified that the heavy minerals reach more than 30 kinds, in which epidote is the most abundant (> 42%). There are many scattering biotite, besides the plate-like or scalar chlorite in light green. In some stations, the content of biotite is 29% and in the other stations is about 20%, which is seldom in the other estuaries in Zhejiang Province. In a word, the basic minerals combination of the bottom sediments in the Qiantang Estuary is epidote, biotite chlorite and ordinary amphibolite, in which biotite is the characteristic mineral (Fig. 4-28).

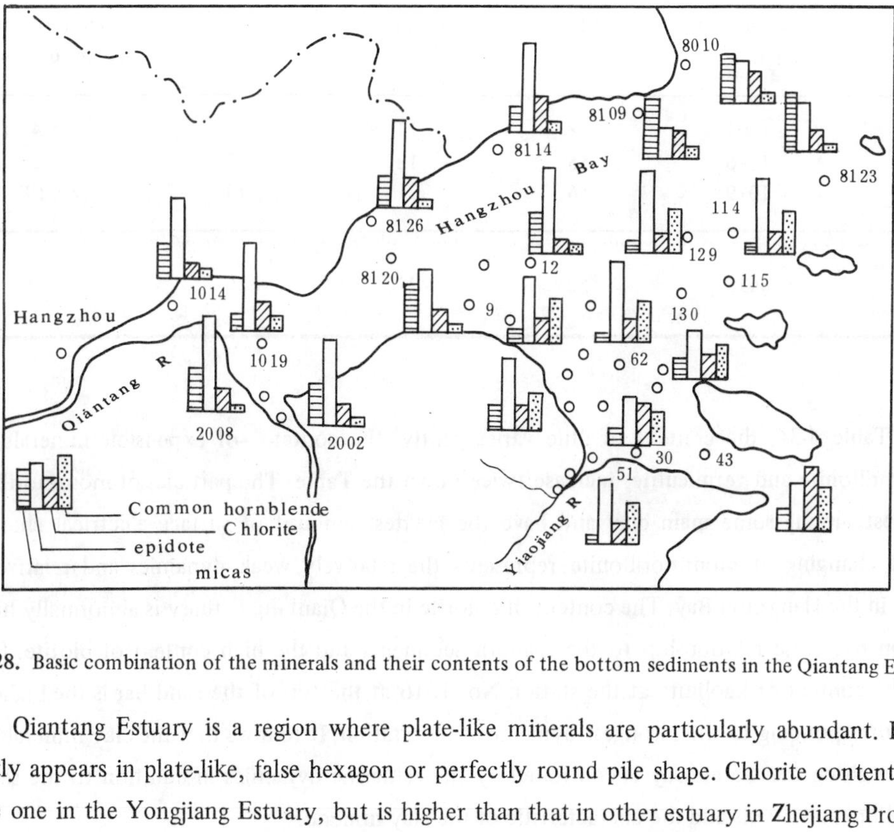

Fig. 4-28. Basic combination of the minerals and their contents of the bottom sediments in the Qiantang Estuary.

The Qiantang Estuary is a region where plate-like minerals are particularly abundant. Biotite frequently appears in plate-like, false hexagon or perfectly round pile shape. Chlorite content is less then the one in the Yongjiang Estuary, but is higher than that in other estuary in Zhejiang Province. The mineral components of the silt sediments, which is the main part of sediments in the Qiantang Estuary, are feldspar, quartz, and calcite. These mineral contents are 80–90% of the light and heavy minerals. The contents of the light minerals are generally about 5%. In the samples taken from the sand bar, not only the cutic-bipyramid and short bar-like zircon but also broken zircon can be found, (Plate 4-21). Zircon is a mineral highly resistant to erosion, with a hardness of 6.5–7.5, stable chemical characteristics. However, zircon in the sample was broken wholly by the tidal bore, we can imagine how strong the tidal bore is Fine silt is distributed over 75% area of the Hangzhou Bay bottom. The particles of biotite observed are finer, which causes the corresponding increase of the contents of clay and the filling in between biotites particles of clay. Therefore, the fine silt is difficult to move in waters of 10 meters depth, because in such deep waters the current speed required to initiate the movements of the fine silt should exceed 2 m/s. So the bottom sediments of the Hangzhou Bay is relatively stable, and the scouring only occurs in some degree at the place where the main tidal stream pass. Although there are abundant seaborn sediments in the Qiantang Estuary, 90% of the suspended sediments are fine sediments. The content of clay grading is less then 10%. Results of the analysis the pure clay samples ($< 2\ \mu m$) by X-ray diffraction, are shown in Fig. 4-29. The semi-quantitative calculation of the X-ray diffraction spectrum of each clay mineral is listed in Table 4-34.

Table 4-34 Clay minerals and their contents (%) in the bottom in the Qiantang Estuary

Reach	Number of Station	Illite	Chlorite	Kaolin	Montmorillonite expensible minerals
Mouth reach	1005	58	24	13	4
	1016	55	18	22	5
	1019	56	21	13	10
Hangzhou Bay	8415	52	21	16	11

In Table 4-34, the content of illite varies gently, the contents of expensible minerals such as montmorillonite and vermiculite, increase twice down the Table. The particles of montmorillonite is the finest among some main clay and have the greatest amount of surface electrical charge. The content changing of montmorillonite represents the relatively weak dynamics and relatively high salinity in the Hangzhou Bay. The content of chlorite in the Qiantang Estuary is abnormally high, this situation has close relationship to the seaborn sediments and the high content of biotite. In Table 4-34, the content of kaolinite at the station No. 1016 at the top of the sand bar is the highest, precisely corresponding to the strongest dynamic at this station. It follows that the clay mineral content changing along the water way is controlled by the estuarine dynamics in addition to the sedimentary medium conditions and the characteristic of the clay mineral.

4. General description of the sediment source of the Qiantang Estuary

Source of sediments in the Qiantang Estuary is briefly described in this section according to grain size and mineral composition of sediments as well as plaeomicroorganisms.

We differentitate river basin-born sediments from seaborn sediments acoording to the material components, the dynamic conditions of sediment transportation or deposition, as well as the general direction of sediment transportation. The former are riverbasin materials, which are transported

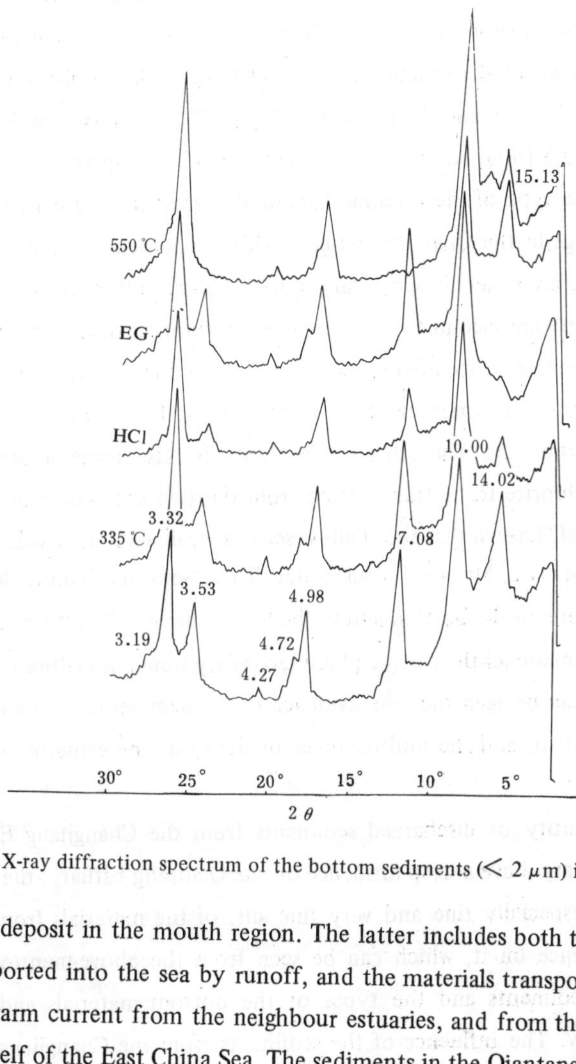

Fig. 4-29. X-ray diffraction spectrum of the bottom sediments ($< 2\ \mu m$) in the Qiantang Estuary.

downstream, and deposit in the mouth region. The latter includes both the riverbasin-born materials which were transported into the sea by runoff, and the materials transported by the coastal currents and the Taiwan warm current from the neighbour estuaries, and from the coastal areas or even from the continental shelf of the East China Sea. The sediments in the Qiantang Estuary move in the forms of suspended sediments and bed load. It is generally considered that runoff discharges sands and tidal flow carries muds. Of coures, the runoff also carries suspended fine sediments away in flood season. The accumulation position of bed load carried by runoff has already been stated. Its minerals components are mainly feldspar, quartz and calcite. They account for 80–90.5% of the bed load in weight. More than 30 kinds of heavy minerals – epidote tribe, amphibolite mica tribe, iron tribe, and ziroon – consistute the high biotite content region characterized by biotete. In heavy minerals, this plate-like biotete is unstable, and its specific gravity is small. So it is easy to be suspended and broken in its movement process. This fact is clearly reflected by the grains of biotite near Zhakou and near the Caoejiang Estuary which is coarser and perfect in its crystal form while the biotite grain in the whole Hangzhou Bay is reduced obviously. The former, therefore, indicates that it keeps the characteristics of biotite inside tuff and granite in the river basin, but the latter relects biotite being transported in a long distance and many times. The problem on the source of the fine biotite in Hangzhou Bay is complex. It looks as if some biotite comes from the river basin, some from the Changjiang Estuary, Zhoushan shallow sea region and the continental shelf. A type of narrow salt foraminifer was found in the light minerals of the sediments of the Hangzhou Bay. Epistominella and Bolivina were found in the Yongjiang Estuary and the Hangzhou Bay in 1971. Afterwords, Wang Pinxian et al. found the type of Sternosaline foraminifer with the shell of 0.17 mm in the Hangzhou Bay and of 0.17 mm near Zhakou. These type of Sternosaline foraminifer grow in an environment of 50–70 m depth, while they now emerge in the Qiantang Estuary. This phenomenon can be explained only by sea-born sediments. Chlorites are generally secondary minerals of alteration, of course, some chlorites are primary minerals too. There are secondary chlorites in the sediments of the Qiantang Estuary. The chlorites we observed under a microscope are precisely secondary minerals, although the crystal forms of chlorites appear to be plate-like, there are two kinds of crystal form in deed. One is similar to columnar crystal form of hornblende. The other one is similar to false hexagon crystal of biotite. So we can consider the above chlorite to be transformed from the two kinds of heavy minerals – amphibolite and biotite. It is well known that the shallow sea is a weak alkaline, oxidizing environment. There are high contents of Mg, Fe, Mn ions in sea water. These ions are favourable to the efflorescence of amphitolite and biotite on the bottom and in the ion exchange. Therefore the chlorites which have two kinds of crystal form are all the marine phase secondary minerals effloresced in sea bottom. From the above analysis, it can be seen that the evidence of the sediments of the Qiantang Estuary derived from the sea is sufficient, and the multi-sources of the Qiantang estuarine sediments in also quite clear.

Although the great quantity of discharged sediments from the Changjiang Estuary, which is adjacent to the Qiantang Estuary, have a deep influence on the Qiantang Estuary, the medium silt and the materials finer than it, especially fine and very fine silt, of the materials from the Changjiang Estuary have principal influence on it, which can be seen from the above-mentioned particle size information of suspended sediments and the types of the bottom materials and its distribution pattern in the Hangzhou Bay. The influence of the sediments from the Changjiang Estuary on the whole Hangzhou Bay is greater than the one on the sand bar. The materials of the sand bar may be from the Zhoushan sea region, even from the eroded materials of farther continental self.

by Zhang Zhizhong
Hangzhou University, Hangzhou

References

Ren Mei-e et al., 1982, Sedimentary Characteristics of tidal mud flat in China. (in Chinese)

Ren Mei-e and Tseng Chenkai, 1980, Late Quaternary continental shelf of East China. Acta Oceanologica Sinica, V. 2, No. 2, pp 91–111. (in Chinese)

Ren Mei-e, 1964, Dynamic geomorphology in the Zhujiang Estuary and problems of beach use. Proceedings of the Symposium of Chinese Society of Oceanology and Limnology in 1963, p50, Science Press. (in Chinese)

Qin Yunshan and Li Fan, 1982, Study on the Suspended matter of the seawater of the Bohai Gulf. Acta Oceanologica Sinica, V. 4, pp 191–200 (in Chinese)

Pang Jiazhen and Si Shuheng, 1979, Fluvial process of the Huanghe Estuary, I. Change in recent history. Oceanologia et Limnologia Sinica, V.10, pp 126–141. (in Chinese)

Pang Jiazhen and Si Shuheng., 1980, Fluvial process of the Huanghe Estuary, II. Hydrographical Character and the region of sediment sitting, *ibid*, V. 11, pp 294–305.

Pang Jiazhen and Si Shuheng., 1982, Fluvial process of the Huanghe Estuary, III. Influence of the evolution of the Huanghe Estuary on the lower reach, *ibid*, V. 13, pp 218–224.

Pang Jiazhen and Si Shuheng, 1980, Fluvial process of the Yellow River Estuary. International Symposium on River Sedimentation Preprints. $C_{3-1}-C_{3-10}$.

Qian Ning and Zhou Wenhao, 1965, Fluvial process of lower Huanghe River, Science Press. (in Chinese)

Chen Lirong, Luan Zuofeng, Zhen Tiemin, et al., 1980, Mineral assemblages and their distribution patterns in the sediments of the Bohai Gulf, Oceanologia et Limnologia Sinica. V.11, pp 46–64. (in Chinese)

Wang Pinxian, Min Qiubao et al., 1982, Quaternary layers from land to sea in East China, Proceddings of Institute of Marine Geology, Tongji University. (in Chinese)

Liu Tungsen, Chang Tsunghu, et al., 1962, The loess of China. Acta Geologica Sinica V. 22, No. 1, pp 7–12. (in Chinese)

Qian Ning et al., 1980, The source of coarse sediment in the middle reaches of the Yellow River and its effect on the siltation of the Lower Yellow River. International Symposium on River Sedimentation Preprints, $A_{2-1}-A_{2-10}$.

Zhong Gongfu, 1981, On the Characteristies of Utilization of land and water resources in the coastal zone of the Zhujiang Delta. Tropical Geography, No. 3, pp 1–8. (in Chinese)

Chen Qinde and He Yashou, 1981, The Characteristics of the tidal and tidal stream in the Zhujiang Estuary, Tropical Geography, No. 3, pp 29–35. (in Chinese)

Xu Junliang, Li Yongxing, et al., 1981, The patterns of the invasions of saline water and the activities of the wedge in the Zhujiang Estuary. Tropical Geography, No. 3, pp 36–44. (in Chinese)

Mai Qiaowei, Zhao Yean, et al., 1980, Sediment problems of the lower Yellow River, International Symposium on River Sedimentation Preprints. $C_{1-1}-C_{1-10}$.

Huang Sheng, Han Naibin, et al., 1980, Analysis of siltation at mouth bar of the Yangtze River Estuary, International Symposium on River Sedimentation Preprints, $C_{6-1}-C_{6-10}$.

Zhang Zhizhong, Huang Wensheng, et al., 1977, A preliminary study of some characteristics in the fluid mud of Changjiang Estuary, Journal of Hangzhou University. No. 1, pp 83– 96. (in Chinese)

Zhang Zhizhong, 1981, The sedimentary characteristics of South Channel of the Changjiang Estuary, Channel Regulation Research of Changjiang Estuary, No. 1, pp 49–61. (in Chinese)

Zhang Zhizhong and Xu Zhigang, 1983, Suspended sediments and its transport in the Changjiang Estuary, Marine sciences, No. 5. pp 6–11. (in Chinese)

Zhang Zhizhong, Wang Yunju and Xu Zhigang, 1982, Test researches on the flocculation in fine sediments of the Changjiang Estuary. Estuary and Harbor Research, Hangzhou University. (in Chinese)

Zhang Zhizhong, Wang Yunju and Xu Zhigang, 1983, Some flocculation characteristics of fine sediments of the Yangtze Estuary. Proceeding of the Second International Symposium on River Sedimentation, pp 274–285.

Feng Huaizhen and Zhang Zhizhong, 1964, Sedimentation of shallow sea along the coastal, Zhejiang Province, Collected Oceanic works, No. 2, pp 25–44. (in Chinese)

Chin Yunshan, 1963, The continental topograhy and sediments of South and East China Sea, Oceanologia et Limnologia Sinica, V.5, pp 71–86. (in Chinese)

Chen Jiyu et al., 1981, Development model of Changjiang Estuary from 2000 Years, Channel Regulation Research of Changjiang Estuary. No. 1, pp 87–93. (in Chinese)

Ren Rushu et al., 1981, Carrying power of sediments in the water of Changjiang Estuary, Channel Regulation Research of Changjiang Estuary, No. 1, pp 104–117. (in Chinese)

Han Naibin, 1981, Analysis of the invasions of saline water in Changjiang Estuary, Channel Regulation Research of Changjiang Estuary, No. 1, pp 93–103. (in Chinese)

Gu Weihao, 1982, Forecast of null point in the Changjiang Estuary, Journal of Water Transport Engineering, pp 24–29. (in Chinese)

Zhang Shaoying, 1982, Present condition and prospects for regulation of Qiantangjiang Estuary, Marine Science, pp 16–19. (in Chinese)

Qian Ning, Li Quanpin et al., 1964, The fluvial processes of the big sand bar inside the Qiantangjiang Estuary, Acta Geographica Sinica. V. 3, No. 1, pp 124–141. (in Chinese)

Zhu Yongkang, 1964, Structures and its development of big sand bar inside the Qiangtangjiang Estuary. Proceedings of the Symposium of Chinese Society of Oceanology and limnology in 1963, pp 53–54. (in Chinese)

Feng Huaizhen, Bi Aohong and Zhu Yankang, 1964, Source problem of the sediments in Hangzhou Bay, Proceedings of the Symposium of Chinese Society of Oceanology and limnology in 1963, pp 59. (in Chinese)

Chen Jiyu, et al., 1964, The formation and historical development of the big sand bar inside the Qiantangjiang Estuary, Acta Geographica Sinica, V. 30, No. 2, pp 109 –123. (in Chinese)

Feng Huaizhen and Zhang Zhizhong, 1983, Sedimentary characteristics of the estuaries of Zhejiang Province, Proceedings of International Symposium on Sedimentation on the Continental Shelf, With Special Reference to the East China Sea, China Ocean Press.

Zenkovitch, V. P., 1962, Basic theory on the coastal development, Science Press USSR, pp 427–449, 643–648. (in Ressian)

Niino, H., Emery, K.O., 1961, The sediment of shallow portion of South and East China Sea. Geological Society of America, Bulletin, V. 72, No. 5. pp 731–761.

Shepard, F. P., 1959. The Eeath Beneath the Sea. pp 81–83.

Allen, G. P. et al., 1976, Transport and deposition of suspended sediment in the Gironde Estuary, France, Estuarine Processe, V. II, pp 63–81.

Gallene, B., 1974, Study of fine material in suspension in the Estuary of the Loire and Dynamic Grading. Estuarine and Coastal Marine Sicence, No. 2, pp 261–272.

Meade, R. H., 1972, Tranport and deposition of sediments in estuaries. Geol. Soc. Amer.

Postma, H., 1976. Sediment transport and sedimentation in the estuarine environment. Estuaries, AAAS, Washington.

Krank, K., 1973, Flocculation of suspended sediment in the sea, Nature, V. 246, pp 348–350.

Krank, K., 1975, Sediment deposition from flocculation suspensions, Sedimentology, V. 112, No. 1, pp 111–123.

Ariathural, R. et al., 1978, Erosion rates of cohesive soils. ASCE, V. 104, No. HY2r, pp 279–283.

Edzwald, J. K., et al., 1974, Coagulation in estuaries, Environmental Science and Technology, V. 8, No. 1, pp 58–63.

Satoshi Yamamoto, 1980, Sedimentology of Suspended Material in Ocean. (in Japanese)

Satoshi Yamamoto, 1979, Size distirbution of delrital mineral grains suspended in surface water of the Yellow Sea and East China Sea. (in Japanese)

Chen Jiyu, et al., 1982, Discription of the morphology and sedimentary structures of the mouth bar in the Changjiang Estuary. Estuarine Comparisons. pp 667–675.

V. Transportation and Diffusion of the Suspended Matter from the Changjiang River into the Sea

It is As the third largest river in the world, the Changjiang is 6380 km in full lenth. According to the 32-years data got at Tatong station from 1950 to 1981, the average runoff of the river is 28,200 m^3/s, which is eighteen times that of the Huanghe River, and the annual mean discharge amounts to 889,500 million m^3 which makes up 90% of the total runoff of various rivers emptying into the East China Sea. The annual sediment discharge is 480 million tons which is about one third of the Huanghe River. It has significant influence on sedimentation and geomorphological development in the river mouth and contributes greatly to the renovation of waterways, maintainance of the port area and exploitation of the coastal zone. The study of the estuarine system of the large river with enormous volume of water and huge quantity of silt is of great importance to the enrichment of the theory of sedimentology and the development of comparative sedimentology.

Based on the results of previous measurements and researches as well as the data from the China-US. joint study on sedimentary process in the Changjiang Estuary and its adjacent East China Sea continental shelf, this section gives a brief introduction to the transportation and movement of water mass, solid and disolved matter of the River.

1. Transportation and movement of the Changjiang River water mass

The Changjiang Estuary belongs to deltatic river mouth with many islands in terms of geometry features. Because of the branching action of islands, banks and sandbars, the estuary is charaterized by a triple-order bifurcation, emptying into the sea through four outlets. From north to south they are the North Branch, the North Passage, the North Channel and the South Channel. The North Branch tends to be silted up now and almost has no water discharge therein. According to investigation data of recent years, the runoff ratio between the South and North Passages has changed from 6 : 4 to 4 : 6 or even 3: 7. As to the runoff of the South and North Channels, the latter has absolute superiority over the former. All thses have resultel in an obvious regional character in the discharge of the river into the sea. Fig. 4-30 is a vertical profile of salinity obtained in a series of large-scale observations in summer at the river mouth gate, which shows that the minimum salinity regions correspond well with the main discharge passageways, i.e. the North Passage and the North Channel.

The amount of water of the Changjiang River emptying into the sea varies greatly from month to month, and flood and dry seasons can be distinguished. The flood season (May – October) makes up 71% of the annual runoff while the dry season (October – April) only 29% (Fig. 4-31). This makes the transportation and diffusion of the runoff present obvious seasonal differences after getting into the sea. Generally speaking, the surface upper water expands strongly toward Chejudo Island in summer, forming a low salinity water tongue stretching northeast as shown in Fig. 4-32 owing to the influence of the Taiwan Warm Current, the northern-Jiangsu coastal water and the Kuroshio surface

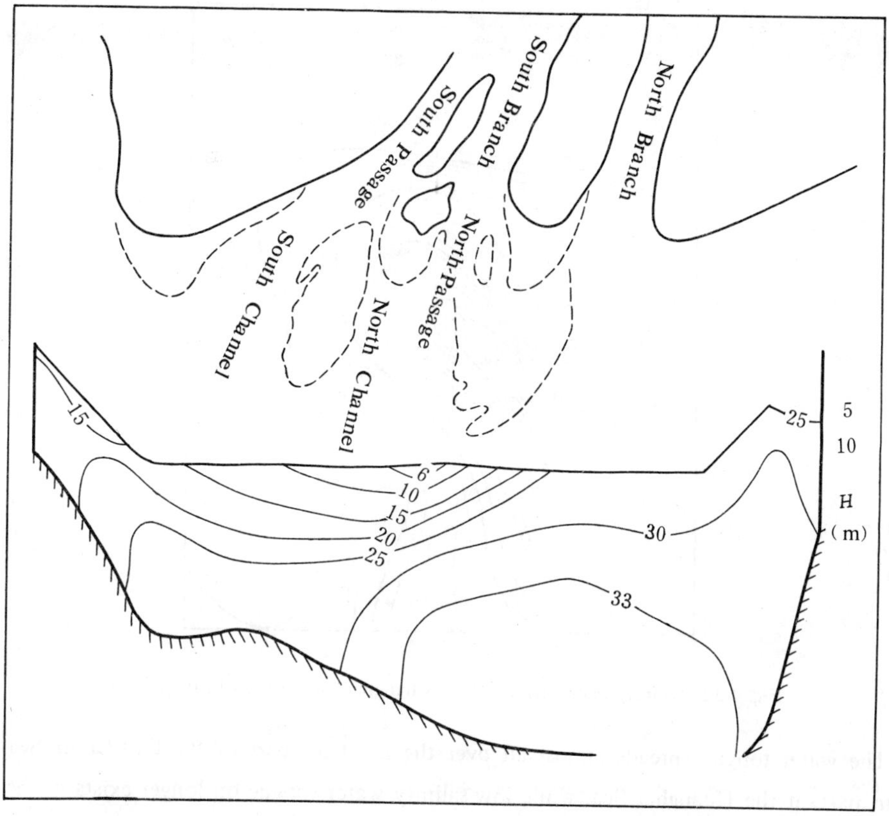

Fig. 4-30. Salinity distribution at the U-Shaped cross-section of the Changjiang River mouth.

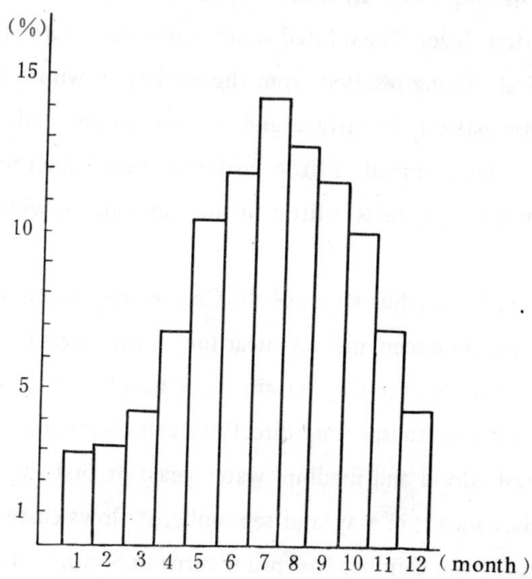

Fig. 4-31. The ratio of the average monthly discharge in the Changjiang River.

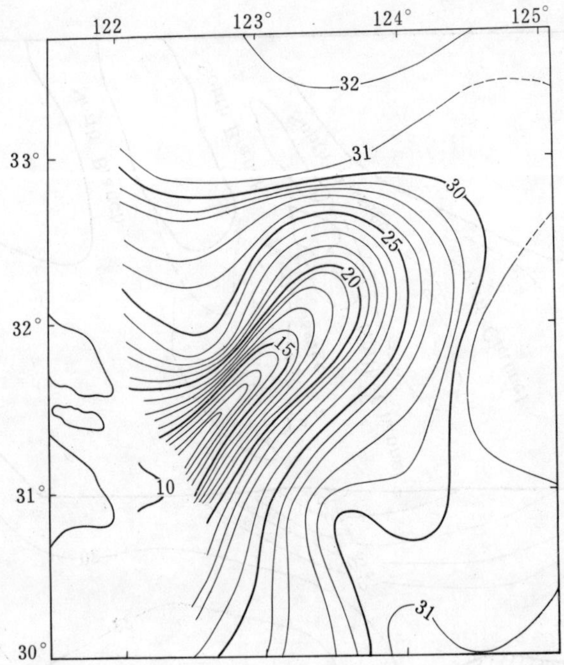

Fig. 4-32. Salinity distribution off the Changjiang River mouth, June 1975.

water. The water tongue spreads almost all over the northern part of the East China Sea and the southern part of the Huanghai Sea, while low-salinity water tongue on longer exists in the bottom layer, and the low-salinity water mass, limited in a narrow area west of 122°20′ E, continuously mixes and exchanges with the Taiwan Warm Current and the coastal water off northern Jiangsu, forming mixed water and moving away. In winter no low salinity water tongue occurs either in the surface layer or in the bottom layer. The diluted water flows southward alongshore in a narrow band off Zhejiang. Strong vertical mixing resulted from the cooling in winter and the strong north wind makes a similar distribution pattern in surface and bottom layers. Only the spreading area of the diluted water in the bottom layer spreads a little farther eastward than in summer because the main flow of the Taiwna Warm Current shifts a little further outward in winter, which is shown in Fig. 4-33.

From the above-mentined spreading states of the Changjiang diluted water, the spreading path of the river runoff into the sea can be seen, namely, near the mouth (west of about 122.5° E) the runoff flows straight ahead to southeast due to the restraint of coastal topography and the action of Coriolis force; afterwards, three different transporting directions can be distinguished, that is, from June to August it turns to northeast (flood and medium water year) or first turns to northeast then to due east at 123° E and stretches ahead; in May and september it flows eastward to about 123°–124° E and then turns to southeast: from October to April it turns to South-Southwest and flows alongshore off Zhejiang.

Fig. 4-34 shows the average monthly route of the Changjiang diluted water.

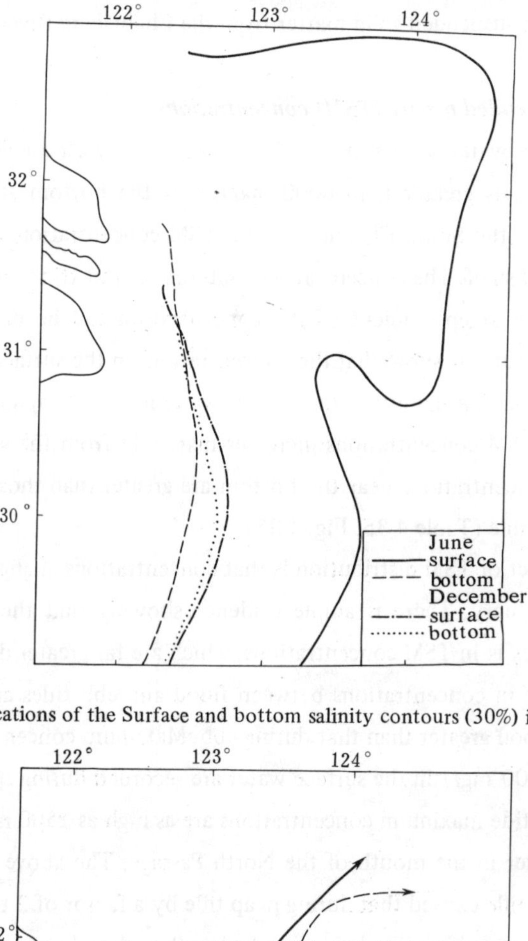

Fig. 4-33. Locations of the Surface and bottom salinity contours (30%) in June and December 1975.

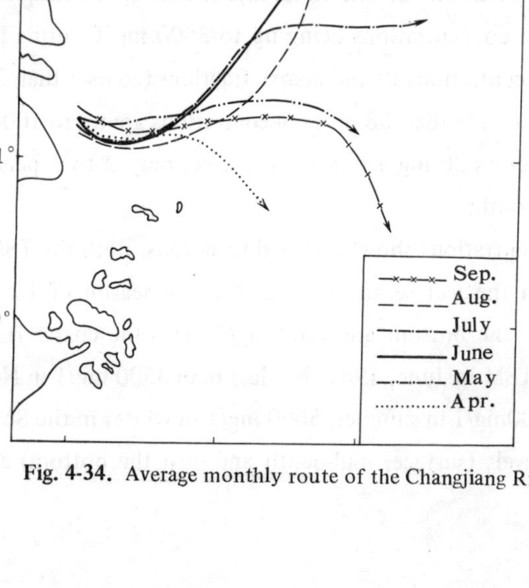

Fig. 4-34. Average monthly route of the Changjiang River Plume.

2. Discharge and transport of suspended solid matter from the Changjiang River

The Changjiang River enters the sea with an average total sediment discharge of 4.86 X10 tons/yr. Due to the lack of data of saltatorial and bedload discharges, this section will only discuss the discharge and transport of suspended solid matter from the Changjiang River.

(1) ***Distribution of total suspended matter (TSM) concentrations***

The high turbidity of the water occurs in the Changjiang Estuary all the year round. During the flood, maximum concentrations greater than 8000 mg/1 near the bottom and 2000 mg/1 in the surface water are measured in the South Channel mouth. TSM concentrations decrease rapidly from the mouth to the continental shelf. The concentration contours are out to the sea southeastward corresponding to the isobaths. A steep gradient of the concentration can be noted in the area from 122°00′ to 122°50′ E. Observations show that the concentrations in the surface water decrease from over 100 mg/1 to 5 mg/1 while the salinity of the surface water increases rapidly from 15‰ to 32‰. The vertical distribution of TSM concentrations increases markedly from the surface to the bottom. These results suggest that concentrations near the bottom are greater than those in the surface water by about one order of magnitude (Table 4-35, Fig. 4-35).

Another interesting aspect of TSM distribution is that concentrations change markedly with tidal period in the Changjiang Estuary. There is ample evidence showing that the greatest discrepancy between spring and neap tides is in TSM concentrations, which are far greater during spring tide than during neap tide. Differences in concentrations between flood and ebb tides are more obvious with TSM concentration during flood greater than that during ebb. Maximum concentration of about 3500 mg/1 near the bottom and 800 mg/1 in the surface water are recorded during spring tide (November, 1981), whereas during neap tide maximum concentrations are as high as 1500 mg/1 near the bottom, 260 mg/1 in the surface water in the mouth of the North Passage. The above facts show that TSM concentrations during spring tide exceed that during neap tide by a factor of 2 to 3. In addition, TSM concentrations during flood are half-one or one times higher than that during ebb of both spring and neap tides, with the maximum concentrations being up to 3500 mg/1 during flood and 2500 mg/1 during ebb. Differences in concentrations of the coarse fractions (coarser than 38 μm) are even more obvious. The coarse-fraction (coarser than 38 μm) concentrations are up to 1000 mg/1 during spring, 40 to 50 pencent of TSM. whereas 20 mg/1 during neap tide, only 2 to 3 pencent of TSM near the bottom in the North Passage mouth.

Distribution of TSM concentrations shows seasonal variations. Both the TSM concentrations and vertical gradient are greater in the wet season than in the dry season (Table 4-35). For example, Maximum concentrations near the bottom are 4200 mg/1, with the coarse fraction concentrations accounting for 35 to 40% of TSM in June, 1981, but less than 3500 mg/1 in November, 1981 in the North Passage mouth and 8000mg/1 in summer, 5000 mg/1 in winter in the South Channel. Average concentrations at the three levels (surface, mid-depth and near the bottom) are greater in summer

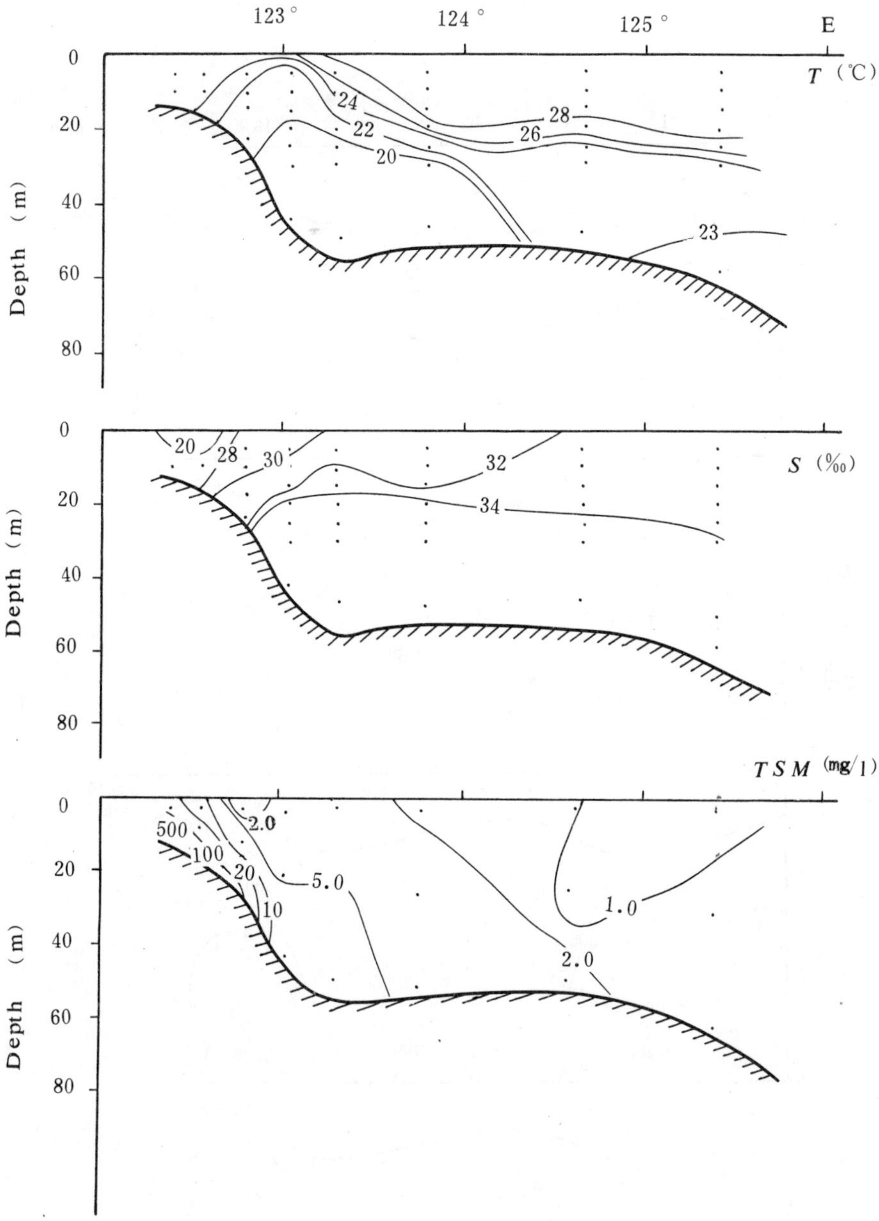

Fig. 4-35. Distribution of temperature, salinity and concentration of suspended matter over the continental shelf off the Changjiang Estuary (30° 50′ N, Aug. 1981).

than in winter (Table 4-35, Fig. 4-36). The reasons are mainly that the water and sediment discharges of the Changjiang River are concentrated in summer months (May – October).

According to spacial variations, TSM concentrations are greater in the South Channel than in the North Passage and North Channel. Maximum concentrations up to 8000 mg/1 during flood in summer occur near the South Channel mouth, and are one times higher than that of the North Passage. However, the grain-size of the suspended matter is finer in the South Channel than that in the North Passage. The coarse fractions (coarser than 38 μm) of the suspended matter are only 2 to 5 pencent of TSM in the South Channel whereas up to 40 perceht of TSM in the North Passage.

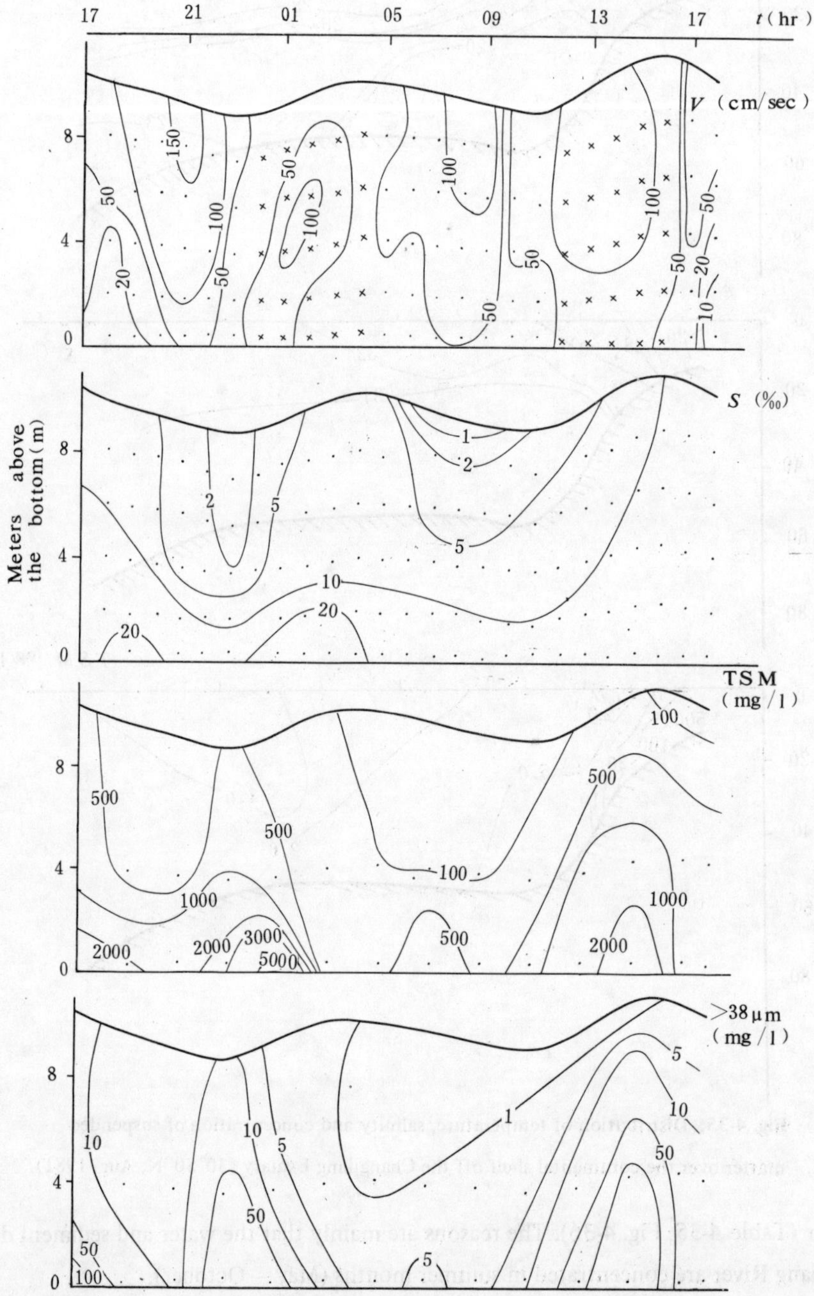

Fig. 4-36. Variation of velocity, salinity, TSM and coarse fraction with time and water depth in the mouth of South Channel, Nov. 1981.

Table 4-35 Average concentrations in the Changjiang Estuary (mg/1)

Location	Layer	Wet season		Dry season	
		spring	neap	spring	neap
Within North Passage	surface		205.0	405.4	
	mid-depth		373.3	689.3	
	bottom		666.4	1176.2	
North Passage mouth	surface		223.4	478.1	117.3
	mid-depth		426.9	1049.6	210.3
	bottom		1518.3	1651.0	663.7
Off North Passage	surface		44.3		
	mid-depth		209.4		
	bottom		720.3		
South Channel mouth	surface	440.0			187.2
	mid-depth	776.5			573.6
	bottom	2648.4			1686.9

(2) ***Composition of suspended matter***

The suspended matter is mainly composed terrigenous material finer than 38 μm which can be divided into combustible and non-combustible components. The non-combustible component consists of at least three different kinds of materials: detrial minerals, clay minerals and skeletal debris. The concentrations in percentage of the non-combustible components decrease rapidly from the estuary to the continental shelf. The distribution of the non-combustible is similar to that of TSM concentrations, and a steep gradient occurs in the area from 122°00′ E to 122°50′ E. Both the TSM concentrations and the content of terrigenous in percentage are lower in the east of this zone. The facts above indicate that most of the suspended matter from the Changjiang River is entrapped in the Changjiang Estuary, west of 123°00′ E. Compositions of the suspended matter are identified with a microscope, X-ray diffraction and scanning electron microscope for some samples. These results are discribed as follows.

(a) *Detrial minerals (quartz, mica)*

Quartz and mica in suspension are concentrated in the Changjiang Estuary, decreasing rapidly seaward. Quartz grains are coated in a kind of ferric oxide membrane and have a subangular shape. The distribution of quartz grains show a seasonal variation. In summer (e.g. june, 1980; August,

1981) quartz grains (coarser than 50 m) from the Changjiang River can reach to the area of 122°00′–122°40′ E, but only to the area of 122°00′ E or so in winter (e.g. November, 1981; Figs. 4-37. 4-38). Quartz discharge in the North Passage is larger than that in the South Channel. Mica grains constitute one of the components of the suspended matter. The distribution of mica grains coarser than 10 μm in size is similar to that of quartz grains.

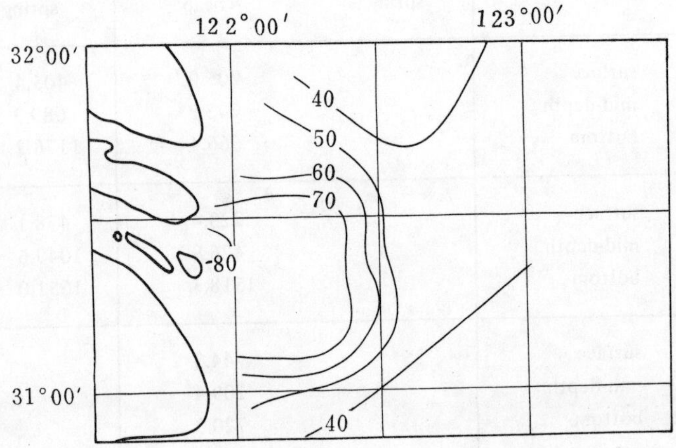

Fig. 4-37. Distribution of the quartz mean grain size (μm) of the suspended matter, Aug. 1981.

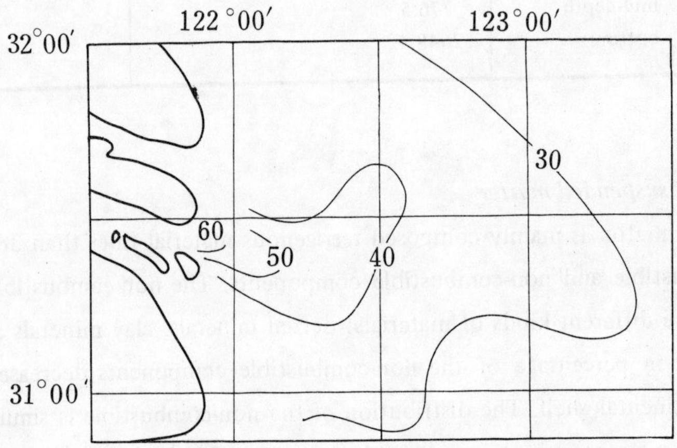

Fig. 4-38. Distribution of the quartz mean grain size (μm).

(b) *Clay minerals*

The concentrations of clay minerals in percentage decrease from the Changjiang Estuary to the continental shelf. The assemblage of clay minerals is illite, montmorillonite, kaolinite and chlorite. Table 4-36 shows that the assemblage of clay minerals in both suspended matter and surfacial sediment off the Changjiang Estuary is similar to the sediments in the Changjiang channel, and is characterized by higher content of illite and lower content of montmorillonite.

(c) *Flocculent materials*

Table 4-36 Concentrations of clay minerals in the Changjiang Estuary and adjacent shelf(%)

Sample	Location	Montmorillonite	Illite	Kaolinite	Chlorite
Suspended matter	off the Changjiang Estuary	5.0	71.4	9.5	14.0
	nearshore of northern Jiangsu	8.5–8.6	56–62	28.5–35.5	
	nearshore of Zhejiang	4.5–6.3	63–64.5	32.0	
Surface sediment	off the Changjiang Estuary	6.4	72.3	10.8	10.7
	nearshore of northern Jiangsu	20.5	63.7	7.7	8.3
	nearshore of Zhejiang	5.0	72.6	9.8	12.5
Bed-sediment	Changjiang River	6.0	68.4	25.6	
	Huanghe River	23.3	59.0	8.4	8.1
	abandoned Huanghe River	24.0	59.0	9.0	9.2

Flocculent materials consisting of sheet mineral, quartz grains clay mineral and organic ormainder are believed to be a production of flocking fine-grained materials they are concentrated in the Changjiang Estuary and decrease rapidly seaward. Such materials take irregular ellipse in shape, with their diameters generally coarser than 10 μm The concentrations and vertical gradient of flocculent materials are higher near the North Passage and South Channel mouths. It is considered that the presence of flocculent materials plays an important role in the development of a broad sand bar system at and near the Changjiang Estuary.

(d) *Combustible component*

A combustible component is defined here as the weight loss of the ashed half cut from a certain amount of filtered suspended matter after its 2-hours combustion in an oven at 550° C. The percentage of a combustible component in the suspended matter turns out to be contrary to the distribution of TSM concentrations, i.e., it increases seaward (Fig. 4-39), and decreases from the surface to the

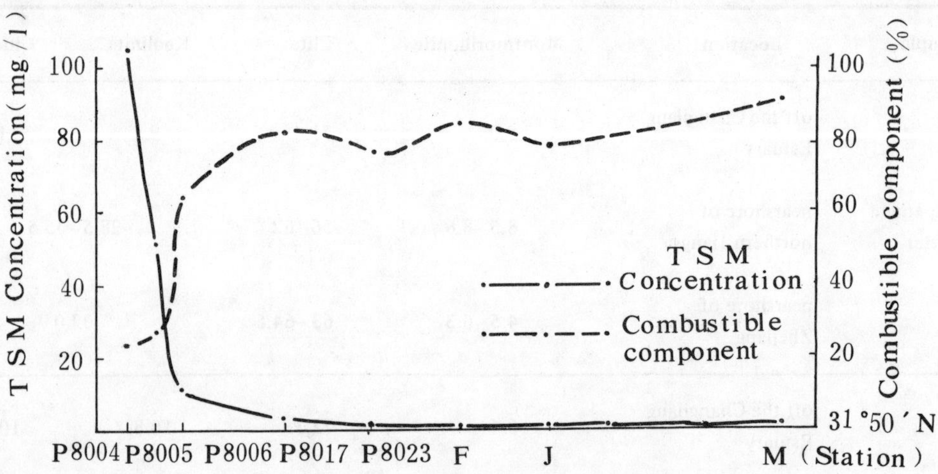

Fig. 4-39. Concentrations and combustile component of suspended matter in the surface water from the Changjiang Estuary to the shelf, June, 1880.

bottom. Its content in percentage in the surface varies over a wide range, with less than 10% at and near the mouth gate, and less than 25% off the gate.

On the basis of N.C.H. analysis, concentrations of particulate organic carbon (POC) may exceed 60 mg/1, and the average percentage of POC ranges from 3 to 1.7 percent, decreasing with the increased suspended matter concentrations in the Changjiang Estuary. Therefore, the Changjiang River transports annually an estimated 10 $\times 10^6$ tons of POC and 1.2 $\times$ 10^6 tons of POC to the sea. The above value is bigger than the average value of POC transpored by rivers throughout the world. In general, this may be associated with the intense human activeties in the Changjiang basin.

On the other hand, organisms and their remainders are found in the suspended matter of the Changjiang Estuary, whose concentrations in percentage increases seaward. It should be noted that the remainders of benthic foraminifera and diatom are present in the suspended matter near the Changjiang Estuary. Benthic foraminifera remainders of *Ammania beccarii* living in the inner-shelf environment occur in the suspended matter within the North Passage. Diatom remainders are present in the suspended matter at 1 m above the bottom off the Changjiang Estuary (Plate 4-22). These results can powerfully prove that the surficial sediment can be reworked in resuspension by hydraulic regime. Under the action of tidal currents, a part of the remainder and resuspemded matter initially deposited in the Changjiang Estuary have participated in the retransport processes of the suspended matter in the Estuary.

(3) ***Exportation and movement of suspended matter***

According to calculations of uni-width silt discharge of several time-series stations at the Changjiang River mouth gate, a diagram of transportation of suspended matter in the Changjiang Estuary is plotted (Fig. 4-40). It shows the transporting path of the suspended matter as follows:

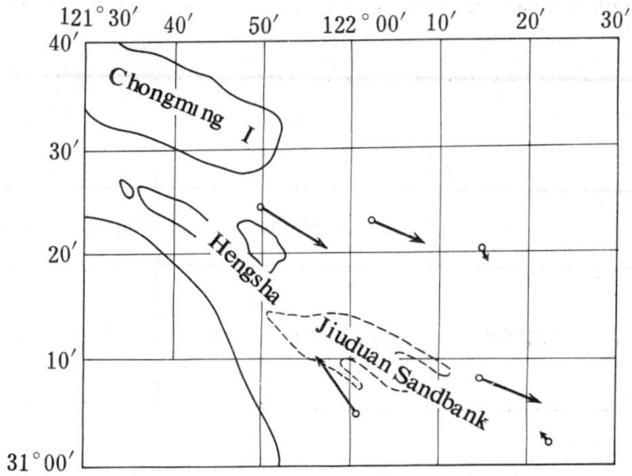

Fig. 4-40. Sketch map showing transport of suspended matter in the Changjiang River mouth, Aug 1981 and June 1980.

(a) Suspended matter in the North Passage is exported in a direction of 120°. At the exit of the passage, it changes to 100°. After getting at the beach outside the river mouth, great changes occur. While it remains southeastward in the surface layer, that in the middle and bottom layers turns to SSW or SSE (190°–165°). The silt discharge decreases outward remarkably further east of 122°30′. Net uni-width silt discharge is only 0.2 ton/day.m on the innershelf (31°30′ N, 123°30′ E) (June, 1981).

(b) Suspended matter in the North Channel is exported all toward to the south-east in the surface, middle and bottom layers.

(c) Except for winter when the suspended matter is exported southeastward, net single-with silt discharge in the South Channel is imported toward the upper reaches of the river (about 325°) during the other seasons. Outside the exit of the Channel (31°02′ N, 122°18′ E) the surface layer moves northeastward and the bottom layer still moves toward the upper reaches of the River (Table 4-37).

According to the analyses of the distribution of TSM concentration its composition, the calculations of uni-width silt discharge, and the satellite images of the same period, it can be seen that most of the suspended loads deposit near the river mouth gate after amptying into the sea, and decrease rapidly in quantity toward the sea, thus exerting their influence only on an area west of 123°00′ E. These can be indicated from the following reasons and facts:

(i) Both the concentration of suspended load and distribution of detrital matter have a turning belt between 122°00′ E and 123°00′ E where the content of suspended matter decreases from 1000 mg/1 to 10 mg/1 and further to less than 5 mg/1 east of 123°00′E.

(ii) Calculations of uni-width silt discharge show that silt discharge decreases rapidly toward the sea. Take the North Passage as an example, the silt discharge at its exit is 75% that in the passage and,

Table 4-37 Net uni-width sediment transport quantity and direction at six time-series stations in the Changjiang Estuary (wet season)

Location		Quantity (ton/day.m)			Direction			Data
ϕ	λ	Surface	Middle	Bottom	Surface	Middle	Bottom	
Inside of North Passage	31°25′ N	121°50′ E	38	105	86	117°	126°	121°	Aug. 1981
Middle of North Passage	31°22′ N	122°02′ E	28	76	41	93°	97°	113°	Aug. 1981
Outside of North Passage	31°20′ N	122°18′ E	8	10	18	132°	188°	166°	Aug. 1981
Inner Shelf	31°30′ N	123°30′ E	0.1	0.3	0.2	30°	179°	111°	Jun. 1980
Mouth of South Channel	31°06′ N	122°00′ E	36	114	76	238°	325°	352°	Aug. 1981
Cutside of South Channel	30°51′ N	122°22′ E	2	29	22	69°	319°	335°	Jul. 1981

outside the exit, it is only 15% that at the exit (Table 4-38).

(iii) A comparison between the satellite images and the surface suspended matter measurements shows that 64% of the suspended matter sink at the river mouth gate and the concentration at 122°30′–45′ E is only 6% of that at the gate (Fig. 4-41).

(iv) The distribution of the muddy sedimentary provinces agrees with the suspended matter spreading area. The southern part of the Changjiang Estuary is a high-rate sedimentary province. ^{210}Pb dating shows that its sedimentation rate can be up to 5.0 cm/yr, which agrees with the distribution of TSM concentration.

Main silt discharge of the Changjiang River is carried to the south by flow except part of it spreading eastward. All the isopleths of TSM content protrude to south-sowtheast. Outside the river mouth, silt discharge in both the middle and bottom layers moves southward. Especially under the action of the north wind in winter a large amount of suspended matter is carried by the Changjiang

Table 4-38 Net uni-width sediment transport and deposition at six time-series stations in the Changjiang Estuary

Position			Net uni-width content of sediment transport		Direction of net sediment transport		Deposit content of nearby two stations	
	ϕ	λ	flood	ebb	flood	ebb	flood	ebb
Inside of North Passage	31°25′ N	121°50′ E	231	134	123°	111°	39.7	34.6
Middle of North Passage	31°22′ N	122°02′ E	139	87	106°	111°		
Outside of North Passage	31°20′ N	122°18′ E	34		165°		75.5	
Inner Shelf	31°30′ N	123°30′ E	0.2		445°		99.4	
Mouth of South Channal	31°06′ N	122°00′ E	184	176	325°	325°		
Outside of South Channal	30°51′ N	122°22′ E	22		335°			

diluted water to the south, becoming a main source of the silt flow alougshore off Zhejiang, which results in muddy sediments in this area.

3. Exportation of disolved matter of the Changjiang River

Measurements of eight main ions in the water mass with salinity less than 0.02‰ show that the water mass of the Changjiang Estuary belongs to the $Ca(HCO_3)_2$ type, that is, $Ca^{2+} > Na^{+} > Mg^{2+} > K^{+}$, $HCO_3^{-} > Cl^{-} > SO_4^{2-}$. Concentration and total output of these ions are shown in Table 4-39.

Annual exports of the disolved matter amount to 200.4 million tons, which approximates to 178 million tons obtained by Le Jiaxiang[1)] et al in 1963. The annual exports ratio between disolved and

1) Le Jiaxiang & Wang Dechun, 1963, Chemical characteristics of the rivers in China, Journal of Geography.

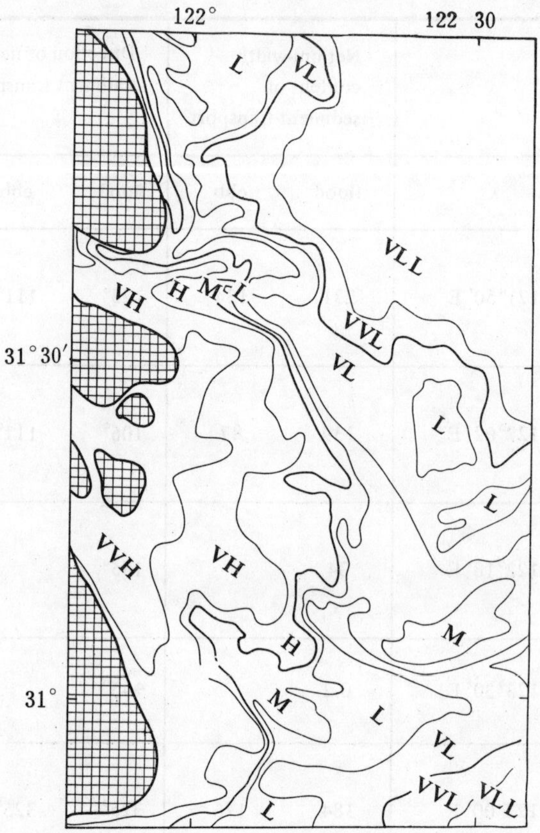

Fig. 4-41. Concentration distribution of the suspended matter off the Changjiang River mouth (July 15. 1981).

Table 4-39 Concentration and output of the major species of ions

Item \ Elements Quantity	Na^+	K^+	Ca^{2+}	Mg^{2+}	SO_4^{2-}	SiO_2	HCO_3^-	Cl^{-1}	Total count
Concentration (ppm)	9.1	2.2	34.5	5.6	13.3	6.2	125.7	20.0	216.6
Output (ton/yr) ($\times 10^6$)	8.4	2.0	31.9	5.2	12.3	5.7	116.4	18.5	200.4

solid matter is 1:2.4 which is larger than the world average 1:4.3. This is related to the middle latitude location of the Changjiang River basin and the high human development.

With the variations of strength of the river runoff and tidal current, the effective area of disolved matter varies greatly. In flood seasons the water inside the gate is of fresh type, that near the gate is of early-mixing type and that outside the gate is of strong – mixing type. The strongest chemical actions and variations occur near the gate.

4. Summary

1. Being carried by huge water mass, the transportation and diffusion of the sediment loads in Changjiang Estuary occur with the exportation of disolved matter and thus should be analysed in combination with water mass, solid matter and disolved matter, of which the quantitative ratio is approximately 4400 : 2.4 : 1. With regard to spreading limits, the water mass is larger than the solid matter, and the solid matter is larger than the disolved matter.

2. Sediment loads of the Changjiang Estuary are chiefly controlled by two main factors, that is, the flood and dry water seasons and the tidal period with which concentration, composition and transportation change remarkably. Its transportation after emptying into the sea agrees with the current system and the growth and decline of hydrodynamical conditions.

3. Suspended loads of the Changjiang Estuary spread and move southeastward all the year round. The gate area of the Estuary is an area where the silt content has a high value and the mix of fresh water with seawater occurs. It decreases toward the sea with its east boundary to be 123° E and north boundary at the exit of the North Branch. To the south it can be transported all along by the along-shore current.

4. Regional differences exist among the passages and channels of the Estuary. The North Passage which has a large grain-size of silt and high content of coarse-grain composition > 38 μm and terrigenous detrital matter is the main outlet of silt discharge emptying into the sea with a constent southeastward transporting direction. The South Channel has a relatively fine grain size and high concentration of suspended loads. A back-turning tendency occurs here in the net transporting direction, which results in the development of river mouth bars and suspended mud layers.

by Jin Qingming

Second Institute of Oceanography, National Bureau of Oceanography, Hangzhou

VI. Sedimentation on the Zhujiang River Mouth Region

1. A survey of the river mouth region

The drainage of the Zhujiang River basin abounds with precipitation, its annual mean runoff modulus takes the first place among those of the large rivers in China (23.6 kg/sec·km^2). Zhujiang River is characterized by an abundance of water(the total annual mean runoff is 3,400 × 10^8 m^3, and the annual mean discharge is 11,000 m^3/sec),a sparsity of sediment (the mean sediment content is 0.126 – 0.334 kg/m^3), and a considerable load-discharge (the annual mean suspended load-discharge is 8,336 × 10^4 ton). After receiving the water from Xijiang River, Beijiang River, Dongjiang River, Liuxihe River, and Tanjiang River, Zhujiang River enters the region of a river network of the Zhujiang Delta, and then passing through eight distributary mouths, namely, Humen, Jiaomen,

Hongjili, Hengmen, Modaomen, Jitimen, Hutiaomen, and Aimen, separately, finally flows into the South China Sea.

The Zhujiang River mouth consists of the above-mentioned eight distributary mouths. The Zhujiang River mouth as a whole is characterized by a strong runoff and relatively weak tides (the mean tidal range is 0.86 – 1.63 m). But there is a significant difference among the these rivers and distributaries in the aspect of discharge and load-discharge. For example, the discharge from Xijiang River is up to 2,460 $\times 10^8$ m^3, accounting for 72% of the total runoff of the entire Zhujiang River mouth region and the suspended load-discharge – 7,250 $\times$ 10^4 ton, accounting for 87% of the total suspended load-discharge from the Zhujiang River; however, the load-discharge from either Tanjiang River, Liuxihe River, or Dongjiang River is very small or rather small. In addition, topography, tides, and other marine dynamic conditions are not the same here and there in the Zhujiang River mouth region as a whole. Therefore, the sedimentation is rather complicated with a variety of sedimentary types. Fig. 4-42 is a scope of the Zhujiang River mouth region compiled with reference to the method proposed by Huang Sheng and Ge Zhijin (1963) for dividing types of an estuary[1)]. As shown in Fig. 4-42, the region between the upper and lower limits of the mean salt water boundary is the center of the Zhujiang River mouth region, or the river mouth section (or transitional section); that above the upper limit up to the boundary of tidal area in flood period is the inshore mouth section; that below the lower limit down to the area where the surface sea waters are directly influenced by the dispersion of runoff is the nearshore beyond the mouth. Fig. 4-42 also shows that the configuration of the Zhujiang River mouth possesses the following features:

(1) The eastern and western parts of the river mouth region are two funnel-shaped bays, that is, Shiziyang-Humen-Lingdingyang Estuary in the east and Yinzhouhu-Aimen-Huangmaohai Estuary in the west;

(2) The coastal zone between these two estuaries is an arc-like siltation zone which bulges outward with Modaomen as its apex. The main distributary mouth of Xijiang River and Beijiang River, i.e. Jiaomen, Hongjili, Hengmen, Modaomen, Jitimen, and Hutiaomen are all involved in this siltation zone.

Forms of the estuaries mentioned above indicate that features of the main distributary mouths of the Zhujiang Estuary are different. Li Chunchu and Yang Ganran (1981) have pointed out that two different types of the Zhujiang River mouth region are developed here.

First, river mouths with a strong runoff and a weak tidal current, i.e. those which are fluvial-dominated river mouths e.g. the main distributary mouths of Xijiang River with Medaomen and Hongqili as the typical ones. They are characterized by the strong runoff, as a great amount of fresh water discharges, so that salt water can only intrude shortly inward, and the variability of runoff is small, so the range within which salt water boundaries (or salt water wedge) shift is small (only 20–30 km); by a weak tidal current action and the smallest tidal range; also by a highly stratified mixing between salt water and fresh water.

---

1) Huang Sheng and Ge Zhijing, 1963, A discussion on the types of tidal estuaries, Nanjiang Institute of Water Conservancy Science.

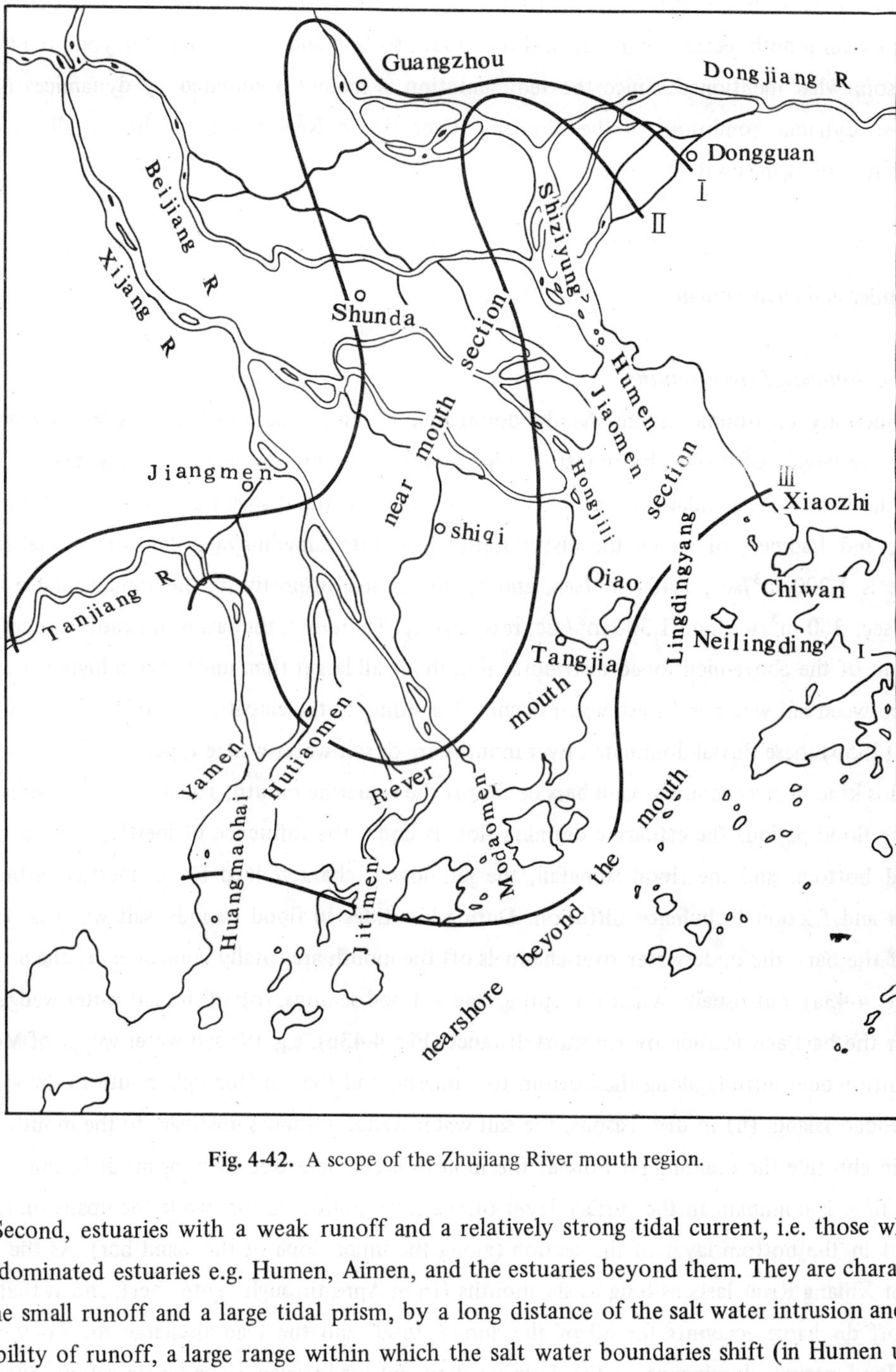

Fig. 4-42. A scope of the Zhujiang River mouth region.

Second, estuaries with a weak runoff and a relatively strong tidal current, i.e. those which are tidal-dominated estuaries e.g. Humen, Aimen, and the estuaries beyond them. They are characterized by the small runoff and a large tidal prism, by a long distance of the salt water intrusion and a large variability of runoff, a large range within which the salt water boundaries shift (in Humen up to 80 km, and in Aimen up to 40 km); by a relatively tidal current throughout the estuary and the largest tidal range in the Zhujiang River Mouth (the maximum can reach 3.64 m), which is increasing from outside inward; and by a partially mixing between salt water and fresh water (a sufficient mixing may occur in dry seasons).

The following is mainly to analyze dynamics and environments, sedimenttation and features of the sediments, sedimentary types and facies in the Zhujiang River mouth region. Emphasis will be put on the river mouth section, while the inshore mouth section and the nearshore beyond the mouth will be somewhat mentioned. Since the sedimentation is mainly dominated by dynamices of the estuary, so dynamic conditions of the Zhujiang River Mouth Region will be first of all discussed briefly in the following context.

2. Dynamics and environments

(1) *Fluvial dominated river mouths*

Sedimentary environments are mainly dominated by river actions, such as Modaomen and Hongqili, are largely controlled by runoff. For Modaomen, the annual mean runoff discharge is 2,500 m^3/sec, the annual mean tidal prism is 1,850 m^3/sec; and for the distributary mouths of Hengmen, Hongqili, and Jiaomen, of which the distributaries flow into Lingdingyang Bay, the annual runoff discharge is 1,200 m^3/sec, 1,400 m^3/sec, and 2,000 m^3/sec, respectively, the annual tidal prism is 280 m^3/sec, 360 m^3/sec, and 1,300 m^3/sec, respectively. Therefore, the ratios of runoff discharge vs. tidal prism of the above-mentioned distributar mouth are all larger than unity, and a highly stratified mixing between salt water and fresh water occurs, According to the classification of estuaries by D.W. Prithard (1955) these fluvial-dominated river mouths are of salt water wedge type.

In this kind of river mouths, sand bars lie far off the estuarine mouth, The dynamic featuress are: (i) during flood periods the estuarine draining flow is under the influence of inertia, friction of the river bed bottom, and the flood slopefall, the jet flow is characterized by an inertial turbulence diffusion and friction turbulence diffusion. During low tides in flood periods, salt water is washed away off the bars, the underwater river channels off the mouth are totally controlled by the draining flow (Fig. 4-43a), but usually when it is spring tide in flood seasons, top of the salt water wedge may pass over the bars and intrude over a short distance (Fig. 4-43b), e.g. the salt water wedge of Modaomen mouth would intrude along the bottom to Dongshi, and that of Hongqili mouth to the vicinity of Damaodao Island, (ii) In dry seasons, the salt water wedge advances upstream to the mouths (Fig. 4-43c), in ebb tide the draining jet flow at the mouths is characterized by buogant diffusion and the draining flow is dominant in the surface layer of the river mouth section, while the upstream flow is dominant in the bottom layer of the section (along the inner slope of the sand bar). As the flood period of Xijiang River lasts as long as six months (from April through September), and at that time the runoff discharge accounts for 80 of the annual total, and the load-discharge for 81–95%, so during flood period, dominance of the draining flow and load-transport under its actions exert the most important influence on sedimentation of the river mouths in question, However, in dry seasons with the intrusion of the salt water wedge, the river mouth section may be affected by sedimentation of suspension and load coming from the sea and salt water.

The above-mentioned facts are basical characteristics of flood and ebb dynamics during flood and dry seasons for the fluvial-dominated distributary mouths. Nevertheless, because of the differences in runoff, tides, reception basins, history of development, as well as other factors, dynamics and environments of these distributaries mouths are not quite the same. For example, at the Jiaomen mouth the draining runoff was rather small, the runoff force was comparable to tidal force, but of these decades under the effects of the evolution of the deltaic river channels in the upper reaches the draining runoff discharge has been increasing at the distributary mouth and leading to a series of changes in dynamics, sedimentation, and geomorphology (for example, the sand Jiposha Bar has been enlarging rapidly; as during flood period a great amount of flood waters has had to be drained, lateral branch channels have been formed one after another; in siltating development, the bar has also been cut into sand islets by the lateral branch channels). Differences in reception basins of the distributary mouths are all the more pronounced. Development of the Jiaomen mouth terminal is limited by the Humen channel and pressed by the Humen downstream, while the Hutiaomen terminal is affected by Yamen downstream. The reception basin off the Hengmen mouth is a semi-enclosed Tangjia-Qi ao shallow bay. On the left of Hongqili is a relatively open Lingdingyang Bay, it is favourable for developing the lateral branching. Only the Modaomen distributary mouth has advanced to the northern margin of the South China Sea continental shelf, facing the open sea and affected significantly by the waves; moreover, as the nearshore beyond the mouth is under the action of the westward longshore current, a lot of suspended loads is transported along directing by west.

(2) ***Estuaries dominated by tides***

The general dynamic conditions of both the eastern and western estuaries at the Zhujiang River mouth region have been mentioned above. Of late years, not a few authors gave detailed analyses of the hydrographic characteristics of Lingdingyang Estuary with contents including the discharging and load-discharging from the upper course, the stagnation and exportation in the estuaries; tidal influx and load influx from downstream and the transport towards the upper reaches[1,2]; features of tides and tidal current in the estuaries; intrusion of salt water and salt water wedge; mixing between saft water and fresh water; positions of the predominant velocity and null point[3], as well as some features of the jet flow at the river mouth, etc. No doubt, these studies have deepended the understanding on features of the estuarine dynamics of Lingdingyang. However, it should be pointed out that some important factors and processes that influence dynamics, environments, and sedimentation in the Zhujiang Estuary fail to attract attentions so far.

First, influence of the special geomorphological conditions on the estuarine flow structure should be attached importance to. Width of Yamen Throat mouth is 700 m, beyond the inlet is the broad Huangmaohai Sea, inside the inlet there was once a relatively broad sea waters – Yinzhouhu; at Humen Mouth two hills stand on either side with a 2,000 – 3,000 m wide water area lying in be-

1) Mo Rujun, Yan Lianhe, 1981, Hydrography of Lingdingyang Bay.

2) Guangzhou Institute of Geography, 1981, Hydrographical features of Lingdingyang Estuary at Zhujiang River mouth.

3) Hu Jiamin, 1979, A preliminary analysis of siltation tendency in Lingdingyang Bay at Zhujiang River mouth.

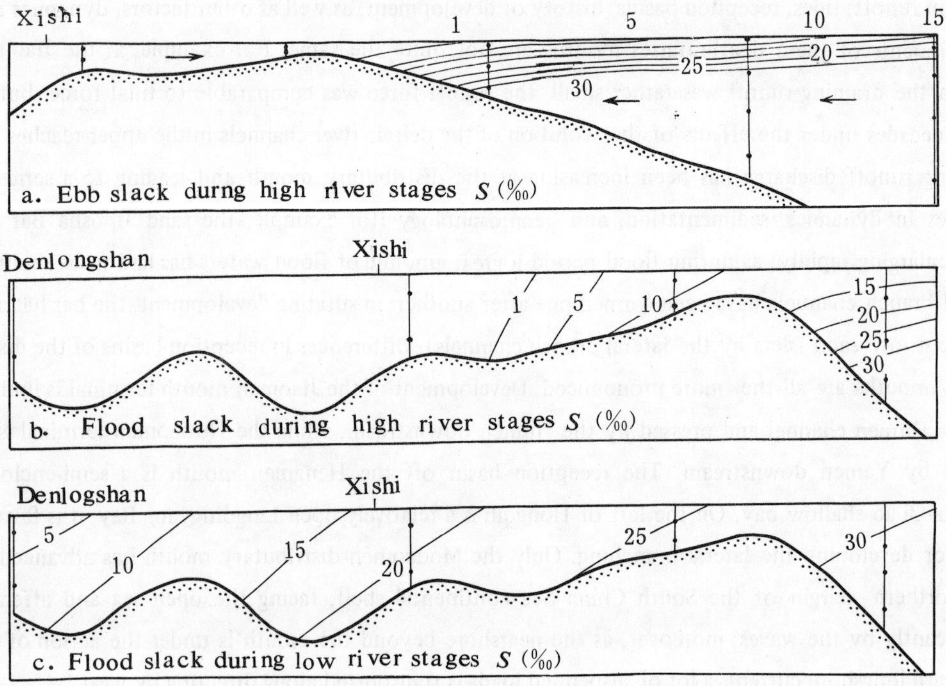

Fig. 4-43. The longitudinal distribution of salinity at the Modaomen mouth.

tween, beyond the inlet is Lingdingyang Bay, inside the inlet it was once a relatively broad sea waters – Shiziyang Bay which was called "immense sea" in Song Dynasty, and was still a bay in Yuan and Ming Dynasties. Therefore, the Yamen Throat and Humen Throat possess geomorphological features of tidal inlets, correspondingly, they possess hydrographical conditions of tidal inlets. Though there is basically on longshore sand drifting here, velocities of flood and ebb currents at the inlets are significant (the maximal velocity is over 2 m/sec), tidal currents wash and erode the bottoms at throats and the channel upstream and downstream, producing a certain amount of sediment sources. Thus, with the formation of tidal scoured troughs at the inlet and in the channel upstream and downstream, in sea areas inside and outside the inlet, tidal accumulative bodies similar to "flood tide deltas" and "ebb tide deltas" may also be developed.

Second, it should be pointed out that the nearshore upwelling of the "deep layer continental shelf water" with high salinity from the continental shelf area off Zhujiang River Mouth exerts significant influence on the sedimentary environments and water body structure of the estuary. In summer the northern part of the South China Sea is dominated by the South China Sea water mass with high temperature and high salinity, nuder the effects of NE-flowing sea current and upwelling from the bottom of continental shelf, the highly salty South China Sea water mass advances to press the waters offshore Zhujiang River mouth. According to the findings in the hydrographical investigation on the continental shelf area offshore Hong Kong by Y.K. Chau and C. S. Wong (1960), in that area the deep continental shelf water is of high salinity (34.5‰), while the surface continental shelf water has a bit lower salinity; the deep continental shelf water may exist all the year round in the area 60 nautical miles off Hong Kong with a water depth of 50 m. In summer when the surface continental shelf water advances seaward as a result of the great amount of fresh water discharging from Zhujiang River, the deep continental shelf water moves onshore, producing upwelling with its front extending to Zhujiang Estuary, which is a compensation resulted from the offshore movement of the surface fresh water runoff. The actual measurements of Lingdingyang Estuary in summer indicate that the compensation and the sea water onshore movement resulted from upwelling from sea floor may impact and extend into the estuary. It is shown in Fig. 4-44 that water body of Lingdingyang Estuary

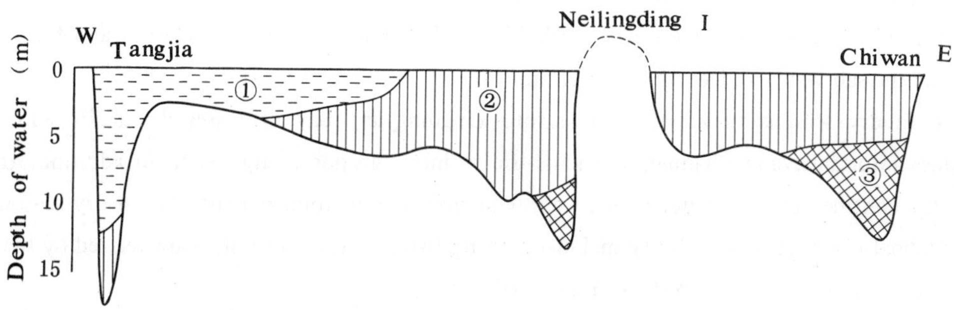

Fig. 4-44. The three water masses dominating Lingdingyang Estuary in summer. (cross-section of Tangjia-Neilingdingdao Island-Chiwan)

1–Fresh water mass; 2–Dilute water mass; 3–Shelf water nass.

consists obviously of three water masses, namely, the fresh water runoff from the northwestern distributary mouths, the initially mixing "dilute water" in the estuary with a salinity of 2–25‰ and the highly salty (the actually measured maximal salinity is 32.86–34.16‰) continental shelf water at the deep trough bottom in the southeastern part of the estuary. It is suggested that the dynamic processess of Lingdingyang Estuary are basically interactions and movements of these three water masses, accompanying the rise and fall of runoff, flood and ebb of tides, the intrusion and the retreat of the continental shelf water from the bottom. Based on such knowledge, the dynamical and environmental characteristics of Lingdingyang Estary can be summed up as follows:

1. Humen Throat is of tidal inlet nature, the south of the channels of the inlet prevail by ebb tide, effecting a range involving the Chibi and Fanshi channels beyond Humen inlet; the north channels of the inlet are prevailed by flood tide when Shiziyang Bay was still an open sea bay, but now Shiziyang Bay has been siltated and getting narrower and become a river section, there is no more flood tide prevailing at some branch channels; nevertheless, at a portion of the river trough of the Shiziyang channel stem flow today (such as the western trough in the Lianhuashan shallow bank area), the flood tide prevailing is still very apparent.

2. The deep trough area south of the line along Xiaochan-Qiao (the south section of the Ling-

ding channel and Anston channel) is a channel through which the continental water resulted from the offshore tidal current and continental shelf upwelling moves onshore, entering the estuary. Here, the bottom layer water is dominated by the highly salty continental shelf water and is mainly upstream.

3. In flood period the runoff discharge in Lingdingyang Estuary increases, at the same time the continental water from the bottom through the estuarine mouth wedges in. Thus, the uppermost and deepest water masses with different densities (the fresh water mass and the continental shelf water mass) both get stronger in summer, but they move in opposite directions. As a result, there occurs a large-scoped pycnocline, and concentration of the "dilute water mass" lying between the two water masses increases, longitudinal distribution of salinity shows somewhat being stratified, with small horizontal gradient and large vertical gradient, as well as an obvious estuarine circulation (Fig. 4-45a). At that time in the north area of the line along Xiaochan-Qiao, the downstream prevails, in the middle by upper part of the bay north of that line is an area where salt water and fresh water mix and the salt water wedge top moves to and fro.

4. In dry seasons runoff that flows into Lingdingyang Bay decreases drastically, salt water intrudes into Shiziyang Channel, sediments are mainly transported upstream. In addition, at that time the continental shelf water which wedges in from the bottom retreats. So, in dry seasons the longitudinal distribution of salinity in Lingdingyang Estuary is, in general, characterized by the large horizontal gradient and small vertical gradient (Fig. 4-45b).

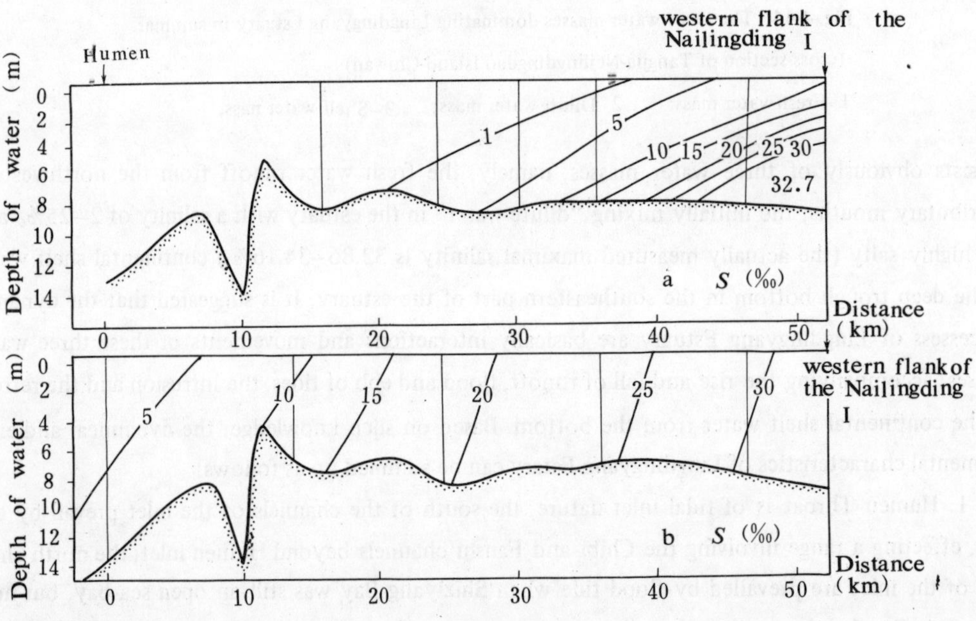

Fig. 4-45. In flood and dry seasons the longitudinal distribution of salinity in the Lingdingyang Estuary.

5. The western boundary of the estuary is relatively unstable, the estuarine deep trough and shallow bank (mainly Lingding Channel and Fanshi Bank) are influenced considerably by diffusion and siltation of the water and sediment discharge coming from the western distributary mouths.

Dynamics and environments of he Yamen-Huangmaohai Estuary are much similar to those of the Humen-Lingdingyang Estuary mentioned above. For example, the bottom layer current in the southern deep trough of the estuary is mainly dominated by the continental shelf water (once its salinity was measured as 31.14‰), and is mainly upstream; the middle part of the estuary is an area where salt water and fresh water mix in flood period; the area beyond Yamen Throat to Chibi possesses the hydraulic characteristic dominated by ebb tide current prevailing on the outer side of a tidal channel; in dry season salt water may intrude into Yinzhouhu Channel.

3. Sedimentation and characteristics of sediments

Estuarine sedimentation is completely decided by estuarine dynamics[13]. Material sources, grain size, composition and distribution, as well as chemical, mineral, and biological features of the sediments in Zhujiang River mouth all represent obviously the dominance and influence of the estuarine dynamics mentioned above.

(1) *Sources and movement trend of materials*

There are three major sources of the Zhujiang River mouth sediments, namely, sands from the river basin, sediments from the sea, and sediments produced through scouring and eroding the estuarine bed (trough) by tidal currents.

(a) *Sediments from the river basin*

Sediments from the river basin that have entered into the estuary are mainly fine sand and the more finer-grained sediment. The fine sand is mostly bed load. Based on our investigations on the bed sediment composition and distribution of the Zhujiang deltaic channels, fine sand coming directly from upper reaches (i.e. Tanjiang, Liuxihe, and Dongjiang) and entering the two estuaries is scarce, But at the mouths of the major distributaries of Xijiang and Beijiang River, which are dominated by river aciton, a considerable amount of fine sand enters the river mouth region. The Modaomen near mouth section of Xijiang River has a shallow channel bed, the medium grain size of bed sediemnts is 0.293–0.150 mm, the mean content is 78%, in some places it is up to 90–95%. The fine sand is mainly discharged from the river basin during flood period, and moves along the river bed to the tidal channel section (i.e. near mouth section), and then influenced by tides, so that the velocity is slowed down and a massive fine sand thus accumulates. As mentioned above, the distributary mouths which are dominated by river actions, of Zhujiang River mouth are controlled by downstream flow during flood period. Therefore, fine sand that has accumulated on the near mouth section river bed may be continuously transported foreward to the river mouth section, and then is stopped in front of the salt water wedge; in flood period the top of salt water wedge is located on crest of the sand bar or in front of the barrier beach, where is the final stop of the bed load motion. When flood encounters spring tide, the salt water wedge intrudes crossing over the sand bar crest, the downstream-shifting fine sand changes its direction and discharges through some lateral branches, such as those in Hengmen mouth, Jiaomen mouth, and crevasse channels (e.g. Hongqili mouth), which is similar to the situation of Mississippi River. In dry season as the salt water wedge intrudes to the mouth, and under the influence of siltation of the sea suspension carried in by the upstream flow, sediments in the river mouth section commonly get finer.

At Zhujiang River mouth fine-grained sediments that are finer than fine sand (including part of extremely fine sand and silt and clay) are mainly transported in suspension. Part of the suspended sediment discharging through Modaomen mouth is deposited on the front slope of the bar, while another part is transported by the longshore current in a somewhat west direction and then deposited; suspended load discharging downstream through the distributary mouths of Hengmen, Jiaomen and Hongqili, and through Hutiaomen is the main source of sediments in these two estuaries, respectively.

(b) *Sediments coming from the sea*

There are two aspects concerning sediments coming from the sea. According to the results of investigations on the sediments on the northern continental shelf of the South China Sea, offshore Zhujiang River mouth on the inner continental shelf with a water depth of 30–60 m are distributed sediments of coarse gravel and sand, which is not coordinate to the fact that on the inner continental shelf of the northern South China Sea are distributed mainly sediments of clayey fine grains. Li Cuizhong et al. (1980) suggested that the composition of this patch of coarse material is the same as that of the Quarternary ancient fluvial gravel layer which is widespread distributed on the basement of the Zhujiang delta; their exposure indicates that here is a local scoured area, formed by the upwelling from the bottom of the continental shelf, which is resulted from the interchange between the complex water masses offshore Zhujiang River mouth, and the obstacle to covering on it by the modern Zhujiang River fluvial load. From this area northward up to Zhujiang River mouth, material gets finer and finer, when water is 20 – 30 m deep, there distributed in the area is fine sand, and when it is in Lingdingyang Estuary mouth front (e.g. the southern section of Lingding Channel), sediments are coarser than those in the middle section of the estuary (see Fig. 4-46), with a 30–40% of silt, indicating that nearshore beyond Zhujiang River mouth, accompanying the upwelling of the continental shelf bottom water and its intrusion into the eatuary, exists a process of sediment sorting, transport and deposition from sea onshore, The other "sediment coming from the sea" refer to such load either of suspension originally carried offshore by rivers, or produced through flocculation and then with the fine-grained material (mud) brought back to the eastuary by the circulation, waves, and tidal current and deposited there; usually they are carried to nearshore tidal flats and to the areas which are under the influence of estuarine salt water wedge by flood tidal currents and then deposited there.

(c) *Load produced through scouring the estuarine bed by tidal currents*

There are not a few troughs coured by tidal currents inside the estuary at Zhujiang River mouth region with a water depth of up to 10–20 m, e.g. Urmston Course, the southern section of Lingding

Channel, and Humen and Yamen tidal inlets. Of these, Humen and Yamen Throats are the deepest ones, with the depths of up to 34 m and 27 m, respectively. In the Zhujiang deltaic area, the Quaternary loose sedimentary layers are, in general, 20–30 m thick with the lower portion consisting of mainly fluvial gravel layers and clay of miscellaneous colours (red, yellow, and white). Thus, currents in the scoured troughs mentioned above can get to and erode this sequence of materials, thereby, producing sources of a certain amount of load, of which the coarse-grained load, such as gravel, coarse sand, and medium sand are not common in sediments of other estuaries. Such coarse load is the material base for development of sand bodies formed by tides inside the two estuary.

Hence, sediments at Zhujiang River mouth region possess a complixity in their material sources and movements.Summing up, sedimentation at the distributary mouth where fluvial actions are dominant are mainly controlled by sediments coming from the river basin; while the estuary is under the action or influence of the abovementioned three kinds of load sources. From the viewpoint of long-ranged time scale (historical processes), load from the second and the third sources (excluding those coming from the river basin and then carried back by flood tide current and deposited) once was the important material source for the estuarine sediments at Zhujiang Rigver mouth region, sedimentation of which laid down the foundation of the present estuarine sediments. However, with the extension of siltation at the entire Zhujiang River mouth and the change of geomorphology of the estuary, the importance of these two sources is decreasing, on the contrary, the developing modern estuarine processes resulted from the extension of siltation at Zhujiang River mouth and the superimposing sedimentation of sediments coming from the river basin and deposited in the estuary become more and more important.

(2) *Grain size composition and its distribution of sediments*

According to the analyses of shallow strata exploration surveys and drilling data, thickness of sediments at the modern Zhujiang River mouth ranges from several meters to 20 m, materials consist of loose sand, silt, and clay with the underlying sediments of relatively hard clay or clayey soil. Usually, they lie 10–30 m below the theoretical depth base level of charts, i.e. the undulated ancient base level of stream (Late Pleistocene sandy soil, clayey soil, or ancient fluvial gravel bed).

According to the characteristics of sedimentary environments and material composition, for the sediments and sedimentation at the modern Zhujiang River mouth region, several regions may be divided as follows (Fig. 4-46):

(a) *Sandy sediments at the river mouth sections of distributary mouths of Xijiang River and Beijiang River*

The river mouth sections of Modaomen, Hongqili, and Hengmen, as well as Jiaomen, which are dominated by river actions, are major areas for accumulation of the coarse grain-sized sand coming from the river basin. Sediments consist of fine sand and silt with a medium grain size ranging from 2 to 5ϕ (0.25–0.32 mm), saltation population is dominant (80–90%), sorting is, in general, good or very good. at the frequency curve a pointed peak may be seen, slope of the possibility accumulation

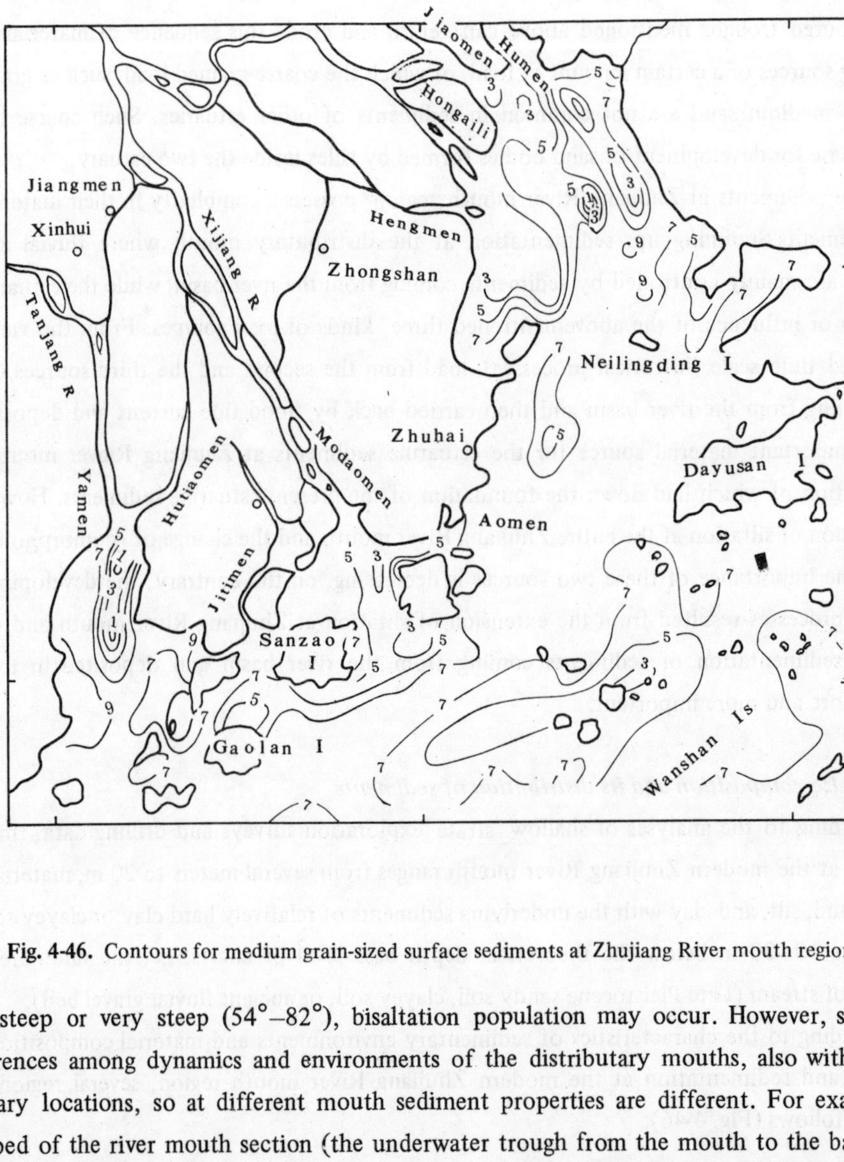

Fig. 4-46. Contours for medium grain-sized surface sediments at Zhujiang River mouth region.

curve is steep or very steep (54°–82°), bisaltation population may occur. However, since there are differences among dynamics and environments of the distributary mouths, also with different sedimentary locations, so at different mouth sediment properties are different. For example, the channel bed of the river mouth section (the underwater trough from the mouth to the bar crest) is mainly sections where bed load from the river basin is transported and deposited; sediments in the underwater trough section by the mouth area well-sorted fine sand with a Md ϕ ranging 2-3; sand waves are present, there are also cross beddings; but as it is intruded by salt water and influenced by density current of salt water wedge, materials in the underwater trough section along the inner slope of the bar are comparatively fine-grained, which usually are clayey sand or sandy clay (Mdϕ = 3–4), virtually are interbeddings of sand and mud with cross beddings and horizontal beddings. Another example is the crevasse-splay sediment on the left side off Hongqili mouth. It is a fan-shaped accumulative body formed in the event which occurred in 1947 when the Zhujiang River highest flood (once happened in two decades) encountered a spring tide, flood and load carried by it broke the subaqueous natural levee in the rear part of Wanqingsha. Sediments by the crevasse channel are digitate crevasses bed sand consisting of bed load and with a thickness of two meters or so, Md ϕ = 2.37–3.02, sorting coefficient is usually smaller than unity; they become finer outward like a skirt, farther present are clayey sand and sandy clay, similarly, showing interbeddings of sand and mud, features of such composition and structure are quite similer to those of the crevasse sediments at the Mississippi River mouth. Of the past two decades (1954–1974), the accumulated load on the crevasse splay sedimentary body has been totally 67,060,000 ton; Again, Jiaobeisha Bank in Modaomen mouth, and the shallow bank area south of it with a water depth of less than two meters are influenced significantly by waves, so sediments there are well-sorted into pure fine sand with a content of over 90%, Md ϕ= 0.15 mm. As a result of sorting and zonal transport by waves, on the shallow bank developed are 3–4 sand levees or beach ridges. Besides, on one or either side of the underwater troughs off Modaomen and Hongqili mouths are also developed subaqueous natural levees; sediments are mainly extremely fine sand and silt.

(b) *Tidal sand sediment in the tidal inlet system outside and indide the Humen and Yamen gorges*

Inside and outside the Humen and Yamen Throats are developed tidal inlet sediment systems formed by the Late Pleistocene loose material from the channel bed scoured by tidal currents. The system is composed of inlet nel bed sediment and the "flood tide delta" and "ebb tide delta" outside and inside the throats. Of this system those that are relatively complete are at the Humen Inlet mouth and in the areas inside and outside – Shiziyang Channel and Lingdingyang Bay. Characteristics of material composition are described below.

Gravel, pebble or such eroded matter as clay of miscellaneous colours which is directly exposed from the cauce bed are commonly seen at the inlet and channels inside and outside scoured by flood tide and ebb tide (e.g. the deep troughs inside and outside Humen Throat), but at the and of the channel (e.g. Fanshi Course) the sandy sediment consists of eroded matter transported here by tidal current from the inlet; the radiant tidal sand bodies (ridges) on the "flood tide delta" and "ebb tide delta" in the bays outside and inside the inlet are mainly composed of medium and fine snad, coarse sand and gravel, as well as the like coarse-grained sediments with fairly good or intermediate sorting. There are some "ebb tide deltaic" sand bodies, such as Shanbansha, Lingdinglanjiangsha, Jiaoyisha, Gongsha, and Fanshisha, outside the Humen Throat, and Chibisha outside Yamen Throat; as for "flood tide deltaic" sand bodies, they are Haixinsha, Jiangousha, Xidatansha, Dasha, Xianwusha and Shazai, which show a radiant distribution inside the Humen Throat (Fig. 4-47). Similar tidal sand bodies may be seen in some estuaries abroad, e.g. in Chesapeake Bay and in Delaware bay in the northeastern United States the "tidal sectors of tidal sedimentation" inside the mouth are "inlet-assoriated shoals". The superimposing sedimentation of modern estuarine accumulatants (silt and clay) has been continuously changing the composition of the inlet sediment system. For example, inside the Humen Throat the flood tide sand body has been gradually siltating and enlarging, and since the end of Qing Dynasty exposing bit by bit, forming plains or sand islands; outside the Yamen Throat the tidal sand ridges have been somewhat buried by modern estuarine mud (e.g. the two sand ridges west of Chibisha); outside the Humen Inlet the siltating development of tidal sand bodies also like that (e.g. Fanshi Shoal). As for the inlets proper, since the tidal hydraulic conditions of the whole channel system tends to be weakening, so siltation occurs in different degree everywhere, Taking the channel inside the Yamen Throat as an example, the course bed is entirely buried by modern estuarine silty mud (or clay); results of shallow strata exploration survey on Fanshi Course indicate that on the original sandy bed foundation the overlying modern estuarine mud matter has a thickness of, in general, 3–5 m, and the maximal thickness of 7–10 m.

(c) *Silty clay sediments on the fore-slope of the river mouth*

Of the fore-slopes at Zhujiang mouth that of Modaomen bar is the most obvious one with a slope of 1.3‰, others are all gentler or much genlter than this one. The range of a fore-slope is roughly from a bar crest (corresponding to the boundary of salt water or the lower limit of the salt water wedge top) below to the area with a water depth of 15 m or so (i.e. offshore the river mouth). Since the bed load (mainly fine sand) discharging from the river basin has basically stopped at the salt water wedge top, and there is not much coarse sediments coming from the continental shelf area; in addition, the latter only influences some deep troughs controlled directly by the intrusion of continental shelf water, therefore, there is not much deposited coarse sediments. As shown in Fig. 4-46, in this region sediments are mainly silty clay with a medium grain size of 7–8 ϕ (0.016–0.004 mm), and a poor sorting ($Qd\phi>1$), mostly being negative skewness. Clay content (grain size less than 0.004 mm) in Lingdingyang Estuary amounts to 70 %. Having studied the distribution and generation of clay mineral in James River Estuary, Virginia of the United States, J.P. Fenillet (1980) pointed out that various sinking factors, flocculation, and diagenesis exert no or very little influence on the distribution of clay mineral, while the estuarine circulation exerts decisive influence on it. Distribution of estuarine fine-grained sediments at Zhujiang River mouth is obviously influenced by the estuarine circulation, too, as the region in question is just situated in the range of salt water wedge in flood period, the surface water is influenced by the diffusion of flood runoff, which is dominated by downstream flow, while the middle and bottom layers are dominated by upstream flow. Therefore, in flood period, suspended load carried downstream by surface water can flow upstream in the middle and bottom layers under the action of estuarine circulation and returns into the salt water wedge, and then becomes deposited.

(d) *Miscellaneous sediments in deep troughs at estuarine front*

Sediments in deep troughs at Lingdingyang and Huangmaohai estuarine fronts are influenced by the intrusion of continental shelf water and the upstream flow from below the bottom, so, they are, in general, somehow coarse. They are either sand (or sand gravel) mixing with mud with poor a sorting (e.g. at the southmost end of Lingding Courss, and the Sanjiaoshan strait section at the Huangmaohai estuarine front), or Late Pleistocene sand-gravel layer and hard clay, or even base rock directly exposing from under the channel bed (e.g. from Urmston Course to Jishuimen of Hong Kong). In the deep troughs in the northern section of Lingding Course and on the west side of Huangmaohai estuarine front there are rather thick sediments of modern mud.

(e) *Silty clay sediment on nearshore shoals or mud flats*

This is another fine grained load sedimentary region at the Zhujiang River mouth region. Medium size of the sediment is about 8 ϕ or finer, frequency curves tend to show polypeak type, sorting is poor and mostly negative skewness. Regions distributed include estuarine tidal flats (Huangmaohai Estuary both sides and the shoal on the east side of Lingdingyang Estuary), interdistributary bays(e.g. Baitungwan Bay, and Tangjiawan Bay), and beaches that are in the deltaic plain stage and beconing lands (e.g. Wanqingsha tail, Denglongsha tail, and Jiposha). According to our investigations, for these tidal flats or beaches, most show an obvious return siltation in dry period, the sediment is mud that temporarily deposits on the fore-slope at river mouth in flood season; as tidal currents strengthen in dry season, it is eroded and suspended again, and then carried by flood tidal currents together with the intruding salt water back to the nearshore tidal flats and beaches in the river mouth sections, finally deposits there.

(3) ***Chemical composition, mineral, and biological characteristics of sediments***

In the aspects of chemical composition, mineral, and biological characteristics, sediments at the Zhujiang River mouth region reflect either the processes of dynamics and environments, or the influences by geological and geographical conditions of the river basin.

(a) *Chemical composition*

Distribution of organic content, pH, and other chemical factors of sediments at the Zhujiang River mouth region shows a tendency of gradual decrease from the river mouth sea-ward. For example, in the distributary mouth sections of Xijiang River and Beijang River organic matter content is larger than 2.0%, in most regions of these two estuaries organic matter content is about 1.6–2.0%, while offshore the mouths it is ususally 1.2–1.6%. Hence, organic matter content at the Zhujiang River mouth is apparently higher than that at Changjiang Estuary. This may be related to the facts that the river basin is situated in sub-tropics, and sources of bio-organic matter are abundant. It is suggested that 690,000–810,000 ton organic matter per year is transported by the Zhujiang drainage from land to the estuarine area. pH value of bottom load near the river mouth section of the distributary mouths of Xijiang River and Beijiang River is less than 7.4, in most regions of the estuaries 7.4–7.6, offshore the mouths 7.6–7.8. Sediments contain mainly Ca, Mg, K, Na, Fe, Mn, Ti, Al, and P. Owing to the influence of human activities, chemical pollutants carried by Zhujiang River also deposit in the river mouth regions. Having studied the distribution and movement of such heavy metal matter as Cu, Pb, and Zn in the water body and suspension at Zhujiang Estuary, Li Feiyong et al. (1981) demonstrated that owing to the interaction of runoff with tidal current, and of salt water with fresh water in the river mouth regions, it favors the chemical pollutants discharged from the river basin to accumulate here. Analyses of Cu, Al, Zn, As, Cr and other pollutants in bottom load[1)] indicate that the highest content is in the river mouth section where it is a medium-degree polluted region, while it is decreasing from the section sea-ward. Guo Wenying and Wu Ping (1981) analysed the geochemical indexes of phosphates deposited near the mouths or in the river mouth sections of Yamen, Modaomen, and Humen at Zhujiang River mouth[2)], their results indicate that the ratio of phosphate components [$\left(\frac{Ca}{Fe + Ca}\right) PO_4 \times 10^{-6}$] is 0.08–0.33 in the near mouth section , and 0.4–0.7 in the river mouth section, that is, from the river mouth sea-ward with the increase of salinity in the water body, content of ferric phosphate in sediments is decreasing, while that of calcium phosphate is increasing. This is due to the fact that when the terrestrial environment changes into a marine one, the migration behaviour of calcium element is different from that of ferric element.

(b) *Mineral*

Mineral deposited at Zhujiang River mouth consists mainly of quartz and clay mineral. Contents of other components are usually under 10%, light and heavy minerals are commonly seen mica, feldspar, magnetite, hematite, zircon, epidote, diopside, titanotite, tourmaline amphibole, barite, niobite-tamtalite, samarshite, tremolite, sillimanite, and limonite, etc. There is a small amount of monazite, garnet, apatite, anatase, cyanite, xenotime, fluorite, wollastonite, titanite, cassiterite, topoz, malachite, rutile, hausmannite, chlorite, psilomelane, and florencite as for authigenic mineral it is mainly pyrite in grained nodules. Most of the above-mentioned minerals reflect the regional geological conditions in which widespread distributed are the intermediate-acidic igneous rocks of Yenshan Period, Lower Palaeozoic group metamorphic rocks, as well as Paleozoic and Mesozoic sandstone, shale, and carbonatite in the Zhujiang River basin. According to the recent studies[3),4)] of interest, mineral composition and distribution of sediments at Zhujiang River mouth are mainly described as follows:

1. Minerals are mainly terrigenous. In the river mouth sections of the distributary mouths of Xijiang River and Beijiang River, as well as in the northern sections of the two estuaries, quartz content in the load with grain size coarser than that of silt usually amounts to 85–90%, while the load which is finer than silt contains more mica, and load of clay order is mainly clay mineral. Ter-

1) Investigation Group of Pollution at Zhujiang River mouth, 1978. Investigation report on pollution in the sea area at Zhujiang Estuary.

2) Guo Wenying & Wu Ping, 1981, Geochemical indication of phosphates deposited at Zhujiang River mouth.

3) Zhu Sulin et al., 1982, A preliminary study on heavy minerals deposited on the surface layer at Zhujiang River mouth and along the adjacent coasts.

4) Wu Wenzhong et al., 1978, A report on the preliminary investigation on the sources of Lingdingyang Estuary load.

rigenous heavy minerals are mainly such components, as magnetite, titanotite, hematite, and zircon, next come tourmaline, epidote and amphibole. In the river mouth section which is under the action of to-and-fro tidal currents and waves heavy mineral contents are relatively high, generally up to 3–4%; in one or two sections (e.g. Hongqili crevasse-splay shoal) the contents are very high, they may reach 17–18%; while those in the offshore section usually are only 1%.

2 In the southern section of the estuary exist minerals indicating sediment from the sea, but with a small amount. Sediments coming from the sea is mainly indicated by the appearence of marine authigenous minerals and marine organism remains. The relatively commonly seen marine authigenous mineral is the spherunellitic authigenic pyrite; next comes alauconite which is also spherulitic. Marine organism remains are commonly seen coral fragments, foraminifera shells, sponge spines, diatom and other biotic debris.

3. In different environment the property and morphology of minerals are different. For example, at Zhujiang River mouth in nearshore regions with water depth less than 5 m, which are dominated by tidal current and waves, magnetite and titanotite and so on are all of isometric grains, whereas zircon shows a semirounded or spheroidic shape, and a great amount of ferric rounded matter appears The latter possesses a core of detritus (or other mineral), and a ferric coating, showing a rounded shape, rich in luster. Its origin is as follows: in the South the fluvial aqueous solution contains plenty of Fe^{+} ions and Fe(OH) molecules; when such soluble ferric matter of low grade enters into the estuarine sea area, under the action of electrolyte it coats detritus or other minerals and precipitates: moreover, the grains are always influenced by runoff, tidal current, and waves, and in a motion manner, so that they become rounded and rich in luster. Such ferric rounded grains may also be commonly seen on the nearshore shallows along the coast of South China, which may be a characteristic of sediments at the Zhujiang River mouth, mouth of the large river in the southern part of China. It differs from that of sediments at the river mouths in the central and northern parts of China. However, in regions with water depths of 10 m or so, shapes of heavy mineral are rather irregular, for example, zircon shows a prismatical, or macro-prismatical shape, or in fragments with clear crystal, and undamaged crystal edge and crystal plane; while titanotite appears in a thick-tabular form or plate-like; prismatic minerals, such as diopside, sillimanite, and tremolite, are obvious prismatice fragments with clear cleavage face. All these indicate that in regions with a water depth of 10 m or so, the modification of minerals by water flow and waves is rather weak.

(c) *Biota*

Bio-sedimentation at Zhujiang River mouth involves that of animals and that of plants. In animal sediments there is a great amount of foraminifera, ostracoda and other micro-paleaobiotic remains, while in plant sediments the buried mangrove remains at the nearshore tidal flats are commonly seen. According to the previous studies[1)], composition and distribution of foraminifera and ostracoda

---

1) Geomorphology and sedimentation investigation branch of the integrated investigation team on the coastal zone and mud flat resources of Guangdong Province, 1982, A report of the investigation on geomorphology and sediments at Zhujiang River mouth.

buried in sediments at Zhujiang River mouth are characterized by the following facts:

1. Foraminifera in the river mouth section where fluvial processes dominate are "Mississippi River type" with low quantity (in Modaomen river mouth section, on an average, only 271 individuals/50g sediment) and monotonous species, normal size of shells. And mainly are the euryhaline compoment buried in situ. For example, in Modaomen river mouth section it is the assemblage of *Elphidum nakanokawaens* and *Ammonia beccarii* Var., accounting for 83.4% of the total individual number of the whole group; as for ostracoda, they are of either terrestrial or marine facies, an assemblage of *Sinocytheridea latiovata* and *Candoniella.* In the western part of Lingdingyang Estuary at the river mouths where river processes dominate foraminifera are an assemblage of *Ammonia beccarii* and *Nonion sinensis.* Offshore Modaomen mouth, 50g sediment may contain nearly a thousand foraminifera individuals. Obviously, they increase in quantity, and a great amount of stenohaline species appears, being an assemblage of *Brizalina striatula* and *Ammonia annectens*; as for ostracoda, they are of an assemblage of *Sinocytheidea latiovata* Hou et Chen and *Neomonoceratina dongtaiensis* Chen and *Leguminocythersis hodgii* (Brady), and nearly all are of marine facies.

2. In the northern section of the estuary where tidal processes dominate, e.g. Fanshi Course of Lingdingyang Estuary, the upper section of Lingding Course, in the deep troughs in the northeastern part of Jinxingmen Throat and Huangmaohai Bay, foraminifera are mainly of an assemblage of *Scutulovis* sp. and *Nonion sinensis,* indicating the environment there is a sea-river interaction zone. In the southern section of the estuary they are mainly of an assemblage of *Ammonia anneatectens-Brizalina striatula,* being euryhaline. Brackish-water and marine benthic species are in symbiotic association, and accompanied by the neritic and stenohaline foraminifera coming from the continental shelf, such as *Miliolina, Nodosaviacea,* and *Buliminacea.* Shells of *Nodosaviacea* and *Buliminacea* are largely well preserved without damages; obviously, they are authchthonous in situ. This differs from the fact at the Qiantang-River mouth (strong tide type) where foraminifera are mainly marine components carried by tidal currents to the mouth and are allochthonous, since the southern section of Zhujiang Estuary is directly influenced by the intrusion of continental shelf water.

Mangroves grow everywhere on nearshore beaches at Zhujiang River mouth. They are mainly distributed on the tidal flats of the two estuaries, and in the interdistributary bays. This is another apparent feature of Zhujiang River mouth, which differs from those of other large river mouths in the country. Major species are *Braguiera gymnorrhiza, Kandelia candel, Avicennia marina, Aegiceras corniculatum, Excoecaria agallocha, Myoporum bontioides, Heritiera littoralis,* and *Acanthus ilicifolius.* Other concomitant plants are *Derris frifoliata, Phragmitis Communis, Cyperus malaccensis* and *Paspalum distichum.* Prof. Zeng Zhaoxuan et al. (1964) pointed out that in sediments of the coastal plain at Zhujiang Delta, where mangroves once grew, are distributed the buried mangrove remains "wood fragment layer". Li Chunchu et al. (1965)[1)] studied the evolution process of mangrove beach and plain on the east side of Huangmaohai Estuary and suggested that: As a baby mangrove beach develops to a period when other mangrove species (e.g. *Aegiceras corniculatum*) have replaced the

1) Li Chunchu et al. 1965, The mangrove coast of Xijiang Delta.

pioneer mangrove plants (e.g. *Avicennia marina*), sedimentary structure of the beach begins to change significantly, that is, with succession of mangrove vegetation and the continuous siltation and heightening of the beach, the eliminated mangrove disappears from the ground of the beach, but their roots are still buried in mud, forming a concentrated distribution of mangrove remains burial layers. Therefore, buried mangroves are resulted from natural siltation and development of mangrove beaches. Having studied the evolution stages of mangrove tidal flats in Ord River Estuary, West Australia, B.G. Thom (1975) et al. also drew the similar conclusion.

(4) *Sedimentary types and facies*

There are not many studies on the problem concerning sedimentary types and facies of the Zhujiang River mouth so far. In early 1960s, Prof. Ren Mei-e(1964) studied and classified the beach types of Zhujiang River mouth. In 1982, Wang Wenjie classified the sedimetns of Zhujiang River mouth into four major types, namely, river mouth sediment of runoff type, river mouth sediment of tidal current type, river mouth sediment of transitional type, and continental shelf neritic sediment[1)]. We suggest that sedimentation at any river mouth, either dominated by fluvial processes or by tidal processes., must finally result in the forming of a delta and establishing a deltaic sedimentary system. Therefore, the study on river mouth sedimentary types and facies should be combined with the study on delta sedimentary types and facies. Based on the above discussion on the characteristics of Zhujiang River mouth sediments, referring the classification of delta and river mouth sedimentary types and facies, commonly adopted in the world, the present study has preliminarily summed up several major sedimentary types facies and sub-facies for the Zhujiang River mouth sediments (Table 4-40, Fig. 4-47). Properties and characteristics of the major river mouth sedimentary facies and sub-facies, listed in the Table have been described in context above and will not be repeated here.

Table 4-40 Sedimentary facies and sub-facies of Zhujiang River mouth region

Facies / Sub-facies / Types	Delta-Plain (I)	Delta-front (II)	Prodelta (III)	Shelf (S)
Distributary mouths (M)	delta plain MI_1 subtidal shoal MI_2 subaqueous levee MI_3 sacond channel bed MI_4 main channel bed MI_5 crevasse splay MI_6 sandy island MI_7 beach ridge and Beach MI_8 interdistributary bay MI_9	Bar MII_1 (back, crest, front, distal) flood-tidal trough MII_2	prodelta MIII	
Estuaries (E)	dalta plain by tide EI_1 tidal flat EI_2 flood-tidal Channels EI_3 channel bed EI_4 sandy island by tide EI_5	ebb-tidal sand body EII_1 ebb – tidal channel EII_2 flood-tidal trough EII_3	prodelta $EIII_1$ bay front trough $EIII_2$	

1) Wang Wenjie, 1982, Shallow strata structure and sedimentary characteristics of Zhujiang River mouth.

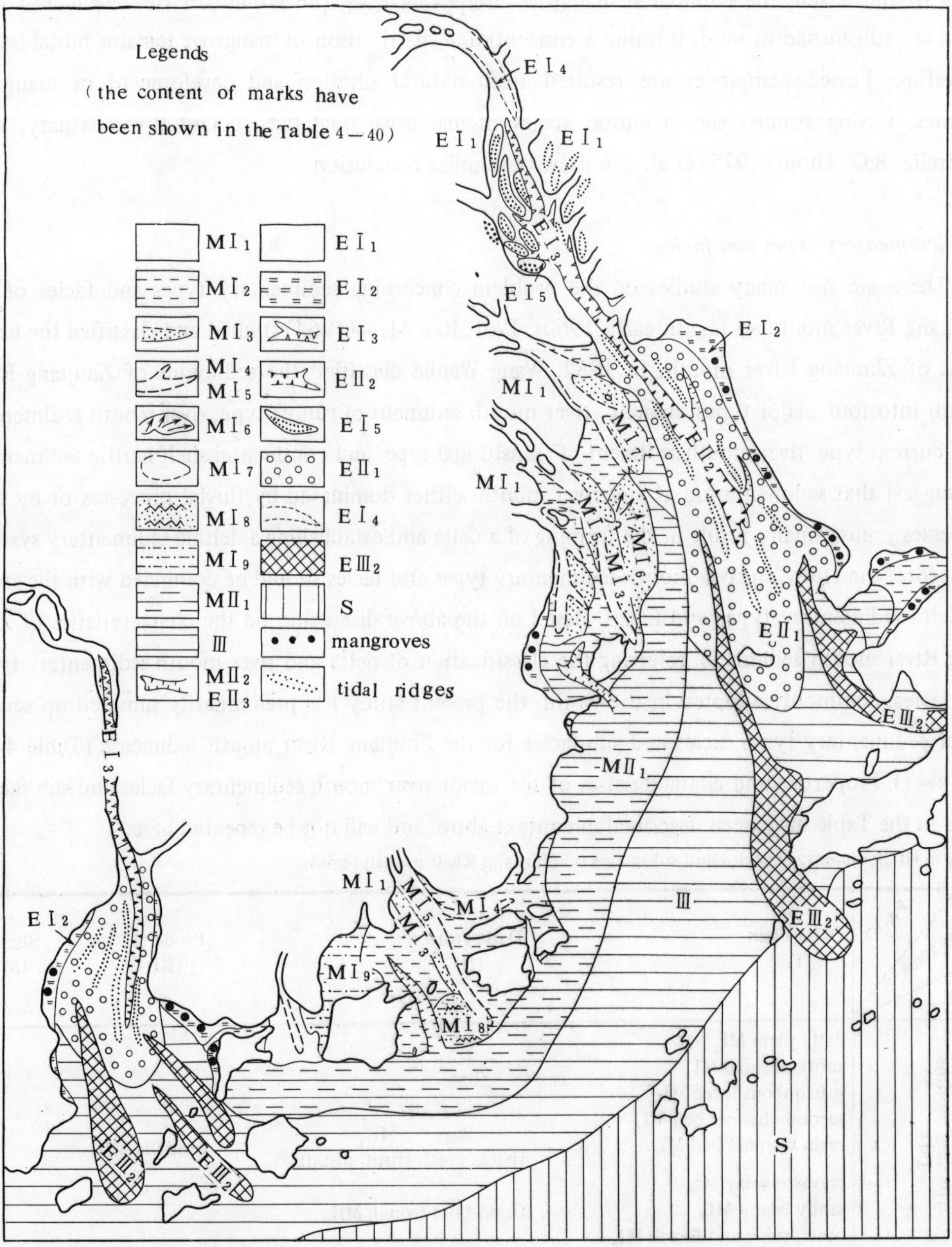

Fig. 4-47. Sedimentary facies and sub-facies of Zhujiang River mouth region.

by Li Chunchu
Zhongshan University, Guangzhou
and Wang Wenjie
South China Sea Institute of Oceanology, Academia Sinica, Guangzhou

Reference

[1] Li Chunchu and Yan Ganran, 1981, Characteristics and evolution of the channel network in the Zhujiang Delta, Proceedings of the 1977 Symposium of Geomorphology, Geographical Society of China, pp 67–75. Science Press. (in Chinese)

[2] Prithard, D. W., 1955, Estuarine circulation patterns, Proc. Amer. Soc. Civil. Eng., 81, No. 717.

[3] Li Chunchu, 1983, Dynamics and sedimentation in the Xijiang Distributary Mouth (Modao-men), Tropical Geography, V. 3, No. 1, pp 27–34. (in Chinese)

[4] Chen Qinde and He Yashou, 1981, the characteristics of the tide and tidal stream in the Zhujiang Estuary, Tropical Geography, V. 1, No.3, pp 29–35. (in Chinese)

[5] Xu Junliang et al., 1981, The patterns of the invasions of saline water and the activities of the wedge in the Zhujiang Estuary, Tropical Geography, V. 1, No.3, pp 36–44. (in Chinese)

[6] Ying Zhifu and Chen Shiguang, 1983, Notable features of the mixing in the Lingdingyan Bay, Acta Oceanologica Sinica, V.5, pp 1–10. (in Chinese)

[7] Fan Xinping, 1982, On the Relationship between efflux structure and deposit types in the Zhujiang Estuary, Tropical Geography, V.2, No. 1, pp 38–45. (in Chinese)

[8] Wang Wenjie, 1982, Recent depositions in the Lingdingyan of Zhujiang Estuary, Tropical Geography, V. 2, No. 2, pp 43–48. (in Chinese)

[9] Hsu Chunming, 1964, Canton in the Sung Dynasty, Acta Scientiarum Naturalium Univeristatis Sunyatsemi, No. 2, pp 248–258. (in Chinese)

[10] Zhou Yuanhe, 1980, Historical changes in the river system of the Zhujiang Delta, Fudang Journal (Social Science), Suppl. pp 85–94. (in Chinese)

[11] Williamson, G. R., 1970, Hydrography and Weather of the Hong Kong fishing ground, Hong Kong Fisheries Bulletin No. 1.

[12] Chan, Y. K. and Wong, C. K., 1960, Oceanographical investigations in the Northern Shelf Region of the South China Sea off Hong Kong, Hong Kong Univ. Fish. U. (3), pp 1–25.

[13] Berthols, L., 1978, Estuarine sedimentation, In the encyclopedia of sedimentology, ed. Faribridge, R. W. and Bourgeois, J., pp 288–293.

[14] Li Chunchu et al., 1985, Features of the bed material and its laws of the sediment transport in the lower reach of the Xijiang River (from Wuzhou to Modaomen), Papers on the Processes of Dynamic, Morphology and Sedimentation in Coastal-Esturine Region, pp 111–119, Science Press. (in Chinese)

[15] Wright, L. D., and Coleman, J. M., 1974, Mississippi river mouth processes: effluent dynamics and morphologic development, J. Geol., V. 81, pp 751–778.

[16] Sedimental Group of Marine Geology Department, South China Sea Institute of Oceanology, Academia Sinica, 1980. Characteristics of the surface sediments on the north shelf of the South China, Nanhai Studia Marina Sinica, pp 50–56, Science Press. (in Chinese)

[17] Marine Geology Department, South China Sea Institute of Oceanology, Academia sinica, 1980, The Quaternary geology along the coastal zone in the South China, pp 84–90, Science Press. (in Chinese)

[18] Li Chunchu and Yan Ganran, 1981, Proceedings of Oceanology and Limnology, pp 115–122, Science Press. (in Chinese)

[19] Gagliano, S. M. and Van Beek, J. L., 1970, Geologic and geomorphic aspects of deltaic processes, Mississippi

delta system. Hydrologic and geologic studies of coastal Louisiana, Report No. 1, pp 140. Centre for Wetland Resources, Louisiana state Univ.

[20] Swift, D. J. P. et al., 1972, Holocene evolution of shelf surface, Central and Southern Atlantic Shelf of North America, In shelf sediment ransport process and pattern, ads. Swift, D. J. P. et al., Dowden, Huthinson & Ross, Inc., pp 499–574.

[21] Ludwick, J. C., 1972, Migration of tidal sand waves in Chesapeake Bay Entrance, In shelf sediment transport process and pattern, eds. Swift, D. J. P. et al., Dowden, Huthinson & Ross, Inc., pp 377–410.

[22] Fenillet. J. P., 1980. Estuarine circulation: controlling factor of clay mineral distribution in James River Estuary, Vinginia. J. Sed. Petr. V. 50, pp 250–267.

[23] Chesterman, W. D. and Wong, H. K., 1971, Bottom sediment distribution in certain inshore waters of Hong Kong, The international hydrographic review, vol. SLV III (2).

[24] Li Congxian et al., 1979, Prelimianary study on sedimentary facies and sequence of the Changjiang Delta, J. of Tongji University, Marine Geology Issue, No. 2, pp 1–14. (in Chinese)

[25] Yun Zukeng, 1982, Characteristics of underground water in the Zhujiang Delta, Tropical Geography, V.2, No. 1, pp 46–50. (in Chinese)

[26] Li Feiyong et al., 1981, Distribution and behaviour of Cu, Pb, Zn dissolved in the water phase and suspended particulate phase in Zhujiang Estuarine Area, Acta Oceanologica Sinica, V. 3, pp 432–435. (Chinese)

[27] Wang Yivou et al., 1979, Application of some geochemical indicators in determining of sedimentary environment of the Funing Group (Paleogene), Jin-Hu Depression, Jiangsu Province, J. of Tongji University, Marine Geology Issue, No.2, pp 51–60. (in Chinese)

[28] Wang Pinxian et al., 1980, Characteristics of the foraminiferal and ostracod thanatocoenoses from some river mouths of China and their geological significance, Papers on marine micropalenotology, pp 101–111, China Ocean Press. (in Chinese)

[29] Zeng Zhaoxuen et al., 1965, Topographical changes of mangrove beach and its agricultural conditions in the estuary: for example in the Zhujiang Delta, Selected Papers of the 1963 Symposium of Geomorphology, Geographical Society of China, pp 90–91, Science Press. (in Chinese)

[30] Thom, B. G. et al., 1975, Mangrove ecology and deltaicestuarine geomorphology: Cambridge Gulf – Ord River, Western Austalia, J. Ecology, V. 63, pp 203–232.

[31] Ren Mei-e, 1964, Morphodynamic characterisitcs and beach-used problems in the Zhujiang Estuary, of Nanjing Univeristy (natural Science), V.8, No.1, pp 135–173. (in Chinese)

[32] Reading, H. G. (ed.), 1978. Sedimentary environments and facies, Blackwell Scientific Publications, Osney Mead: Oxfore, pp 97–142.

[33] Wrigth, L.D., 1978, River deltas, In coastal sedimentary environments, ed., Davis, R.A., New York, pp 5–63.

Chapter Five
Deltaic Sedimentation

As a river-fed depositional system, delta is formed in the area where river debouches into sea and is characterized by deltaic sedimentary sequences. Nearly all the major plains along the coasts of East China are formed by large deltas. The continental shelves of China are one of the broadest shallow-sea shelves on the world, in which the sedimentary strata were formed when the deltas of great rivers moved to and fro frequently during the processes of transgression and regression. It is of great importance, therefore, to study deltaic sedimentation for the recognition of depositional characters and distribution of sediments along the coastal areas and in the shelf regions. Since 1960's it has been proved by the exploration practice for oil and gas that many large oil fields were born in ancient deltaic sediments, hence, to study deltaic deposits has become one of the important research topics of coastal sedimentology. We study ancient deltaic deposits in order to seek for petroleum, natural gas as well as other sedimentary mineral resoures; while we study modern deltas in order to provide identifying criteria for the recognition of ancient deltas. Moreover, shallow sedimentary strata in river-mouth areas and continental shelves are formed by modern deltaic sedimentation. The particular structures of delta sequences bring about the unstability of strata, which often causes serious accidents during the construction of submarine engineering works and brings hazards to some marine constructions.

I. Factors and Processes Influencing the Development of Deltas

There are many factors and processes influencing the formation and development of deltas, such as, river mouth flow, wave, tide, climate, river discharge, sediment load and the compound action of above-mentioned factors, as well as the relative water densities of river and sea water bodies, and regional tectonics where the delta lies in. However, the effects of these factors during the formation and development of deltas are different, in which the river-mouth flow is the leading factor that determines the formation of a delta, causes essential characteristics of deltaic sedimentation and the generality of all kinds of deltas. River-mouth flow, wave and tide are the three factors and processes capable of forming delta sand bodies, and the combination of which determines the various types of deltas and the distribution pattern of sedimentary facies in the delta. The other factors either provide the conditions and backgrounds for the formation of deltas, such as tectonics, or cause certain depositional units of delta to have some changes, such as climatic changes etc.

At the same time, the current slows down further due to bottom friction. Taking the river-mouth of Changjiang River as an example, no matter in flood or dry seasons, the average velocites in vertical and on channel cross-section tend to gradually decrease toward the lower reaches. However, as the Changjiang River mouth is a tidal estuary and the flood current can go upstream as far as the Jiangyin County, Jiangsu Province, about 150 km from the river mouth, the real flow-field of the Changjiang River is the result of superimposition of runoff and tidal currents. The flow-field becomes more complicated after superimposition of runoff and tidal currents. Moreover, it makes the flow-speed increase or decrease locally that the local changes on the cross section of the river channel and the exchange of water among distributaries as well as between distributaries and shoals on either sides.

The changes of the flow at river mouth possess two-fold significance for the formation of a delta. The flow changes at river mouth in decreasing flow-speed downstream results in the sedimentation of a large amount of sediments with no exception; hence, a delta forms. Meanwhile, the flow at river mouth is one of the factors and processes to form sand bodies, which constitute the framework of delta and may determine the type of deltas and the assemblage of deltaic sedimentary facies.

1. River discharge and sediment load

Discharge and load of a river is the material basis in the formation of a delta. Generally speaking, the larger the discharge and load of a river is, the more extensive its delta will be. The average discharge of the Changjiang is approximatly more than 0.9×10^{12} m^3 per year, while the average load is about 4.7×10^8 ton per year, and the whole area of the Changjiang delta is about 50000 km^2. Those rivers with small discharge and load have small area of deltas. Taking the Hanjing River as an example, the average load is 7.24×10^6 ton and the delta area is about 1200 km^2, respcetively about 1.4 percent and 5.2 percent of that of the Changjiang River. However, there often occur exceptions and the Huanghe Delta is one of them. The load of the Huanghe River is the largest in the world, about 2.3 times as much as that of the Changjiang River, but the area of modern delta (5400 km^2) is only one tenth as large. The factor of time is the main cause of such exceptions. The modern Huanghe Delta began to form from 1855 when the Huanghe River broke up its bank at Tongwaxiang, and changed its route into the Bohai Gulf via north part of the Shandong Province. In other words, the modern Huanghe Delta was formed in less than 150 yr[4]. Actually, almost all the Holocene sediments in the North China Plain and the North Jiangsu Plain are related to, in some degree, the sedimentation of deposits carried by the Huaughe at the river-mouth area, estimat.

Usually, the change of load corresponds to that of discharge, but the ratio of the two may vary greatly with the changes of climate, geomorphology and geological conditions. There are about 5000 rivers in China, and the total drainage area is more than 1000 km^2, the total discharge is 2.6×10^{12} m^3 in which 1.86×10^{12} m^3, making up 71.5 percent of the total, flows into the Bohai Sea, the Huanghai Sea, the East and South China Seas. As showed in Table 2, the total discharge of the major rivers in China makes up 84 percent of the discharge of all rivers flowing into the four seas[6], hence, representing the basic situation of the rivers flowing into seas in China. It is obvious in Table 5-1 that the Changjiang River serves as a demarkation line, to the north of it the ratio of sediment load to discharge is usually over 0.5, the maximum one being up to 27.85, and the deltas extending rather rapidly are usually formed at those river-mouth; to the south of the line the ratio is mostly under 0.24, the standard proposed by Moore in 1909[7], which is in favour for the developing of estuaries

Table 5-1 Load and Discharge of Major Rivers Flowing Into Seas off China*

River	Load (10^4 ton)	Discharge (10^8 m^3)	Load/Discharge (10^6 ton/km^3)
Liaohe	2418	42.2	5.72
Dalinghe	3230	37.1	8.72
Ruanhe	1900	45.5	4.18
Haihe	600	51.9	1.15
Huanghe River	111800	422.8	25.25
Changjiang	47200	7250.0	0.52
Qiantanjiang	540	320.0	0.17
Oujiang	245	194.0	0.13
Minjiang	751	623.7	0.12
Hanjiang	724	297.1	0.24
Zhujiang	8218	3492.0	0.24
Moyanjiang	80	84.3	0.09
Jianjiang	191	85.8	0.22
Nandujiang	52	61.2	0.08

* Based on References [6] and other related data.

2. Tides

The influences of tides on deltaic sedimentation are from many aspects. Tides make the gradient water surface at river-mouth change unceasingly, leading to constant accumulation and erosion there. Tidal waves at river mouth are called standing waves, when the flow speed is minimum during the high or low waters. The fine matters deposit and that is why clay-borne intercalated beds are often found in the deltaic sediments. The tidal currents at river mouths are reciprical currents; therefore, bidirectional cross stratification and reactivation surfaces are often developed in deltaic deposits. In addition, tidal currents are one of the three dynamic factors capable of producing sand bodies at river mouth area, and tidal sand bodies have important influences on the distribution patterns of depositional facies in deltaic area.

Tidal range at river-mouth reflects the intensity of the tide. The river-mouths, therefore, are often classified according to the tidal range, i.e. microtidal river-mouth having the tidal range less than 2 m; mesotidal river mouth, 2–4; macrotidal one, over 4 m[8]. The changes of tidal range along the coasts of China are of evident regularity, namely, the tidal range is within 1–3 m to the north of the Changjiang River mouth , and the river mouths there belong to microtidal or mesotidal ones. (Table 5-2) From the river-mouth of the Changjiang to that of the Minjiang, the tidal ranges are mostly over 4 m; therefore macrotidal and mesotidal river-mouths are developed [9] , in which the tidal range of the Qiantang River, being about 8.93 m, is the largest in China and famous as a macrotidal estuary in the world. To the south from the Mingjiang River the tidal range is about 1–2 m along the coasts of the South China Sea, and most river-mouths belong to microtidal ones.

Table 5-2 Tidal ranges of major rivers in China

River mouth	Tidal Range (m)
Liaobe River	2.7
Ruanhe River	1.0
Haihe River	2.5
Huanghe River	0.5–0.9
Changjiang River	2.6
Qiantang River	>5.0
Oujiang River	4.0
Mingjiang River	4.0
Zhujiang River	1.0

The macrotidal river mouths often have subaqueous tidalbodies elongation parallelly with tidal currents, such as the case in the Hangzhou Bay (Plate 5-1); tidal currents at mesotidal river mouth tend to make sand bars connect with subaqueous tidal sand-bodies, which results in the huge volume of river mouth sand bars, such as in the Changjiang Delta. Generally, it is difficult to build tidal sand bodies at microtidal river mouths. However, marine microfossils might be presserved in deltaic plains because of the process of tides.

Tidal flats, of which width depends on the flat slope and tidal range, usually develop along the marginal zones of tidal delta. The thickness of vertical sequences, formed under the seaward extension of tidal flat, depends mainly on the tidal range. The river mouths of the Changjiang and the Luanhe belong to meso-and microtidal mouths respectively.The thickness of plain facies in those deltas are separately 3–4 m and 1–2 m.

Tidal ranges controll the distance of tide-level fluctuation and the limit of tidal flood currents intruding upstream in lower reachs of a river. If the other conditions are similar, the limits of tide level fluctuation and that of flood tidal currents are quite long at macrotidal river mouth, such as in the Changjiang River, but rather short at microtidal river-mouth such as in the Huanghe River (Table 5-3).

For rivers running in steep mountainous districts, the distance influenced upstream by tide is small although the tidal range is quite large, the distance of the tide level fluctuation of the Yalujiang River, for example, averages 30 km. Moreover, the changes influenced by tides in flood and dry seasons are evident. As a rule, the distance that tides retrogress in dry seasons exceeds that in flood seasons (Table 5-3). Marine microfossiles can often be found in river channels intruded by tidal flood currents. What is more, several species of Foraminifera have been found in modern channel-sediments of central Yangzhou district, about 200 km away from the Changjiang River mouth[10]. At a microtidal river mouth where the slope is quite steep, the transporting distance of marine microfossiles carried by tides is rather short; for example, no microfossills 1) have been found near Lianhuachi village, 7–8 km from the Ruanhe River mouth. Hence the distance of reaches influerced by tides is of certain significance in the study of deltaic sediments.

Table 5-3 Upstream limits of tide level fluctuation and tide-flood currents in lower reachs of the major rivers in China*

River	Tidal range	Upstream limit of tide-level fluctuation (km)			Limit of tide-flood current (km)		
	(m)	Flood season	Dry season	Average	Flood season	Dry season	Average
Changjiang	2.6	80	389	156	500	700	624
Zhujiang Xijiang	–	12	190	–	28	280	–
Beijiang	1.0–1.5	0	116	–	43	127	–
Dongjiang		10	58	–	40	90	–
Huanghe	0.5–0.9	–	–	2–3	–	–	30
Haihe	2.5	–	–	–	–	–	100
Liaohe	2.7	–	–	–	34	100	–
Yalujiang	–	–	–	–	–	–	30
Mingjiang	4.0	–	–	–	65	76	–
Oujiang	–	–	–	–	–	76	–

* Based on Physical Geography of China; Marine Geography, and Xu Mingfeng (1973).

3. Waves

Wave often plays an active role in the formation and development of a delta where it is not only one of the factors forming sand bodies, but also an important energy to transform abandoned deltas when the river has changed its course. The principal function of waves in a deltaic area is to transform sediments brought by rivers, and the magnitude of wave action on delta depends upon its energy, which is mainly related to wave heights. The general trend of annual mean wave-height in coastal areas of China appears to be small in the north and large in the south; to be more exact, it is about 0.3–0.6 m along the coast of the Bohai Gulf; 0.6–1.2 m , in the south part of the Shandong Peninsula, the North Jiangsu Province, the Changjiang River mouth and the coasts of Zhejiang Province, as well as Fujian Province; and about one meter along the coast of Guangdong Province [9]. It

1) As usural, "fossils" refer to those lithified test of microorganisms preserved in sedimentary rocks. In this chapter, we sometimes call those unlithified tests remained in modern sediments "fossils" just for the convience of writing.

should be pointed out that sea waves can only be regarded as the possible agents to transform deltas; its real action on deltas and their sedimentary units is also affected by the quantity and property of fluvial sediments and the geomorphology where the delta lies in as well as other conditions. Although the river-mouths of the Huanghe River and the Luanhe River are all located in the Bohai Gulf with lower energy waves, the two rivers are different in the formation and development of deltas. The Huanghe River is very long and rich in sediments mainly consisting of fine-grained substance, and the wave action on the delta is considerablly weak. As a result, nearly no wave-built sandbodies have been found during the progradation of the delta. After the distributaries migrated or the river completely changed its course and flowed into the Huanghai Sea by way of North Jiangsu Province. Large–scale scouring and shell chanier present in the old river mouths and in the area of abandoned deltas may be resulted from wave action. The Luanhe River is quite short carring larger amount of coarse grained deposits in sediments. The seaward gradient from the Launhe is relatively steep, and the wave action on the delta is quite strong. Barrier islands, which are one type of important sand bodies constituting deltaic framework, were formed during the progradation of the delta. The Zhujiang delta is located in the coast of the South China Sea with strong wave action; however, the Zhujiang flows into the bay being protected by bed-rock islands, the wave energy is low and sandbodies formed by waves on the deltaic area are not well developed. Therefore, the Zhujiang delta is a fluvial dominanted one.

Seasonal changes of wave heights are obvious. Enormous waves possess important significance for the erosion of deltas as well as for the transportation of sediments. Some papers emphasize the simultaneous or separate processes of giant waves and flood seasons for the formation of barrier bars in a deltaic area, and hold that when the seasons of wave differ from that of flood, sandy sediments transported by rivers to the sea will form barrier in offshore area after being carried landwards by waves. Therefore, a barrier lagoon system appears. In contradiction to this case, it is impossible to develop a barrier-lagoon system in delta areas[11]. It seems that this point of view overemphasizes the process of time difference between wave and flood seasons. Taking the Bohai Sea as an example, the flood season of major rivers is from July to September when waves are quite calm and average wave height is less than 0.9 m. The barrier bars have been formed in the Luanhe Delta but not in the Huaughe and Haihe Deltas. This shows that the formation of some sedimentary facies doesn't depends on a single factor or the mutual relation of a few factors, but on a combination effects of numerous factors.

4. Climate

Climate, in addition to exerting influence on a delta through waves and currents caused by winds, has certain restrication to the assemblage of deltaic plain-facies. The coastal areas of China stretch over temperate, subtropical and tropical zones. Furthermore, deltaic plain-facies vary with climate zones to a large extent. Taking the most common peat layers on deltaic areas as an example, these layers are quite different in respective climate zones. The tropic and subtropic regions in South

China are characterized by warmth, moisture, rich rainfall, and dense vegetation, and these layers consist mainly of woody plants, in the delta areas, belonging to xylophytaherg peats of low moor carex type. The peats in the Zhujiang Delta contain a large amount of tall *deciuou argores Glyptostrobus Pensilis,* of which thick stems are not carbonized yet[1)]. In general, the nearer to the tropical, the more the components of woody plant in the peat layers. For instance, the peat layers of Klan Delta in Malaya are made up, almost wholly, of woody plants[13]. Northwards it is characterized by less rainfall and sparse vegetation mainly consisting of herbal plant. Low-moor carex peats in the deltas of these areas are largely made up of reed and other kinds of herbage. For an example, in the Changjiang Delta, the stems of reed are commonly preserved in the peat layers. The peats of the Luanhe Delta, locating in North China, are composed almost wholly of herbal plants. It is thus obvious that woody plants in peat layers decrease gradually, but herbal plants become dominant from south to north in China.

5. Tectonic movement

The structure patterns of NNE uplift belts alternated with depression belts along the coastal East China have controlled the distribution of deltas, or in other words, all deltas of large rivers take place within depression belts. The deltas of the Huanghe River, Liaohe River and Changjiang River, for example, are not only vast, but also thick in deltaic sediments. For rivers flowing into seas in uplift belts, even the large river with abundant rainfall as Zhujiang is no exception, their deltas mosaic in hills and the thickness of deltaic sediments is limited. It is thus clear that only continued tectonic depression can be as prerequisite for the continuous supply of river-derived sediments and for the formation of very thick deltaic deposits.

Based on drilling data, deltas of large rivers in depression belts all possess thick sediments. Quarternary strata on coastal areas of the Changjiang Delta, for instance, amount to 300–400 m, in which four or five marine beds intercalated with continental beds between them, are found (Fig. 5-1). Similar phenomena can be found in the deltas of the Luanhe, Liaohe, Haihe and Huanghe rivers etc (Fig. 5-2). But the number of marine beds in these deltas are not well-matched one another, which may result from the drilling holes being located in different places of the deltas. It indicates the sea level changes in Quarternary and that marine beds intercalate with continental ones. However, it is the result of regional depression that all those beds, marine or continental facies, can be preserved in the strata in turn. Those deltas located in uplift belts, even if the regional depression has taken place, tend to be no polycycle superimposition of marine beds. For instance, there is only one marine bed in the modern Zhujiang Delta[15]. Thus, continual depression is the prerequisite for the development of a delta as well as the essential conditions to form superimposition strata orderly in deltaic sediments. Furthermore, only in a continual depression region can deltaic sediments be preserved

1) Department of Geography, Zhongshan University, 1977, The formation, development and evolution of the Zhuiiang Delta.

in strata.

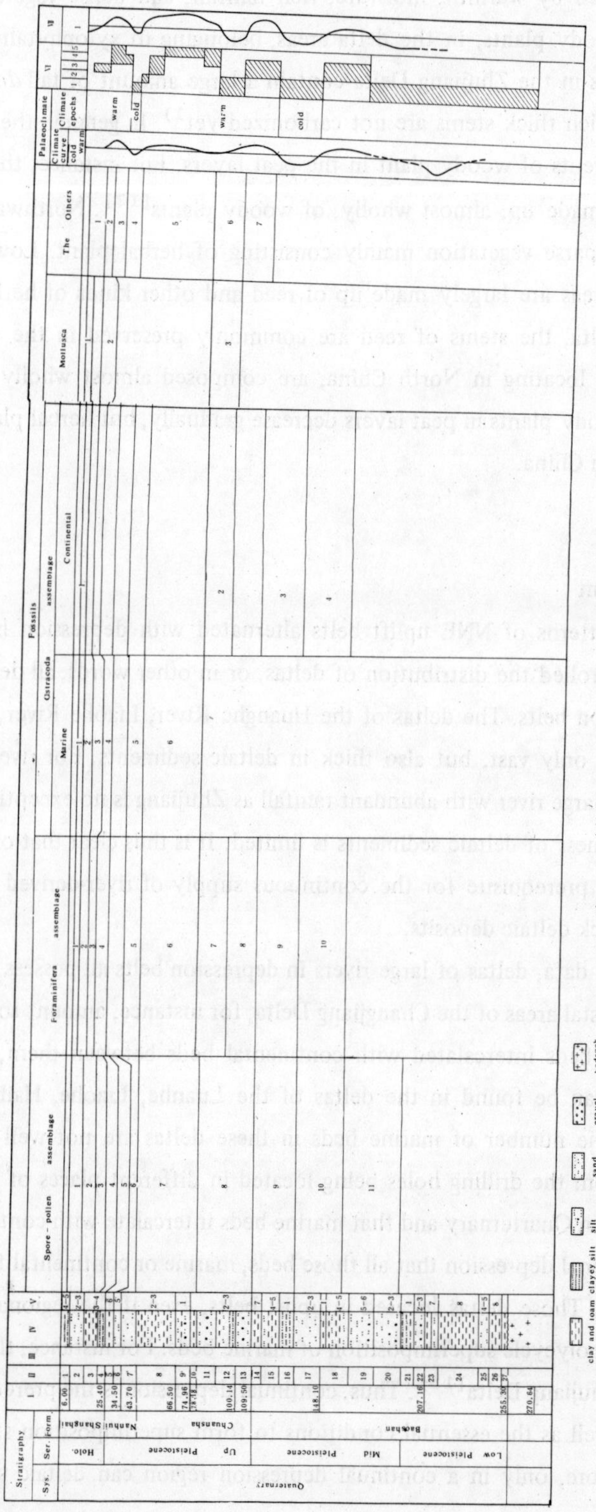

Fig. 5-1. Section of drilling-hole at Shanghai Flour Mill, Changjiang Delta.

I–Depth of bottom (m); II–Number of beds; III–Depth (m); IV–Lithoprofile; V–Color; VI–Sedimentary facies; VII–Land-sea curve.

Spore-pollen assemblage

1 – *Pinus-Quercus-Qramineae;*

2 – *Castanopsis-Quercus glauca-Polypodiaceae;*

3 – *Castanea-Quercus aliena-Pinus-Gramineae;*

4 – *Castanea-Quercus-Pinus-Chenopodiaceae;*

5 – *Gramineae-Cupressaceae-Quercus-Artemisia;*

6 – *Leguidambar-Quercus aliena-Ceratoperis;*

7 – *Castanea-Quercus glauna-Ulmus-Palypodiaceae, Abies, Pinus* and a lot of herbage pollen present in bed of 66m;

8 – *Leguidambar-Parkeria-Quercus-Ulmas;*

9 – *Pinus-Quercus-Abies;*

10 – *Quercus-Ulmus-Cupressaceae-Graminere;*

11 – *Pinus-Abies-Picea-Quercus.*

Foraminifera assemblage

1 – *Ammonia beccarii* var.;

2 – *Epistominella naraensis-Ammonia convexidorsu,*

3 – *Ammonia beccarii* var. *Elphidium advenum;*

4 – *Ammonia confertitestu;*

5 – *Ammonia beccarii* var.;

6 – *Feorilus decorus;*

7 – No foraminifera;

8 – *Ammonia beccarii* var.*Epistominella naraensis;*

9 – No foraminifera;

10 – A little foraminifera.

Marine

1 – *Neomonoceratina dongtaiensis;*

2 – *Sinocytherídea latiovata-Neomonoceratina dongtaiensis;*

3 – *Echinocythereis bredyi-Sinocytheridea latiovata;*

4 – *Sinocythereidea bredyi-Sinocytheridea latiovata;*

5 – *Sinocythereidea bredyi;*

6 – *Munseyella-Cytheropteron.*

Continental

1 – *Ilyocypris;*

2 – *Ilyocypris-errabundis-Condoniella sulacensis;*

3 – *Ilyocypris errabundis-Condoniella albicans;*

4 – *Ilyocypris.*

Mollusca

1 – *Mactra veneriformis;*

2 – *Nassarius* sp.-*nuculana* sp.-*Aloidis* sp.-*Omala* sp.;

3 – *Venerupis philipinarum.*

The others

1 – Spines of *echnoid;*

2 – Spines of *echnoid,* fragments of *ophiuroid;*

3 – *Coscinodiscus radiatus, Coscinodiscus marginatus, Cyclotella stylorum;*

4 – *Chorophyta,* fish-bone;

5 – Spines of *echnoid,* fragments of *ophiuroid,* sponge skeleton;

6 – *Chorophyta,* fish-bone;

7 – Spines of *echnoid,* sponge skeleton.

VI (Sedimentary Facies):

1 – Lacustrine; 2 – Alluvial; 3 – Estuarine; 4 – Littoral; 5 – Neritic.

VII (Land-sea curve):

1 – Land-Sea.

II. Sedimentary Characteristics of Deltas

1. Major types of deltas in China

Fisher(1969) proposed the stratigraphic classification of detals[16], and considered that sand bodies and their assemblages that control the types of delta and the distribution patterns of sedimentary facies. It is appropritate, therefore, to adopt the essential principle of stratigraphic classification from the angle of deltaic sedimentation.

Besides, the effects of landform, tectonics as well as other factors may also make the distribution of deltaic sediments and the assemblage of sedimentary facies change. In the piedmont with active tectonics, rivers tend to form alluvial fans, and then deltas. Fans and deltas constitute a unified organic sedimentary system, the distribution patterns of sedimentary facies, the forming process and evolution of the fans and deltas restrict each other and correlate closely. Because of the wide spreading of the fan-delta along the coasts in China, and of the depositional mineral resource in the ancient analogue, it is necessary to distinguish the fan-delta as an independent type, to which the Luanhe Fan-Delta is very typical.

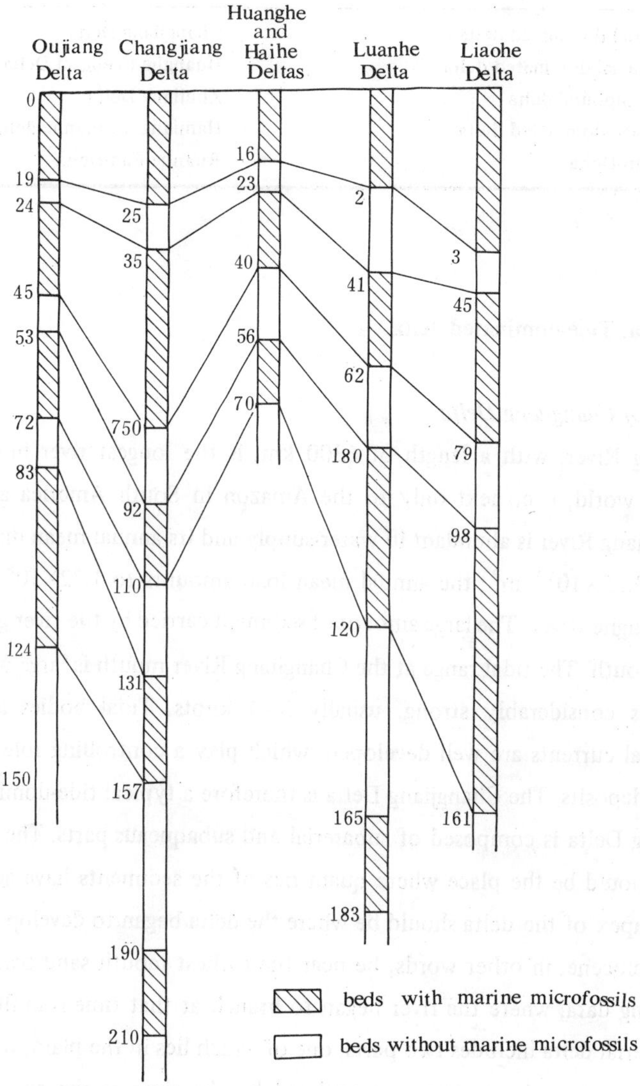

Fig. 5-2. Stratigraphic correlation in delta areas.

Thus, main delta types in China are as follows: tide-dominated, fluvial-dominated, wave-dominated and fan-deltas (Table 5-4). There are a series of transitional types among the main types. In addition, the Zhujiang River Delta belongs to fluvial-dominated type on account of its deltaic sedimentary characteristics and developing processes, but the deltas formed by Xijiang, Beijiang, Dongjiang Rivers, which are independent from one another, are crisscross and very complex, therefore the Zhujiang Delta as a whole may be named as a compound delta which is shaped when Xijiang, Beijiang and Dongjiang, the three isolated rivers, empty into the same gulf.

Table 5-4 Main delta types in China

Delta type	Example
Tidel-dominated delta	Changjiang Delta
Fluvial-dominated delta	Huanghe (Yellow) Delta
Compound delta	Zhujiang Delta
Wave-dominated delta	Hanjiang, Futuanhe deltas
Fan-Delta	Ruanhe Fan-Delta

2. Changjiang Delta: Tide-dominated delta

(1) *General setting of Changjiang Delta*

The Changjiang River, with a length of 6300 km, is the longest river in China, and the third longest one in the world, i. e. next only to the Amazon in South America and the Nile River in Africa. The Changjiang River is abundant in water-supply and its annual mean discharge tops in China by amounting to 9.25×10^{11} m^3; the annual mean load amounts to 4.72×10^8 ton, which is second only to that of Huanghe River. The large amount of sediment carried by the river gives rise to a gigantic delta at the river-mouth. The tidal range at the Changjiang River mouth is large with an average of 2.6 m; tidal current is considerably strong, usually 3–4 knots. Tidal bodies and the sand bodies transformed by tidal currents are well developed, which play a controlling role over the distribution patterns of deltaic deposits. The Changjiang Delta is therefore a typical tide-dominated one.

The Changjiang Delta is composed of subarerial and subaqueaus parts. The area of the Holocene Changjiang Delta should be the place where quantities of the sediments have acaumulated since the Holocene and the apex of the delta should be where the delta began to develop during the maximum transgression in Holocene, in other words, be near the earliest mouth sand bars of the Changjiang . Based on the drilling data, where the river began to branch at that time is in the vicinity of Yizheng County. The subarerial delta includes two parts, one of which lies in the plain, which possesses deltaic sequences and is known as delta plain, accumulated by the river at the ancient river mouth. The north boundary line of the plain is through Yangzhou, Taizhou, Qutang, Haian and Lushi, while the south line through Zhengjiang, Jiangyin and the south bank of the present river. Therefore the whole plain is in the form of a long and narrow belt. The other part is littoral plain, whose sediments are from Changjiang and deposited under the processes of ocean factors. Although its sedimentary sequences differ from the deltaic ones, the two parts can correlate each other because both underwent the similar history of sedimentation and development. The south boundary line of the littoral plain is approximately through Jiangyin, Taichang, Shongjiang, Jinshan; while the north one through Yangzhou, Taizhou, Haian, Rudong and the south of Lushi. The subaqueous delta of Chang-jiang River extends as for as to 33.5°N north-wards, 30°N southwards and 123.5°E eastwards, in term of hydrodynamic factors, characteristics of sediments and the distribution of micro-fossils.

The statistical figures for the above-mentioned scope of the Changjiang delta in Holocene are showed on Table 5-5. The total area of the delta is 51,800 km^2, in which the subarerial part is 22,800 km^2, making up 44 percents of the total; and the subaqueous part about 29,000 km^2, making up 56 percents of the total. Therefore the area ratio of the two parts, the subarerial versus the subaqueous, is 0.78.

Table 5-5 Statistical figures of the area of Changjiang Delta

Subarerial part			Subaqueous part
Delta plain	Littoral Plain		
	South of the river	North of the river	
15,000 km^2	5,800 km^2	2,000 km^2	29,000 km^2

(2) *Sedimentary characteristics of subaqueous part in the delta*

(a) *The river mouth outline and underwater topography*

There are two points, Nanhui in the south and Qudong in the north, at the Changjiang River mouth. The topography at the mouth is quite complicated, numbers of sand islands and shoals paralleled to the currents scattered in the distributary channel, among them the Chongming Island is the largest mouth sand bar. The place, upstream from the Chongming Island, where Changjiang begins to branch, is the starting point of modern Changjiang River mouth. The Chongming Island branches the Changjiang River into two distributaries, the north one of which is called Beizi or Haimen distributary which is being and will further be abandoned. In the middle of Nanzi (south distributary) there exist Changxin Island, Hensha Island and Laotongsha Shoal which are linked together one another; therefore they divide the Nanzi into Beigang and Nangang (distributaries). Furthermore Nanggang is subdivided into Nancao and Beicao (distributaries) by Jiuduansha. The outlet of the Nancao is met with Xintongsha Shoal, which is a river-mouth sand bar being formed, and is the main obstacle of navigation on Changjiang. At present, Beizi, Beigang, Beicao and Nancao are four main distributaries at the Changjiang River mouth (Fig. 5-3).

The water depth off the river mouth gradually increases from northwest to southeast, with a change from 5 m to 40 m. In the area with a contour between 5–10 m, the submarine morphology is flat and constant, having the slope only about 1/1000, and may be called a subaqueous flatform, which is usually known as subaquatic delta-plain or delta-front platform. The slope in the area off the contour of 10 m becomes steep, being about 1/1000 , a slope zone with the width about 30 km and surrounds the river mouth in the form of inverse "S" shape. Seawards from the slope, the morphology of the north part becomes flat once more; while that of the southeast part reveals the troughlike relief. There are two troughs on the zone, one extending to the Changjiang River mouth

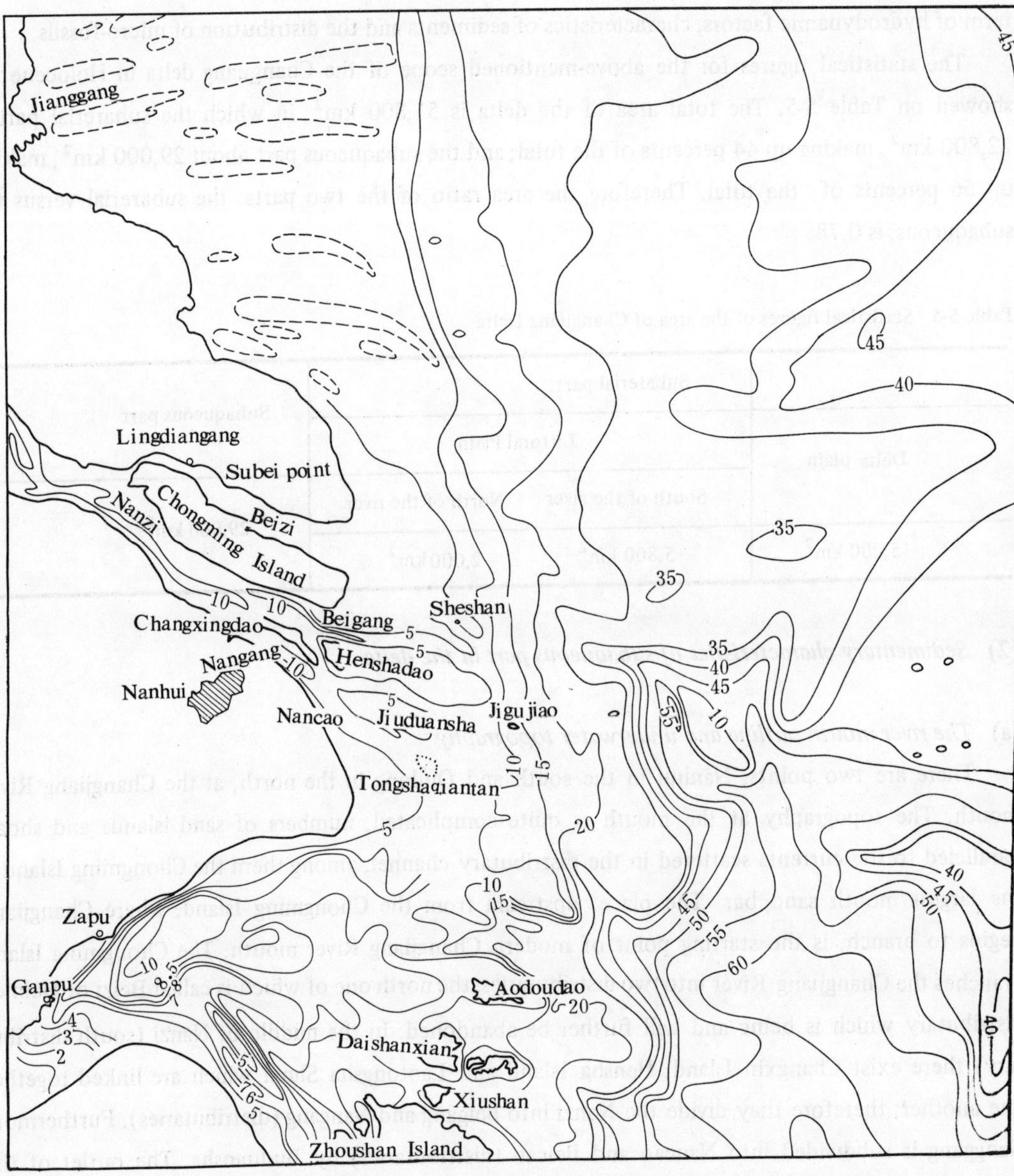

Fig. 5-3. The river mouth outline and underwater topography.

and the other to the Hangzhou Bay. Southeastwards, the two troughs merge together and then extend as for as the continental margin near the north end of Taiwan Island. The troughs may be the ancient Changjiang valley during the glaciation with low sea level. Chengshi and Zhoushan Islands, located in the mouth of the Hangzhou Bay, are directly affected by the silts and clays transported by Changjiang. There are generally deep troughs among islands and the water depth is changeable.

(b) *Grain size of sediments*

The hydrodynamics at river mouth is the dominant factor in controlling of the distribution of sediments. The river flow velocity at the mouth gradually decreases and the deposits become finer and finer seawards (Fig. 5-4).

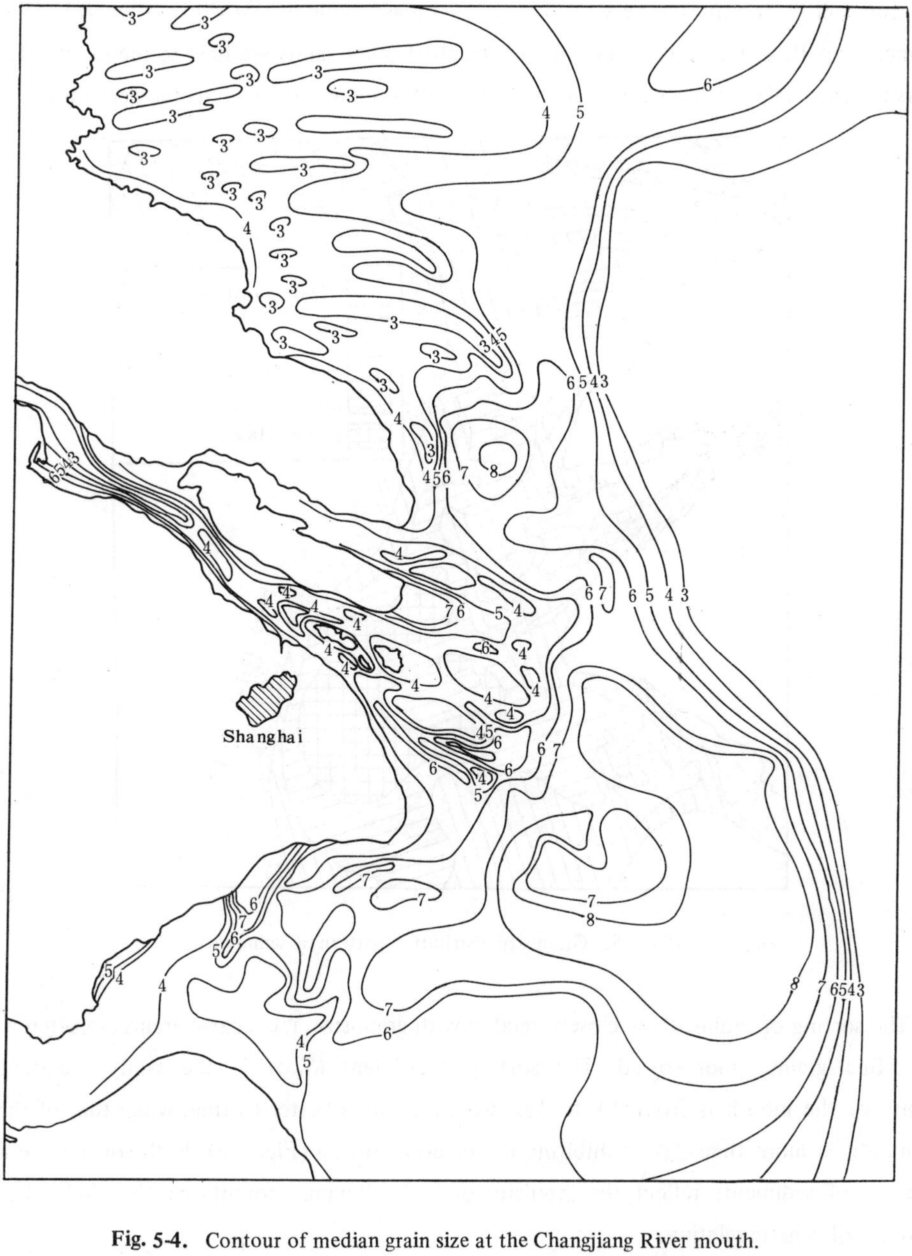

Fig. 5-4. Contour of median grain size at the Changjiang River mouth.

There is an area of sediments with median grain size less than 14 in the vicinity of the mouth, in which fine sands account for more than 80 percents of the sediments. Off the Nanzi (south distributary) mouth, the area of sediments with the median grain size greater than 8 surrounds the mouth in the shape of a horseshoe, in which the contents of fine sands are less than 10 percent.

Because the current velocity at the mouth decreases gradually, the sediments change little by little, that is seawards from the mouth the deposits are successively fine sands, silt, clayey silt and silty clay. There are relict sands seaward from the silty clay (Fig. 5-5). In addition, as mentioned above the south distributary of the Changjiang River is the main channel, therefore the fine grained sediments can be transported very far, the farthest place being in 123°30′E; whereas the sediments at the mouth of Beizi are soon deposited and the shelf relict sands are near to the mouth. This further shows the close relationship between sediment distribution and hydrodynamic conditions.

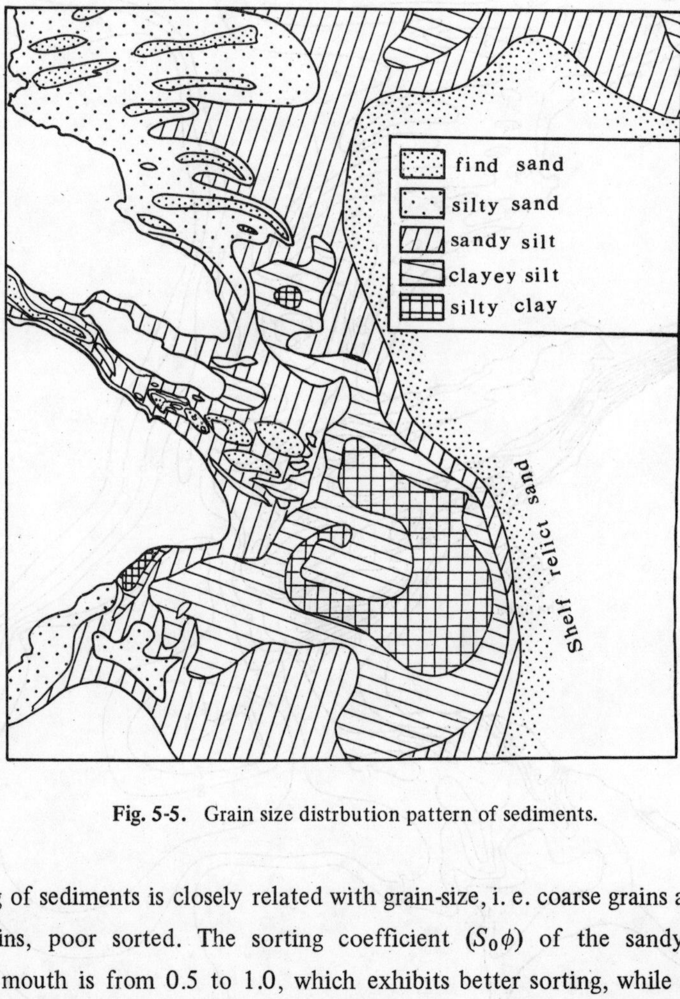

Fig. 5-5. Grain size distrbution pattern of sediments.

The sorting of sediments is closely related with grain-size, i. e. coarse grains are often well sorted while fine grains, poor sorted. The sorting coefficient ($S_0\phi$) of the sandy sediments in the vicinity of the mouth is from 0.5 to 1.0, which exhibits better sorting, while that of the clay off the mouth is more than 2.0, exhibiting rather poor sorting (Fig. 5-6). Both sorting coeficient and grain-size of sediments reflect the intensity of hydrodynamic conditions; therefore, the variation of both is of genetic relation.

The CM pattern taking the form of "S", similar to that of the typical river, is drawn based on the analytical data of 108 samples from the reaches of Nanjiang to the area off the mouth. There are no data about the reaches of NO and OP because all samples were collected from the lower reaches of Changjiang. There are two distribution areas of fine-sediments in the form of rectangular and rhombus on the lower left of the CM pattern for normal river sediments. The two areas represent the

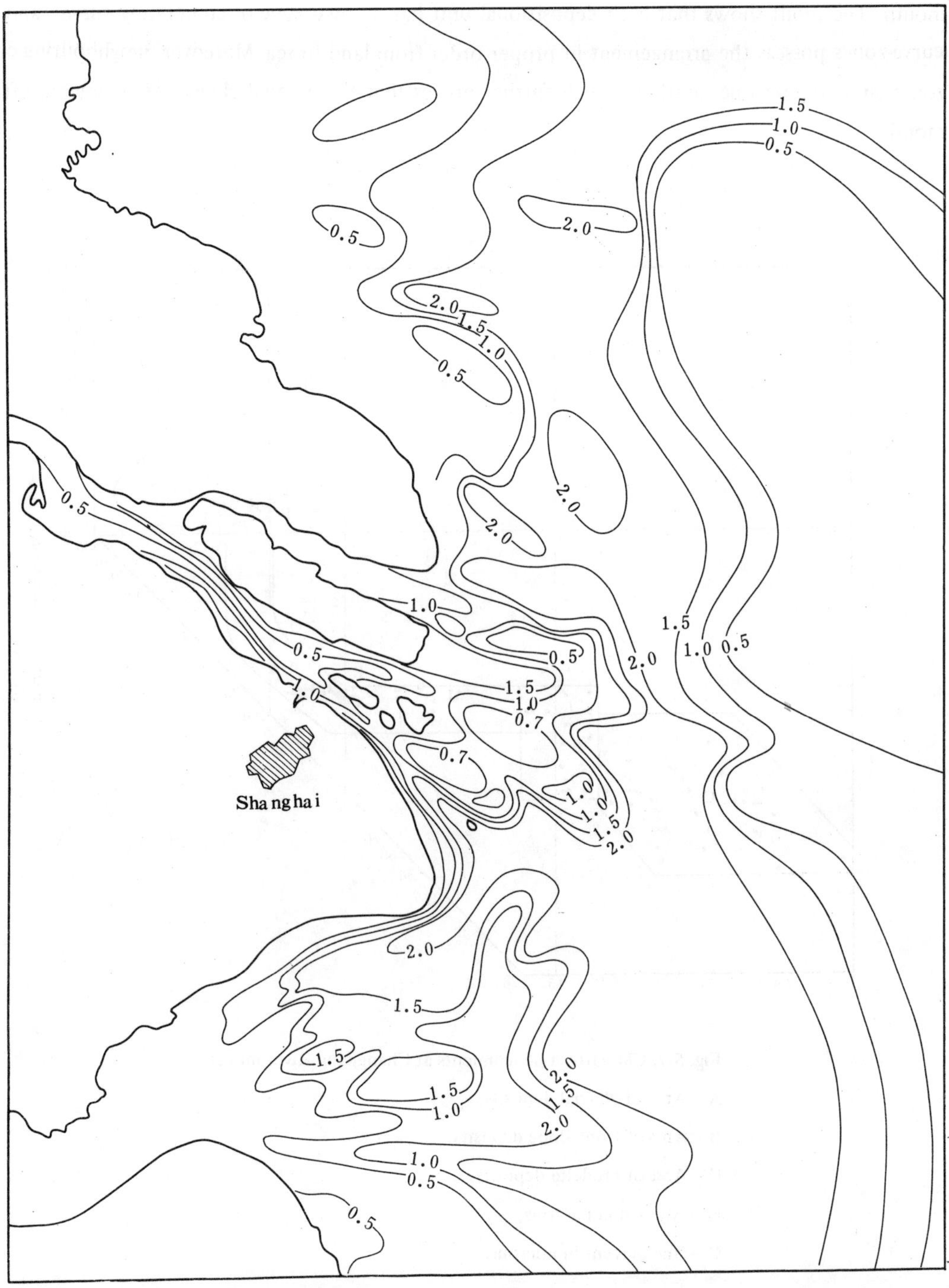

Fig. 5-6. Contour of sorting coefficient ($S_0\phi$) of sediments.

delta-front slope and front delta respectively and they either superimpose or connect closely each other, which reflects the gradual transition of the sediments from land to sea (Fig. 5-7). Fig. 5-8 is the cumulating curves of the coarsest and finest sediments sampled in various depositional units off the mouth. The result shows that each depositional unit has its own zone of cumulating curves, and any curve-zones possess the arrangement in proper order from land to sea. Moreover, neighbouring curve-zones superimpose one another, which further proves that the gradual change of sediments off the mouth.

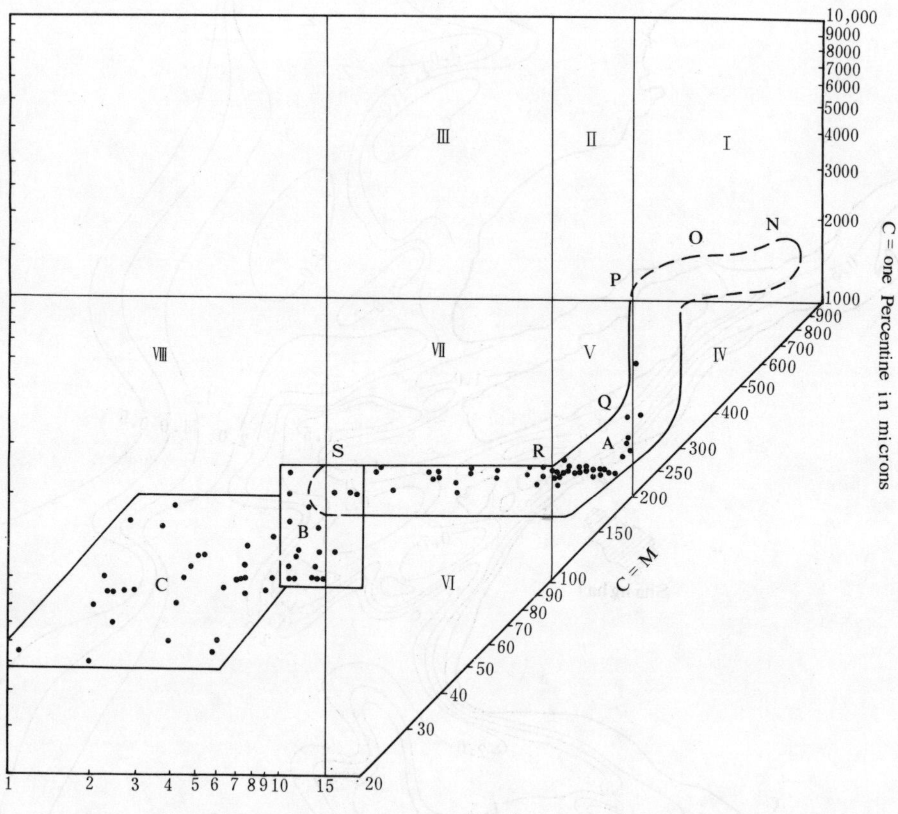

Fig. 5-7. CM pattern of sediments at Changjiang River mouth

A – Area of alluvial deposits;

B – Area of front slope deposits;

C – Area of prodelta deposits;

M – Median in microns;

C – One percent in microns.

(c) *Chemical properties of sediments*

Hydrodynamic condictions act as the important control over the distribution and migration of main chemical elements in river mouth sediments. If the hydrodynamics is strong, upper and lower water can exchange continually and fully mix, which makes the water bodies rich in oxygen arrive at the sea bottom and tends to produce an oxidation environment. To the contrary,

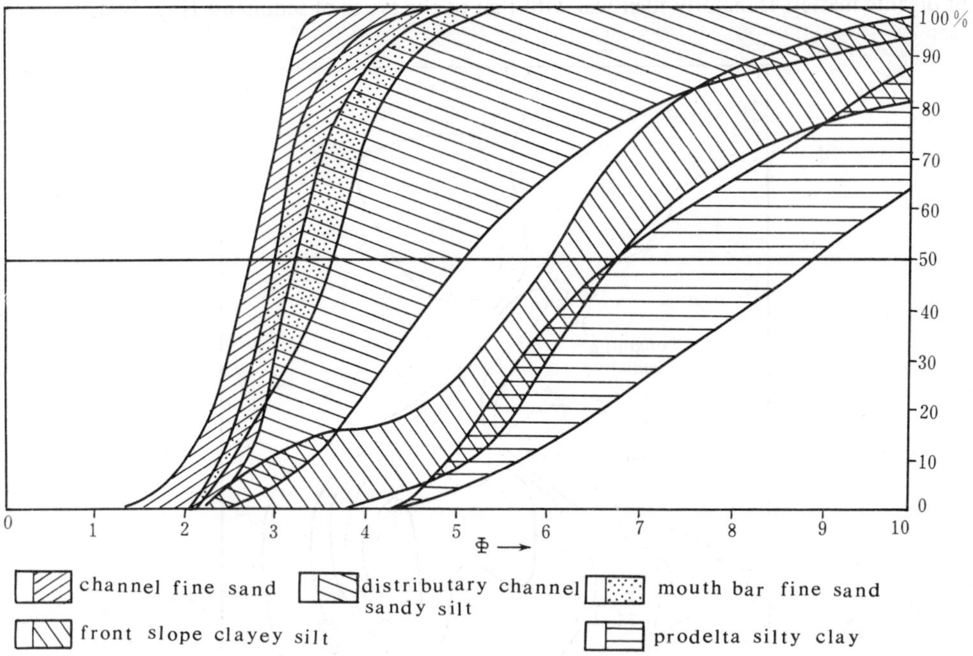

Fig. 5-8. Cumulating curves of various depositional units at Changjiang River mouth.

if the hydrodynamics is weak, the sedimentary enviroments are relatively stirless and the water rich in oxygen can not reach the bottom, the submarine zone will then be in favour for the formation of reducing environments. Off the river mouth the hydrodynamics gradually decreases with the increase of water depth. As a result, the environment in the bottom changes slowly from oxidation into reduction The hydrodynamics in the distributaries and at the river mouth sand bars of Changjiang is strong, the ratio of Fe^{+++}/Fe^{+} is low, usually less than 0.5 and Eh value varies greatly, usually from +100mV to+400mV; therefore, this area belongs to an oxidation environment. Seawards, on the argillaceous sediments area with water depths from 30 to 50 m, ratio of Fe^{+++}/Fe^{++} is less than 0.5 and Eh value is less than +100 mV[1)]. Hence, this area belongs to the reducing environment.

(d) *Organic matter*

There is obvious relationship between the contents of organic matters and the grain-size of sediments in the subaqueous part of Changjiang Delta. The sediments in the vicinity of the distributary channels and river mouth sand bars are mainly composed of fine sands and fine silt, in which the contents of organic matters are rather low, generally less than 0.5% and the smallest only 0.09%, while the contents of organic matters in the sediments composed of silty clay off the Changjiang

1) Tan Hueifang, 1981, Preliminary study of geochemical characterestics off Changjiang River mouth.

Estuary are fairly high, the largest being about 1.5%. Moreover, the distribution pattern of the content of organic matters is majorly identical with that of the types of sediments (Fig. 5-9).

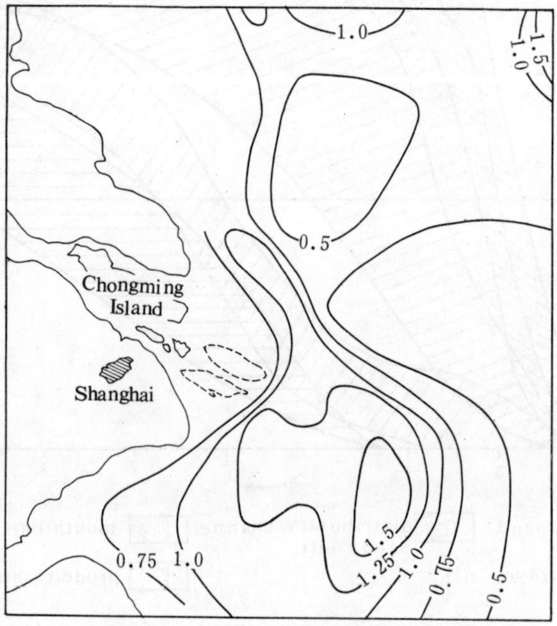

Fig. 5-9. Distribution of organic matters.

The distribution of organic matters in the subaqueous part of Changjiang delta is related to many factors. Clay minerals possess absorptionability, and they can absorb not only trace elements but also organic matters. The area with high contents of clay usually has high contents of organic matters. However, clay minerals can absorb organic matters only when the sources of organic matters are abundant. The distribution of organic matters, therefore, are closely related to the sources of organic matters. Parts of organic matters in the superficial sediments at river mouth area are brought by the river while the other parts are autochthonous. The organic matters from land consist mainly of plants but that of the autochthonous, of planktoic microoganisms. The ratio of organic carbon/organic nitrogen in the organic matters mainly composed of advanced plants is relatively high, for instance, from 18 to 35 in littoral swamp-sediments; whereas the ratio of the two in the organic matters mainly composed of planktonic microoganisms is rather low, for example, 8–12 in marine sediments on the shelf[19]. Based on the calculation of 27 samples collected at Changjiang River mouth, the above-mentioned ratio is from 4.0 to 14.4. Thus it can be seen that organic matters are largely from planktonic microoganisms. When the fresh water mingles with the saline water at Changjiang River mouth, the organic matters in suspension carried by the river is in favour for the growth of planktonic organisms, which can directly absorb the dissolved salt. The inorganic suspension matter transported by the river might convert into organic colloid and solvable organic matters through the processes of hydrolysis and the action of bacteria, and becomes nutritional matter easily absorbed by planktonic microorganisms such as nitrogen and phosphorus. Moreover, abundant suspension matter at the Changjiang River mouth makes planktonic microoganisms reproduce and grow in large quantities. Fig. 5-10 shows the distribution of planktonic plants.

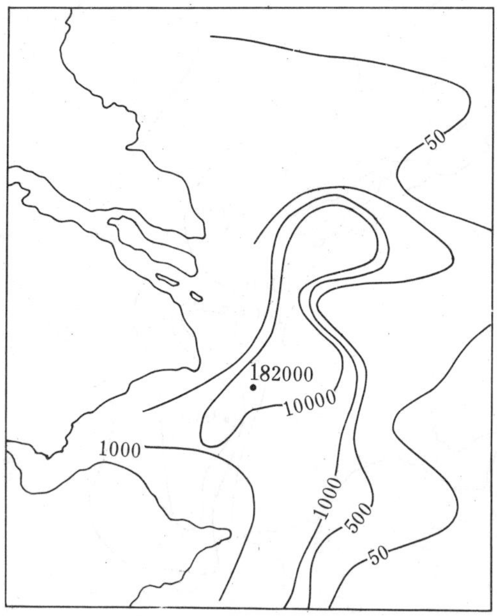

Fig. 5-10. Dstribution of phytoplankton (10^4 specimen/m^3) in Changjiang River mouth area.

It can be seen that the contents of the plants in the sea area off the mouth are the highest, as high as to 1.82×10^9 specimen/m^3, but both landwards and seawards the contents decrease evidently. This is simlar to the distribution pattern of organic matters in sediments. The area with high contents of planktonic organisms is near the mouth, of which the highest content amounts to 4300 mg/m^3 (Fig. 5-11), which is slightly different from the distribution pattern of phytoplankton as well as that of organic matters. It should be pointed out that the contents of organic matters in the sediments at river-mouth area depend not only on the sources of organic matters, but also on the sedimentary rate, oxidation-reduction conditions and the intensity of hydrodynamics along with other factors. For example, the area near the mouth, compared with the offshore, is characterized by high sedimentation rate, strong oxidation-reduction and intensive hydrodynamics; therefore, the contents of organic matters are still low even if the quantity of organic matters provided by planktonic organisms is slightly large. Dead planktons sink to the bottom of the sea, which either becomes the source of organic matters or provides abundant food for benthonic organisms and makes them reproduce in great quantities. The distribution pattern of benthonic organisms is on the whole in agreement with that of planktonic organisms and organic matters. That is to say, the contents off the mouth are the highest, amounting to 69.3 g/m^3 (Fig. 5-12). The benthonic organisms are supplementary sources of organic matters in submarine sediments, and the activity of the organisms often disturbs bedding structure and leave lots of organic burrows, which become the assistant criteria in the recognition of sedimentary facies of deltas.

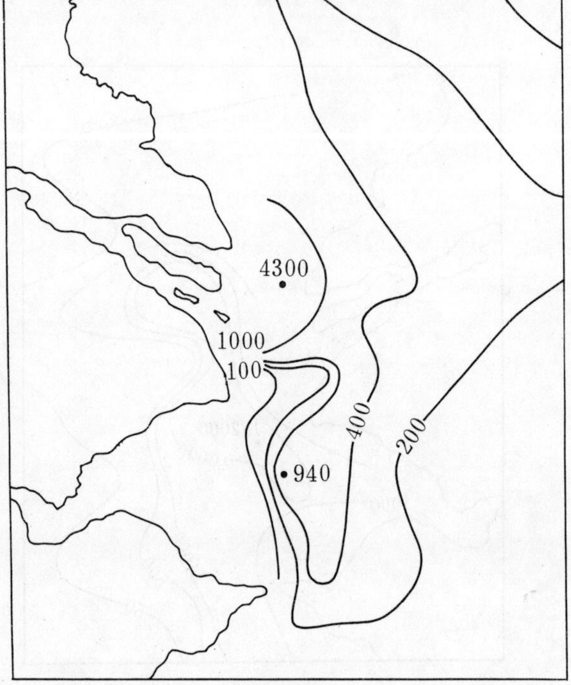

Fig. 5-11. Distribution of zooplanklon (mg/m^3) in Changjiang River mouth area.

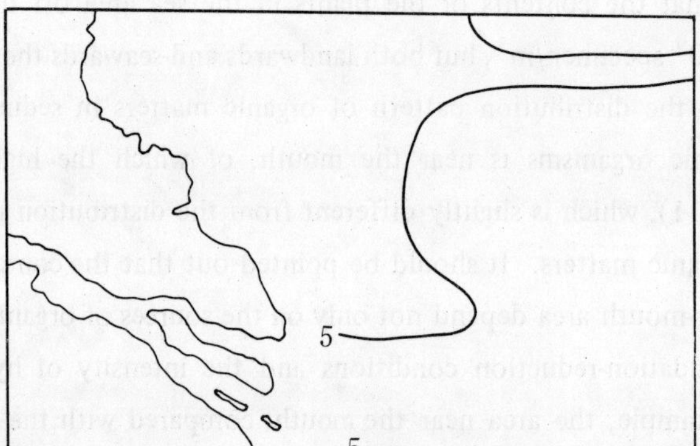

Fig. 5-12. Distribution of benthonic organisms (10^4 specimen/m^3) in Changjiang River mouth area.

(e) *The assemblage of microfossils*

Changjiang River mouth belongs to mesotidal one and the distance that tidal currents retrogress is very long. Because huge amount of flood discharge empties into the sea, and the processes of the sea and river interact each other, Changjiang River mouth becomes a typical environment of marine continental transitional facies, and the assemblage of foraminifera and ostracoda is evidently characterized by marine-continental transitional facies.

The marine and trrestrial ostracoda, foraminifera and *Thecamoebina* are buried together in the sediments at Changjiang mouth. There are not only euryhaline species such as *Ammonia beccarii* var, but also a lot of marine stenohaline species such as *Epistominella naraensis* and planktonic foraminifera. The contents of the latter species amount even to 12.4 percents. Because of the mixing of a large number of stenohaline species, the contents of euryhaline ones decrease and the superiority degree is not high, being only 22.7 percents, which is remarkably different from that at microtidal mouth [20].

There are 30 to nearly 100 species of foraminifera in the sediments of normal neritic facies, to which every 50 grams of dry samples contain several thousand to several ten thousand foraminifera and the average test-diameter is over 0.22 mm; while there are only 30 to 40 species of foraminifera off Changjiang River mouth, to which every 50 grams of dry samples contain several hundred to several thousand foraminifera and the mean test diameter is about 0.22 mm. Furthermore, there are only 2–20 species at Changjiang River mouth area, ten to several hundred specimen in 50 grams of dry samples, and the mean diameter is usually less than 0.17 mm. Marine ostracoda follows the similar law of change. Take the maximum test-diameter for example, it is 0.6–0.8 mm in neritic areas, about 0.5 mm off Changjiang mouth and 0.3–0.4 mm in the mouth area. (Fig. 5-13)

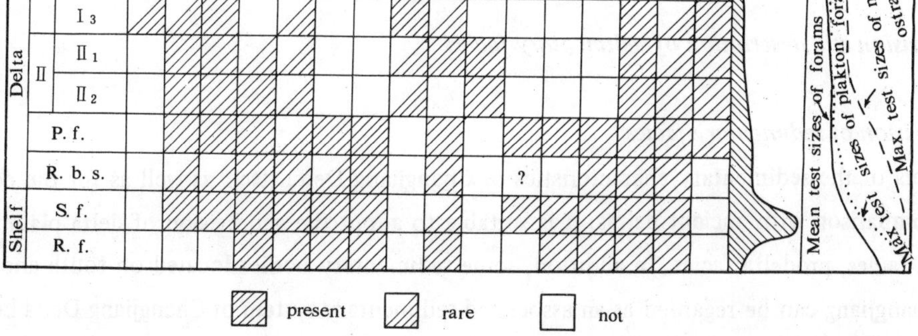

Fig. 5-13. Distribution of the fossil assemblage (thanatocoenosis) in sedimentary facies, Changjiang Delta. (After Wang Pingxian, 1978)

T – Thecamoebena;

Foraminifera

1 – *Globigerinidae;* 2 – *Epistominella;* 3 – *Miliolidae;* 4 – *Elphidium;* 5 – *Arenaceous foraminifera;* 6 – *Ammonia annectens.*

Ostracoda

C.O. – Continental Ostracoda.

Marine ostracoda

1 – *Sinocytheridea;* 2 –*Neomonoceratina dongtaiensis;* 3 – *Neomonoceratina triangulata;* 4 – *Munseyella;* 5 – *Leguminocythereis;* 6 – *Echinocytheries.*

E – Echinodermata.

1 – *Echinoidea;* 2 – *Ophiuroidea;*

P. d. – Plant debris.

I – Delta plain facies; I_1 – Freshwater lake subfacies; I_2 – Tributary subfacies; I_3 – Changjiang channel facies; II – Delta front facies; II_1 – Distributary channel subfacies; II_2 – Front slope subfacies. P. f. – Prodelta facies; R.b.s. – Radiating bar subfacies; S. f. – Shelf facies; R. f. – Relict facies.

The micro fossils of delta from the offshore area to Changjiang River mouth shows that the mariness of sedimentary facies exhibits evident progressive change, and this phenomenon had a close connection with hydrological factors. The distribution of above-mentioned foraminifera and ostracoda is controlled by salinity and the isosaline curve of 30% is taken as the boundary line in neritic and offshore areas; while the boundary line of foraminifera is the isosaline curve of 18% in and off the Changjiang mouth area. The boundary area where foraminifera can live at Changjiang mouth is in the vicinity of Hensha Island, which is nearly identical with the moving scope of salt-water wedge. Upstream from the reach of Nantong, all foraminifera found are dead and mainly composed of *Ammonia beccarii* var and *Epistom nella naraensis.* The species of foraminifera decrease to about 10 and there are only dozens of specimen with an average test-diameter of 0.05 in 50 grams of dry samples. Further upstream from Taipinzhou of Yangzhong County, about 200 km from the mouth, the flood-tide current disappears and no marine microfossils have been found.

(3) *Subdivision and description of sedimentary facies*

(a) *Subdivision of sedimentary facies*

In terms of the sedimentary characteristics of Changjiang Delta itself as well as for the convenience of comparison with ancient deltas, it is suitable to adopt the subdivision of delta plain facies, delta front facies, prodelta facies. In addition, some sedimentary bodies located on south and north sides of Changjiang can be regarded as an associated sedimentary system of Changjiang Delta because their sediments are largely from Changjiang and their existence as well as size is closely related to Changjiang Delta, where sediments are transported and accumulated by marine factors. As shelf sedimentary facies is closely associated with deltas and can be compared each other, it will be discussed together with the deltas in the following paragraphs.

(b) *Delta plain facies*

Delta plain facies is the subaerial portion of Changjiang Delta and consists of marine-alluvial plain and alluvial plain.

(i) *Marine-Alluvial plain*

The marine alluvial plain consisting of all the Holocene subdeltas is the major component of delta plain facies and is gradually shaped from subaqueous river mouth into present subaerial zone. The plain distributes in the north of Changjiang and spreads in the form of a belt (Fig. 5-14). The topography of the whole plain is relatively high except the area of paleo-distributary channels. Generally the sediments of the plain are rather coarse and mainly composed of tawny and yellow-greyish silty clay and clayey silt. In some areas such as Huangqiao zone silt-fine sand outcrops, which are usually the ancient river mouth sand bars without being overlaid. There are several scattered lakes and swamps on low-lying zones of ancient distributary channels, where sediments consist of mud and clay rich in plant remains.

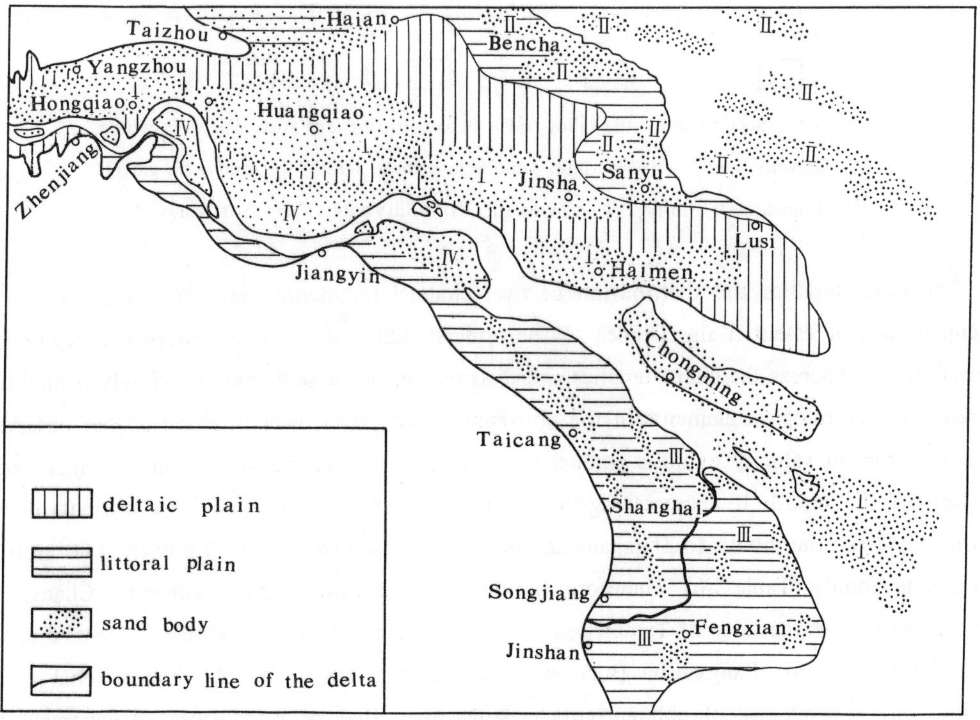

Fig. 5-14. Main sedimentary units of Changjiang Delta plains.

I – River mouth bar; II – Radiating bar; III – Littoral sand bar; IV – River channel sand body.

(ii) *Alluvial plains*

Alluvial plains are built when a delta progrades continually and the river transforms the delta plains. In Changjiang Delta, alluvial plains spread out along the river banks. Consequently, original deltaic sequences have been totally or partly transformed into river channel sequences possessing two sedimentary units–flood plain and river channel, and superimposed river channel sedimental formed on different stages are developed in some reaches of the river (Fig. 5-15).

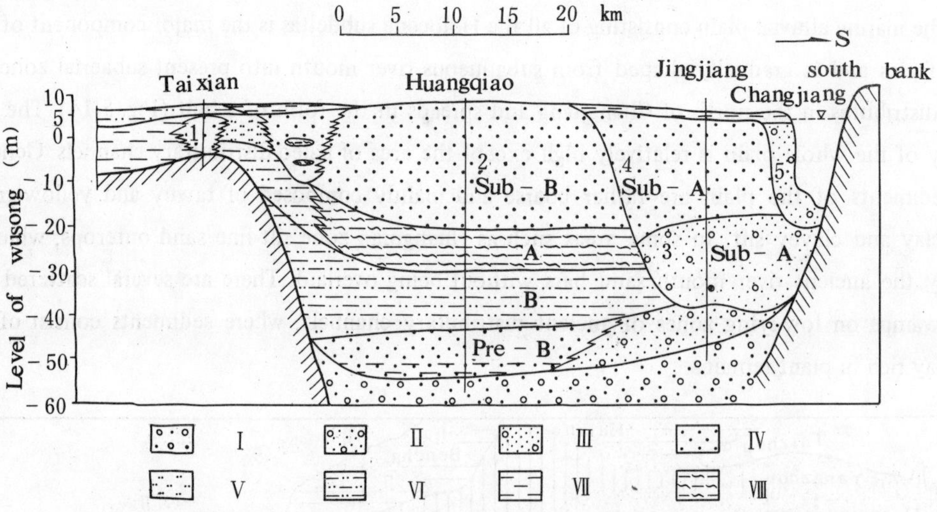

Fig. 5-15. Section across the valley of Changjiang in Jingjiang area, Jiangsu Province.

I – Gravelly sand; II – Medium sand with gravels; III – Fine sand with gravels; IV – Fine sand; V – Silty; VI – Clayey silt; VII – Silty clay; VIII – Soft sandy clay.

The characteristics and distribution of river channel sediments from the mouth upstream to Zhenjiang are of essential significance to the understanding of the sedimentary characteristics of alluvial plains, whereas the characteristics and distribution of the sediments are closely related to the variation of hydrological elements in the above-mentioned river reaches. Based on our observation data, ebb current velocity exhibits gradual decreasing and increasing again from Zhenjiang to the mouth, off the mouth it decreases again. On the contrary, velocities of flood currents decrease continually from the mouth to Zhenjiang and becomes zero upstream from Taipinzhou (section DF), Yangzhong County (Table 5-6). Therefore, the upstream limit of flood tidal current in Changjiang in dry seasons is in the vicinity of Taipinzhou, and further upstream the flood-tidal currents disappear. From Taipinzhou, to Jiangxinsha (section D-4), Nantong, flood and ebb bidirectional currents coexist; but the velocity of ebb currents is larger than that of flood tides. Downstream from Jiangxinsha, the currents in channel are still bidirectional, but the velocity of flood currents is larger than that of ebb tides.

In reaches downstream from Zhenjiang the change of salinity is distinct. To be specific, from Zhejiang to Jiangxinsha near Nantong water is fresh, of which salinity is 0.015–0.018‰ and does not vary with tides. From Jiangxinsha to Hensha Island fresh and salt water mingle, and the salinity is 0.43–0.83‰ in dry seasons, only 0.1–0.2‰ in flood seasons. The salinity increases to 10‰ seawards from Hensha Island (Table 5-7). The bottom isosaline of 2‰ is in the vicinity of Hensha Island (121°45′E) where it might be the front edge of salt-water wedge in dry seasons. Therefore, the bottom isosaline of 2‰ of Changjiang in the south trough is adjacent to Hensha Island, while that in the north distributary is at NW end of Chongming Island (121°10′E). The boundary between salt and fresh waters is located near Jiangxinsha of Nantong. From Jiangxinsha (section D-4) to Taipinzhou water of Changjiang River is fresh, but sometimes affected by flood tidal currents, whereas upstream from Taipinzhou water is fresh and never affected by flood currents.

Table 5-6 Mean current velocity in vertical direction of Changjiang River beds from Zhenjiang to Hensha Island

Section	Velocity of flood current (m/sec)	Duration (hr)	Velocity of ebb current (m/sec)	Duration (hr)
Xiaojiuduan	128.2	5	126.7	8
D-1	75.5	4	70.6	9
D-4	53.5	3	91.7	10
D-5	41.3	4	66.3	9
D-6	34.0	4	75.4	9
D-10	20.0	2	63.8	11
D-7	20.8	3	89.2	10
D-8	0	0	107.2	13
D-9	0	0	132.8	13

Table 5-7 Maximum salinity (‰) in the reach from Zhenjiang to Hensha Island, Changjiang River

Layer \ Section	D-9	D-8	D-7	D-10	D-6	D-5	D-4	D-1	Hensha Island
Superficial layer	0.015	0.015	0.014	0.016	0.016	0.016	0.017	0.43	7.12
Middle layer	0.017	0.014	0.015	0.014	0.015	0.016	0.018	0.69	10.93
Bottom layer	0.015	0.015	0.015	0.014	0.015	0.017	0.016	0.83	11.64
Mean value	0.016	0.015	0.015	0.015	0.015	0.016	0.017	0.65	9.92

Deposits in Changjiang channel consist mainly of sandy sediments, of which variation downstream is closely related to hydrological factors. In river reaches from Zhenjiang to Taipingzhou, upstream from the limit of flood currents, the sediments are relatively coarse and composed of gravel-bearing fine-medium sands and medium-fine sands but not of argillaceous sediments, in which the contents of the grain size being coarser than medium sands amount to 70%. The log probability curves possess typical characters of channel sands (Fig. 5-6). In reach from Taipinzhou to Jiangxinsha, Nantong, the channel sediments are chiefly made up of fine sands, accounting for 60–90% of the total content. Besides, this river reach is controlled by bidirection flow, and in log probability curves of the sediments there are mostly three subpopulation curves so the sediments possess the characters of channel sands. However, there are two suspension subpopulation on a few log probability curves because of the existence of flood tides. Downstream from the Jiangxinsha, Nantong, Changjiang is branched into several distributaries by Chongming Island, Changxin Island and Hensha Island etc. The sediments in this area are fine, mainly composed of silt-fine sand and with high contents of clay, though hydrodynamics is strong (Table 5-6).

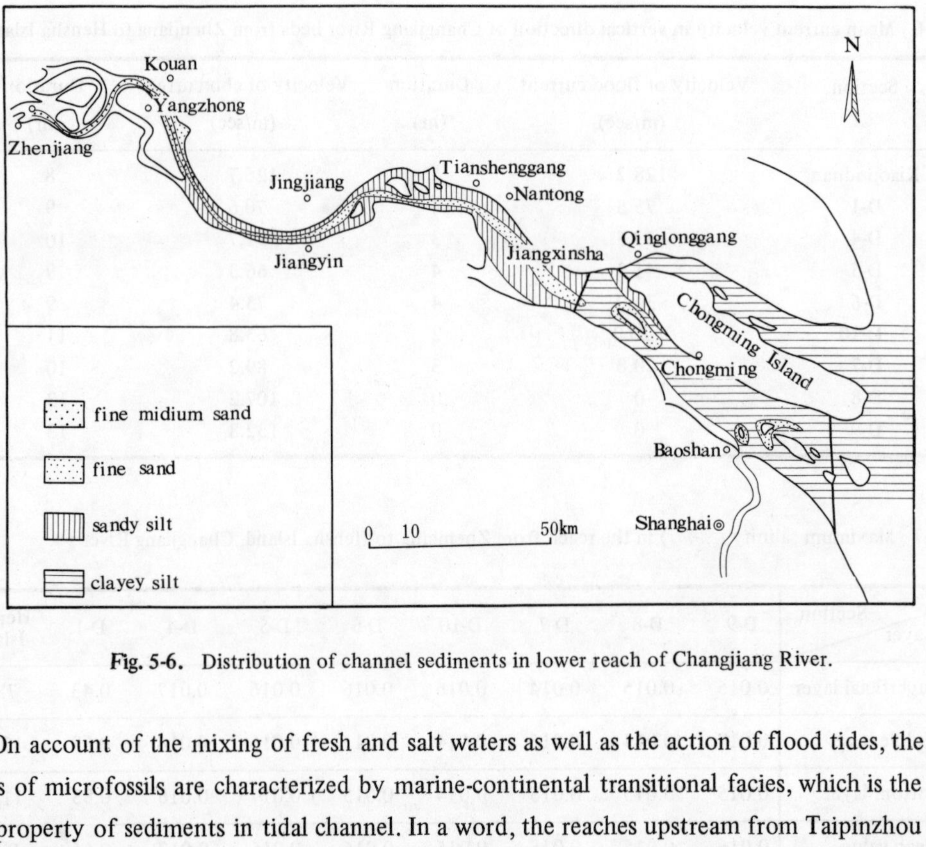

Fig. 5-6. Distribution of channel sediments in lower reach of Changjiang River.

On account of the mixing of fresh and salt waters as well as the action of flood tides, the assemblages of microfossils are characterized by marine-continental transitional facies, which is the important property of sediments in tidal channel. In a word, the reaches upstream from Taipinzhou belong to a continental facies environment, and those from Taipinzhou to hensha Island belong to a brackish environment, but not their ecological environment of marine microoganisms. In fact all microoganisms were brought in by flood currents and the area seaward from Hensha Island is truely the ecological environment of some marine microorganisms.

(c) *Delta front facies*

Delta front is situated on the broad area from distributary channel to delta front slopes where sea and river interact violently. Subaqueous landform of this area is changable and the distribution of sediments is very complex. The sediments are made up mainly of yellow-greyish or grey fine sands, sandy silt and clayey silt, as well as of silt clay on local areas. There are a variety of sedimentary structures such as a great many kinds of cross-bedding, and on the surface the ripple marks are well developed. In addition, abundant plant remains often form horizontal laminae, in which the content of organic matters amount to 0.3–0.5%. There exist marine microfossils such as foraminifera and echinoidea, as well as continental ostracoda in this area. Delta front facies, moreover, can be subdivided into distributary channel, mouth bar, frank marginal bank, inter-distributary shoal and front slope subfacies (Fig. 5-17).

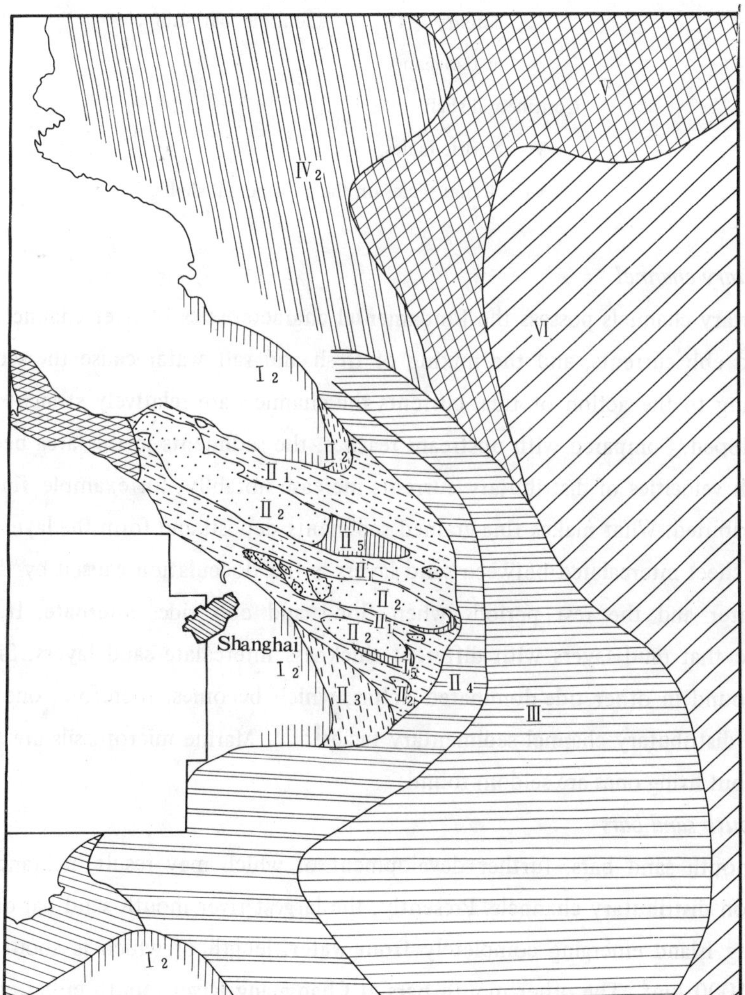

Fig. 5-17. Sedimentary facies in Changjiang Delta.

Delta system

I – Delta plain facies;

I_1 – Main channel subfacies; I_2 – Strandplain subfacies.

II – Delta front facies;

II_1 – Distributary channel subfacies; II_2 – Mouth bar subfacies, II_3 – Frank marginal shoal subfacies; II_4 – Front slope subfacies; II_5 – Interdistributary shoal subfacies.

III – Prodelta facies

IV – Associated facies

IV_1 – Sand bar subfacies in Qiangtang Estuary; IV_2 – Radiating bar subfacies in Jianggang.

Continental shelf

V – Neritic facies.

VI – Relict facies on the continental shelf.

(i) *Distributary channel*

Distributary channels possess the fundamental characteristics of river channel, but the existence of flood and ebb currents, and the mixing of fresh and salt water cause the sediments to change greatly. Owing to the action of tidal currents the channels are relatively straight and meanders are poorly developed. Compared with upstream reaches, the sediments in this area become remarkablly fine although velocities of distributary currents increase notablly; for example, fine silt-fine sand are patchly distributed. What makes fine-grained sediments deposit and form the layer of surface mud as well as clay sheet intercalated between sand layers is the flocculation caused by the mixing of fresh and salt water and the rest periods when flood and ebb tides alternate. It is the important characteristic that mud layers with different thickness intercalate sand layers. Similar phenomena have been found in other tide-dominated deltas, which becomes, therefore, one of the criteria in recoginizing distributary channel sedimentary facies[23]. Marine microfossils are distributed widely in this area, but living ones are seldom found.

(ii) *River mouth sand bars*

River mouth sand bars, further development of which may result in branching of channels, are located on distributary channels. Presently, the largest river mouth sand bar of Changjiang River is Chongming Island emerging completely from water, length of which is about 70 km, and area, more than 1000 km^2. The other mouth bars at Changjiang River mouth either emerge partly from water such as Changxin Island and Hensha Island or submerge completely in water such as Jiuduansha. Water depth of subaqueous mouth sand bars varies in range from 2 to 3 m or 8 to 10 m. Besides, the superficial sediments, mainly consisting of fine sands, in sand bars with shallow water depth are coarse and possess well developed wave ripples. On log probability curves these sediments exhibit two suspension subpopulations, but miss rolling subpopulation. For the deeper mouth bars their sediments contain a considerable amount of mud. The content of organic matters in the mouth bars is lower, being usually less than 0.5% and the lowest only 0.08%. Marine microfossils species in this area are similar to those in distributary channels; whereas the quantity decreases and a few *Neomonoceratina* and *Bosquetina* are present.

(iii) *Frank marginal bank*

Frank marginal banks, which are similar to subaqueous levees at certain river mouths, are attached to Shubei and Nanhui Points of Changjiang. The frank marginal banks are rather giant because of abundant matter brought by Changjiang. Sediments at this area consisting mainly of clayey silt, in which horizontal bedding is developed mainly. Moreover, sand belts parallel to the coast line appear because transportation and accumulation of sediments are chiefly subject to wave actions in this area.

(iv) *Interdistributary shoals*

Interdistributary shoals distributed in the area of stagnant currents near the end parts of river mouth bars. Except for low tidal flat where sandy doposits are dominant, sediments in interdistributory shoals are composed of clayey silt and silt clay, which are often overlaid by the superficial layer consisting of tawny mud in the form of semi-fluid. The sediments in this area are poorly sorted and characterized by the parallel bedding of sand-clay interbedding.

(v) *Front slope*

Front slope, with a gradient of 0.7‰ in rough, is located on the outmost of delta front. This region looks like a slope zone, wide in the south part and narrow in the north part surrounding the river-mouth in the form of inverse S-shape. Front slope sediments consist of clayey silt, in which the content of fine sands is less than 10‰. The sedimentary structures are characterized by horizontal bedding and wavy bedding. In the sediments the content of plant remains is still high and that of *Echinoidea* increases; what is more, *Ammonia beccarii* is the dominant species of foraminifera and neritic foraminifera species increase but *Epistominella naraensis* which is typical of Changjiang mouth greatly decrease. Ostracoda are mainly composed of *Neomonocertina dongtaiensis* and *Neomonoceratina latiovata.*

(d) *Prodelta facies*

Prodelta facies is distributed in the zone seaward from the delta front and of which the south part is wide; while the north part is narrow. Because the Nanzi is a principal distributary of Changjiang to debouch water and sands into the sea, the prodelta seaward from Nanzi protrudes in the shape of a lobe, while that off the north distributary mouth is rather narrow. Prodelta facies is a widespread facies in Changjiang deltaic sedimentary system, and these sediments, consisting mainly of silty clay, are rich in organic matter and blue-grey in colour. Besides, water content of the sediments, amounting from 60% to 80%, exceeds the liquid limit; therefore, the sediments are in the form of semi-fluid state. Horizontal bedding is well developed and the stratification is usually alternated with thick clay beds and thin silt beds; of which thickness is 5–10 cm and 0.1 cm respectively. There are many benthic organisms in the prodelta, which bring forth numerous burrows and cause partial original bedding structures to be damaged. Foraminiferas in this region are similar to those in delta front slopes, except for numbers of neritic species, such as agglutinant foraminifera,

Spiroloculina, Ammonia annectens; while ostracoda in this area are composed of *Cythereis, Echinocythereis, Munseyella, Neomonoceratina triangulata* and *N. cicatricosa* etc. Moreover, remains of *Echinoidea* are widespread, but plant remains are seldom in the sediments.

(e) *Associated sedimentary facies*

The existance and figuration of associated sedimentary facies in Changjiang Delta is closely related to the development of the delta. The associated sedimentary system can be subdivided into Qionggang radiating-bars, sand bars of Qiantang Estuary and littoral plains.

(i) *Qionggang radiating-bars*

There exist series of subaqueous bars in the north sea area off Changjiang mouth (Plate 5-2). They spread radiately from Qionggang as an apex. The occurrence and development of radiating bars is associated with the Changjiang River. It is the cause to build radiating bars that the northward sediments from the Changjiang River mouth meet near Qionggang area with the southward ones from the Huanghe abandoned delta and spread seawards.Thus, radiating bars should be regarded as an associated sedimentary system of the Changjiang Delta. The bars are composed mainly of fine-grained sands, which are well sorted and with a standard deviation (σ) of 0.5–0.6. The sediments between bars consist of sandy silt and silty sands. The landward sediments of radiating bars become fine-grained and gradually transit into the sediments of flat; while seaward sediments gradually change into marine silty clay. The assemblage of organisms in this zone is similar to that in normal neritic area, but presents a small mumber of common species typical of the Changjiang River mouth.

(ii) *Qiantang sand bars*

Qiangtang sand bar located in the Hangzhou Bay is sandy sedimentary body with exceedingly giant volume being up to 4.25×10^8 m^3. The sand bars are made up chiefly of sandy silt and the content of clay is less than 3%. Because only part of the sediments are from the Qiantang River and the most are brought in by the southward currents from Changjiang,[26,27] it is reasonable to regard Qiantang sand bar as an associated sedimentary system of Changjiang Delta. Presently the sand bar are still scoured by violent swell waves and incessantly reconstructed; to be specific, strong tides can form a trough, which might be filled with sediments afterwards, with the depth about 15 m on the sand bars.[28] Besides, unceasing transformation of tides makes the sandy sediments pure and gradually develops a series of sand bodies extending in parallel with tides. Foraminifera in the sediments consist mainly of *Epistominella naraensis,* but of which test diamters are smaller than that in the Changjiang River.

(iii) *Strandplains*

Strandplains, built when the coast continually progrades, are located on the north and south flanks of Changjiang Delta. The strand plain north to Changjiang extends in the form of a belt parallel to the coast. The coast line of the strandplain south to Changjiang is flat, on which there are several series of littoral sand bodies that represent the ancient coastlines in different stages, to which the farthest line is more than 70 km from the present coastline (Fig. 5-16).

Strandplains are converted from tidal flat, on which the sediments of the uppermost 3 or 4 m are characterized by the vertical sequences of tidal flat. The upper part of the deposits are composed of tawny silt clay containing iron and manganese as well as rust spots, and even peat layers and bog ores in some parts. The downward sediments are made up of greyish-yellow or grey clayey silt and silt. In addition, there are buried radiating sand bars to the north of Changjiang under sandy-clay layes, of which the sediments are similar to the same sedimentary facies in this sea area; while in the south frank of Changjiang Delta there are many rows of sandy bars and shell bars. The material distribution of littoral sand bodies in the south part of Changjiang exhibits evident regularity, i. e. sediments in the area adjacent to the mouth are mainly comopsed of terrigenous clastics, such as that in north part of Songjiang, Naihui, Fanxian etc, and the littoral bars in the area distant from the mouth are mainly composed of shells, in which terrigenous clastics are well sorted. Quartz is the major mineral constituent of sandy bars, in which the contents of heavy minerals are high, the largest one being more than 20% (Fig. 5-18).

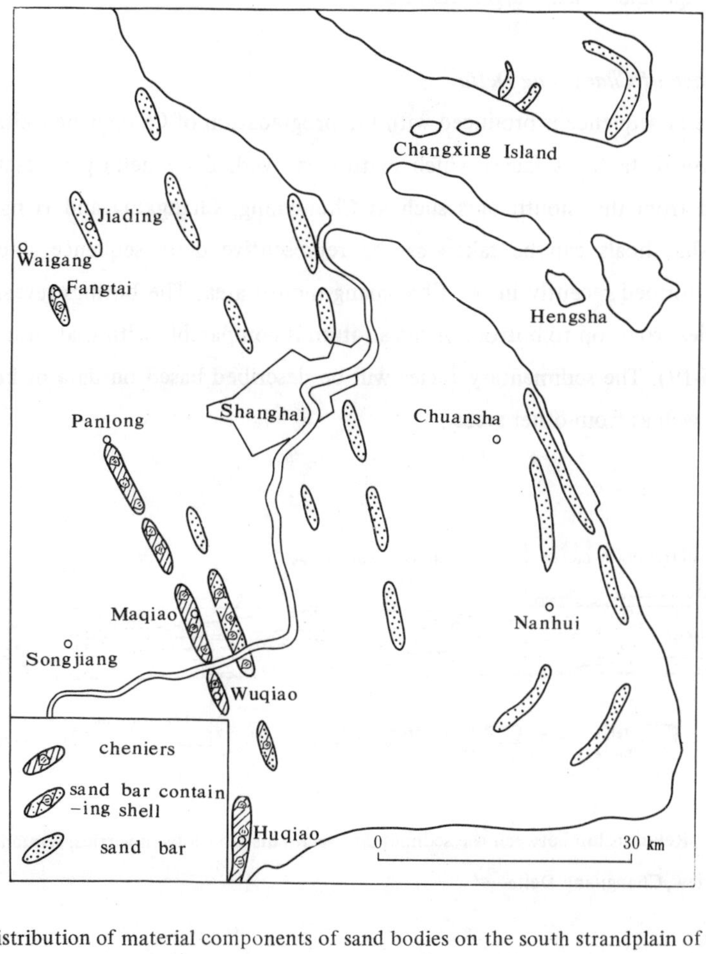

Fig. 5-18. Distribution of material components of sand bodies on the south strandplain of Changjiang Delta.

Microfossil assemblages of strandplains, in which the assemblage of the supratidal zone can be regarded as the representative, are similar to those of tidal flats. According to the researches of the recent years, the assemblage of foraminifera in littoral swamps and marshs on the supratidal zone possesses distinct characteristics, because these areas are still affected by the tides, and the salinity of water is in the range of 1.2% to 6.3%, even up to 16% in some creeks leading to the sea. Besides, these areas are overgrown with plants, suitable for the growth and reproduction of certain foraminiferas. All the foraminiferas living in this environment are stenohaline and euryhaline species such as *Ammonia beccarii* var, of which mean content in samples is 47%, and the highest, 98.6%. The species of foraminifera in the supratidal enviroments are few, usually less than 10, and only 2–3 in some parts. The dominancy is high. Moreover, the tests of foraminifera are rather small, while the range of foraminifera quantities per 10 cm^3 is great. The ostracoda assemblage in littoral swamp environments of the supratidal zone exhibits the characteristics similar to foraminifera and is composed of saltwater species such as *Sinocytheridea latiovata* Hou et Chen, *Pontocypris hyposaline* n. *sp.*, *Tanella opima* Chen etc. The specimen quantities of ostracoda, being up to several hundred per cm^3, are quite large. The dominancy is high and usually larger than 50%.

(4) *Vertical sequence of Changjiang Delta*

A complete delta sequence is produced with the progradation of Changjiang Delta when prodelta facies is overlaid by delta front facies which in turn is overlaid by delta plain facies. The vertical sequence obtained from the mouth bars such as Chongming, Changxing and Hengsha islands and Tongsha, Jiuduansha shoals can be taken as the representive delta sequence since they are the sedimentary units formed recently in the Changjiang mouth area. The borings reveal that the order in sedimentary facies from top to bottom of the stratum is comparable with that from land to sea in a plane view (Fig. 5-19). The sedimentary facies will be described based on data of borings drilled in Hengsha Island, as well as from other areas.

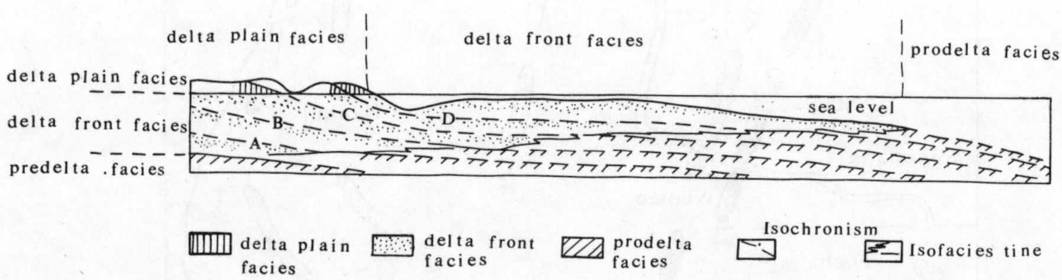

Fig. 5-19. Relationship between the sedimentary facies distributions in vertical sequence and in plane view, Changjiang Delta.

Delta plain facies is made up of intertidal and flood sediments deposited subaerially. The layer 3–4 m thick covers most part of the islands in the Changjiang mouth and the plain north of the Changjiang. The sediments of the upper part of the layer are yellowy brown silty clay with brown terrigenous spots and those of the lower part are yellowy grey clayey silt. The primary sedimentary structures are mainly lamination partly destroyed by the well-developing plant roots. Undecayed leaves and stems of plant and thin layers of peat are locally discovered. The sediments are rich in plant fragments, but poor in small sized foraminfera and fragments of ophiuroid . This layer is deposited in distributary channels and on shoals such as Tongsha shoal (Fig. 5-20).

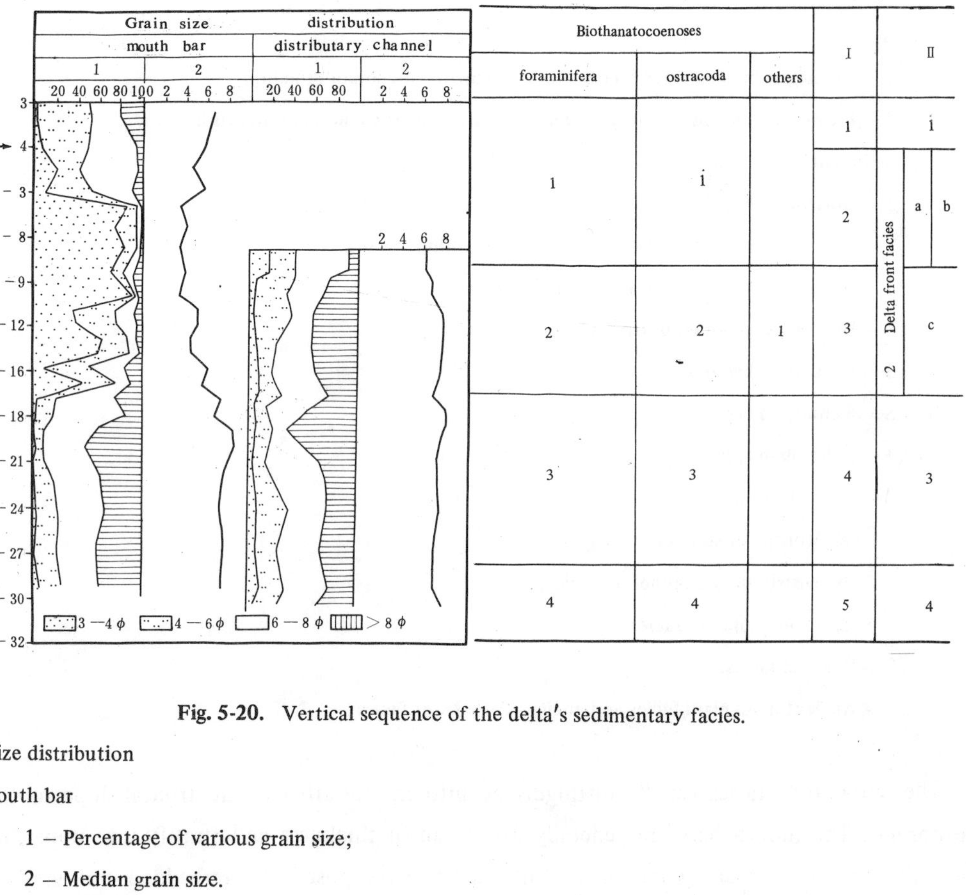

Fig. 5-20. Vertical sequence of the delta's sedimentary facies.

Grain size distribution

Mouth bar

1 – Percentage of various grain size;

2 – Median grain size.

Distributary channel

1 – Percentage of various grain size.

2 – Median grain size.

Biothanatocoenoses

Foraminifera

1 – Test sizes are small, 0.1–0.15 mm in diameter, dominantly *Ammonia beccarii, Epistominella naraensis,* secondly *Nonion, Protephidium, Grobigerina* etc.

2 – Test sizes are middle, about 0.15 mm in diameter, dominantly *Ammonia natitongensis, Quinqueloculina seminulum, Elphidium advenum.*

3 – Test sizes increases, 0.15–1.0 mm in diameter, dominantly *Ammonia nantongensis, Elphidium advenum, Quinqueloculina seminulum.*

4 – Dominantly *Ammonia, Nonionidae.*

Ostracoda

1 – Test sizes are relatively small, test numbers are rare, dominantly *Neomonoceratina, Cushmanidea,* containing continental ostracoda.

2 – Dominantly *Neomonoceratina.*

3 – Abundant in test number, test sizes are normal, dominantly *Senocytheridea, Cythereis.*

4 – No fossils are found.

Others

1 – Abundant in plant debris, containing fragments of echinodermata.

2 – Fragments and species of *echinodermata* are numerous, no plant debris is found.

I – Sedimentary structures

1 – Horizontal bedding;

2 – Horizontal bedding, wavy bedding, cross bedding, ripple;

3 – Horizontal bedding, wavy bedding, burrows;

4 – Horizontal bedding, burrows;

5 – Horizontal bedding.

II – Sedimentary facies

1 – Delta plain facies;

2 – Delta front facies;

a. Mouth bar subfacies;

b. Distributary channel subfacies;

c. Front slope subfacies.

3 – Prodelta facies;

4 – Ancient delta plain facies and ancient delta front facies.

The delta front facies can be distinguished into the mouth bar and frontal slope subfacies in the borings. The mouth bars are generally 10–15 m in thickness and stratify into upper, middle, lower three parts. The upper part is 2–3 m thick and composed of interbedded dark grey silt and yellowy brown silty clay with distinct lamination or inclined bedding and lebensspuren; the middle part 7–9 m thick, is made up of well sorted grey fine sand, bearing shell and shell fragments, which may be as thick as 1 m with thinly layered yellowy brown clayey silt intercalated, in primary sedimentary structures being lamination and inclined bedding and small cross-bedding whose bed sets are 5–8 cm thick; the lower part is interbedded bluish grey fine sand and downward thickening clayey silt with lamination and wavy bedding. In borings drilled in Chongming Island it was discovered that intercalated in mouth bar sediments there were some clay lens which may be the remains of smaller distributaries once developed in mouth bars. The sediments of the front slope facies are interbedded grey silt and grey sh yellow silty clay and 4–5 m in thickness with lamination and wavy bedding in which there are thin layers and small lens of fine sand. More species of foraminifera, *Ammonia beccarii. Epistominella naraensis. Ammonia annectens, Elphidium advendum* etc., are found in the sediments of delta front slope facies, the first two being the majority. The tests are found becoming larger from top to bottom. The size is 0.1–0.15 mm in upper part and larger than 0.15 mm in lower part. The bottom surface is generally about 14 m below Wusong Level.

The sediments of prodelta-neritic facies are thicker than 10 m, composed of fairly homogeneous bluish grey silty clay with thin silty striation, rich lebensspuren and poor plant fragments more neritic and stenohaline foraminifera are found besides those common in delta front facies and the normalized test size ranges 0.15–1.0 mm. The bottom surface of the prodelta-neritic sediments is generally at the depth 28 m below Wusong Level.

Below the prodelta-neritic sediments lies the interbedded compacted greyish yellow silty clay and grey silt and fine sand. The sediments are deposited in flood plain landward of river mouths influenced by sea water in transgression on the basis of the disappearence of neritic and stenohaline species and the dominance of the monotonous foraminifera of euryhaline species, mainly *Ammonia beccarii* which are few in number, samll in size, with some agglutinated ones.

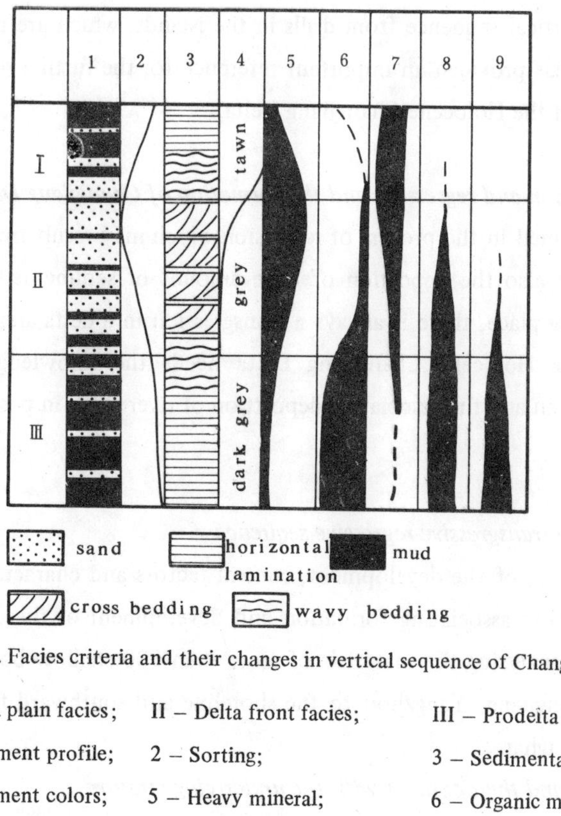

Fig. 5-21. Facies criteria and their changes in vertical sequence of Changjiang Delta.

I – Delta plain facies;	II – Delta front facies;	III – Prodelta facies.
1 – Sediment profile;	2 – Sorting;	3 – Sedimentary structures;
4 – Sediment colors;	5 – Heavy mineral;	6 – Organic matter;
7– Plant debris;	8 – Marine organizm;	9 – Bioturbation.

By comparison and analysis of boring cores, general characteristic of delta sequence can be summerized as in ascending order: the change in grain size of the sediments is fine-coarse-fine; in sorting poor-good-poor; and in primary structure lamination-wavy and all sorts of cross bedding-lamination bedding; in colour dark-light; there are less marine fossils and more continental ones in the upper part than in the lower part; the drilling structures such as borings and burrows and plant roots are common respectively in lower part and on the top; content of organic matters becomes lower upward but turns to be high again in the swamp sediment of delta plain facies. It should be pointed out, however, that the changes of all the facies criteria are gradual, without any visible demarcation (Fig. 5-21).

Compared with the world's deltas formed by large rivers such as the Mississippi and the Niger, Changjiang Delta's vertical sequence shows no difference in large [23,30]. The fact that there are some variances in detail is not in support of the conclusion that the delta sequence obtained from any delta can not serve as the model for comparing with the others[11]. Investigators have been working for the purpose to summerize the general characteristics in the facies criterion changes of delta sequence through studying all types of delta to obtain a criterion for determining the range of a delta and a model for recognizing ancient ones. The satisfactory application for petroleum and natural gas in ancient delta has proved that the common characteristics of all deltas, some and foreign, modern and ancient, are centralized in the vertical sequence which is the essential indicator of the existence of an ancient delta. The vertical sequence from drills in the islands, which are recent emergent mouth bars of the Changjiang, has provided an important reference for the further research on distribution, range and development of the Holocene Changjiang Delta.

(5) *Postglacial transgression and regression and the formation of Changjiang Delta*

All the delta are formed in the process of regression which may result from not only the relative lowering of sea-level but also the depostion of large amount of sediments brought by river. Since before a regression taking place, there is always a transgression in a delta area, a full understading of the development of the Holocene Changjiang Delta needs the knowledge of the processes of transgression and regression and the associating deposttion of river loads in post-glaciaticn recorded in the strata of the area.

(a) *Characteristics of the transgressive-regressive sequence*

A fuller understanding of the development, control factors and characteristics of transgressive-regressive sequence and the associating formation and development of the delta has been acquired through comparing and analysing the materials of more than 500 shallow borings drilled in the vast area eastward from Zhengjiang, Yangzhou to the shoreline and southward from Taizhou, Haian to Taihu Lake area and Shanghai.

(i) *Postglaciation strata and their contact with the underlying stratum*

A layer of dark green loam is widely distributed in Changjiang Delta area. It belongs to the late-Pleitocene stratum which was deposited during the last lowstand of sea level since it distinctively differs from the overlying sediments in lithology, sedimentary facies and sporo-pollen assemblage. The loam is characterized by its compactness, the presence of yellowy brown spots, vertical white stripes, iron and manganese nodules, siderite, plant debris and root system and Charophyta and the absence of marine fossils, all of which undoubtly verify the continental origin of the loam. The layer

3–5 m thick is overlaid by the marine fossil bearing and grey littoral-neritic sediments which are loose comparing with the dark green loam though they change insignificantly from place to place in lithology.

The spores and pollen in the sediments from the boring drilled in Shanghai Flour Factory indicates that the temperature of the period when the dark green loam formed was 6–8°C lower than that of present days and the climate was drier and colder, since there is a *Graminerae-Cupressaceae-Quercus-Artemisia* assemblage with some other spores and pollen fond of cold climate, such as *Abies* and *Larix*[31]. The climate reflected by the spores and pollen is similar to that of the last glaciation in the area. In the overlying layer the spores and pollen fond of humid and wet climate increase upward until their dominance, reflecting the optimum climate of the Atlantic Stage. Since the climate change reflected by the spore-pollen assemblages of the strata is very near to that in the middle and lower reaches of the Changjiang since the last glaciation[32] it is reasonable to refer the dark green loam to the sediments of the Dali Glaciation of late Pleistocene and the overlying marine layer to those of post-glaciation.

The depth where the dark green loam occurs varies in different region. It is about 10 m in Jiadin, Haian and amounts to 20--25 m in Shanghai and Rudong area and locally as deep as 30 m in Shanghai, showing an eastward dipping and undulating incline with a slope of 0.18–0.31%. and a relief about 10 m at the most. The dark green layer is absent in the valley where yellow and yellowy grey sand and gravelly sand layers are detected. And local area near the mouth there are only marine sediments which belong to one of the underlying units of the dark green loam. The disappearence of the loam layer in the area may be the result of local alluvial erosion and cutting. Pre-existing rock (carbonate rock, or the strata of Xiashu Formation is encountered in the small valleys in delta apex area which once discharged into the old Changjiang valley[33]). It is obvious that an undulating land surface of the last glaciation of late-Pleitocene when sea level was at its lowstand is represented by the dark green loam overlaid uncomformably by the covering layer deposited in the postglacial transgression due to the sea-level rising. Perpendicular to the shoreline, the land surface is not isochronous as the farther part from the shoreline, the later is the arrival of the transgression. And again, although the sediments on the land surface are comparable to each other in stratigraphy, they are not an isochronous stratigraphic unit[34].

(ii) *The structure of postglacial transgressive-regressive sequence*

Old valley and interfluve are differentiated in the vast area of Changjiang Delta according to the presence or absence of the dark green loam and they significantly differ to each other in sedimentary environment. But since different regions of the delta have experienced the similar change from transgression to regression, the sequence of delta and strandplain are also comparable.

A) Strandplain

The strandplain is situated in the vast area east of the line linking Taican, Songjiang, Jinshan south of the Changjiang and north of the line linking Taizhou, Chitang, Haian and Lusi north of the Changjiang where widely exists the dark green loam layer. The boring drilled in Shanghai Flour Factory is representive of the postglacial sedimentation in strandplain (Fig. 5-22). From bottom to top two sequences are met.

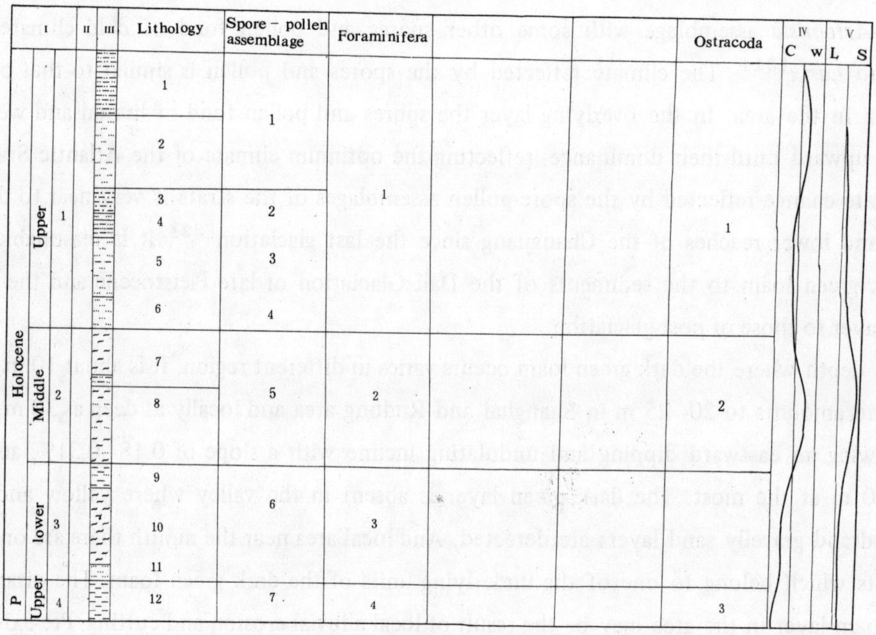

Fig. 5-22. Section of post-glacial sediments from Shanghai Flour Factory.

I – Geological epoch

Holocene

Upper

1 – Subatlantic epoch.

Middle

2 – Subboreal-Atlantic epoch.

Lower

3 – Boreal-Preboreal epoch.

P – Pleistocene

Upper

4 – Wurm (Dali) Glaciation.

II – Depth (m)

III – Profile

Lithology

1 – Greyish brown, tawny sandy clay and clayey silt;

2 – Greyish brown silt intercalculated with thin sandy clay beds;

3 – Greyish sandy clay;

4 – Greyish brown clayey silt;

5 – Greyish yellow and greyish brown silt intercalated with thin clay beds;

6 – Greyish brown silt;

7 – Dark grey soft sandy clay with thin clayey silt beds;

8 – Greyish brown soft sandy clay with thin silt beds;

9 – Greyish brown sandy clay, containing a small amount of fine sand;

10 – Greyish brown sandy soft sandy clay with calcareous concretions;

11 – Dark grey silt;

12 – Brown sandy clay.

Sporo-pollen assemblage

1 – *Polypodiaceae Saluinia* etc;

2 – *Pinus Polypodiaceae;*

3 – *Onychium;*

4 – *Gibotium;*

5 – *Pinus Leguminosae Polypodiaceae Quercus* etc;

6 – *Pinus Abies Polypodiaceae* etc;

7 – *Abies Pinus Picea*.

Foraminifera

1 –*Epistominella naraensis, Ammonia coniexidorsa, Ammonia tepita,* less number, with small test size;

2 – *Cribrononion subincertum, Ammonia coniexidorsa, Ammonia tepita,* rich in number, with middle test size;

3 – *Epistominella naraensis, Ammonia coniexidorsa,* number and test size decrease downwards;

4 – *Epistominella naraesis* etc. a small amount of test number.

Ostracoda

1 – *Tanella opima, Sinnocytheridea latiorata, Sinocythere sinensis,* number and size of tests are small;

2 – *Neomonoceratina dongtaiensis, Sincytheridea latiorata, Albileberis sinensis, Leguminocythereis hodgii* rich in test number and test size is normal;

3 – Fragments of *Cyprididae* occasionally present.

IV – Climate curve

C –Cold, W –Warm.

V – Land -sea curve

L –Land, S – Sea.

a) Transgressive sequence

The sediments of the transgressive sequence overlying the dark green loam are composed of yellowy grey and greyish brown loams, in which calcium nodules are rich and siderite, pyrite are common, with calcium content about 10%. Vertical reed roots 1–2 cm in diameter are also found. Microfossils are monotonous *Ammonia confertifesta* assemblage and *Sinocythere* with some other continental and marine transitional fossils such as shieldform fossil, *Coscinodiscus* and fragments of fish bone etc., indicating a coastal swamp environment. The spores and pollen are found to be *Castanea-Quercus aliena-Pinus-Graminea-Chenopodiaceae* assemblage, reflecting that the climate then was cool warm. Fine sand lens 3–5 m thick are often encounted in the transgressive sequence which are remains of old littoral barriers and peats deposited behind the barriers are locally discovered. Upward, the barrier-lagoon and coastal swamp facies are gradually substituted by neritic facies and marineness is enhanced. The thickness of the transgressive sequence generally is 7–8 m and the thickest amounts to 20 m (Fig. 5-23).

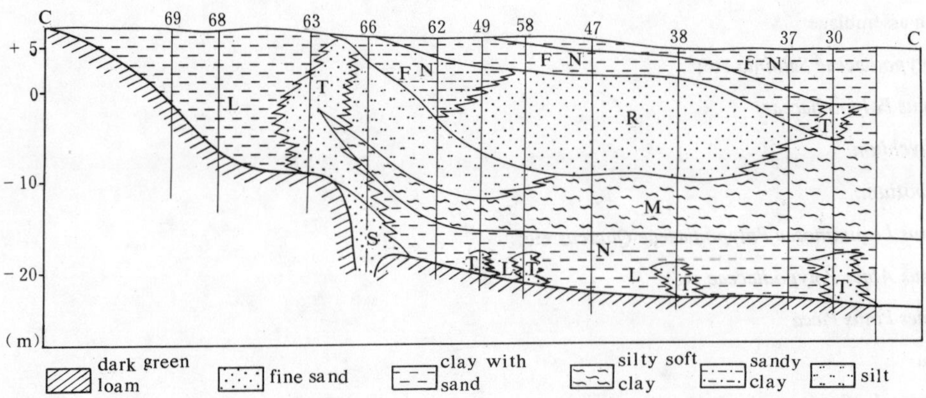

Fig. 5-23. Profile of the strandplain north of the Changjiang River.
L – Lagoon; T – Littoral bar; N – Nearshore shallow water deposits; M – Littoral-neritic deposits; R – Radiating sand bar; F – Flood-tidal deposits; S – Channel-filling sand body.

b) Regressive sequence

This sequence stratifies in two. The sediments of the lower part are homogeneous yellowy grey silty clay with a few siderite and relatively higher calcium content, amounting to 16.4%. Foraminifera are found rich in species and number, mainly normal sized neritic species, and of *Ammonia beccarii-Elphidium advenum* assemblage. The rich ostracoda is *Echinocytheris bradyi-Sinocytheridea latiovata* assemblage. And other neritic fossils such as spines of echnoidea, bryozoans and triceratium etc. are common. Spores and Pollen of *Castanopsis-Quercus glauca-Polypodiaceae* assemblage reflects a warm and humid subtropic climate. Therefore, the lower part belongs to neritic facies.

The upper part is mainly grey silt containing rich foraminifera of *Epistominella naraensis, Ammonia convexidor* assemblage, which are small sized in majority, and ostracoda, mainly *sinocytheridea latiovate, Neomonoceratina dongtaienses,* and some marine and continental fossils such as spines of echnoidea and Charophyta. The 3–4 m thick upper layer of the part is referred to littoral facies as it is composed of yellowy grey loam containing iron-manganese nodules and many plant roots and fragments, with some peat layers locally intercalated and the foraminifera in it become poorer with the relative increase in the euryhaline species. The upward decrease and disappearence of pollen *Chenopodiaceae* in the sporopollen analysis of the sediments from the depth 2–3 m in Songzi area reflects the weakening of sea water influence with time and freshwater intruding of coastal swamps. A relatively visible regional change of the upper part is shown by the dominance of grey loam with sand lens south of the Changjiang (Fig. 5-24) and of fine and silty fine sand north of the Changjiang (Fig. 5-24).

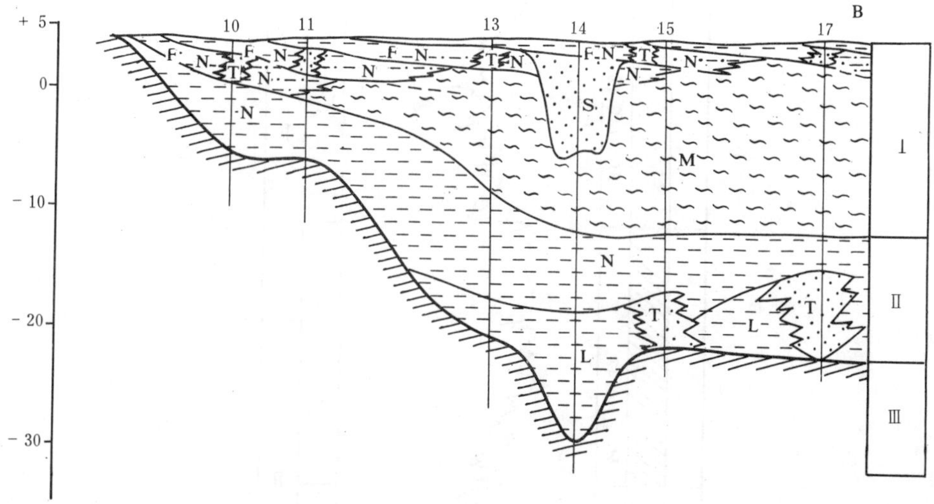

Fig. 5-24. Profile of the strandplain south of the Changjiang.

I – Regressive stage of Middle and Late Holocene;

II – Transgressive stage of Early Holocene;

III – Late Pleistocence.

B) Delta plain

In a strip north of the south Changjiang bank and south of the line linking Yangzhou, Taizhou, Chutang, Haian and Lusi ranges the delta plain. The area in which six sub-deltas of the Changjiang develop in the Holocene is about the same one of the old Changjiang valley cut in the lowstand of the last glaciation. The characteristics of the transgressive-regressive sequence of the delta plain are to be described in ascending order.

a) The transgressive sequence

The transgressive sequence is differentiated into two parts. The sediments in the lower part are grey fine sand with a thin layer of gavelly sand at the bottom containing gravelly sand at the bottom containing gravel 0.5 to 1.0 cm in diameter. And the sedimentary structures are inclined bedding and cross-bedding. The sediments in the upper part are grey loam with many thin intercalations of sand. The saw-tooth shaped electric log which damps upward matches the change in lithology in large (Fig. 5-25). This clearly reflects the characteristics of fluvial deposition which onlaps the layer of yellowy brown sandy gravel. In the area of lower reaches such as Nantong, the sand layers in the lower part of the sequence become thinner and the loam layers in the upper part become thicker.

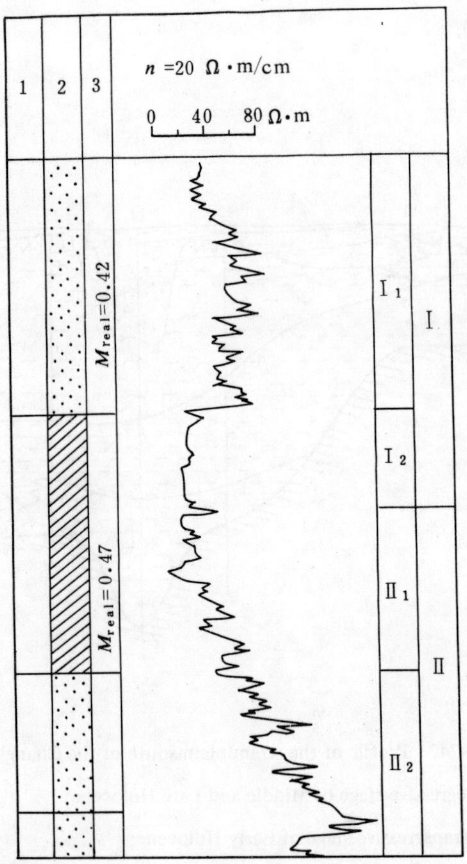

Fig. 5-25. Cross-section showing borings in Kouan.

1 – Depth (m); 2 – Lithological profile; 3 – Grand-water minenalization.

I – Regressive sequence of Middle and Late Holocene,

I_1 – Delta plain-delta front facies,

I_2 – Prodelta-neritic facies.

II – Transgressive sequence of Early Holocene,

II_1 – Littoral-flood basin facies,

II_2 – Channel facies.

The foraminifera of *Epistominella naraensis-Ammonia covexidorsa* assemblage found in the sediments are small in size and few in number. With the occurence of freshwater species *Thecamoebina,* they are characteristic of river channel deposition under the influence of sea water which enhances upward in the sequence. A ^{14}C dating of 10,100±160 BP from the counterpart of the upper part of the sequence, sampled from the core at the depth of 28–31.5 m below the datum level[35], is near the value of ^{14}C dating from the sediments of the East China Sea, representing the arrival of transgressive influence at the frontal margin of present Changjiang delta[36]. A set of swamp sediments deposited successively in some smaller valleys entering the ancient Changjiang valley and underlaid by pre-existing rocks and the "Xiashu" Formation are intercalated with several layers of peats. It belongs to post-glaciation sediments since the oldest of which dates 14,650±400 BP (Fig. 5-26). The paleoclimate reflected by sporo-pollen indicates that during the period from the formation of the swamps to 10,000 BP, temperature was lower than that of today, though it fluctuated, and that of the period following was 2°C higher than that of today which was approached grodually since 6000–7000 BP.

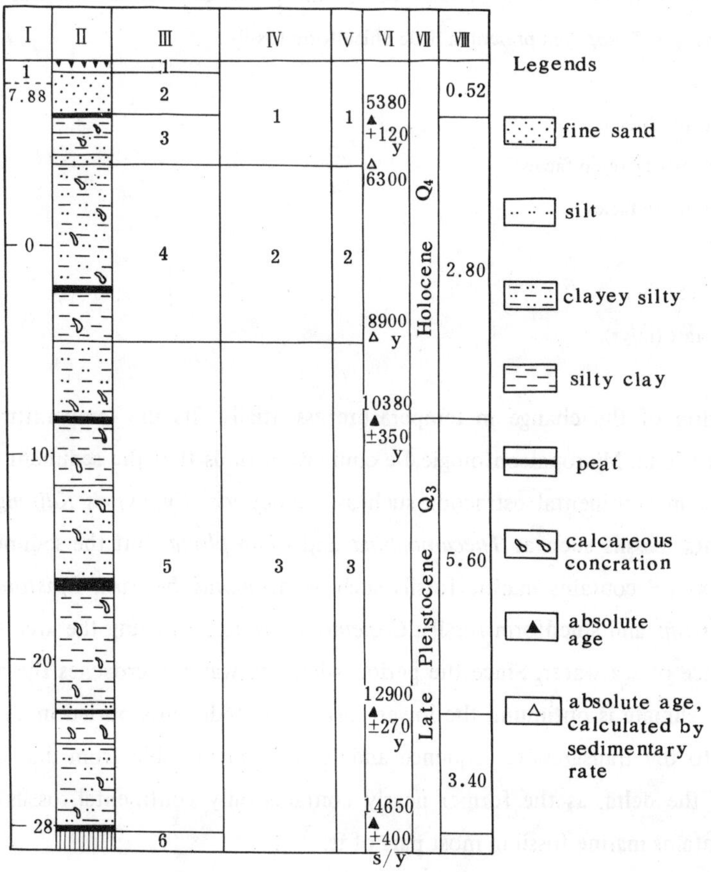

Fig. 5-26. Cross-section showing borings drilled in Zhenjiang and Dantu area.

I –Absolute height

1 – Level of Changjiang.

II –Profile

III – Lithology

1 – Soil;

2 – Grayish yellow fine sand, well solted;

3 – Gray, blue-gray clayey silt intercalated with fine sand, clay and peat beds;

4 – Gray, blue-gray silty clay, rich in plant debris, roots stem, branches of trees, fruit, calcareous concretions and peat beds present;

5 – Gray, blue-gray silty clay and clayey silt with parallel lamination calcareous concretions, plant debris, roots, leaves, branches of trees, roots and leaves of reed, several peat beds present;

6 – Bed rock.

IV – Microfossils

1 – A little shiedform fossils;

2 – A small amount of forams: *Ammonia beccarii* (vay.) , *Cribrononion visreum, Elphidium magellanicum, E. kangsuensis, Epistominella naraensis. Chorophyta,* Shieldform fossils and *Coscinodiscus* present;

3 – Continental ostracoda: *Ilyocypris errabundis, I. kaifengensis, Candoniella sutacensis* etc. *Thecamoebian* (*Centropxis* sp.) *Chorophyta* present, a little Shieldform fossils.

V – Sedimentary facies

1 – Freshwater marsh facies;

2 – Littoral-estuarine swamp facies;

3 – Freshwater marsh facies.

VI – Absolute age

VII – Geological epoch

VIII – Sedimentary rate (m/yr)

The undulation of the change in temperature essentially fits the temperature curve of post-glaciation in East China. Micropaleontological examination finds that the sediment older than about 9000 BP are rich in continental ostracoda such as *Ilyocypris. Ilyocypris Kaifengensis* and many sorts of continental fossils such as *Thecamoebina* and *Charophyta,* but the sediments dating from 6000 BP to 9000 BP contains marine fossils such as *Ammonia beccarii. Epistominella naraensis, Cribrononion vitreum* and shiedform fossils, *Cosicinodiscus* etc, implying the area was then already under the influence of sea water. Since the period when sea water encroaches the area is matchable to that when the climate is optimum, the lower part of the sediments buried in the old valleys also can be referred to the transgressive sequence and it is distinguishable from that developing in the marginal part of the delta, as the former nearly contains only continental fossils in its lower part and the latter contains marine fossil in most part of it.

b) Regressive sequence

The regressive sequence stratifies in two. The lower part is composed of homogeneous silty mud with low sand content. It manifests itself in electric log as a straight line of low resistance (Fig. 5-25). Its counterpart in Nantong area is rich in foraminifera of *Ammonia convexidorsa-Ammonia tepedas-*

Cribrononion subincertum assemblage and ostracoda, mainly *Albileberis subeinia, Cushmanidea,* with some marine spines of echnoid, ophiuroid etc, belonging to prodeltaic-neritic facies.

The upper part is made up of well-sorted grey fine sand mixed with a certain amount of silt. Its resistance amounts to a maximum in the middle of the electric log, typical of distributary-filling sequence (Fig.5-25). But the electric log from the borings drilled in Huangqiao and Panjin area are of the shape of a triangle upside down, reflecting the coarsening-upward sequences of a mouth bar. In the counterpart of it in Nantong there are foraminifera of low diversity, mainly *Epistominella naraensis* which are numerous and small sized, and some spine of echnoid and *Thecamoebina.* The upper part belongs to delta front facies. In some borings, however, the upper part is clayey sediment. It may be the sediment filled in the distributary channels. The top of the upper part which is silty clay with very few small-sized foraminifera in it is of delta plain-intertidal facies.

In the old valleys of delta apex in Zhenjiang area which once discharged water into the Changjiang River, the mud layers change gradually from the lower part containing marine fossils to the upper part without typical marine fossils. Thus a regressive sequence is undoubtedly presented.

(iii) *Rhythmatism and regional change of the transgressive-regressive sequence*

The comparison of all the borings in Changjiang Delta shows that there is a complete cyclic sedimentation from transgression to regression in which the grain size change is coarse-fine-coarse, resistance change high-low-high, sedimentary facies change littoral-neritic-littoral and climate change reflected by sporo-pollen analysis cold-warm-cool upward. All the changes in the sequence mentioned keep step essentially with each other, indicating that the sedimentation occurred by stages.

The regional change of the transgressive-regressive sequence is very notable, especially the layer of neritic facies, which thins westward in the lower part of the regressive sequence. The marine facies on the north and south flanks of the Changjiang transforms to littoral facies respectively west of the line linking Weigang and Maqiao and noth of the line linking Taizhou and Haian (Fig. 5-29, Fig.5-25); and the delta plain facies to river mouth facies in Hongqiao area. In the area where neritic facies disappears the strand sediments deposited in transgression and regression superimpose on each other. And in delta plain area the muddy neritic sediments over it is thin in the same direction (Fig. 5-27). The thickest post glaciation strata (50–60 m) are formed in the old valley of the Changjiang, but the strand plains developing on both sides of the valley only acquire a thickness of 10–25 m.

Neverthless the strata in the old valley and those in strandplain are comparable to each other. The thickness of transgressive sequence is thinner than that of regressive one, the former being only a half as thick as the latter.

The transgressive-regressive sequence of Changjiang Delta onlaps unconformably the continental deposits made up of sandy gravel alluvial and erosion surface of bedrocks formed during the last low-stand in Late Pleistocene, only in local area being on the old marine layer of late Pleicene, and graduates into continental sediments of flood-plain and freshwater swamp. Therefore the transgressive-regressive sequence is sandwiched in between the upper and the lower continental sediments.

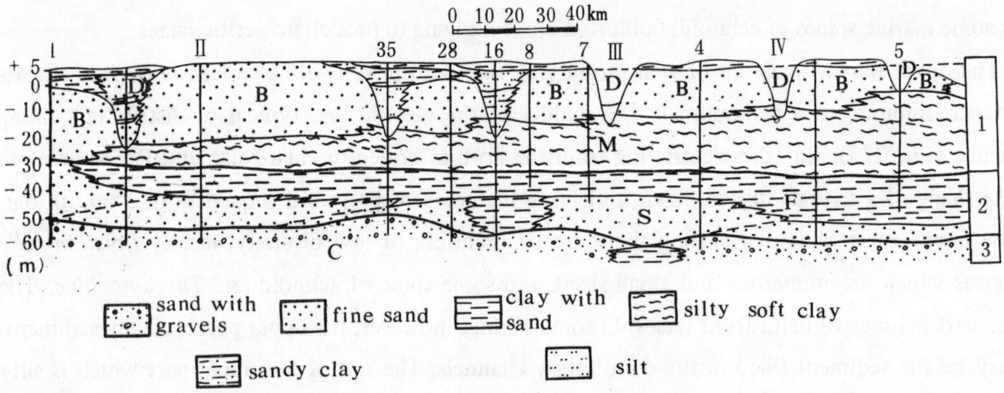

Fig. 5-27. Profile of Changjiang Delta plain normal to shoreline.

C– Late Pleistocene deposits;

S– Transgressive river channel-filled sand bodies;

F– Tide-flood basin clay deposits;

M– Littoral-neritic clayey deposits;

B– Mouth sand bar;

D– Distributary channel.

I– Hongqiao; II– Huangqiao;

III– North distribution of Changjiang;

IV– North breach of Cahngjiang.

1– Regressive sequence of Mid and Late Holocene;

2– Transgressive sequence of Early Holocene;

3– Late Pleistocene.

(6) *The development of Changjiang Delta*

(a) *Process of developing*

It is verified by synthetic analysis of the information from stratigaphy, sedimentary facies, history and archaeology that Changjiang Delta develops by stages notably. The sixstages since the full stage of transgression have borne six sub-deltas which make up the Changjiang Delta system. The six stages during which the subdeltas develop are named Hongqiao stage, Huangqiao stage, Jinsha stage, Haimen stage, Chongming stage and Changxing stage after the names of the locations where locate the nuclei for the developing of the sub-deltas (Table 5-8). The youngest of the six are Chongming stage and Changxing stage.

The basement of Chongming subdelta is the Chongming Island itself. From Tang Dynasty to Song Dynasty all parts of the subaqueous sand body, namely the basement of the sub-delta, emerged one by one to become sand islands which were latter connected to each other and near the trun from Ming Dynasty to Qing Dynasty the configuration of today's Chongming Island came into being (Fig. 5-28).

Table 5-8 Stages of the sand bodies in the Changjiang Delta system

Absolute age (BP)	International stage	Archaeological stage	Stage of sand bodies: River mouth sand bar	Littoral sand barrier: South of the lower reaches of the Changjiang River	Littoral sand barrier: North of the lower reaches of the Changjiang River	Radiating bar	River channel sand bank
	Subatlantic	Iron Age	Changxing stage 700	Donghai-wanxiang		Jianggang radiating bar	
–1000			Chongming stage 200–1700	Wangjianqiao-Jiangzhen Jiangwan-zhoupu		Sanyu radiating bar	Changyinsha
–2000			Haimen stage 1200–2500	Luodian-Qianqiao		Bencha radiating bar	Matousha
–3000	Subboreal	Bronze Age	Jinsha stage 2000–4500				
–4000				Jiading-Zhelin	Taizhou-Haian		
–5000		Neotithic Age	Huangqiao stage 4000–6500				
–6000	Atlantic			Waigang-Caojing			
–7000			Hongqiao stage 6000–7500				

The coastal line change is complicated in Chongming Stage. The coastal line north of the Changjiang maintains in the position of Fangongdi and Shengongdi (names of two sea dykes) built in Song Dynasty. The coastal line south of Changjiang is about the line linking Gaoqiao, Chansha, Huinanzhen, made up of several lengths of ridges upon which the Lihutang was built in 1172. When the former was under destruction on a large scale, the latter maintained essentially steady and in early Qing Dynasty it was silted up greatly and advanced to the line linking Huangjiawan and Wanxiang where distribute discrete coastal ridges upon which sea dyke were built later[43-46].

Three new channel bars of large scale named Changying Island, Dongxing and Nang Island formed in Qing Dynasty with the emergence of small channel bars in the Changjiang channel proper. The three eventually incorporated into the south bank of the Changjiang. The mouth bars of Changxing subdelta include Changxing Island, Hengsha Island and Tongsha Shoal which are separated by

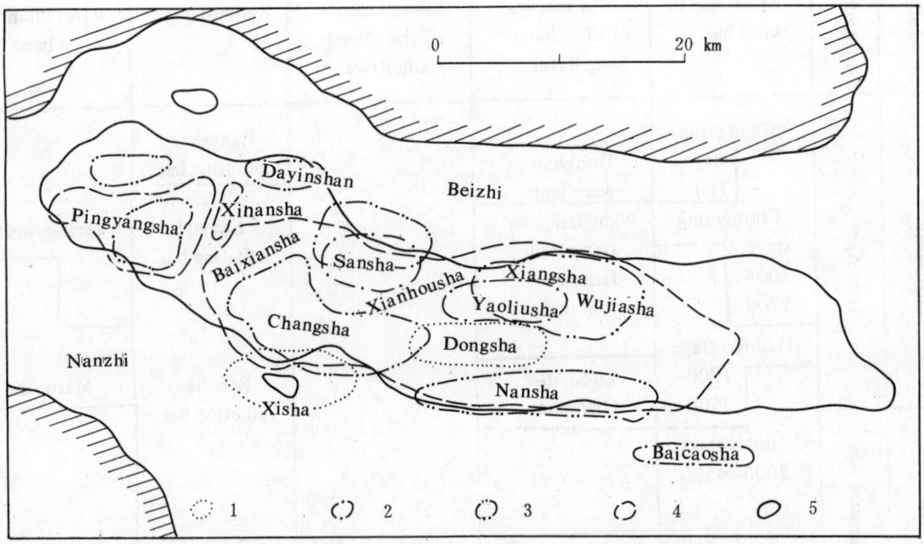

Fig. 5-28. Schematic diagram of the Chongming Island development (from Historical Geograph Laboratory, Fudan University. 1975).

1– Sand island emerged in Tang-Song Dynasty;

2– Sand island emerged in Song Dynasty;

3– Sand island emerged in middle Ming Dynasty;

4– Sand island emerged in Late Ming and Early Qing;

5– Present Chongming island Dynasty.

small distributaries but are connected to each subaqueously. The islands of SE strike were formerly mouth bars in the south distributary of the Chongming subdelta. The mouth bars of Changxing stage came into being before the year 700 BP, and Sodid Chongbaosha Island, Yawocha Island, Shitongsha Island and Yuanyuansha Island etc. which emerged from the same basement during the last 200 yr later. The Hengsha Island appearing in 1858 is separated only by a narrow distributary from the Changxing Island, and the Tongsha Shoal is still a subaqueous one now (Fig. 5-29). The south distributary mouth bar of the Changxing subdelta was surveyed for navigation chart in 1900, and later it grows gradually to be Jiuduansha Island (Fig. 5-30).

A the same of the Island development (from the comparison of historic navigational charts), Changxing subdelta advanced the coastal line further, producing muddy intertidal flat of 2–3 km in width, which eventually turned into a coastal plain with aggration.

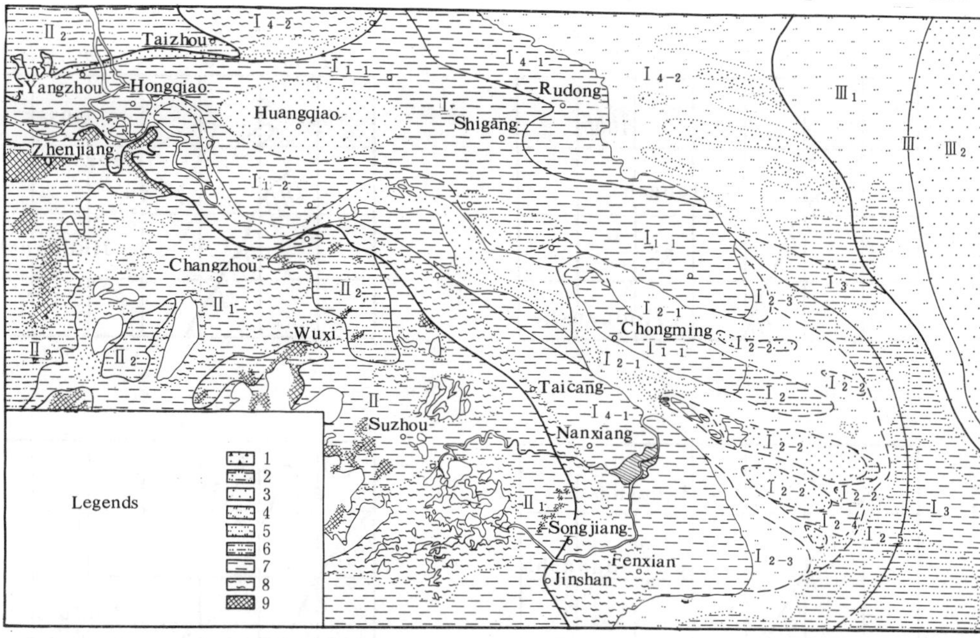

Fig. 5-29. Schematic diagram of facies distribution of modern Changjiang Delta.

I– Deltaic sedimentary system;

I_1 – Delta plain facies;

I_{1-1} – Marine-alluvial plain subfacies;

I_{1-2} – Alluvial plain facies;

I_2 – Delta front facies;

I_{2-1} – Distributary channel subfacies;

I_{2-2} – Mouth bar subfacies;

I_{2-3} – Frank marginal bank subfacies;

I_{2-4} – Interdistributary shoal subfacies;

I_{2-5} – Front slope subfacies;

I_3 – Prodelta facies;

I_4 – Association facies;

I_{4-1} – Strandplain subfacies;

I_{4-2} – Radiating sand ridge subfacies;

II– Continental sedimentary system;

II_1 – Alluvial-lacustrine plain faies;

II_2 – Ancient alluvial plain facies;

III– Marine sedimentary system;

III_1 – Neritic facies on continental shelf;

III_2 – Relict facies on continental shelf;

1– Breccia; 2– Gravelly clay sand; 3– Fine sand;

4– Fine silt; 5– Silt; 6– Clayey silt;

7– Silty clay; 8– Muddy silty clay and mud;

9– Bese rock.

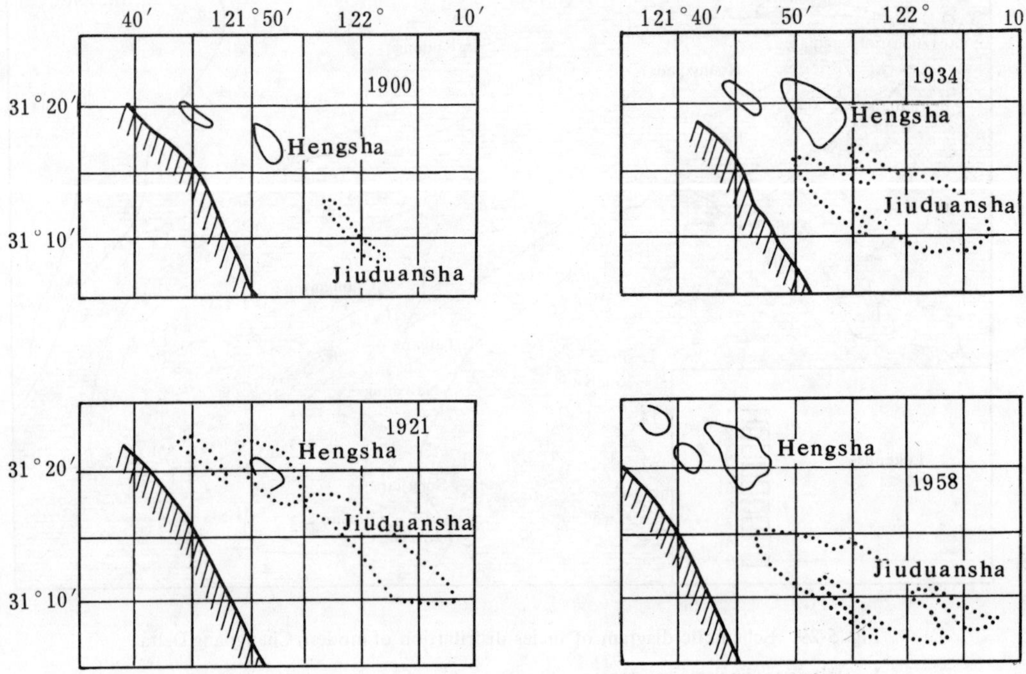

Fig. 5-30. Schematic diagram of Jiuduansha.

(b) *Development of Changjiang Delta*

From the full stage of transgression of post-glaciation the Changjiang has borne six sub-beltas since 7500 BP. Therefore, Changjiang Delta is different from Mekong Delta and Niger Delta in that the former is a complex system made up of many sub-deltas and other associating deposit bodies but the latter are individul deltas[47,48]. The sub-deltas making up the system are gradationally arrayed in order of generation instead in a chaotic like Mississippi Delta and Huanghe Delta[49,50], which is also taken as one of the specialities of Changjiang Delta[51].

(i) *Model of development*

The developing of a delta depends on the change of the flow at the mouth. The embryo and the fundamental elements for a delta are the mouth bar. The process of Changjiang Delta building is actually the formation and development of mouth bars. The formation of a mouth bar forces the stream to bifurcates, producing the south and the north branches. The north branches' mouth bars of the six subdeltas are small in size and short in existence, being incorporated into the north bank one by one with the dying away of north branches, as in the mesotidal mouth of the Changjiang, under the influence of the Coriolis force, there always exists stronger flood current in north branches and stronger ebb current in south ones. Therefore, it is the Coriolis force that flourishes a south branch and makes it become the main distributary for water and sediment discharge where a new and large mouth bar is deposited rapidly. The new mouth bar is to become the main part of a new generation of sub-delta and forces the branch to bifucate into south and north distributaries again. The new distributaries result in the new mouth bars, preluding the formation of a sub-delta still younger (Fig. 5-31). The mouth bars of the six stages array themselves gradually from NE to SE in an en-echelon pattern in

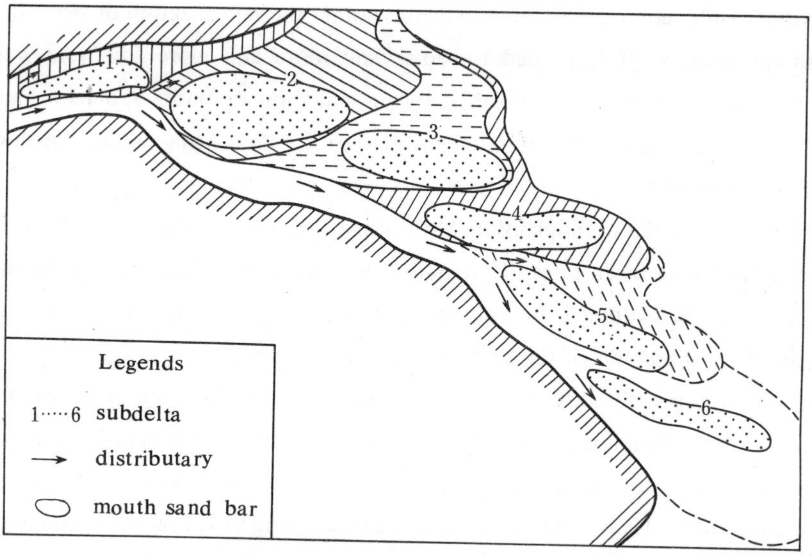

Fig. 5-31. Developing model of Changjiang Delta.

plane view. And it is clear that sub-deltas regressively offlop one another orderly not only in space but also in time without notable destructive phase in delta development. These features in the development of Changjiang Delta are seldom found in other deltas in the world. Of course, erosions of short period have occurred locally in the building up of some sub-deltas in the form of shoreline receding, as is seen in Chongming sub-delta, but they are too short in duration and too local to negate the generized pattern of the delta development.

(ii) *Rate of delta progradation*

The rate of delta progradation is figured about 40 m/yr averagely from the information of history and archaeology and ^{14}C dating and the development of Changjiang Delta. The rate is larger than those of 27m/yr and 24m/yr estimated by Ting, V.K. and Heidenstan, H.V. respectively[52,53].

The rate of delta progradation has a tendency of becoming larger with time in the building up of Changjiang Delta (Table 5-9), e.g., the rate of delta plain progradation increases from 40 m/yr in Hongqiao and Huangqiao stages to 75–90 m/yr in Haimen stage. The reason for this is that the large estuary with deeper water in the earlier stages needs more sediments than the mouth of shallower water in later stages for the emergence of a subaqueous sub-delta.

Table 5-9 Progradation-rates of the sub-deltas in the Changjiang Delta system

Stage	Rate of delta plain	Progradation (m/yr) strand plain of the Changjiang
Hongqiao stage	45	2.5
Huangqiao stage	47	3.0
Jinsha stage	37	5.0
Haimen stage	90	11.2
Chongming stage	75	20.2

And another important reason for the speeding up of the Changjiang Delta progradation is that with the enhancement of man's economic activities in historic times the vandalization of vegetation in the Changjiang drainage basin went from bad to worse and more and more soil were washed away[54,46]. The exceptional drop in the rate of delta progradation in Jinsha sub-delta may be assinged to the diminished water discharge and sediment load of the Changjiang because of the dry and cool climate during that period.

The rates of progradation of the strandplains on both sides of the Changjiang are notablely lower than those of the delta plain. The mean rates of progradation of the north and south strandplain are 12 m/yr and 8 m/yr respectively. Although the rate varies greatly in the stages, it also grows higher with time, e.g, the rate of progradation of the south coastal line is only 2–3 m/yr in Hongqiao and Huangqiao stages but it amounts to 20 m/yr in Chongming stage. The Nanhui point outgrows as fast as 45 m/yr averagely in last one hundred years. The reasons are that the Changjiang continuously shifts south-ward, supplying the south strandplain with more and more sediments, that the enhancement of man's economic activities leads to the increase in the Changjiang's load capacity, and·that the sea dyke building for reclaimation also tends to speed up the advancing of the south strandplain.

(iii) *Sedimentary rate*

The sedimentary rate of a delta is an improtant parameter for describing its growth and high sedimentary rate is the main characteristic of delta sedimentation. It is know from the information of navigation engineering that the sedimentary rate of a growing mouth bar, Xingtongsha shoal, in the south branch of the Changjiang reaches as high as several tens of centimeters per year when there are unusual floods. Same reports are received from the Mississippi and the Rhone, 30 cm/yr[55] and 35 cm/yr[56]respectively. It should be pointed out that the high sedimentary rates as such are only representives of those of main distributary mouths in the early stage of delta building, not the mean rates. In·face, the rate differs from place to place in the direction perpendicular to the shoreline. It drops gradually off shore. By comparing the sea charts of 1842, 1865 and 1973 outside the Changjiang River mouth, the sedimentary rates of mouth bar, delta frontal slope and prodelta were respectively 12 cm/yr, 5 cm/yr and 0.3 cm/yr, near the data from Mississippi Delta[39]. In the direction parallel to shoreline, the sedimentary rate in the region where the main distributary is active is the highest but it decreases latee away from the mouth. In the Changjiang River mouth area the sedimentary rate of the mouth bar is higher than that of the north and the south strandplain, as is found in the Po, Italy. A pronounced variance also exists among the rates of different stages. It is discovered that higher sedimentary rate belongs to earlier stages and the lower to later stages, as for example, the formation of the mouth bar's subaqueous basement 10–15 m thick in Chongming stage takes only several hundreds of years but the process of the Chongming Island formation in which all parts of the basement emerge successively lingers as long as more than one thousand years. Thus the mean sedimentary rate calculated from the overall thickness and the duration from the beginning to the end of a sub-delta formation is much lower. The averages of six sub-deltas' sedimentary rates are 1.14 cm/yr for mouth bar, 0.54 cm/yr for delta frontal slope and 0.31 cm/yr for prodelta (Table 5-10). And it can be concluded from Table 5-10 that the variance of sedimentary rates of the sub-deltas is visible though not pronounced, of all the sub-deltas, the Jinsha stage one has a lowest rate, which may have a bearing on the dry and cool cilme leading to the lower water discharge and sediment load of the Changjiang River, and the Hongqiao stage one has the highest, which may be attributed to not only the warm and humid climate which causes a high water discharge and sediment load, but also the deeper water of the Changjiang Estuary which takes large amount of sediments. The later factor is far from the negligible one for the high sedimentary rate in Hongqiao stage.

Table 5-10 Sedimentary rates of the sub-deltas of the Changjiang Delta system

Stage	Sedimentary rate of the sub-deltas (cm/yr)		
	Mouth bar	Frontal slope	Prodelta-neritic zone
Hongqiao stage	1.66	0.18	0.40
Huangqiao stage	1.20	0.36	0.22
Jinsha stage	0.72	0.30	0.16
Haimen stage	0.15	0.65	0.42
Chongming stage	1.00	0.57	0.37
Average	1.14	0.54	0.31

If the regional difference is not taken into account, the mean sedimentary rate of 0.53 cm/yr is obtaied from the overall amount of sediment of the Changjiang Delta system. And the mean sedimentary rates of Rhone Delta and Niger Delta are respectively 0.53 cm/yr and 0.56 cm/yr[23,56], comparable to that of Changjiang Delta.

The sedimentary rates strandplains in south and north sides of Changjiang, being 0.20 cm/yr and 0.28 cm/yr respectively, are less than those in Changjiang Delta plain.

(7) *Characteristics and distribution of sand-bodies in Changjiang Delta*

Sand body is the main sedimentary unit in a delta, it forms the framework of a delta determining its shape, refecting sensitively its evolution, and thus is the key for recognition of the delta. The delta sand bodies lie close to argilloceous sediments containing abundant organic matters and are good resevoirs for oil and natural gas in stratigraphical sequences and form major objectives of exploration. Sand bodies are very impotant in Changjiang Delta area because of their large area, wide distribution, numerous types and complete system, they constitute a complicated complex (Fig. 5-32). There are no deltas in the world that can be compared with the Changjiang Delta. With alternative transgression and regression sand bodies of the Changjiang Delta are distributed in two-storeyed beds.

The external geometry, internal characteristics, direction of arrangement and contact with overlying and underlying beds of sand bodies closely related to hydrodynamics. Therefore, it is reasonable to classify the sand bodies of the Changjiang Delta according to main factors and processes forming them. Generally speaking, the sand bodies in Changjiang Delta may be grouped into river mouth sand bodies, channel-filling sand bodies and marine sand bodies according to their oringins (Table 5-11).

Table 5-11 Characteristics of various sand bodies in Changjiang Delta

Characterictics	Mouth sand body		Marine sand body		Channel sand body	
	Mouth bar	Distributary Channel sand body	Littoral bar	Radiating bar	Regressive channel sand body	Transgressive channel sand body
Length (km)	60–70	10–15	2–30	10–100	10–30	–
Width (km)	10–15	2–3	0.2–2	5–10	5–10	2–30
Thickness (km)	15–30	10–15	1–10	5–10	10–20	5–15
Area (km^2)	800–1000	20–30	–	–	50–300	–
Volume ($b.m^3$)	15–35	0.2–0.6	0.0001–0.5	–	1–5	–
Major hydrodynamic factors	River flow and tidal current woth wave action	River mouth flow	Wave	Tidal current	Channel unidirectional flow	
Distribution	Perpendicular to coastline, an echelon	Oblique to coastline, distributed as a string of beads	Parallel to coastline in shoelace shape	Oblique or perpendicular to coastline, radiating shape	Perpendicular to coastline or no strict direction, elongated shape	
Sediment	Fine sand, well sorted occationally with clayey lense	Silt-fine sand, intercalated with clay beds	Fine sand with shell detritus and whole shell	Fine sand and sand with shell	Fine sand and fine sand with gravels	Fine sand and sand with gravels intercalated by clay beds.
Sedimentary sequence	Coarsening-upwards	Fining-upwards	Coarsening-upwards	Coarsening-upwards	Fining-upwards	Fining-upwards
Sedimentary structures	Horizontal lamination-wavy bedding-cross bedding in ascending order	Trough cross bedding-wavy bedding-horizontal lamination in ascending order	Cross bedding with low angar	Main horizontal lamination, small medium cross bedding in middle and upper parts	Trough cross bedding-oblique bedding-horizontal lamination in ascending order	

(to be continued)

Table 5-11 (*continued*)

Characterictics	Mouth sand body		Marine sand body		Channel sand body	
	Mouth bar	Distributary Channel sand body	Littoral bar	Radiating bar	Regressive channel sand body	Transgressive channel sand body
Contact	Bottom: gradational, underlying prodeltaneritic sediments; top: gradational, or sharp, overlying tidal flat and floodplain	Bottom: sharp, erosional lower contact; top: gradational, overlying tidal flat and floodplain deposits	Bottom: gradational or sharp, underlying continental or neritic beds; Top: gradational, overlying tidal flat or neritic deposits	Bottom: gradational, underlying neritic deposits; Top: gradational, overlying tidal flat deposits	Bottom: sharp, erosional Top: gradational, overlying floodplain deposits	Bottom: sharp, erosional Top: gradational, overlying littoralneritic deposits
Microfossils	Foraminifera are relatively small *Ammonia beccarii-Epistominella naraensis* assemblage, main ostracoda are *Neomonoceratina*, *Sinocytheridea*, containing continental ostracoda, fragments of *Echinoderimata*, rich in plant debris		Test size of foraminifera is normal, *Ammonia beccarii-Elphidium* assemblage; ostracoda: *Lequnimocythereis*, *Albileberis*, fragments of *Echinoderimata* are common, plant debris rare		A small amount of species and tests of foraminifera, *Ammonia beccarii-Ammonia convexidorsa* assemblage, continental and marine ostracoda mixed, containing plant debris and echinoidea	
Electric log (resistivity)	From low to high upwards	From high to low upwards	From low to high upwards		From high to low upwards	

These sand bodies can further be subdivided. For example, marine sand bodies are divided into littoral sand bars and radiating sand bars, and river mouth sand bodies into river mouth sand bars and distributary channel sand bodies etc.

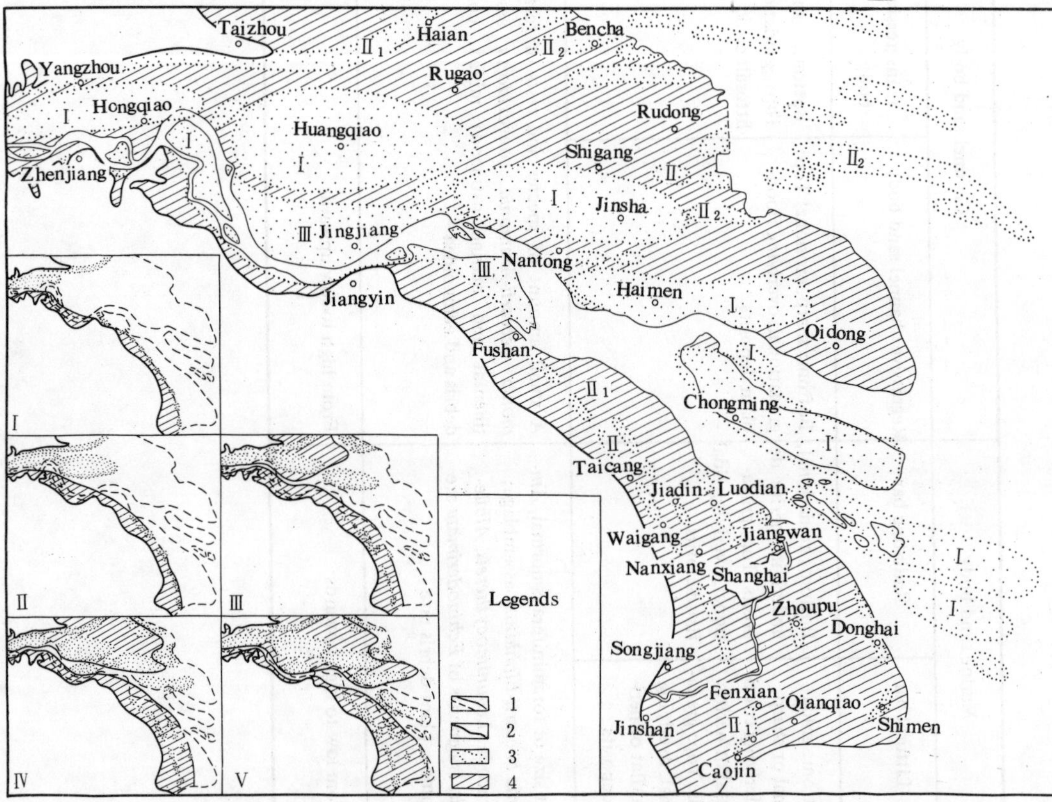

Fig. 5-32. Types and assemblage of sand bodies in Changjiang Delta.

I– Mouth sand body;

II– Marine sand body;

II_1 – Littoral sand bar,

II_2 – Radiating sand body;

III– Channel sand body.

1– Present coastline; 2– Boundary of delta;

3– Sand body; 4– Land.

(a) *Characteristics of sand bodies in Changjiang Delta*

(i) *River mouth sand bodies*

a) Mouth sand bar

In tidal river mouth, tidal sand ridges are generally present seawards from the mouth sand bar, between which is the deep water area[16]. In Changjiang Delta this kind of tidal ridges has not been discovered, but the length of mouth bars is unsually great. This may be resulted from the combination of the mouth sand bar and underwater tidal ridges as a united sand body. The typical characteristics of the river mouth sand bars of the Changjiang Delta are large bodies, each of them becomes the nucleus and the main part of the subdeltas. Usually the mouth bars are 60–70 km in length, 10–15 km in width, and the thickness of the sand beds in 15–30 m, with maximum up to 40 m. The area of each sand bar is about 800–1000km^2, its volume 3X10^9m. (Table 5-12). The total area of the mouth sand bars is 5894 km^2, which makes up 39.3% of the delta plain.

Table 5-12 Parameters of mouth bars in Changjiang Delta

Stage of mouth bar	Length (km)	Width (km)	Average thickness (m)	Area (km^2)	Volume (10^8m^3)
Hongqiao	63	13	25	819	204.75
Huangqiao	72	16	30	1152	345.60
Jinsha	67	15	18	1005	180.90
Haimen	71	13	15	923	138.45
Chongming	70	15	18	1050	163.00
Changxing	72	9	14	648	90.72
Jiuduansha	33	9	10	297	29.70
Amount				5894	1158.32
Average	64	12.9	18.3		

The mouth bars of the Changjiang Delta arrange chornologically an echelon from the northwest to southeast with axis perpendicular or oblique to the coastline (Fig. 5-32).

The vertical sequence of mouth bar in Changjiang Delta is coarsening-upwards. Apparent resistivity increases from bottom to top (Fig. 5-33). Their underying beds are deltaic-neritic sediments. Their contacts are gradational. The overlying tidal flat and flood plain facies, 3–5m thick, also have a gradual changing relationship (Fig. 5-34) with the mouth bars.

b) Distributary channel sand bodies

This type of sand bodies is formed in distributary channel, and usually arranges like a string or beads. Their strike is perpendicular or oblique to the coastline. The differeces between the distributary channel sand bodies and mouth sand bars are in vertical sequence, the former is fining-upwards and in the electric log its resistivity is from high to low upwards (Fig. 5-35), and its contact with underlying beds is sharp and erosional.

(ii) *Marine sand bodies*

Marine sand bodies may be subdivided into littoral and radiating bars which are located in the southern and northern flanks of the Changjiang Delta, respectively.

a) Littoral sand bodies

The littoral sand bodies in the southern part of the Changjiang Delta spread arcuately from Fushan towards southeast, subparallel to ancient coastline. They are several or more than a dozen 1 km long, about 200–300 m wide and the sand layer is about 0.5–0.6 m thick but the maximum thickness reaches 2.0 m. The buried depth of the sand bodies is 1–2 m from the terrene, the maximum 2.75 m, the minimum only 0.4 m, and some of them even are exposed on the surface. The sand bodies of tidal flat normally are characterized by irrgularity of buried depth and their thickness. The width of the so-called "Gangshen" (series of bars) from Jiading to Caojing is about 8 km in the north, consisting of 5 littora bars, but getting norrower southwards while (1.5–2.0 km) with 3 littoral sand bars. The shell content in these littoral bars generally increases from north to south. All these show that sandy materials of the littoral bars came mainly from the Changjiang River. The littoral sand bodies are underlain by marine argillaceous beds and overlain by tidal flat deposits. The contacts with underlying and overlying beds are gradational. The littoral sand bars in the northern frank of the Changjiang Delta are well developed along the belt from Taizhou to Haian where they are about 1 km wide and 10 m thick. The underlying sediments are continental and the overlying ones are tidal flat deposits. They began to accumulate in the maximum period of postglacial transgression and finished at the end of Huangqiao stage, lasting 3000 yr (Table 5-6).

b) Radiating bars

The radiating bars are a specific associated sedimentary system in Changjiang Delta, which are distributed in the strandplain in the northern part of the Changjiang Delta and the offshore area of Jianggang. The sequence of radiating bar is coarsening-upwards, and the resistivity in electric log changes from low to high in ascending order (Fig. 5-36). A similar vertical sequence and electric log of the tidal sand ridges are present in the Corolado Delta[58]. The radiating bars are underlain by neritic deposits and overlain by tidal flat deposits. The contacts between them are gradational. The existence of the two groups of radiating bar system in Bencha and Shayu areas indicates that the sedimentary environments similar to recent ones in Jianggang area appeared several times during development of the Changjiang Delta. Obviously, the radiating bar system is not an occasional phenomena.

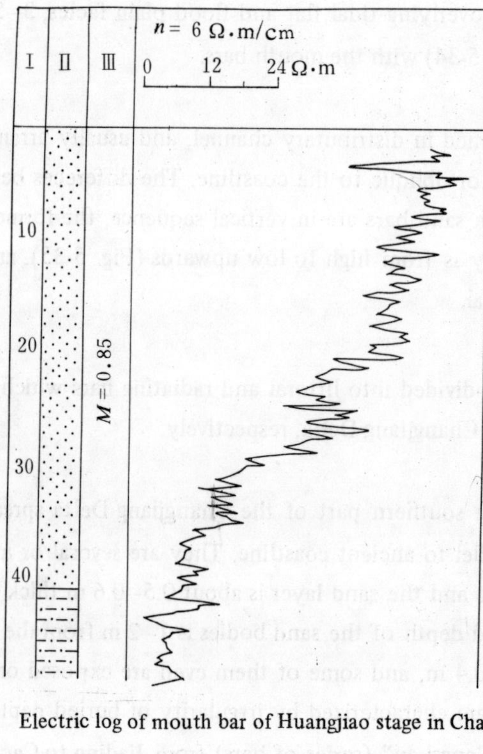

Fig. 5-33. Electric log of mouth bar of Huangjiao stage in Changjiang Delta.

I– Depth (m); II– Lithoprofile; III– Groundwater minerallization (g/1).

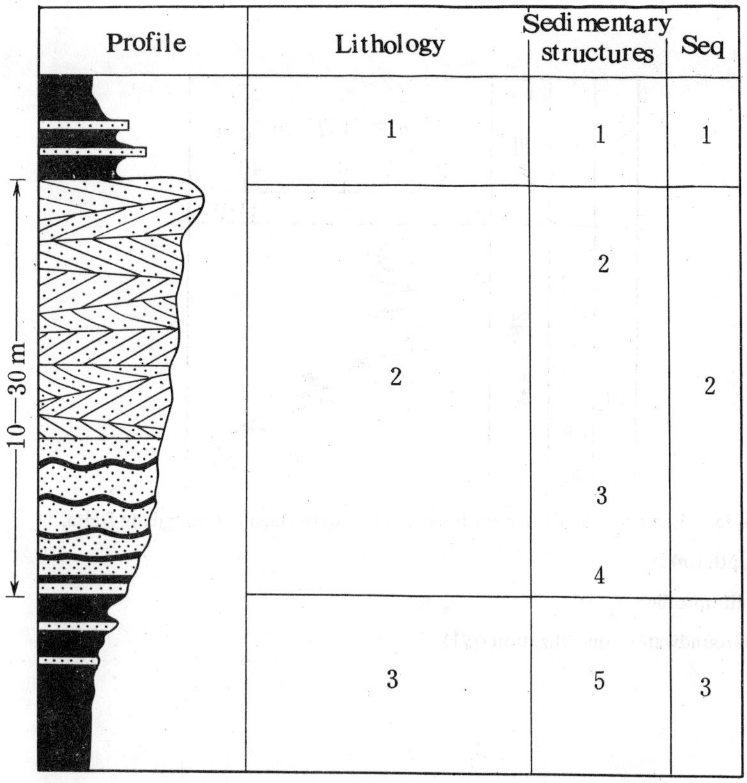

Fig. 5-34. Vertical profile of mouth sand bar in Changjiang Delta.

Lithology

1– Tawn and yellowish gray silty clay with thin beds of silt, Root, ferriginous concretion and occasional peat present;

2– Gray fine sand, well solted in upper part with clay beds in lower part, shell and plant debris present, contacts with overlying and underlying beds are gradational;

3– Dark gray silty soft clay.

Sedimentary structures

1– Parallel bedding consisted of clayey and sandy laminae;

2– Cross-bedding with low angle and cross-bedding;

3– Wavy bedding;

4– Parallel bedding;

5– Parallel bedding consisted of sandy and clayey laminae.

Sequence

1– Delta plain facies;

2– Delta front facies (mouth bar);

3– Prodelta neritic facies.

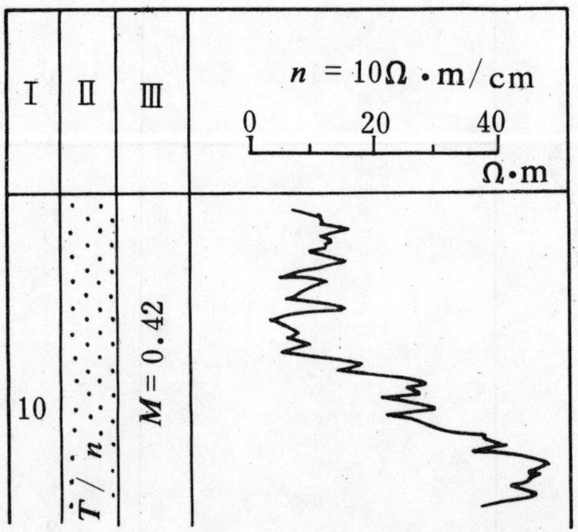

Fig. 5-35. Electric log of the mouth bar of Hongqiao stage in Cahngjiang Delta.

I– Depth (m);

II– Lithoprofile;

III– Groundwater mineralization (g/1).

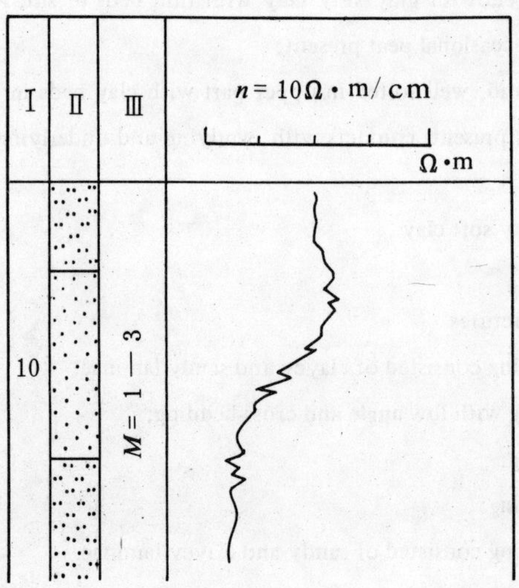

Fig. 5-36. Resistivity log of radiating bar.

I– Depth (m);

II– Lithoprofile;

III– Groundwater mineralization (g/1).

(iii) *Channel sand bodies*

In the progradation of the delta, the main channel of the Changjiang extented, reworking the delta sequences and forming channel sand bodies in the process of the channel evolution. The sand bodies in the Changjiang River are mainly formed by the various kinds of the bars in the centre of the river, the biggest one being Yangzhong Sand bar. Some of sand bodies have connected with the river bank, forming into land, such as in Matuosha of Jingjiang area and Changyinsha of Shazhou County. Moreover, the tributary channels entering into the Changjiang River also form their channel sand bodies; for instance there are channel sands, about 2 km, wide and 7–8 m thick in the meander belt of the Suzhouhe River in Shanghai. The vertical sequence of sand bodies is finingupwards. As examplified by that in Jingjiang area, gravelly sands are present in the lowermost part, bceoming medium sand, then fine sand and silty sand upwards. The contacts of the channel sand bodies are sharp, erosional with underlying beds and gradational with overlying floodplain deposits.

(b) *Transgressive sand bodies in Changjiang Delta area*

During postglacial period the sea level rose and transgression occurred. As a result, modern Changjiang Delta area was progressively submerged and different types of transgressive sand bodies were formed in ancient valley and strandplain.

(i) *Transgressive channel-filling sand bodies*

The transgressive channel-filling sand bodies, about 10–15 m thick, are nearcontinuously distributed along the ancient Changjiang Valley, On the bottom of the transgressive sand bodies there is a gravel sand bed about 10 m thick. The sequence is fining-upwards and therefore, resistivity in the electric log changes from high to low in ascending order (Fig. 5-37).

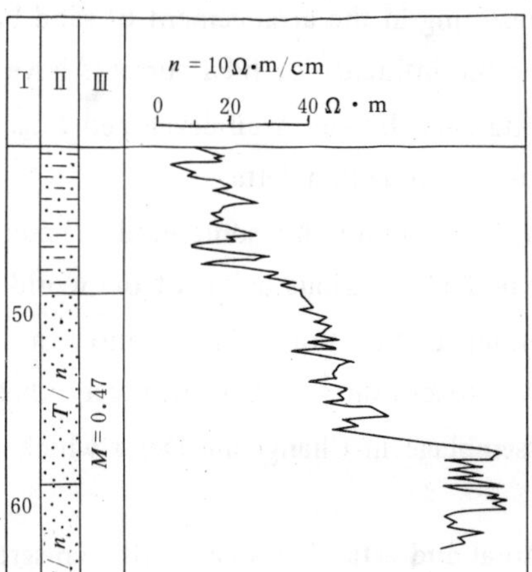

Fig. 5-37. Electric log of the transgressive channel-filling sand body in Changjiang Delta.

I– Depth (m); II– Lithoprofill; III– Groundwater mineralization (g/1).

The transgressive channel-filling sand bodies overlapped discordandly on the continental gravel sand bed during the last glacial low sealevel. The contacts with overlying littoral-neritic muddy deposits are gradational. The transgressive channel-filling sand bodies are discovered in Zhujiang (Pearl) River[59], Mississippi[30], Lavaca[60], and Brasos[61] rivers. Therefore, it is evident that the distribution of the transgressive channel-filling sand is quite popular.

(ii) *Transgressive littoral sand bodies*

The transgressive littoral sand bodies in Changjiang Delta are generally 3–5 m thick but its maximum thickness may reach 10 m. Their forming factors and internal characteristics are similar to the regressive ones. The main difference between them is that the transgressive littoral bars overlap discordantly on the continental beds. Because the amount of sands from source is relatively smaller in the interfluve than that in the channels, the sand bars are discontinuely distributed from land to sea as an echelon, no sand sheet is formed.

(c) *Distribution of sand bodies in Changjiang Delta area*

Although physical factors in the delta area are numerous, only river flow, wave and tidal current can form sand bodies, and, therefore, the delta sand bodies include the river mouth sand bodies, wave sand bodies and tidal sand bodies. All these types are present in Changjiang Delta area. The channel sand bodies form an important part in the delta during the process of progradation of the Changjiang Delta. As these sand bodies are within the same estuarine hydrodynamic system, they are interrelated each other, and may be comparable in forming time as well.

The Changjiang River mouth is a mesotidal estuary. Tidal current is comparatively strong and tidal sand bodies are well developed. Firstly, the mouth bars link with tidal ridges, so that with the growth of bars, their volume become very great. Flood and ebb currents run up and down stream in the distributary channels, making them seldom silting up. The river mouth is progressively moving southwards by the Coriolis force, resulting in the arrangement of sand bodies an echelon. Secondly, the radiating bay systems reflecting the influence of tidal currents have appeared several times and become important units of the delta sand bodies. Well-developed tidal sand bodies in Changjiang Delta reveals the characteristics of tidal dominanted delta.

Recent attempts have been made to summarize some models based on the complicated distribution and assemblage of the sand bodies in various deltas of the world in terms of the need of oil and natural gas exploration. According to the study of many modern deltas, J.M. Coleman (1975) obtain six models of the sand body distribution[11]. Unfortunately, these models have not included the sand body distribution and assemblage in Changjiang Delta which are more complicated than these six models.

The sand bodies formed in littoral and estuarine areas. With transgression and regression as well as reciprocation of the coastlines, two-storeyed sand bodies are formed in which the transgressive sands are covered by regressive ones. In vertical sequences of the ancient Changjiang Valley, the delta sand bodies are located at the upper part, while transgressive river channel-filling ones at the lower part (Fig. 5-38). In the strandplain, the regressive littoral bars or radiating bars are in the upper beds and transgressive littoral ones in the lower part (Fig. 5-39). The shallow marine argillaceous sediments are present between them, showing sharpenoid shape, and pinching out westwards. Where sand bodies of upper and lower beds are superimposed directly, thick sand layer is formed as in Hongqiao and Haian areas. The overlying and underlying beds of these two-storeyed sand bodies are continental. Similar two-storeyed structure of sand bodies can be found in deltas of the Mississippi[30], Niger[23], Rhone[56], Lavaca[60], Brasos[61] etc.

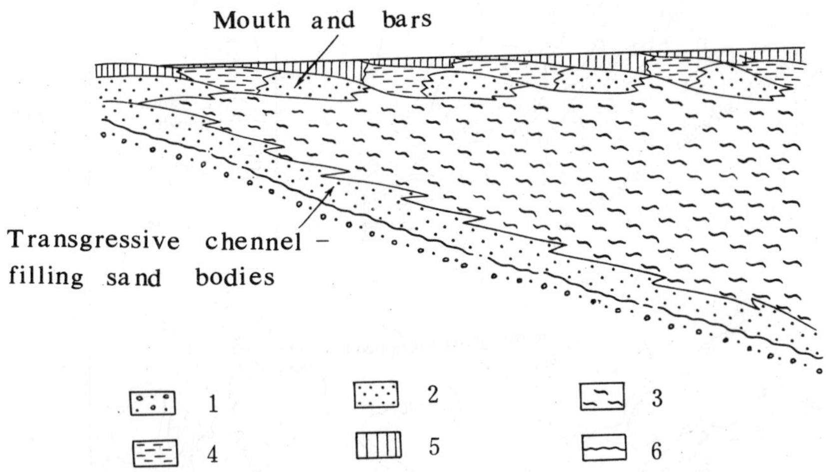

Fig. 5-38. Vertical distribution of sand bodies in Changjiang Delta. Plain.

1– Tawn sandy grvels;

2– Gray sand;

3– Littoral-neritic sediments;

4– Distrbutary channel sediments;

5– Tidal flat-floodplain sediments;

6– Erosional suface.

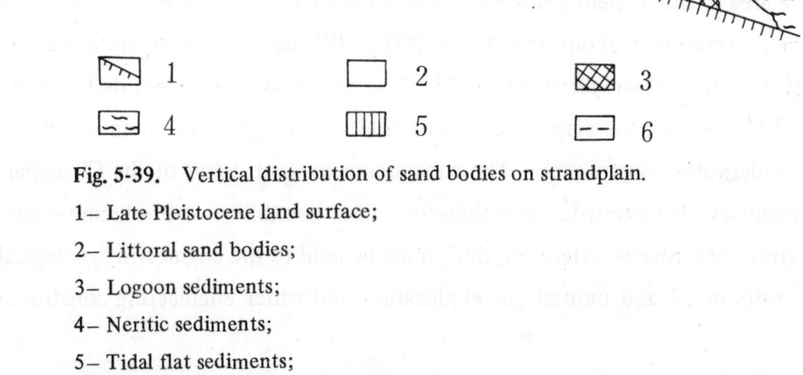

Fig. 5-39. Vertical distribution of sand bodies on strandplain.

1– Late Pleistocene land surface;

2– Littoral sand bodies;

3– Logoon sediments;

4– Neritic sediments;

5– Tidal flat sediments;

6– Marsh peats.

(8) ***Subaqueous ancient delta of the Changjiang***

From the modern Changjiang prodelta to water depth of 70–75 m, there is a fan-shaped shallow water platform, known as Great Yangtze Bank in foreign literatures (Fig. 5-40). Recently, investigation in the East China Sea shows that this is an ancient Changjiang Delta. Along its marginal bathymetric contours are concentrated. The pointed part of the delta is located in the north of the modern Changjiang River mouth, at about 32°20′N. Both franks of delta are cut by the troughs and are rather sinuous. The area of this ancient delta is about 40 000 square kilometors according to the ^{14}C dating. The age of its formation is about 15,000–36,000 yr BP suggesting that the delta was formed in the regressive stage prior the last glacial period[62]. The surface of the ancient delta is usually covered by relict sands[63]. Besides, the diapir strctures of modern Changjiang Delta had resulted in some damages to the underwater constructions. The subaqueous ancient deltas of the Changjiang Should have similar sedimentary characteristics, and therefore, have possibility of producing diapir and other deformation structures. Special attention, thus, must be paid to the engineering geological conditions of shallow deposits in oil and natural gas exploration and other engineering construction in the East China Sea.

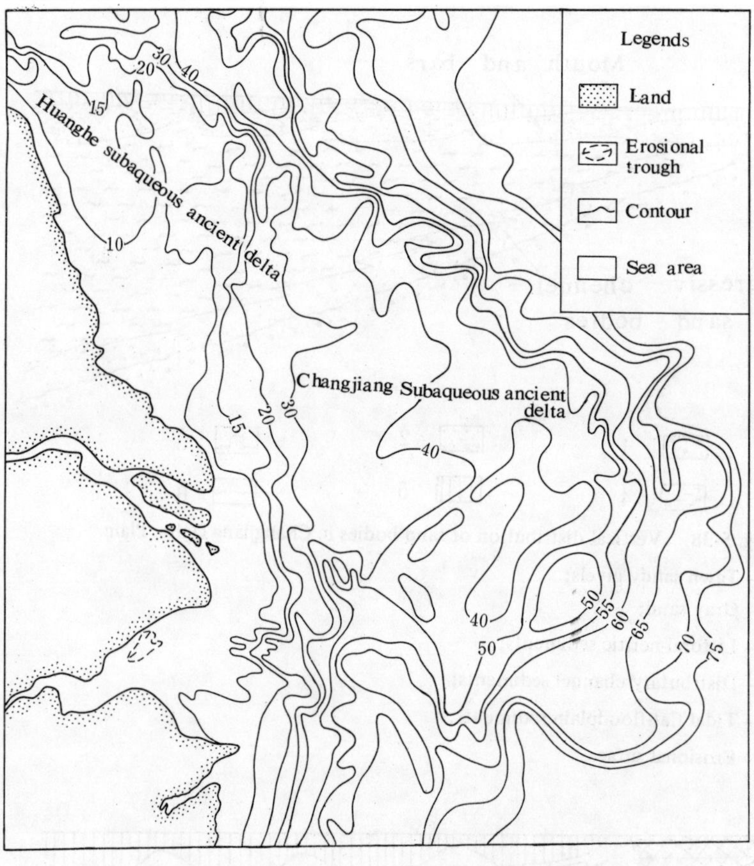

Fig. 5-40. Subaqueous ancient delta of Changjiang.

3. The Huanghe River Delta: A river-dominated delta

(1) *General Situation*

The trunk stream of the Huanghe River is 5460 m long. The annual sediment discharge of the Huanghe River ranks first in the world[65], the average one being 1.118×10^9 tons[64]. Because of the large amount of sediments, riverine processes have been dominating in the formation and development of the Huanghe Delta, which in turn has made the delta a typical one of fluvial type. As known to the world, the Mississippi Delta is also a river-dominated delta and, just like the Huanghe River, it also empties into a gulf with small waves and weak tide. The annual sediment discharge of it, however, is only half of that of the Huanghe River, and that's why riverine processes are more notable in the Huanghe Delta than in the Mississippi Delta.

(a) *Extent and Boundary*

Being between 37°10′ and 38°05′N, and 118°30 E, the present Huanghe Delta is sandwiched in between the Bohai Gulf and the Laizhou Bay. With Ninghai as its apex and taking the form of a fan, the delta spreads from the Douhai River in the north to the Nanwang River in the south. The topography of the delta is high in southwest, low in northeast; high along the abandoned distributary channels of the Huanghe River and low in zones between them. The elevation of it is below10 m (at Dagu zero point) and the average slope of its surface, 0.1–0.15‰. The present principal river channel of the Huanghe River winds towards the east[64].

The water depth of the subaqueous part of the Huanghe Delta can reach 20 m outside the mouth of the main river channel. The offshore slope varies markedly, and the variations are roughly of two types. Outside the main distributary mouth, from the low water line to where the water depth is 10–15 m, the slope is usually over 1‰, such as outside the abandoned distributary mouths of Shenxiangou and Tianshuigou, where the slopes are respectively 1.3‰ and 2.3‰. When the water depth goes larger than 15 m, the slope will become flattened, normally less than 1‰, such as in the offshore areas of Shenxiangou and Tianshuigou, Where the slopes are respectively 0.25‰ and 0.24‰[1)]. The slope-profile-variation of this type differs from normal deltas. Outside the abandoned distributary mouths exists a shallow-water paltform. The outer edge of it is 2 m deep outside the Taoerhe River mouth and its slope there varies between 0.5‰ and 0.7‰. Further outward, the slope will turn gentle again and becomes about 0.3‰ [2)]. The slope-profile-variation of this type is similar to normal deltas. The offshore slope has a tendency of gradually turning gentle outside the abandoned distributary mouths on the both sides of the main distributary mouth (Table 5-13).

The deposition of the large amount of sediments carried down by the Huanghe River has led to the formation of two spits (subaqueous natural levees) at the main distributary mouth, which pro-

---

1) Department of Geography, Nangjiang University, 1966, Report on sediment source and evolution tendency of coast and tidal flat in Xingang, Tianjin.

2) Department of Geography, Nangjiang University, 1966, Report on Sediment Source and Evolution Tendency of Coast and Tidal Flat in Xingang, Tianjin.

Table 5-13 Offshore slopes of different parts of the delta (after the Water Conservancy Committee of the Huanghe River, 1970)

	Chezhigou mouth		Main river mouth		Shenxiangou mouth		Qingshuigou mouth		Songchunronggou mouth		Xiaoqing River mouth	
	Dis. (km)	Slope (‰)	Dis. (km)	Slope (‰)	Dis. (km)	Slope (‰)	Dis. (km)	Slope (‰)	Dis. (km)	Slope (‰)	Dis. (km)	Slope (‰)
5 m isobath	2.5	2.0	0.5	10.0	2.0	2.5	2.5	2.5	3.0	1.6	3.5	1.4
10 m isobath	7.0	1.1	1.5	5.0	5.0	1.6	6.0	1.4	17.0	0.35	23.0	0.25
15 m isobath	20	0.38	2.0	0.9	10.0	1.0	14.5	0.58	–	–	–	–

turdes seaward, making the mouth look like a bird-mouth. The extent of coast influenced by the spits can reach 25–40 km. The zones in between the distributary channels protruding seaward are called "muddy bays", which are in fact interdistributary bays. It is of such protruding mouth spits, relict spits and muddy bays in between them that the whole edge of the Huanghe Delta are composed. And the coastline of the Huanghe Delta, which has thus been made rather zigzag.

(b) *Hydrology*

It should be pointed out that the distributions of the Huanghe River runoff and sedimet discharge within a year vary greatly, and that its runoff in the flood season from July to October usually makes up 59.8% of the yearly total, and the sediment discharge, 83.1%[(64)].

Situated in the marginal part of the mid-latitude monsoon region, the Huanghe Delta is dominated by north wind in winter, and by northward winds in summer, with the maximum wind speed usually occuring in April and May. The winds that can generate waves to affect the delta are generally southward winds.

The Huanghe River mouth is microtidal, the tidal range of it being only 0.6 m at the Shenxiangou mouth, and 1.6–2.0 m at the mouths of the Douhai River and the Xiaoqing River. The tides of most coast sections of the delta are irregular semidiurnal except at the Shenxiangou mouth, where the tide is normal diurnal. The sea area outside the Huanghe River mouth takes Shenxiangou as a rough demarkation, north of which, the tidal currents circle anti-clockwisely with the flood flowing southward and the ebb, northward. Due to the low tidal range, the tideinfluenced reach of the Huanghe River is relatively short, the influenced extent being 15–30 km. And the influenced distance of the river by the water-level-variations induced by meteorological tides can attain 50 km, and it's smaller in flood season and larger in non-flood season. The landward limit of tidal flood current is near the mouth.

(2) ***Sedimentary Characteristics of the Subaqueous Part of the Delta***

(a) *Suspended load*

Running through the Huangtu Plateau, the Huanghe River carries a large amount of fine-grained materials down into its mouth area. According to the observation in the mouth area during the flood season in 1953 to 1955, silt and clay make up over 90% of the suspended load; very fine sand, 4–7%; and fine sand, around 1% (Table 5-14). The bedload is rather small in quantity in the mouth area, being only 0.017– 2.04%[1)] of the suspended load according to the observation by the Water Conservancy Committee of the Huanghe River in 1959.

Table 5-14 Percentage content of the suspended load in flood season

Time	Silty clay	Very fine sand	Fine sand
1953	92.9	6.0	1.1
1954	95.7	3.3	1.0
1955	94.1	5.0	0.1

The sediments coming into the mouth area of the Huanghe River are the material base for the formation of the Huanghe Delta. From 1964 to 1973, the total sediment discharge in these ten years was 11.32×10^9 tons, of which, 2.75×10^9 tons (24% of the total) were deposited on the subaerial part of the delta (above the low water line), 4.5×10^9 tons (40% of the total) on the subaqueous part (below the low water line), and 4.06×10^9 tons (36% of the total) transported into the Bohai Sea[64].

The large amount of sediments transported by the Huanghe River into the sea has inevitably resulted in an area with high content of suspended load outside the river mouth. Observations indicate that, whether in flood season or in dry season, outside the mouth of the Huanghe River always exists a zone with high content of suspended load, the zone that extends seaward 10 to 15 km in a tongue shape. The content of the suspended load is over 100 to 150 mg/1 in flood season and also above 100 mg/1 even in dry season. It decreases progressively seaward to below 50 mg/1. In dry season, the influenced extent of the high content area is comparatively small and the content drops rapidly[66].

Vertically the suspended load in the Huanghe River mouth increases downward from below 50 mg/1 up to above 150 mg/1. The mixed-water tongue of the basal high suspended load stretches northward along the sea floor and that as long as more than 10 km. This shows that turbidity current may form outside the Huanghe River mouth with high content of sediments. The suspended

1) Water Conservancy and Electrical Power Division of the Water Conservancy Committee of the Huanghe River Ministry of Water Conservancy & Electrical Power, 1959, Preliminary analysis on the evolution of the Huanghe River mouth.

load in the Huanghe River mouth is dominated by terrigenous detritus, and next to it, plankton. The terrigenous detritus consist mainly of quartz, mica, feldspar, hornblende, epidote, microparticls of carbonate, and "polyment" of clay minerals etc., while the plankton, of *Sangitta, Labidocera euchaeta, Coscinodiscales* and *Chaetoceraceae* etc. The zooplankton/terrigenous material ratio is about 1/500 (Table 6-15). Even if the phytoplankton is included, the terrigenous materials still occupy the dominate position[66] (Table 5-15).

Table 5-15 Content of the palnkton in the suspended load in the Huanghe River mouth (after Qin Yunshan et al., 1982)

Zooplankton (mg/1)	Suspended load (mg/1)	Zooplankton/Suspended load
0.154	99.0	1/500

(b) *Types and distribution of the sediments*

The sediments distributed in the mouth area of the Huanghe River in a fan shape are comparatively coarse (Fig. 5-41), and they change seaward first from fine to coarse and then finer again. The nearshore area is where silt deposits, the Md of it being 5.6–6.9ϕ and the content, 60–80%. The

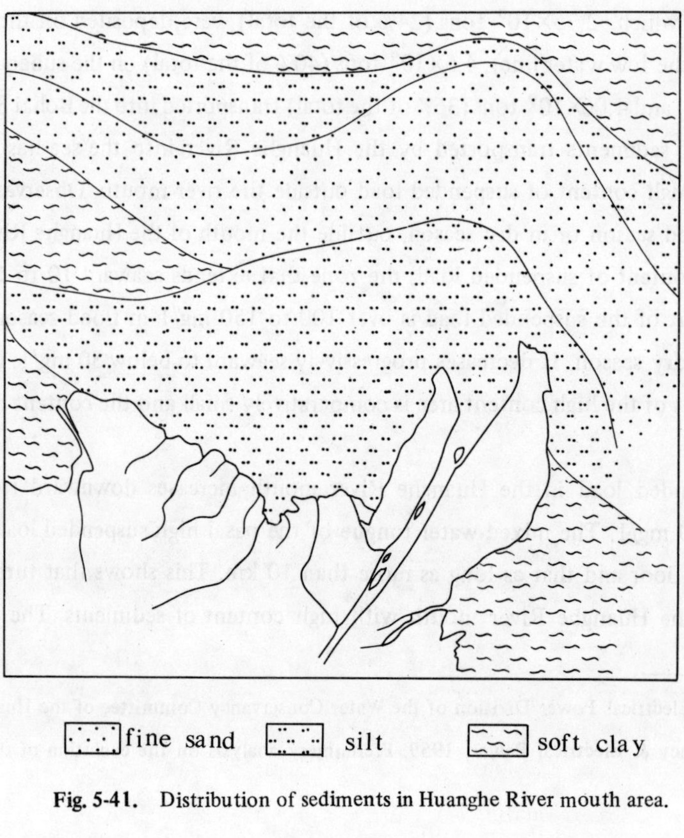

Fig. 5-41. Distribution of sediments in Huanghe River mouth area.

fine sand here is rather low in content, making up only 0.06–0.9%. And there is also a certain amount of clay here. The area around the outlet of the subaqueous river channel is a zone of very fine sand, the Md of which is 3.3–4.3ϕ and the content, as high as 90%. The content of the fine sand here is 1–3%. And there is also a little amount of umd here. This zone of v.f. sand protrudes seaward in the mouth area and at the same time, extends southeastward and westward with the width decreasing gradually. The outer edge of it appears where the water is 10–15 m deep. Towards the two sides, the water depth decreases to 5–8 m. On this outer edge again appears a silt zone. The colour of its sediment is relatively dark, but its lithology, similar to that of the nearshore silt zone. Seaward, this silt zone graduaIy gives way to silty mud with Md of 6.6–7.6ϕ. The fining-outward tendency of the sediments from the outlet of the subaqueous river channel exists.

The "muddy bays" on the both sides of the river mouth are comparatively weak in hydrodynamics and hence, the site for fine-grained materials to deposit. The fine-grained materials deposited here consist mainly of grey silty mud, and also some silt with Md of 7–8ϕ. The relationship between this area and the depositional zone of coarse-grained materials in the mouth area is gradational.

(c) *Mineral association and chemistry of the sediments*

The heavy mineral association in the sea area off the river mouth is similar to those in the distributary channels among which, hornblende and epidotes have the highest content. Some of the stable minerals with larger specific gravity are concentrated in the mouth area, such as ilmenite, garnet and sphene, while the schistose minerals represented by mica, in the "muddy bays" on the both sides of the river mouth and in the deep-water area. The content of the schistose minerals can reach over 50%. All these minerals decrease in content outward from the mouth[68]. The distribution of different types of minerals in the Huanghe River area is a reflection of the hydrodynamics of the area.

The clay minerals in the Huanghe River mouth consist mainly of illite and next to it, montmoril and kaolinite. Mixed in them are a little calcium carbonate, vermiculite, organic materials and non-crystalline ironstone etc[68]. Flowing through the Huangtu Plateau, the sediments of the Huanghe River are characterized by high Ca content, the $CaCO_3$ content reaching over 10%[69]. In grian sizes less than 0.001 mm, the CaO content can go over 8% in the mouth area, but it decreases outward, forming a distinct dispersing halo (Fig. 5-42), and drops to below 3% in the sea area 60 km away from the coast.

(3) ***Division of the sedimentary facies***

(a) *Distribution of the sedimentary facies*

Based on the sedimentary character, the Huanghe Delta can also be devided into three facies: delta plain, delta front and prodelta-shallow sea.

The delta plain owes its origin to the to-and-fro flow diversion and the many times of siltation of the Huanghe River. The sediments are pork-liver-coloured clay and silt, the colour of the silt changing downward from yellow to grey. Horizontal bedding is developed and minor cross-bedding occurs in

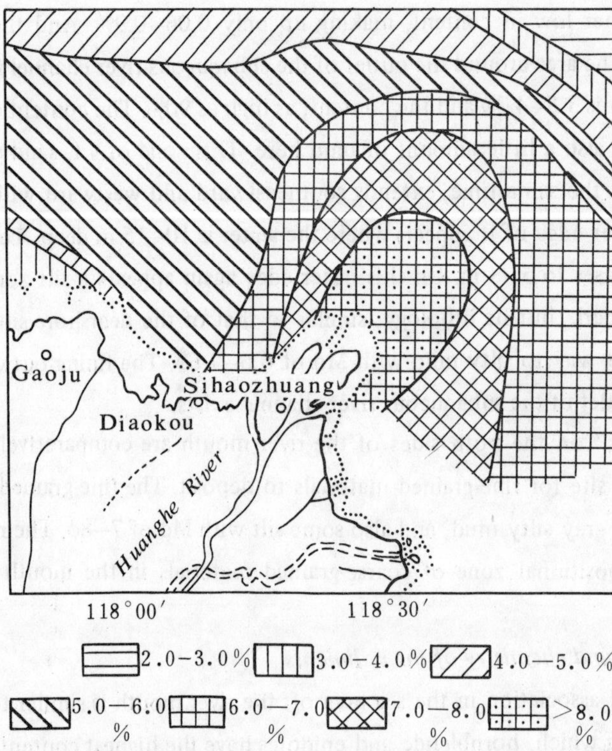

Fig. 5-42. Distribution of the CaO content in grain sizes less then 0.001 mm in the Huanghe River mouth (after Gao Mingde et al., 1980).

the silt beds. With root systems relatively abundant, vertical reedroots very common, and being rich in plant remains, the delta plain sediments are of typical flood plain deposits, the thickness of which is 5–6 m (Fig. 5-43) and inside which there are many paleosol intercalatied beds, reflecting the route-changing and oscillating of the river channel. There are mainly two types of sediments outcropping on the plain: silt and silty clay. The distributary and abandoned distributary channels are comparatively high in relief, the sediments of which are dominated by silt, whereas the zones in between the distributary channels, relatively low and depressed, the sediments of which turn finer and, the farther way from the river channels the sediments are, the finer they turn, i.e., near the distributary channels usually occur interbedded silt and clay and, towards the interdistributary low and depressed zones, thick beds of clay doposits will gradually appear instead[66].

The waves being weak, the interdistributary bays on the marginal part of the Huanghe Delta are favounable depositional sites for fine-grained materials. And the sediments here are silty mud. The tidal flats in the bays can attain widths of 7–10 km. And there are plants like reed and chaste twigs growing above the mean high water line. The interdistributary bays are an important depositional unit of the delta plain.

The delta front facies refers to the sand and silt zone in front of the river mouth, which starts from the low water line and extends seaward until where the water depth reaches 10–15 m. Widest in the mouth area where it can attain widths of 20–25 km, it narrows towards the two sides and

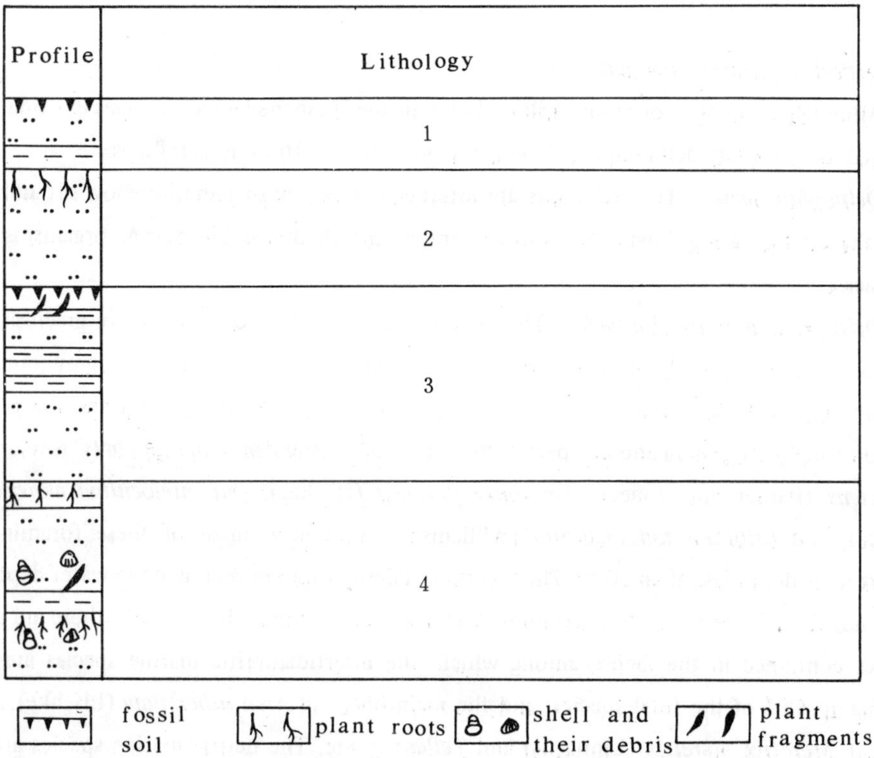

Fig. 5-43. Sedimetary profile of the Huanghe Delta plain (based on the data from Guiyang Institute of Geochemistry, 1972).

Lithology

1– Grayish yellow, yellow silt intercalated by brown clay, soil texture present in upmost part;

2– Grayish yellow silt with small cross-bedding, black plant roots present in upmost part;

3– Brown clay intercalated with yellow silt in upper part, gray, yellowish gray silt in lower part. Clay with soli texture in upmost past. Plant fragments, root (vertical need) present;

4– Brown clay intercalated by thin silt with plant roots in upmost part. Yellow, gray silt with shell depris, plant fragmente in middle and lower part, clay in lowermost part. Grayish yellow silt with marine shells, need roods present in uppermost part.

extends farther away westward, Influenced by the remained flow, the mouth area has the tendancy to shift northwestward, and the maximum thickness of the sediment, which is over 10 m, occurs in the area directly adjacent to the outlet of the subaqueous river channel northwest of the mouth. Seaward of the delta front appears the prodelta, the sediments of which are the finest, high in water content (70–80%) and rich in organic materials. Except the surficial portion, the colour of the sediments is greenish-grey. Further seaward, the prodelta gradually gives way to shallow sea. However, as there is no distinct demarcation between them, they are called the prodeltaneritic marine facies.

(b) *Vertical sequence of the delta*

Although there are not many drilling bores in the Huanghe Delta, the ones existing have already revealed the complete delta sequence, which, from top to bottom, is as follows.

Delta plain facies. The sediments are interbedded yellow or greyish-yellow silt and clay of mild clay, the silt increasing downward both in content and thickness. No marine organisms are found in this facies.

Delta front and prodelta facies. The sediments in the delta front are grey or greenish-grey silt and fine sand containing shell and plant debris, while that in the prodelta, greenish-grey muddy mild clay and silty clay with high water content. Rich in foraminifera, the delta front and the prodelta facies contain totally 21 genera and 58 species, the main ones being *Ammonia tepida* (Cushman), *Ammonia annectens* (Parder and Jones), *Ammonia flevensis* (Hofker). *Quinqueloculina akneriana rotunda* (Gerke), and *Cribrononion incertum* (Williamson). Presently, most of these foraminifera inhabit waters with depth less than 20 m. The layers containing foraminifera living in water depths of 10–15 m or less than 10 m have the maximum thickness. As for the mollusc fossils, there are altogether 20 species contained in the facies, among which, the intertidalneritic marine species are dominating, making up 45% of the total species, and the main ones are *Arca subcrenata* (Lischke), *Aloidis laevis* (Hinds), *Meretrix meretrix* (Linnaeus) and *Tellian sp* etc. The neritic marine species are the fewest, making up only 10%. The deposits of the delta front and prodelta are about 4–16 m in thickness. And they thin gradually from north-east towards southweast and finally taper out[70]. Judging by the marine-continental variation of the deposits, during the postglacial period, complete transgressive-regressive-sequences were developed in the Huanghe River Delta (Fig. 5-44). The vertical sequence of the Huanghe Delta is, however, comparatively thin, usually being 10–15 m, and it thickens towards the sea.

(4) *Development of the Delta*

The Huanghe River is a river with high suspended load. Due to the large amount of sediments, riverine processes have been dominating in the development of the delta, whereas in the accretion and progradation of the delta, the channel-crevasse-and-flow-diversion of the river serves as the most basic process.

(1) *Process of development*

In the Huanghe Delta area, the river gets breached and changes its route very frequently. In the 122 years from 1855 to 1977, there were 87 years when the river actually flowed through the present delta area, and in these 87 years took place 11 times of major flow-diversion (on an average, once every 8 years) and over 70 times of minor crevasse. As the delta develops, the river extends unceasingly and the longitudinal slope of the river channel gets flattened gradually. When the dropping slope

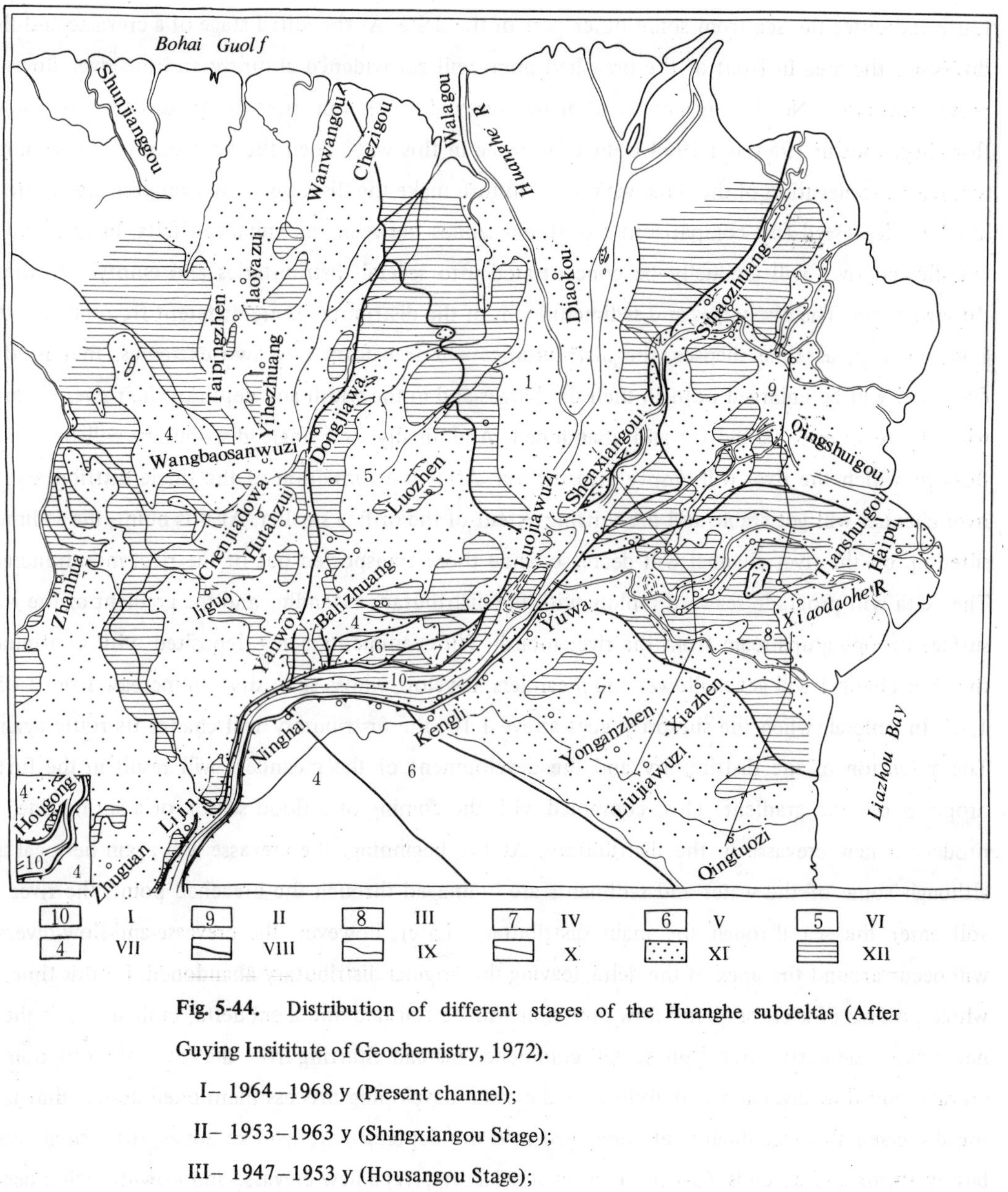

Fig. 5-44. Distribution of different stages of the Huanghe subdeltas (After Guying Insititute of Geochemistry, 1972).

I– 1964–1968 y (Present channel);

II– 1953–1963 y (Shingxiangou Stage);

III– 1947–1953 y (Housangou Stage);

IV– 1934–1938 y (Qiansanggou Stage);

V– 1929–1934 y (Jizhuang Stage);

VI– 1926–1929 y (Balizhuang Stage);

VII–1904–1926 y (Kouzhuang Stage);

VIII– Boudary of subdeltas;

IX– Litholine;

X– Huanghe alluvial profire;

XI– Sand;

XII– Clay.

reaches a certain critical value, the river will, under favourable conditions, be breached, change its route and enter the sea from some other part of the delta. At the initial stage of a crevasse-and-flow-diversion, the area in front of the breached point will get widened abruptly and the river flow, dispersed randomly. Not having a constant main-stream, the river will enter the sea dispersedly. The two flow-diversions in 1954 and 1964 both took place in this way. Then the accretion of the sediments will raise the elevation of the flats, which will in turn make the dispersed flows get converged into the low and depressed (by comparison) trough-like zones between the enhanced flats. In this way the over-flowed river will gradually be concentrated into several distributaries and empty into the sea through them, The discharges and sediment loads of the distributaries are different from one another. Consequently, as time diapses, the distributaries will transform into two extremes, that is, those dominant will be enlarged both in their discharges and in the width and depth of their river channels, while those inferior, silted up and abandoned. And finally one single distributary will be formed, through which the river will empty into the sea. After the formation of the single distributary, the river channel will get stable and the shifting extent of the orifice smaller. At this point, the sediments silted up on the river channel will decrease while those transported out of the river mouth increase. The delta will prograde seaward repidly with the distributary extending and the gradient of the water surface dropping upstream from the river mouth. With sediments being deposited on it continually, the river channel will get shallower and meanders, develop, whose sinuosity can increase from 1.05 to 1.13. In general, when the sinuosity goes above 1.10, the distributary will change its route again[1]. The extention of the distributary and the development of the meanders will result in the further dropping of the gradient. This, combined with the coming of a flood season or a spring tide, will produce a new crevasse in the distributary. At the beginning, the crevasse will again be local and, although some of the water and sediments are drainaged through the breached point, the river will still enter the sea through the main distributary. Later, however, the crevasse-and-flow-diversion will occur around the apex of the delta, leaving the original distributary abandoned. By this time, the whole process of a crevasse-and-flow-diversion will be finished and a subdelta, built up. And then, a new phase, the destructional phase will commence instead. Starting from the front of a new point of crevasse-and-flow-diversion and following the same developing process mentioned above, that is, (i) the dispersed flows gradually get converged into the depressed trough-like zones; (ii) a single distributary forms and extends; (iii) the river channel shifts; (iv) a new crevasse-and-flow-diversion occurs, a new distributary will again be evolved and a new subdelta, built up. With the same process occuring repeatedly, the delta will grow and prograde incessantly.

In response to the alternative flow-diversion of the distributary channels of the Huanghe Delta and the incessant accumulation of the large amount of sediments, the coastline of the delta progades seaward and the area of the delta enlarges progressively. Based on the field investigations, historical literatures, layout of the residential areas and airborne photographs, the location of the coastline in

---

1) Water Conservancy Science Institute, Water Conservency Committee of the Huanghe River, 1972, Analysis on the basic data of the existing Huanghe River Channel & Prediction of its developing tendency.

1855, the total length of which is 128 km, has been roughly delimited. Comparison among the coastlines in 1855, 1954 and 1972 indicates that the area of the delta increased by 1510 km^2 in the years from 1855 to 1954. If the actual time when the distributaries flowed through the delta was 64 years, the mean prograding rate of the delta would be 150 m/year, and the mean accreting rate of the delta area, 23 km^2/year. As is already known, the mean prograding rate of the delta from 1954 to 1972 was 420 m/year and the mean accreting rate of the delta area, 23.5 km^2/year. It can thus be seen that the accreting rate of the delta area in these later years was relatively low, close to that in the years from 1855 to 1954, whereas its prograding rate, rather high, almost three times of that in the early years. This must be resulted from the influence of the human activities[4].

In the development of the delta, the prograding and area accreting rates are unequal at different stages. As can be seen from Table 5-16, the prograding rate and the accreted area in a short period are, on an average, larger than those in a longer period. Probably this is because the mean values in a longer period include the destruction of the delta. Besides, in the years from 1964 to 1971, both the prograding and area-accretion rates of the delta had a tendency to decrease progressively, reflecting the natural variation in the different developing stages of the delta, the variation from rapid progradation in the early stages to slow deposition in the later stages. At the initial stage of a crevasse-and-flow-diversion, the delta can prograde seaward with a rate of 10 km/year[1)].

Table 5-16 Prograding rate and accreted area of the Huanghe Delta (After Pang Jiazhen et al.,1980; and where with, after the Water Conservancy Science Institute, Water Conservancy Committee of the Huanghe River, 1972)

Item / Duration	Accreted area of the delta (km^2)		Prograding rate (m/year)	
	Total	Average/Annually	Max.	Average
1855–1954	1510	23	–	150
1964–Oct., 1965	180.5	90.3	5500	3200
1965,10– Oct., 1966	65.8	65.8	4400	2190
1966,10– July, 1968	105.8	52.9	2250	1490
1968,7– Sept., 1970	87.5	43.8	2700	1410
1970,9– Sept., 1971	18.3	18.3	2300	1390
1971,9– Sept., 1973	20.1	10.0	–	–
1973,9– Sept., 1974	14.0	14.0	–	–

After a crevasse-and-flow-diversion, the original distributary will, as a result of the decreasing or even stopping of the water and sediment supply, be gradually silted up and the subdelta, abandoned, whereas the wave action and the tidal currents off the river mouth, relatively enhanced and the coast and tidal flat, eroded and retreated, the eroding and retreating rate of the delta reaching as high as

1) Geography Department, Nanking University, 1966, On sediment source and evolution tendency of tidal flat in Xingang of Tianjin.

100–300 m/year (Table 5-17). Based on the eroding and retreating rate, the Huanghe Delta can be devided into 4 areas. The mouth area of the main distributary is a area of high accretion and the areas on the both sides of it, areas of high erosion. Farther away from the mouth of the main distributary, stable coasts will occur in response to the decreasing erosion rate.

Table 5-17 Erosion rate of the Huanghe Delta (After Geography Department, Nanjing University, 1966)

Area	Eroding and retreating rate (m/year)	
	From visiting the local people	From comparison with airborne data
Shunjianggou	75–100	150–175
Walagou	125–150	125
Wanwangou	250	175–200
Tianshuigou	330	300

As a river which breaches its banks and diverts its flow most frequently and oscillates its channel most significantly, the Huanghe River has changed its route many times to enter the Huanghai Sea from North Jiangsu and the Bohai Sea from the Shandong and Hebei provinces. When the river changes its route and enters the Yellow Sea from North Jiangsu, the delta formed in the Bohai Gulf will be eroded and destructed large-scalely. And the same will be true to the abandoned one formed in North Jiangsu when the river changes its route again and empties into the Bohai Sea.

As a result, the development of the Huanghe Delta can easily be devided into the constructional and destructional phases, which is very similar to the development of the Mississippi Delta, a typical river-dominated delta. However, in the destructional phase of the Huanghe Delta, no littorial sand bodies are formed, and this is where the Huanghe Delta differs from the Mississippi Delta.

The developing history of the Huanghe Delta over the last 100 years can be traced by consulting the many historical documents, the visiting the local people and the comparison with the historical maps. Totally there are 11 stages of subdeltas.

In 1976, the Huanghe River artificially changed its route at Xikou and entered the Bohai Sea by way of the Qingshui River. This time of flow-diversion has been lasting up to now and the eleventh stage of subdelta is just under development.

(b) *Similarities and differences in development between the Huanghe Delta and the Mississippi Delta*

Both the Huanghe Delta and the Mississippi Delta are river-dominated deltas, and so, the development of crevasse-splays and subdeltas is the basic way for them to prograde and widen. Since crevassing and routechanging are, to a great extent, accidental, either the crevasse splays or the subdeltas of the two deltas are not arranged in definite order. In addition, both of the two deltas have distinct constructional and destructional phases in their development. However, compared with the Mississippi

River, the Huanghe River has much larger sediment discharge and its sediments are finer in grain size, and so it is much easier and hence more frequent for it to get breached and routechanged and for its subdeltas to develop. In the 7000 yr of the Mississippi River, altogether 16 stages of subdeltas have been built up, 7 of them being exposed on the land and the rest 9, buried under the ground[74]; whereas, over the last more than 100 yr, the Huanghe River has been breached for more than 70 times and has got 11 stages of subdeltas built up. The Mississippi Delta usually takes an elongated form in its constructional phase, but will change into a fan shape when reworked after abandoned. Of this, Fisher has already given a detailed description in his stratigraphic classification of the deltas. Elongated spits (natural levels) will also occur on the both sides of the Huanghe River mouth when a new delta first appears after a crevasse-and-flow-diversion of the Huanghe River. With the shape like a bird mouth, the spits (natural level) will plundge themselves rapidly into the sea, representing in fact the embryonic form of a birdsfoot delta. And that up to now no complete birdsfoot deltas have ever been built up is only because each new channel of the Huanghe River has existed too short a period of time (compared with the Mississippi River). Each time, before it can find enough time to build up a complete birdsfoot delta, it will have had its route changed already and the embryonic form of a birdsfoot delta just formed will gradually be reworked and become fan-shaped. What is more, taken as a whole, the Huanghe Delta formed over the last nearly 100 yr also takes the shape of a fan, and this has also resulted from the more frequent crevasse-and-flow-diversion of the Huanghe River and the more active development of its subdeltas (compared with the Mississippi River). Offshore barrier islands will form when the Mississippi Delta suffers from erosion and destruction. The offshore barrier island of Chandeleur is such an example, which is in fact the outcome of Bernard Subdelta under erosion and destruction. It is because of this that Fisher takes it a criterion to determine wave-destroyed deltas whether barrier bars exist or not. Unfortunately, due to the too fine sediments, no barrier bodies will form when the Huanghe Delta is subjected to erosion after abandoned.

(5) ***Subaqueous Huanghe Delta off North Jiangsu***

Presently the subaerial part of the abandoned Huanghe River mouth in North Jiangsu no longer takes the shape of a delta. Subaqueously,however, there still exists a fan-shaped shallow-water platform protruding seaward. The platform is nearly 200 km long from north to south and 100 km wide from east to west, its most seaward-protrusive part being located at 34°N. On the frontal edge of it there is a distinctive slope where the water depth is 30–40 m. This fan-shaped platform in front of the abandoned Huanghe River mouth overlaps the marginal part of the subaqueous ancient Changjiang Delta (Fig. 5-40).

The sediments of the subaqueous Huanghe Delta are silty clay or clayey silt in grey or greenish-grey colour. Similar to the mineral association of the Huanghe River, its heavy minerals mainly include mica, hornblende, apidote, hemitite, ilmenite, etc., which are distributed along the subaqueous river channels[75]. The most striking characteristic of the subaqueous Huanghe Delta is: its sediments are rich in Ca with the mean content of $CaCO_3$ being 11.4% and the max. one over 15%. This content is similar to that in the sediments in the Huanghe River channel, but higher than that in the sediments from any rivers emptying into the Huanghai Sea or the Bohai Sea.

The microfossil assemblage in the subaqueous Huanghe Delta is the *Ammonia beccarii-Senocytherides* assemblage which includes 38 species and the main ones are the nearshore shallow-water species such as *Ammonia beccarii, Ammonia annectens, Shackoinella globosa, Ammonia convexidorsa, Spiroloculina laevigate, Florilus decorus* etc. Averagely, in every 50 g of dry sample can be found about 500 pieces of microfossils, the species diversity of which averages 2.9. Of the foraminifera contained in the deposits of the subaqueous Huanghe Delta, the planktonic type has the highest content and the content tends to increase towards the coast, while the agglutinative type has the lowest content and the content increases in an opposite direction, displaying the basic distributive character possessed by the foraminiferal assemblage in the tidal river mouths in China. In shallow bores, the foraminifera changes little vertically, indicating that the strata are composed of recently accumulated sediments with high sedimentary rate and the character of delta areas. Totally 8 species of ostracoda are found and the main ones are *Senocytheridea, Albileberis sinensis* and *Neomonoceratina dongtaiensis* etc. With the number of the ostracod individuals constituting 12% of that of the foraminifera[1)], its foraminiferal and ostracod assemblages are characteristic of the littorial deltas with water depth less than 20 m.

The spore pollen is dominated by that of *Artemisia Chenopodiaceae* of herbage and there also exists that of *Poto magtonaceae* of aquatic plants.

It can be inferred based on the relationshep of the subaqueous Huanghe Delta with the ancient subaqueous Changjiang Delta, its sedimentary characteristics and paleontological assemblages that this delta system was formed during middle or late Holocene and it may have resulted from the many times of route-changing and the overlapping of several stages of subdeltas on the another.

1) Micropaleontology Research Group, Department of Marine Geology, Tongji University, Distribution regularity of foraminifera and ostracoda in the bottom sediment of the northwest part of the South Huanghai Sea and its geological significance.

4. Zhujiang Delta: a compound delta

Gathering together Xijiang, Beajing, Dongjiang and other small rivers is scalled Zhujiang River, runs into the South China Sea. The Zhujiang Delta consists of all the types of deltas of previous rivers. Xijiang, Beijiang, Dongjiang and some middle and small deltas, which connect and intercalate each other, forming a compound delta. Each delta of the Zhujiang compound delta results from independent rivers and develops simultaneously. They are different from those subdeltas of ordinary delta system.

(1) *Outline of the Zhujiang Delta*

(a) *Area and boundary*

It is generally considered that the apex of Xijiang Delta is situated in Linyang Cayon, Gaoyo County, and apexes of the Beijiang and Dongjiang in Lubao, Sanshui County, and Shilong, Dongwan County, respectively. The northern boundary of Zhujiang Delta is along the line of Lubao, Guangzhou and Shilong, the eastern boundary along Shilong, Dongwan, Humen, and the weastern one Linyangxia, along Jiangmen, Yamen (Fig. 5-45A)[76].

Fig. 5-45. Area and boundary of Zhejiang Delta (After Xu Junming, 1973; Li Chunchu, 1977).

1 – Hills and islands; 2 – Boundary of the delta.

Taking, however, into account of existence of the delta sequence, the area of Zhujiang Delta would greatly decrease. The boundary of Xijiang, Beijiang deltas are on the south of Huangqu, Shiqiao Daliang, Jiangmen. The Dongjiang Delta spreads in west of Xintang, Daojiao[59] (Fig. 5-45B). According to the viewpoints of sedimentalogy, the later should be accepted. Although the fluvial plains located in the north of Xijiang and Beijiang deltas, and in the east of Dongjiang Delta are different from the delta plains in sedimentary characteristics, but their forming process is closely associated. The following desccusion will concentrate on the fluvial plains downstream from Linyangxia, Lubao, Shilong, and illustrate the relationship and comparison with delta plains.

(b) *Topography*

The Zhujiang Delta is franked by the hills in the east, west and north sides, and by the fluvial plains in the northwest and northeast. On the deltaic and fluvial plains scattered more than 160 monadrocks, such as Huangyangsha, Wuguisha etc., accounting for about 20% total plain area. In a funnel-shape the Zhujiang River mouth includs 8 main distributaries running into the sea, namely, Humen, Jiaomen, Hongqili, Hengmen, Jitimen1), Hutiaomen, Yamen, Modaomen. The patterns of the distributary channels, often bifurcating and rejoining, belong to Coleman's type B[11], similar to braided stream.

Off the Zhujiang River mouth, there are islands of Dalingdao, Xiaolingdao, Sanzaodao, Dahengqindao, Xiaohengqindao, Wanshan Islands in various sizes composed of rocks and distributed arcuately, surrounding the Zhujiang Estuary and making it a sheltered bay. The numerous rocky hills in the delta plain restrict the river flow, so the shoals and rapids develop in the channels, which influence greatly the distribution of sediments in the distributary channels of delta.

(c) *Hydrology*

Situated in the subtropic area with a great amount of precipitation the Zhujiang River has abundant water discharge, averaging 302×10^9 m^3 per year, in which the flood season, from April to June, makes up about 90% of annual amount. The discharge in various rivers of the Zhujiang drainage system is extremely differentiated, with a maximum of about 73.5% in Xijiang River, and minimum of only 0.5% in Liuxijiang River (Table. 5-18). The water discharges in various distributaries are not equal. Discharge in Modaomen including Jitimen is the maximum, about 29%, and the minimum, about 8.9% is in Yamen including Hutiaomen.

There are 8 distributary outfalls of Zhujiang River, from which the Humen, Jiaomen, Hongqili Hengmen are four eastern distributary outfalls; the Modaomen, Jitimen, Hutiaomen, Yamen are four southern outfalls. The former discharges about 62.5% of total runoff of Zhujiang River, and the later only 37.5%[78].

The averagc annual sediment load of Zhujiang is 83.36 million tons, among which about 91.7–95.7% discharged in flood season. The maximum sediment load presents in Xijiang River, making up 86.9%, and the minimum in the Liushi, only 0.1%. Although the sediment discharges in the eastern and southern four distributaries are almost equal, both about 50%, the load in each distributary is

1) After embanking in Beitan, Niwanmen was abundoned and has converted into Jitiman[77].

Table 5-18 Discharges in various rivers of Zhujiang drainage system

River	Average discharge of many years		Sediment load of many years	
	$10^9 m^3$	%	10^4ton	%
Xijiang	220.0	73.6	7250.0	86.9
Beijiang	41.0	13.5	515.4	6.2
Dongjiang	22.0	7.3	314.0	3.7
Zengjiang	37.6	1.2	–	–
Tanjiang	–	–	23.0	0.3
Liuxijiang	15.2	0.5	10.2	0.1
Zhujiang	3020.0		8336.0	

(After Huang Zhenguo et al, 1982. slightly abridged)

greatly differentiated. Only the sediment discharge of Modaomen distributary makes up 42.2% of the total, and about 70% of the southern four outfalls. The difference of water and sediment discharges in various discharges result in the differentiated rates of formation and development of subdeltas. The Modaomen distributary has maximum water and sediment discharges so that its development of subdelta is a representative of the Zhujiang Delta.

The waves in the South China Sea is fairly strong. In terms of the data measured in Lidao, Chuandao Islands off the Zhujiang River mouth, average range of wave height is 1.4–1.9 m, with the maximum up to 4.0–4.5[78]. The wave height, therefore, in the Zhujiang Estuary is rather small, becuase of the bedrock islands surrouding the Estuary and scattering within it. The wave effect on the delta is very weak. The Zhujiang River mouth with the average tidal range of various outfalls is about 0.86–1.57 m (Table 5-19), belonging to a microtidal estuary[78]. In the circurumstances of small wave and microtide, the fluvial processes pronounce. Thus, each lobate of Zhujiang Delta is characterized by fluvial-dominated type. As far as the whole delta is concerned, it is a compound delta.

Table 5-19 Average tidal ranges (m) in various distributary mouths

Tidal range \ Outfall	Humen	Jiaomen	Hongqili	Hengmen	Modaomen	Jitimen	Hutiaomen	Yamen
Average	1.47	1.36	1.20	1.11	0.86	0.90	1.20	1.24
Maximum	3.36	3.81	2.56	2.48	2.29	2.36	2.66	2.95

(After Huang Zhenguo et al, 1982)

(2) *Sedimentary Characteristics of the delta*

(a) *Delta plain facies*

The plain downstream from the Liyangxia, of the Xijiang River, Lubao of the Beijiang River and shiloug of the Dongjiang River is generally called the Zhujiang Delta. In fact, it contains a delta plain with deltaic sequences and fluvial plain without deltaic sequences. Although the latter is not included in the delta plain, we still discussed it here for the sake of differentiation and comparison.

(i) *Texture of sedimentary beds*

The Zhujiang Delta locates in a sinking depression basin of Cenozoic and Mesozoic filled with a Cretaceous, Tertiary red rocks covered by the postglacial sediments. Late Quaternary sediments ^{14}C dated over 20000 yr, occur only in the peripheral and depressed areas of the delta. The contact between them is unconformable. The tectorium in the Zhujiang Delta area, therefore, is quite thin, different from that in Changjiang, Huanghe and other large deltas. Based upon the data of over 1200 boring cores, the average thickness of tectorium on the bedrock is 25.1 m, maximum 63.6 m. The area with thickness less than 30 m, covers 74.8% and with that over 40 m making up 8.6% of the total delta. The postglacial beds filled mainly in the ancient valley and overlie on the Mesozoic and Cenozoic red rocks and whethering residuum. The contact between them is sharp due to the distict difference in lithology. There is obviously difference of the postglacial beds between the delta area and the fluvial plain.

a) Fluvial plain

The postglacial beds consist of two superiomposed finingupward sedimentary sequences.

The first (lower) sedimentary sequence can be subdivided into two parts. The lower part consists of yellow, dark yellow gravels, sandy gravels, and sands with gravels. The gravels are made up of quartz, purplish red sandstone and siltstone. The gravels are general 0.5–2.0 cm in diameter and subrounded or subagular in shape. The well-rounded sands are dominantly quartz, with small a amount of feldspar. The mineralization of groundwater in sand and gravel is 0.2–0.5%, of freshwater origin. The sediments become fining-upwards in the vertical sequence, the base of which is marked by the thin (30 cm) gravels. The upper sandy gravels and sands with gravels may be more than 10 m thick and intercalated commonly the lenses of grey and black silty muds, 2–3 thick, with plant fragments and peats. The upper part is the granophyric clay in colours of white, light-yellow, greyish yellow, and Brownish purple. The clay containing ferriferous nodules is pure and stiff, 1–5 m thick, and locally absent. At the uppermost of the sequence is soft mud with decayed woods, uncoalified stem of *Glyptostrobus pensillis* and fragments of tree piercing vertically[59] into muds.

The second sedimentary sequence covers on the first sequence. Between them there is a mud bed with decayed woods. The sequence can also be subdivided into two parts. The lower part consists of dark-yellow sandy gravels, and the upper mottled clay covered by marsh peats and soft mud with decayed woods. At the uppermost there is a light yellow silty clay, of the river flood deposits. As a whole, the second sequence is also a fining-upward alluvial one.

b) Delta plain

The postglacial beds in the delta plain may also be divided into two parts, the lower fining-upward and upper coarsening-upward.

The first (lower) sequence in fining-upward and underlain by the red bedrocks and whathering residual beds. Their lithological characteristics are similar to that in fluvial plain. The difference, however, is only that the mineralization of groundwater in sandy gravel and sands is higher, usually 7–13%, maximum up to 21% in delta plain, belonging to the types of sodium chloride and sodium chloride-sodium bicarbonate[79].

The second sequence can be grouped into the lower and upper parts. The lower part consists of the light black mud and silty mud, including a small amount of graycanite transported possibly by tidal currents from the shallow sea area. The sediments contain the well-preserved mollusca shells, such as *Oystea sp, Aleidis, Trepezium japanicain* Pilsbry, *Arca sp, Anomia sp, Natica macalasa* Lamarck, *Mita ambigns* Swainson etc. occurring in the bay and intertidal zone. This bed is distributed widely and has less change in lithology. The upper part is made mainly up of silty mud and mud, including sand bodies with thickness up to 9–11 m. The sand bodies are dark grey silty sands existing as lens in the cross section and bar-finger in plain, contaction gradationally with overlying and underlying beds. They are a type of the mouth bar-finger[59]. The topmost sediments are light yellow, grey black silty clay, containing usually peat and shell beds, and abounding in plant debris and root systems. Considering this sandy lenses of the mouth bars, the second sequence, which is coarsening-upward, is characterized by a typical deltaic one. The thickness of second sequence progressively increase southwards, 3–5 m in Shunde, 8–12 m in Zhongshaxian, Xinhueixian, 13–20 m in Wanqinsha area and up even to 17–25 m near Baiteng, the southernmost area[59].

Although the sedimentary sequences constituting the daltaic and fluvial plains and their formative time are different in some degrees, their underlying beds and the lower sequences are similar. In addition, in the upper beds the sedimentary facies assemblages of fluvial and deltaic sequences are not alike, they, however, grade laterally one to other and may be correlated in age, regarded as contemporoneous deposits in various facies (Fig. 5-46).

(ii) *Distributary channel*

Distributary cahnnel is an important unit in the Zhujiang Delta, where the distribution of sediments is similar to that in river channels. The grain size of the distributary channels becomes fining from the deep trough to both sides (Fig. 5-47). The ebb currents in the thalweg, which connects the point of maximum depth in successive cross-sections of the deep trough, is about 70–90 cm/sec. The sediments in the main trough are coarse sandy gravels with some hemitite, limnite and rounded ferrifinous sands[80]. The depth eroded by river reaches usually dozens of meters, such as 67.5 m below the sea level in Daxiongkao, 66 m in Linyangxia, 42.5 m in Makaoxia, 46.5 m near Ganzhuxi, and over 20 m upstream from outfall of Niwanmen. The distributaries eroded not only the underlain marine beds, but also the ancient channel-filling sequences, and even down to the whethering

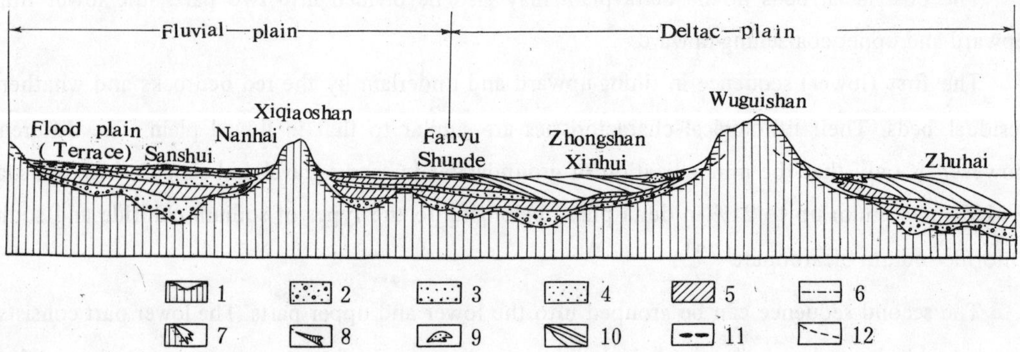

Fig. 5-46. Longintudinal section of Zhujiang Delta (After Li Chunchu, 1981, with slight abridgement).

1– Bedrock and weathering crust;

2– Sandy gravel;

3– Midium-coarse sand;

4– Fine sand;

5– Sandy clay;

6– Longitudinal profile of river channel;

7–Ancient pluvial-fluvial fan;

8– Modern pluvial-fluvial fan;

9– Coastal bar;

10 – Mud, sandy mud and muddy sand;

11 – Mud with decayed woods;

12 – Modern flood deposit (sandy clay).

residum or bedrocks[1)].

The centre bars and islands develop in the distributary channels. The former emerge above the water level during dry seasons or the low tidal water, while the later emmerge only during the flood period or high tidal water. They possess the coarsening-upward sedimentary sequence. The lower part of the sequences consists of sands and gravel sands and the silty clay is at the uppermost where plants grow. The sediments in the distributary channels become progressively finer downstream, changing to fine sands and silty sands near the outfalls.

(b) *Delta front facies*

(i) *Subaqueous distributary channel*

Distributary channels going out the outfall became the subaqueous ones, which are restricted by

1) Qiao Pengnian, 1980, Evolution of the Xijiang Channel in the Zhujiang Delta system.

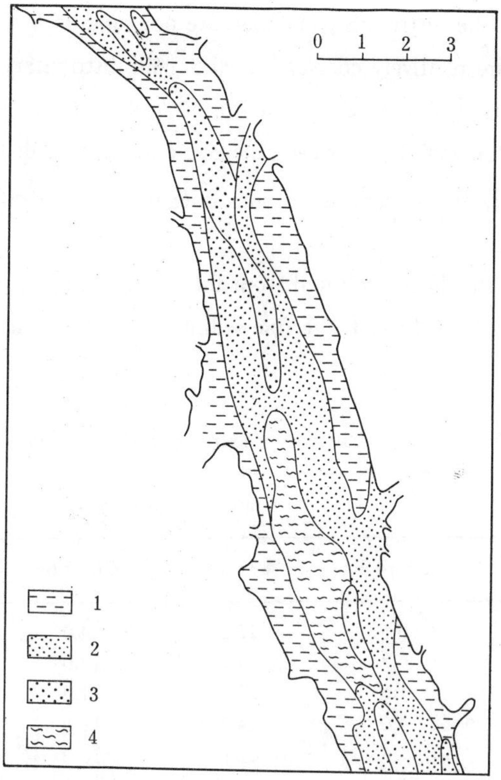

Fig. 5-47. Distribution of the sediments in Humen Distributary (After Zhau Huanting, 1980).

1– Muddy silt; 2– Fine and medium sands; 3– Coarse sand gravel;

4– Sands and gravels mixed with clayey/subclayey cemeutation of oyster shells.

subaqueous levels on both sides, such as the eastern and western troughs of Modaomen, where the subaqueous channels extend over 20 km long, 1–2 km wide, and maximum water depth 8–10 m. In the oufall, of the subaqueous channel, the mouth bar was formed, where the distributary branches off several sub-channel, which extend seawards and finally disappear. The sediments of the subaqueous distributary channels are mainly grey silts and silty sands.

(ii) *Interdistributary shoal*

The interdistributary shoals are the areas where the fine sediments remain in a standstill. The silty clay and clayey silt with Md 7–8 ϕ are deposited, containing, caly (40–50%) and fine sand (less than 5%). The shoal sediments near the mouth exhibit double peak in frequency curve, one of which has the value of 4.3–6.6 ϕ, another 7.6–8.0 ϕ, such as the samples in Denglongsha on south of Modaomen Distribatary (Table 5-25).

The sediments of the shoals far from the mouths are fairly fine, being silty clay with Md of 7.6–8.1 ϕ and contain sands less than 1.0%, but clay up to 50–63%[1]. There is dsingle-peak in frequece

1) Wu Wenzhong, 1975, The sediments of Niwanmen.

curve of the sediments, such as that on both sides of Niwanmen Distributary. When the water discharge and sediment load of the distributary channels are not same in both sides of the interdistributary shoals, the sediments are relatively coarser near the main distributary and here the elevation of the tidal flat is high. For instance, fine sand content is 50%, elevation about –0.4– –1.0m, near Modaomen Distributary of Denglongsha However, near the secendary distribution of the southern frank of this shoal not only the sediments become finer but also the elevation decreases to about –1.6 m.

Some interdistributary shoals are evoluted from the mouth bars, and preserve still the sedimenttary characteristics of mouth bars. In fact, the fine sediment layer covering the shoal, is the overlying bed of the mouth bars.

Table 5-20 Grain size distribution of surface sediments in southern frank of Denglongsha

No. samples	Grain size distrbution (%)					
	1–3.3	3.3–4.3	4.3–6.6	6.6–7.6	7.6–7.9	Md
12	0.5	0.5	22.5	24.0	53.5	7.6
13	4.0	6.5	31.0	13.0	45.0	7.3
20	0.5	1.5	21.5	21.0	56.0	8.0
21	1.5	4.0	24.5	19.0	51.5	7.7
22	2.0	5.5	26.5	16.0	50.0	7.6
23	1.0	5.0	35.6	16.9	40.5	7.1
76	5.5	9.5	31.5	12.5	41.5	7.0
77	3.5	6.5	29.0	16.0	45.3	7.3

(After Wu Wenzhong, 1975)

With extension of subaqueous distributary the interdistributary shoals continualy prograde and their front may spread out to water depth of 5 m. Meanwhile, there are plants growing along nearshore zones, including mainly *Aegiceras Cornicutalum*, *Acanthus ilicifolius*, *Kandelia candel*, *Phragmites communis*, and man planted greass, *Cyperus malaccensis*[1]. The plants promote the sedimentation and make the shoals grow progressively to emerge above water surface. The interdistrbutary shoals are transformed gradually to marshs with peats.

(iii) *Mouth bar*

In the front of the subaqueous distributary there is a mouth bar with water depth of 3–4 m. The slope of the bar is about 0.1–0.9% landwards, and about 0.8–1.0% seawards. The mouth bars are 10–20 km long, and several times largerin width than that of the subaqueous distributary channels.

The sediments of mouth bars are mainly silt and fine sands, about 80–90% of the samples, and get progressively fining seawards from the top of the mouth bars. In the zone of water depth less than

1) Department of Geography, Zhongshan University, 1973, Process and development of tidal flats in Modaomen mouth area.

5 m, the medium grain size is 2.7–3.7ϕ, and in the 10 m water depth silt content is 33–56%.

Microfossil assambleges differ obviously inside and outside the mouth bars. The foraminifera inside the bars are euryhaline ones such as *Elphidium nakanokawaense-Ammonia beccarii* assemblage. Every 50 g of samples may contain 200–300 tests. The continental and marine ostracoda co-exist, being *Senocytheridea latiovata-Cadoniella* assemblage. In the area outside the mouth bar a great amount of the *stenohaline* foraminifera present, being *Brizalina striatula-Ammonia annectens* assemblage. The ostracoda, *Senocytheridea latiovata-Neomonoceratina dongtaiensis-Liguminocythereis hodgii* assemblage, are almost mariné. Every 50 g of dry samples contain more than thousand foraminfera tests (Fig. 5-48).

Table 5-21 Slope gradient on both sides of the mouth bar

Area	Landward slope	Seaward slope
Yamen Distributary	–0.89%	0.75%
Modaomen Distributary	–0.10%	1.00%

(After Zhao Huanting, 1975)

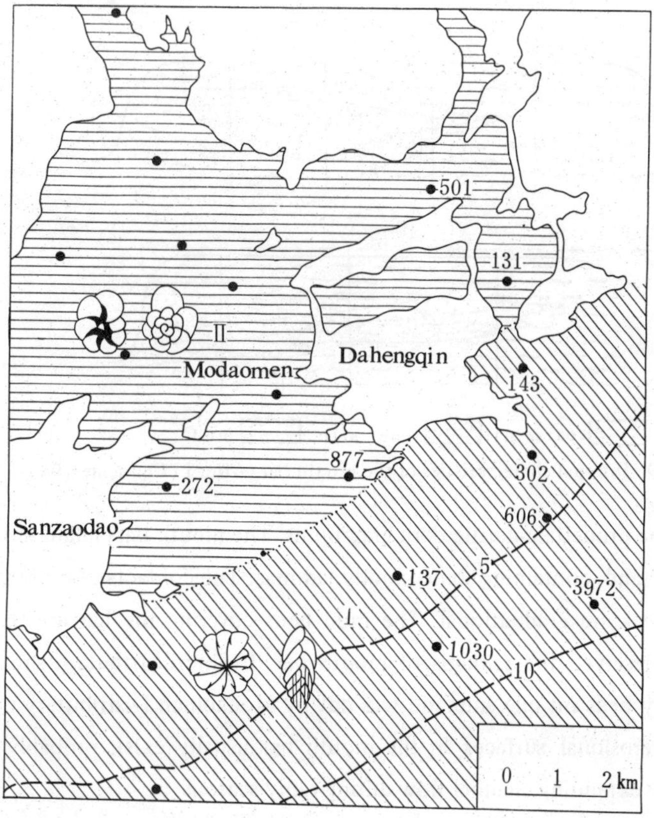

Fig. 5-48. Distribution of foraminifera assemblages in the mouth area of Xijiang River (After Wang Pinxian, 1980).

I– Ammonia annectens-Brizalina striatula assemblage;

II– Elphidiella nakanokawaense-Ammonia beccarii assemblage.

As above-mentioned, the interdistributary shoals were often the mouth bars in the past. These shoals, therefore, still preserve the remains of the mouth bars. For instance, the Denglongsha which was used to be the mouth bar of the Modaomen Distributary forced the channel to bifurcate into two branches by the end of 19th centuary, One of the branches becomes the main distributary, Modaomen, another is the Tianshenhe. Then, the Denglongsha gradually grew into the interdistributary shoal with an extension of the distributary. The shoal still maintains internal sedimentary characteristics of mouth bar (Fig. 5-49). Based on the data of the boring cores the maximum buried depth of the bottom of the mouth bar is –12 m.

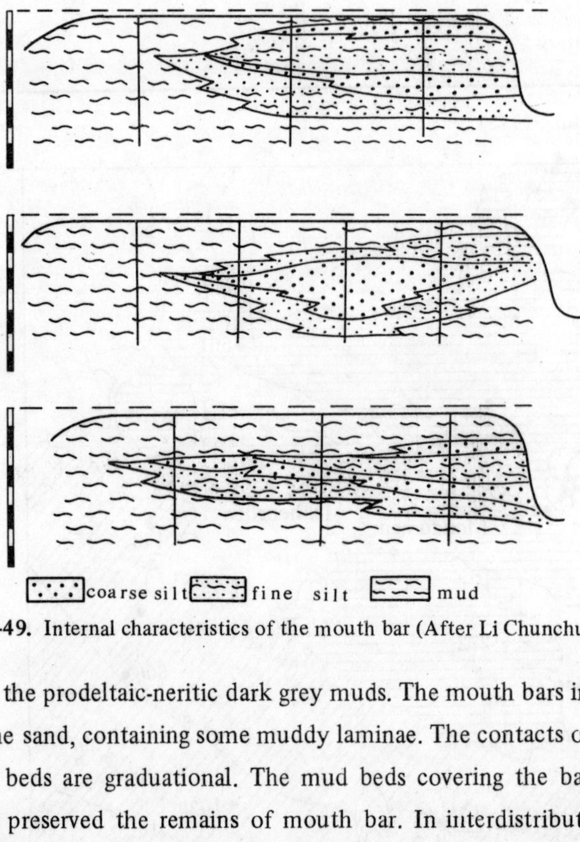

Fig. 5-49. Internal characteristics of the mouth bar (After Li Chunchu, 1983).

It is underlain by the prodeltaic-neritic dark grey muds. The mouth bars inlain in the muddy beds are well sorted silt-fine sand, containing some muddy laminae. The contacts of mouth bars with overlying and underlying beds are graduational. The mud beds covering the bar are 0.5–4.0 m thick. Denglongsha has still preserved the remains of mouth bar. In interdistributary shoal which graded laterally into the muddy beds on one side and was cut by Modaomen Distributary on anouther side (Fig. 5-49 B,C,). The erosional surfaces in the mouth bar usually exist, probably resulting from erosion of mouth bar by extending channel flow during rainy seasons.

It should be pointed that the mouth bar surrounded by muddy deposits serve as a lens in cross-section, by which the fluvial-dominated delta as the Mississippi is characterized[81]. Presence of this kind of mouth bars in Zhujiang Delta further confirms Zhujiang Delta as a fluvial-dominated one.

(c) *The prodelta facies*

The sediments of the prodelta are those of fine grain size brought by the rivers. The sediments of fairly high water-content are generally silty clay and clay, grey and blue in colour, with much orgainc matters, the highest percentage of which ranges 6.2–12.7%[1)]. The heavy minerals in the sediments, with lower percentage of 0.5% at most, mainly are hornblende, limnite, mica, ilmenite and zircon etc, and of terrigenous clastics. And the assemblage of the heavy minerals resembles that in the Zhujiang. Since the hydrodynamic conditions in prodeltaic environment is weak, the content of minerals of slab-shape and of less stability is high, e.g. the percentage of hornblende can be higher than 40%. But the percentage of stable minerals of high[82] special gravity, such as ilmenite and zircon is very low. The sediments of the prodelta are mainly from the Zhujiang, but are also under the influence of sea, which is clearly reflected by the chemical elements in the sediments of the delta. The presence of more than 50 chemical elements in the prodelta sediments and that of fewer than 20, or only several ones, in the sediments of the apex of the delta and of distributary channels present a striking contrast. The microfossils are of *Ammonia annectens-Brizalina stratula* assemblage.

In the region 30–50 km in width, out from the mouth of the Zhujiang River, exists the muddy sediments of the prodelta which ends at the depth about 20–30 m. The outer demarcation of the prodelta is fairly clear because on the sea side of the muddy sediments are the sandy and gravelly sediments of relict origin. The observation data shows that there is a constantly westerly-flowing longshore current outside the mouth of the Zhujiang River, which, carrying a great deal of fine sediments from the Zhujiang, reaches the Leizhou Peninsula directly and turns southernly when it meets the east coast of the peninsula. Of course, the rivers such as Maoyangjiang, Jianjiang contribute their fine materials to the west-ward flowing and sediment-carried longshore current when they pass their mouths, resulting in the widening of the zone of fine sediment which reaches its culmination in width and depth, respectively 100–130 km and 40–50 m, at the east coast of the Leizhou Peninsula[82].

(3) *The development of Zhujiang Delta*

(a) *The deposition history of the delta since the last glaciation*

The region where Zhujiang Delta develops has experienced the same change from transgression due to the higher rate of sea-level rise at the early stage of regression due to the relative decrease in the sedimentary rate of the delta resulting from the lowered rate of sea level rise at the late stage as other coastal deltas in the world since the last glaciation, which can be taken as the background of the development of Zhujiang Delta.

The Zhujiang entered sea at the spots where now belong to the outer continent shelf when sea level was 100 to 130 m lower than that of today at the full stage of the last glaciation. And the vast

1) Zhao Huanting et al., 1975, Preliminary research of the development of the Xijiang River mouth, Zhujiang Delta.

area now covered by the sediments of Zhujiang Delta was then an undulating land of hills with the Mesozoic and Cenozoic redbed and its weathered crust on the surface. Meandering through the area are the Xijiang, Beijing, Dongjiang and Tanjiang rivers, which cut the redbed into so many deep and wide valleys that only in limited patches are preserved the layers of unconsolidated sediments deposited in the last glaciation or prior periods.

The shoreline moves landwards because of transgression when sea level rose rapidly at the early stage of postglacial period. The transgression shows its effect first in the river channels of the area of Zhujiang Delta in the form of backwater and retrogressive aggradation, results of the rise of datum plane with the rise of sea level, which causes filling in the valleys of the lower reaches of the Zhujiang River. The sediments are finer in the lower reaches than in the upper ones. Therefore, a fining-upward sequence as the first periodites in the post-glacial development of the delta is produced while the retrogressive aggradation occurring in the process of transgression makes the finer sediment overlay the coarser. The channel-filling sequence of transgressive origin is different from both that of delta sedimentation and that of river sedimentation when there is no transgression. It is known from model experiment and field survey that the effects of retrogressive aggradation can pass beyond the points the tidal current and backwater can reach. Therefore, channel-filling sequences of transgressive origin which contain no marine fossil will be produced within the reaches where the retrogressive aggradation has its effects and out of the reach of tidal current, and thus alluvium plain will develop. The alluvium plain of the Zhujiang falls into this catagory.

The line linking Jiangmen, Daliang, Shiqiao, Huangpu, Xintang and Daojiao approximately fits the shoreline at the full stage of transgression when what is now Zhujiang Delta is a large estuary in which stand islands of all sizes. The formation and development of Zhujiang Delta begin with the sedimentary rate over that of sea level, which really is the result of the relative decrease in the rate of sea level rise. And then the delta plain with sequence of deltaic structure shows its existence. There are two characteristics unique to the development of Zhujiang Delta. The first is that Xijiang, Bejiang, Dongjiang, Tanjiang river etc. which all enter the estuary develop deltas of their own almost synchronically at the mouths, forming what is called the Zhujiang compound delta in which the small deltas keep independence of every one's own to certain degree though they are overlapped and adjacent to each other in their building up. The second one is that the islands in the estuary become the deposition nuclei around which the process of delta formation occurs and are linked to each other by the sediments, which eventually turned into parts of the delta plain growing outward in an inlaid pattern.

(b) *The Construction of Zhujiang Delta*

Since among the distributaries of the Zhujiang Delta system the water discharge of Modaomen in flood season accounts for 21.47% of all that discharged by the Xijiang and Beijiang, the deltaic deposition near Modaomoen can essentially be taken as an example of the process of progradation of Zhujiang Delta. As is shown in Fig. 5-49, the mouth bars and interdistributary shoals are formed by the accumulation of large amount of load, and the shoals grow vertically and stretch laterally with the extending of distributaries and seaward moving of mouth bars while the middle parts of the bases of the mouth bars are scoured by river, leaving its side parts undestroyed[1]. This process which can also be observed in the building up of Mississippi Delta may be characterized by all fluvial-dominated deltas. With Zhujiang Delta's advancing, the shoal growth causes seaward shifting of contours. The enlargement of the shallower than 2m is much more easily visible than that of the deeper, e.g. of the shallows within the contour of 10 m at Modaomen, the area of that within the contour of 2m only accounted for 1.7% in 1946, but it reached the percentages of 12.7% and 29.6% repetively in 1964 and 1971[2]. The natural accretion and the reclaimation for land speed up the building up of Zhujiang Delta.

(c) *The destruction of Zhujiang Delta*

When the alteration of main branches or the abandoning of some channels in the deltaic system causes decrease in sediment supply and enhances the marine factors the destruction of the delta occurs, and the reports about which are only inferences from the existence of sand layers deposited in coastal region, lacking direct information from field investigation and record, so we considered that the destruction of Zhujiang Delta are only a phenomenon of local scale and are not so easily visible as the deltas built by the Mississippi and the Huanghe etc, which may be attributed to the protection of the islands and the estuary in which Zhujiang Delta stands.

(d) *Rates of deltaic sedimentation and outgrowth*

Like other deltas in the world the sedimentary rate of Zhujiang Delta, which though varies greatly in different parts of the delta and at different stages of developing, is averagely high. It is calculated from the amount of sediments silted up off Modaomen during 1964–1971 that the sedimentary rate averages 6.9 cm/yr, right outside the mouth of main channels, but it drops rapidly down to 1.1 cm/yr seaward away from the mouth. And again, the seidmentary rate of the interdistributary shoals, which averages 0.4 cm/yr from the sediments deposited during 1961–1964, is much lower than that at the mouth of the channels[2]. However, the ratio of the mean thickness of the deltaic sediment during the period when Zhujiang Delta develops ranges 0.2–0.3 cm/yr averagely, much lower than those in a local areas of the delta during a shorter period mentioned above. The mean rate of Zhujiang Delta's sedimentation is in the same order of magnitude of Changjiang Delta though slightly lower. And the mean sedimentary rate in the stage the delta develops is greater than that before the starting of the delta formation, the former being 0.25 cm/yr and the latter only 0.07 cm/yr[3].

The mean rate of progradation is obtained by virture of the dates when dykes are built to re-

---

1) Geography Department, Zhongshan University, 1973, Research of the reclaimation of Modaomen beach, Zhujiang Delta.

2) Qiao Pengnian, 1980, Evolulion of Xijiang River channel in the Zhujiang Delta system.

3) Calcu Lated from reference [78].

claim land since this sort of man activity is widely seen in Zhujiang Delta area. Generally the mouths of main distributaries have a high rate of progradation which is 150–160 cm/yr averagely and 220 m/yr extremely at Modaomen, and the distributary mouths of minor importance have a relatively low rate of progradation which averages 63.3 m/yr, and respectively 55.6 m/yr during 1830–1883, 60.3 m/yr during 1883–1936 and 91–2 m/yr during 1936–1956 at Wanqinsha in the vicinity of Jiaomen, by comparison of nautical charts[78].

5. Luanhe Fan-Delta

(1) *General Situation*

(a) *Tectonic Setting*

The Luanhe River debouches from the Yanshan Mountain and empties into the Bohai Gulf. With the Yanshan fold belt keeping uprising and the Bohai Basin, subsiding since mid-Cenozoic[88,89], the tectonic setting thus formed represents the optimum environment for the fan-delta to develop.

(b) *Hydrology*

The Luanhe River is 1200 km long in total and its drainage area, around 45000 km^2. According to the 1950–1977 data provided by the Luanxian County Station of Hydrology, the average runoff of the Luanhe River is 4.55×10^9 m^3 annually and the average sediment discharge, 19×10^6 tons annually. The flood season of the area starts from July and lasts until the end of September, during which, the runoff of the Luanhe River makes up 70% of the yearly total and the sediment discharge, 93%.

As a microtidal river mouth, the tidal range of the Luanhe mouth area is 1.0–1.5 m and it tends to increase gradually from north to south (Table 5-22). The current system off the Luanhe River mouth is under the influences of the warm currents entering the Bohai Sea, which, once reaching the offshore area of the Luanhe Delta, will go along two separate routes, with one turning to the right to enter the Liaodong Gulf and the other, to the left ot enter the Bohai Gulf, thus forming a longterm steady circulation. And that is why, west of the Luanhe River mouth, the flood flows southwestward and the ebb, northeastward; whereas in the area east of it, the opposite situation appears.

Table 5-22 Correlation of the tidal ranges along the coast of the Luanhe Delta

Area	Mean high tide	Mean low tide	Tidal range (m)
Qilihai Sea	1.82	1.08	0.74
Luanhe River mouth	1.68	0.36	1.36
Daqing River mouth	1.91	0.52	1.39
Nanbao	2.53	1.02	1.51

The Bohai Gulf is a semi-closed bay, where the relationship between the waves and winds is apparent, with the fetch length being relatively short and the swells making up only a small proportion. Data from the Leqing County and Laoyujian indicate that the normal winds are the northward winds and the strong breezes blow in the NEE direction. As a result, the NE and NEE waves are the highest waves in the Gulf.

(c) *Extent and boundary*

In the postglacial period, the Luanhe Fan-Delta took the Luanxian County as its apex and spread southward in a fan shape, with the line running from the Luanxian County to Luannan and then to Anniudian serving as its west boundary and the line from Dafangezhuang – Kanshang – Jingan and then to Huanghouzhai as its east boundary. On the both sides of the fan delta there are steep scarps separating it form the old fan. The steep scarps on the north side can be as high as 10–12 m and so are more distinct compared with those on the south side. The scarps, however, decrease southward in elevation and consequently, becomes merged into the coastal plain when reaching the coastal area. Even though, the boundary lines on the both sides of the Luanhe Fan-Delta are still rather clear as a whole (Fig. 5-50). And this fan-shaped sedimentary body takes the line from Hugezhuang – Yangezhuang – Xiamaozhuang – Huanghouzhai as a demarcation, north of which, it appears as an alluvial fan and south of which, a delta, the two constituting an unified sedimentary body – the fan-delta system though the fan and the delta differ from each other in their sedimentary characteristics.

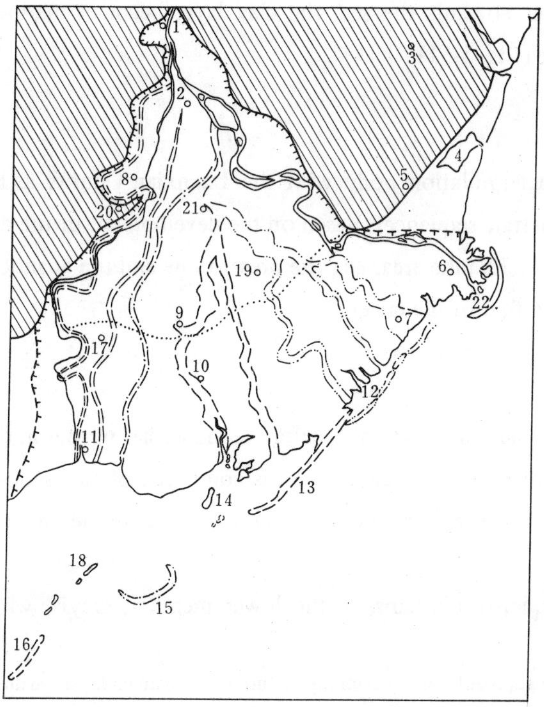

Fig. 5-50. Extent and Boundary of the Luanhe Fan-Delta.
1. Luanxian; 2. Macheng; 3. Changlixian; 4. Qilihai; 5. Liutaizhuang; 6. Jiujiafang; 7. Langwokou; 8. Luanhe River; 9. Jiangjiazhuang; 10. Xinzhai; 11. Matouying; 12. Xihe; 13. Hulinkou; 14. Dawangang; 15. Shijiutuo; 16. Dongkengjianzi; 17. Caofeidian; 18. Tuoli; 19. Getuo; 20. Leting; 21. Luannanxian; 22. Tingliuhe River; 23. Luanhe River mouth.

(2) ***Alluvial Fan***

No marine organisms such as foraminifera are found in the sediment, indicating that the layer is of terrigenous facies. Palaeontological data suggest an late Pleistoncene age[1] profile texture.

Varied parts of the alluvial fan possesse different sedimentary characteristics and profile textures. The Luanhe Bridge in the Luanxian County is the area where the Luanhe River debouches from the mountain and enters the piedmont plain, representing the apex of the alluvial fan. The apex area or the proximal alluvial fan is composed of granule gravel layer. The gravels, with mean grain diameter of 3.56 cm and maximum one of 15 cm, are subrounded or subangular in shape and with medium to fine sands filling in between. A host of mechanicaldash markings can be found on the surfaces of the quartz grains (Plate 5-3), reflecting the strong water-flow effects. The granule gravel layer shows large planar bedding, trough cross bedding, oblique bedding and occasionally, grading bedding. Scour-and-refill structures are very common, revealing the violent rising and falling of the water level and the inconstancy of the water flow. In the top 30–50 cm of the granule gravel layer occur yellow silt-fine sand and clay beds containing gravels and exhibiting minor cross bedding and climbing bedding.

The longitudinal profile running through the whole alluvial fan from the apex to the coast indicates that the lower sedimentary cycle represents a transgressive sequence whose facies changes upstream form sandy clay beds to sand beds and then to granule gravel beds, thus forming a kind of overlapping deposits. The same facies variation also takes place in the upper sedimentary cycle but with the result that a kind of downlapping deposits forms. In the proximal fan, as the clay layer embedded in between the upper and lower cycles has tapered out, the two granule gravel layers have merged into one with greater thickness. Towards the sea, the clay layer becomes marine in facies while the upper cycle, a deltaic sequence (Fig. 5-51).

(3) ***Delta***

The south part of the triangular accumulation body with the Luanxian County as its apex is where deltas develop and so possesses deltaic sequences. Based on the development of the deltas, two types of deltas can be distinguished in the Luanhe area, i.e., the present one and the abandoned type. The area of the former is far smaller than that of the latter.

(a) *Vertical sequence*

The underlying layer of the postglacial deposits in the delta area is similar to that in the alluvial fan area in character. As for the layer of postglacial deposits, it is composed of two sequences: the transgressive channel-filled sequence in the lower part and the deltaic sequence in the upper part (Fig. 5-52).

The transgressive channel-filled sequence contains, in the lower member, greyish-white gravel-

---

1) Zhao Lianbi et al., 1964, Late Pleistocene palaeontological assemblage (mollusca, ostracoda, spore and pollen) in the Zhua Chun Village, Qianan County of Hebei Province and its geological significance.

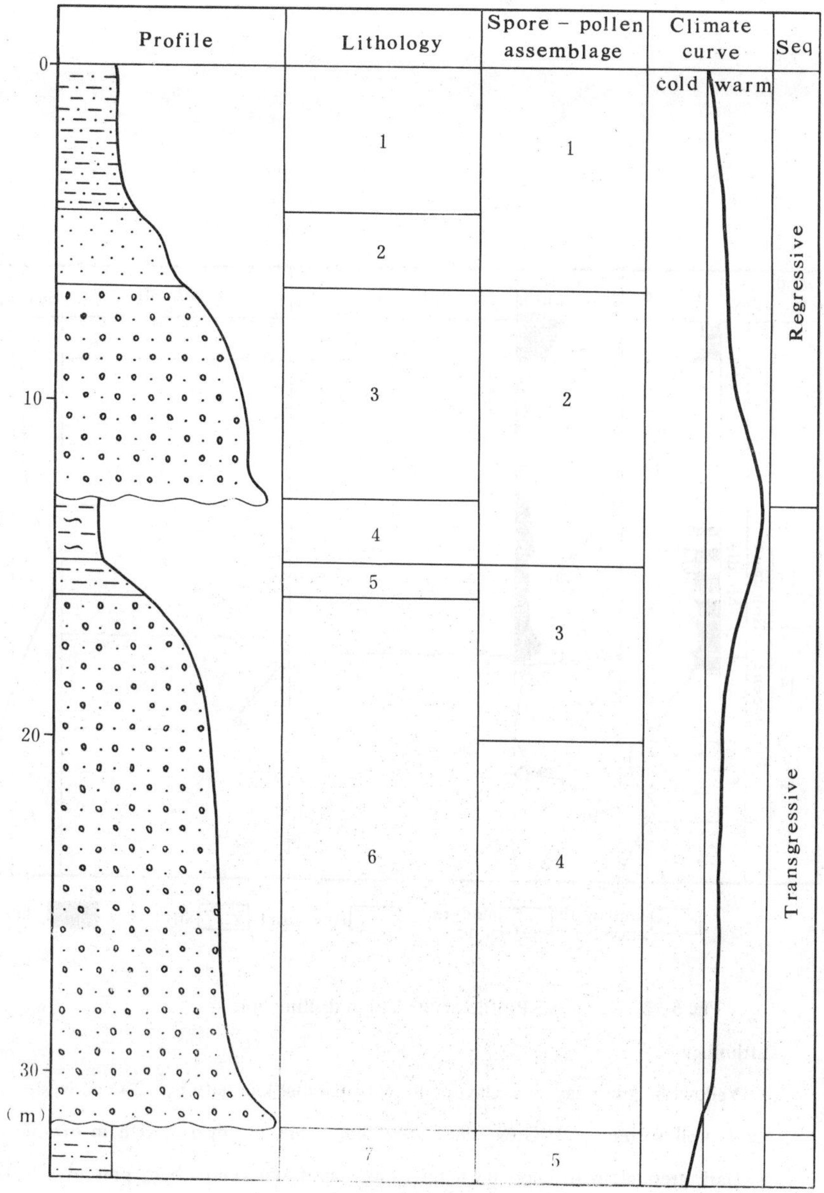

Fig. 5-51. Longitudinal profile of the Luanhe Fan-Delta.

Lithology

1–Yellowish, greyish yellow loam; 2–Grey medium sand;

3–Grey, dotted sandy gravel, lower erosional contact;

4–Black, dark grey soft loam; 5–Yellowish clayey sand;

6–Grey, dotted sandy gravel with lower erosional contact;

7–Tawn Clayey sand.

Sporo-pollen assemblage

1–*Pinus-Betula;* 2–*Betula-Quercus-Ulmus;* 3–*Pinus-Betula-Ulmus;*

4–*Pinus-Betula;* 5–*Pinus-Betula-Picea.*

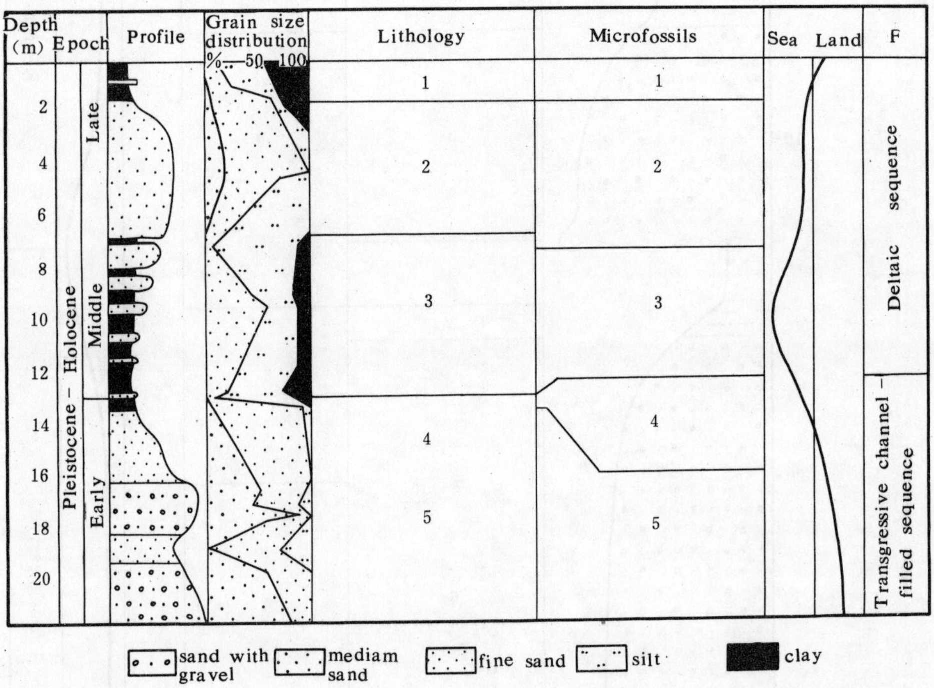

Fig. 5-52. Columnar Profile of the Yutan drilling bore.

Lithology

1–Yellowish, grey clay and clayey silt with parallel lamination; 2–Yellowish grey fine sand, well sorted; 3–Grey, dark grey sandy silt, clayey silt, rich in organic matter; 4–Dark grey silt with clay, tawn ferriginous spot and yellow belts present; 5–Greyish white sand with gravel and median-fine sand, intercalated by silty sand, median sand.

Microfossils

1–A few *Nonion Shansiense, Ammonia beccarii* etc; 2–*Ammonia beccarii,* spines of echnoid, shieldform fossils; 3–Rich in neritic foraminifera, marine ostracoda, spines of echniod, fragments of *ophiuroid* and *bryophyta;* 4–A small amount of *Ammonia beccarii, Cribrononion subincertum, Neomonociratina;* 5–A few foraminifera: *Cribrononion subincertum, Elphidium magellanicum.*

F–Facies interpretation.

bearing sand beds, which change upward into greyish-white medium sand, fine sand and silt-fine sand contained shell debris. In the upper member, it appears as greenish-grey sandy clay, inside which rusty spots and yellow strips can be found and plant roots, often wrapped by brownish-red iron-sheets, are well developed. The minor amount of foraminifera contained in it includes mainly *Elphidium magellanicum, Cribronomin subimertum, Ammonia beccarii* var. and *A.grabosa* etc. Also present in it are *Thecamoebina* and shieldform fossils. With marine and terrigenous biogenic fossils coexisting in it and with very weak marine characters, this layer is of flood-plain facies upstream from the river mouth and influenced by sea water. Correlation with the other drilling bores in the Bohai Gulf area suggests an early postglacial age.

The deltaic sequence can be divided into three members (the upper, middle and lower). The lower member is composed of soft, organic-rich, greenish-grey silty clay, inside which foraminifera is very abundant and with high diversity (2.0). The main genera and species of the foraminifera are *Elphidium magellanicum, Ammonia beccarii, Triloculina Trigonula, Cribronomin subimertum,* and *Ammonia convexidonsa* etc. As for ostracoda, the main ones are *Loxococha ocellata Sinocythere senensis* and *Cushmanidea* etc. Also present in it are a large amount of echinoid spines, ophiuroid fragments and a small amount of Bryozoa. The biogenous assemblage indicates that this member is with strong marine characters and of neritic marine-prodelta facies. The ^{14}C dating of it is 5400–5600 yr[93].

The middle member consists of well-sorted yellowish-grey fine sands containing shell debris, uniform in grain size and relatively rich in foraminifera with normal tests. The foraminifera is dominated by *Ammonia beccarii, Elphidium magellanicum* and *Triloculina triganula* etc., while the ostracoda, by *Loxoconcha ocellata* and *Sinocythere sinensis* etc. In addition, echinoid spines, small Gastropoda, shield-form fossils and plant debris are also relatively abundant in this member, indicating a delta front-nearshore shallow-water facies influenced by fresh water.

The upper member consists of grey and earthy-yellow clayey silt and silty clay which contain rusty spots, iron and manganese nodules and lots of plant debris, and show horizontal bedding. Also present in it is a minor amount of small foraminifera dominated by *Nonion shansiense* and *Ammonia beccarii* etc. With very low mariness, this member is of delta plain-tidal flat facies.

The upper sedimentary sequence of the postglacial sedimentary layer is coarsening-upward and its marine characters weaken upward. All this shows it is a typical deltaic sequence. The stratigraphic correlation with the other drilling bores around the Bohai Gulf indicates a middle-late postglacial age.

(b) *Modern delta*

The Luanhe River has been emptying into the sea from the present position ever since 1915 when it breached the littoral sand dune near the Lianhuachi Village, and in the 65 yr from 1915 to 1980 it formed a modern delta with an area of 69 km^2. The prograding velocity of the delta is 123–138 m/yr and the area-accreting velocity, 1.06 km^2/yr. The subaqueous delta is about 250 km^2 in area and the subaerial/subaqueous area ratio is 0.27. The Luanhe River used to enter sea through many distributaries and mostly through the southwest and northeast distributaries. In 1979, it changed its route to enter the sea from the central part of the delta. On th frontal margin of the Luanhe Delta exist barrier bars, reflecting strong wave action. The tidal inlets within the lagoons and the tidal channels between the bars, however, are the outcome of tidal current action, and, therefore, the Luanhe Delta, just like the Niger Delta, is a transitional delta between the wavedominated and tide-dominated ones.

(i) *Delta plain*

With its natural state preserved for not having been suffered from artifical destructions, the delta plain provides a desired locality for studying deltaic deposits. The delta plain can be further divided into flood plains, distributary channels, abandoned distributary channels, natural levees, lagoons and barrier bars etc (Fig. 5-53).

a) Flood plain

The sediment consists of greyish-yellow and yellowish grey clay-bearing silts, locally with silt-fine sand intercalated beds. Inside the sediment, plant debris are abundant and often form laminae; plant roots develop; and rusty spots and rusty-dyed reddish-brown strips etc. are common. Besides, the sediment shows well-developed structures such as minor cross bedding (plate 5-4), horizontal bedding, climbing bedding and lens-shaped bedding etc., and attains a thickness of 3–5 m. The flood plain is not only drowned by flood, but also under the influences of tide. And that's why its sediment often contains a minor amount of marine microfossils (generally speaking, 3–5 pieces or at least 1 piece every 50 g of dry sample). Molluscan remains are also commonly found in it.

b) Active and abandoned distributary channels

Altogether 8 active and abandoned distributary channels are found in the Luanhe Delta (Fig. 5-54). Although there exist so many distributary channels, only one of them serves as the active channel during each stage with all the rest abandoned. The sediment is well-sorted greyish-yellow fine sands with a mean grain diameter of 3–3.5 ϕ and a standard deviation, less than 0.5. The heavy minerals are dominated by hornblends (high as 50%) and next to it, biotite. In the light minerals, the feldspar content is higher than the quartz content, the ratio between the two being 3.0. With high rock debris content and low maturity of sandy materials, the sediment displays fully the characteristics of proximal-river-dominated deltas (Fig. 5-55). On the surface of the quartz grains, besides the large amount of markings left by mechanical dashes, appear siliceous balls, siliceous films and dissolution-eroded pits, which have occurred in response to chemical processes. There is also a minor amount of foraminifera in the sediment.

In the upstream reaches of the abandoned river channels often develop eolian dunes, whereas in the downstream reaches, argillaceous marsh deposits.

c) Natural levee

Because of the frequent flow diversions of the Luanhe River, the natural levees are poorly preserved, their lengths being varied and their widths ranging from 50 to 150 m. And there is vegetation covering them. The sediments are yellow fine sand and silt-fine sand with light-brown clay inlercalated beds. In the heavy minerals, the mica content is reatively high (about 5.0–7.0%), whereas in the light minerals, the feldspar content instead, and the ratio between feldspar and quartz is over

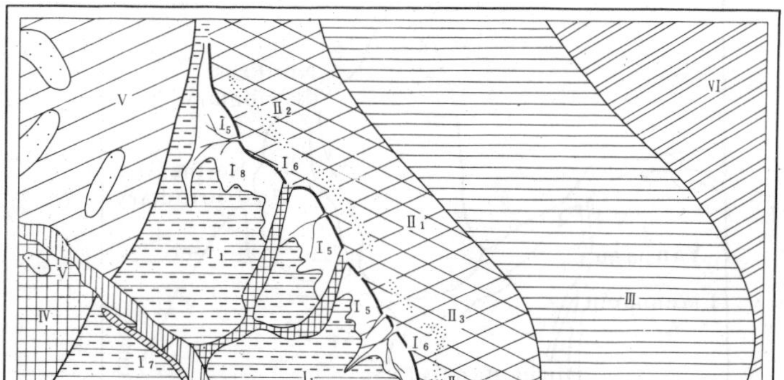

Fig. 5-53. Sedimentary facies of the Luanhe Delta.

Deltaic sedimentary system

I Delta plain

I_1–Flood plain I_2–Luanhe channel I_3–Distributary channel I_4–Abandoned distributary channel I_5–Lagoon I_6–Barrier island I_7–Levee I_8–Ancient barrier island I_9–Tidal Channel

II Detla front

II_1–Front platform-slope II_2–Subaqueous sand bar II_3–Subaqueous mouth bar

III–Prodelta

IV–Ancient delta

No deltaic sedimentary system

V–Coastal sand dune

Interdune marsh

VI–Relict sediments on continental shelf

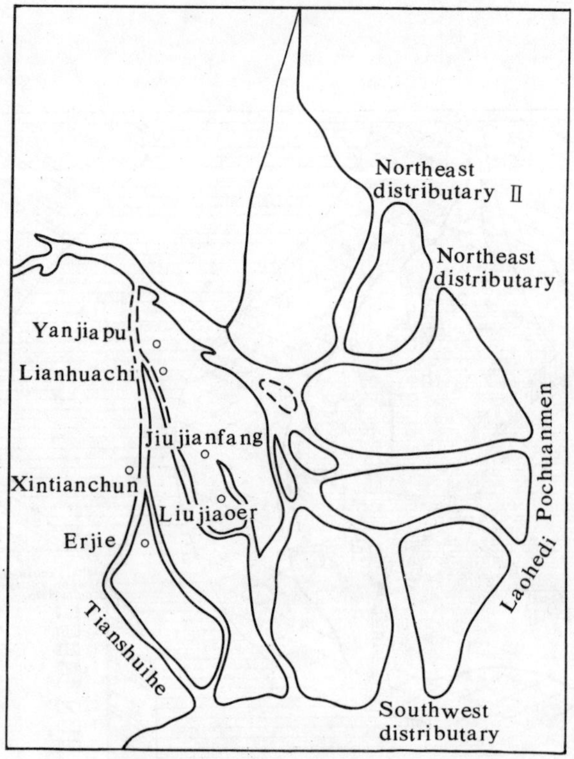

Fig. 5-54. Main distributaries in Luanhe Delta.

2.0. Plant debris are abundant and laminae often form. Argillaceous gravels can occasionally be found which owe their origin to local scouring. As for plant roots, their distribution is more widespread.

d) Lagoon

The lagoons are usually located in the marginal part of the delta and their sediments are ill-sorted grey and greenish-grey argillaceous silt-fine sand and sandy clay, the argillaceous content being 10–26% and the standard deveiation, 1.62–2.37. The heavy minerals are dominated by hornblends and epidots, reaching over 60%. The organic contents are around 0.5%, making up 3–10% of the coarse grain (>0.063 mm). Also present in them are shells, most of which are interact, and there are even some living ones.

Being influenced at the same time by the fresh water from the Luanhe River and by the tidal currents from the sea, the logoons change markedly in their salinities, lower than 20% in flood season, and as high as 32.28% in dry season. Consequently, the foraminifera in the lagoons are dominated by euryhaline *Ammonia beccarii*, just one species of which makes up 73.1% of the total individuals. The individuals of *Ammonia beccarii* are so abundant that over 2000 pieces can be found in every 50 g of dry sample. As for ostracoda, it mainly includes euryhaline *Loxoconcha ocellata*. Also present in it are freshwater organisms such as *Thecamoebina* and brackish fossils such as *Gyrosigma* and shield-form fossils etc. The character that the marine and continental biogenous fossils are mixed together is thus fully displayed.

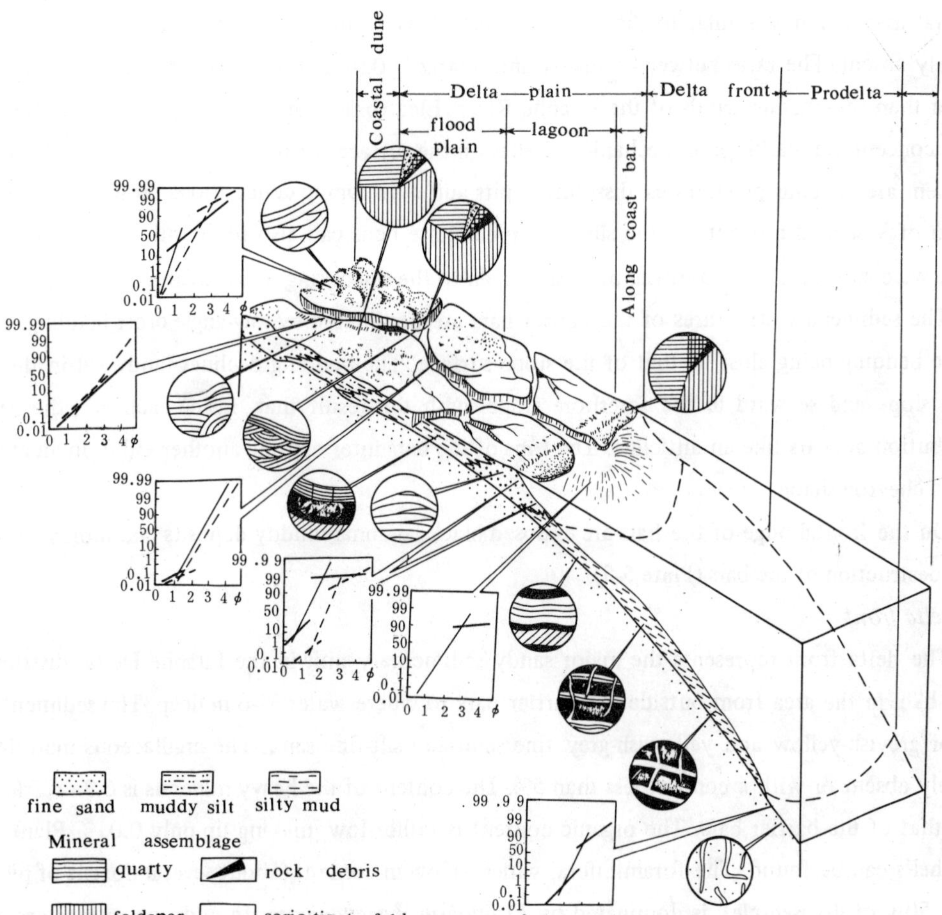

Fig. 5-55. Block diagram illustrating characterristics of different sedimentary environments in modern Luanhe Delta.

e) Barrier bar

In an arc form, the barrier bars distribute along the margin of the Luanhe delta plain. The bars are about 20 km long, 50–100 m wide, and 1–2 m higher than the high water line. Their mean seaward slope varies between 1/50 and 1/100 and the steepest part, around 4°. The sediment is fairly-sorted yellow fine sands with mean grain diameter of 2.23–2.40 ϕ and standard deviation of 0.35–0.39. The average heavy-mineral content is 11.3% and the maximum one, 24.7%. The heavy mineral association is similar to that in the Luanhe River, but with schistose minerals such as mica entirely absent. The ratio between feldspar and quartz is 0.9–1.0. The mature degree is significantly higher than the channel sands of the Luanhe River. Electronic scan microscopic analysis shows that there concentrate markings of mechanical dashes on the surfaces of the quartz grains, and also present on them are siliceous precipitates, dissolution pits and overlapping denudations. Besides, the phenomennon of V-shaped pits cutting the siliceous precipitate beds can also be found, indicating the these grains were subjected first to fluvial processes, then to the reworking of waves and sea water.

The sedimenary structures of the barrier bars are dominated by low-angle cross bedding, the dip of the bedding being close to that of the slope surface. The bedding declines landward in the backshore slope and seaward in the foreshore slope, with the result that, inside each bar, the bedding distribution appears like an anticline. The tops of the bars intersect one another either in an arc form or in a chevron shape.

On the frontal edge of the bars are exposed black lagoonal muddy deposits commonly, revealing local destruction of the bars (Plate 5-5).

(ii) *Delta front*

The delta front represents the major sandy sedimentary unit in the Luanhe Delta, distributing, sheet-like, in the area from outside the barrier bars to where water 5–6 m deep. The sediment consists of greyish-yellow and yellowish-grey fine sand and silt-fine sand. The argillaceous materials are entirely absent or with a content less than 5%. The content of the heavy minerals is 6.3–8.2%, lower than that of the barrier bars. The organic content is rather low, making up only 0.01%. Plant debris and shells can be found. The foraminifera, which is low in content (about several dozens of pieces in every 50 g of dry sample), is dominated by *Elphidium magellanicum.* In addition, minor cross bedding is also recorded.

(iii) *Prodelta*

The prodelta, with its outer edge being 4–8 km away from the coast and its width, 2.5–6.5 km, situates where the water depth varies between 5–6 and 10–12 m. And it is the most-widely-distributed sedimentary unit in the delta, the sediments of which are ill-sorted, semifluid grey and greenish-grey clayey silt and silty clay with water content being 40–60%; Md, 6.0 ϕ; sand content, less than 10%; clay content, as high as 40–50%; and standard deveiation, 2.0–2.8. The heavy minerals, whose content is about 1% (lower than that in the delta front), are dominated by hornblende and epidote. The organic content is relatively high, usually over 0.50%.

Outside the prodelta, the north part is characterized by well-sorted coarses sandy sediment with mean grain diameter being 1.25–1.29 ϕ, sand content over 90% (among which, coarse sand makes up over 20%), standard deviation 0.26–0.38, and heavy-mineral content as high as 15.0% (that's why the sediment is much coarser than that in the presetn Luanhe Delta); while the south part, by stiff greyish-black silty clay with water content being around 20%, containing brown rusty spots, iron and manganese nodules, plant roots and plant debris (making up 20% of the coarse grain class) and organic content being 0.36–0.50%.

(c) *Abandoned delta*

In the south part of the Luanhe Fan-Delta, a bar lagoon system is developing along the coastal zone from the Luanhe River mouth in the east to Caofeidian in the west, and a vast plain, on the land. The whole area is considered as the abandoned Luanhe Delta on th following evidences:

(i) *Deltaic sequence*

The major criterion to determine a delta is the deltaic sequence. And as above described, this area possesses a complete deltaic sequence (Fig. 5-55). The development of modern deltas is usually led by postglacial transgressions. And here,underlain by the deltaic deposits is the transgressive channel-filled sequence.

(ii) *Subsurface texture*

Subsurface acoustic profiles reveal that there is commonly a seaward-oblique layer in the subsurface of the outer sides of the barrier bars, and that the layer is relatively steep near the shore but gets flattened towards the sea, the mean slope being 3–5%, close to that of the subaqueous part of the present Luanhe Delta (Fig. 5-56). Over this layer lies a horizontal marine layer. And so the obligue layer may represent the deltaic deposits while the overlying horizontal layer, the marine layer deposited after the delta was abandoned.

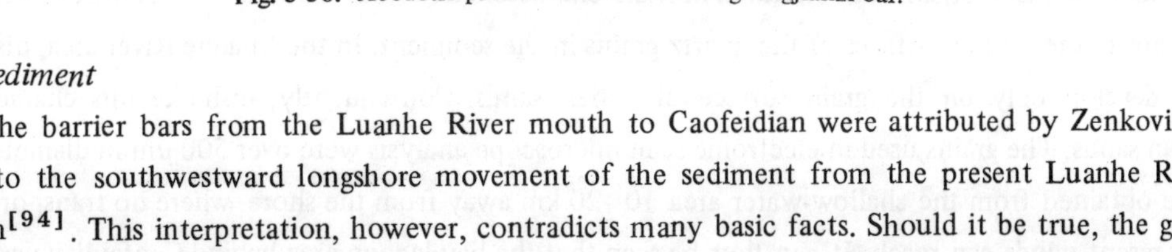

Fig. 5-56. Acoustic profile outside the Dongkengjianzi bar.

(iii) *Sediment*

The barrier bars from the Luanhe River mouth to Caofeidian were attributed by Zenkovitch, V.P. to the southwestward longshore movement of the sediment from the present Luanhe River mouth[94]. This interpretation, however, contradicts many basic facts. Should it be true, the grain sizes above fine sand would get finer outward from the material-source area. The fact, however, is that the sediment of the barrier bars tends to get coarser from the Luanhe River mouth to Caofeidian with the grain sizes above medium sand increasing in content (Table 5-23). The heavy minerals in the barrier bars from the Luanhe River mouth to Caofeidian include mainly magnetite, ilmenite, hemitite,

Table 5-23 Content variation of grain sizes above medium sand in the barrier bars

Area	Content (%)	Number of sample
Luanhe River mouth	10.02	5
Langwokou	10.28	4
Daqing River mouth	34.02	17
Dongkengjianzi	36.50	2
Caofeidian	38.85	10

limnite, hornblende, epidote, actionlite, titanite, garnet, apatite, tremolite, diopside, hypersthene, zircon, tourmaline and mica, etc. The minerals here being similar to those in the Luanhe River indicate that the material source of these barrier bars is no other than the Luanhe River itself. However, the content of the heavy minerals, just like the grain sizes above medium sand, increases from the Luanhe River mouth to Caofeidian (Table 5-24). This shows that most of the sediment forming

Table 5-24 Heavy mineral content in the different parts of the barrier bar

Grain size (mm) / Content (%) / Area	0.25–0.125	0.125–0.063
Luanhe River mouth	0.86	6.74
Daqing River mouth	2.15	16.44
Dongkengjianzi	2.96	56.50
Caofeidian	11.99	85.35

the barrier bars do not come from the longshore transportation of the sediment in the present Luanhe River Mouth but derives from the abandoned Luanhe Delta. The vast shallow-water area in the back side of the Caofeidian barrier bar connects with the open sea through tidal channels. The sediments here and the paleo-organisms contained in it are characteristic of bay-lagoons[95]. A host of dish-like pits are found on the surfaces of the quartz grains in the sediment. In the Luanhe River area, dish-like pits develop only on the grain surface of eolian sands. Consequently, dish-like pits characterize eolian sands. The grains used in electronic scan microscope analysis were over 500 μm in diameter and were obtained from the shallow-water area 10–20 km away from the shore, where no transportation by present winds can reach. It can thus be seen that the bay-lagoon area behind Caofeidian was once subaerial and with eolian sands developing on it.

In the Caofeidian bay-lagoon area develop a series of shell-accumulated bodies such as Hatuo, Caomutuo, Nangoutuo and Beigoutuo etc. These shell-accumulated bodies, whose shell content is over 80%, owe their origin to the wave transportation (Plate 5-6). As for the shells, they are of three types. The first type is the molluscs inhabiting river mouth areas, such as *Ostrea rivularia* Goula, the shells of which are most abundant and most severely eroded due to friction and dissolution. Because of weathering, some of the *Ostrea rivularia* Goula shells have already peeled off in small pieces, indicating that they have already existed for a long time. The second type refers to those living in present shallow lagoonal environments such as *Ostrea gigas* Thunberg and *Meretrix* Linne etc. Generally speaking, this type is comparatively fresh. The third type inhabits nearshore neritic environments. *Ostrea denselamellosa* Lischke is an example of this type. The presence of large quantities of shells of river mouth-inhabiting molluscs indicates that the Caoferidian bay-lagoon area was once a river mouth area.

(iv) *Dynamics of barrier bars*

Many traces indicating wave-erosion are left in the barrier bar area west of the Luanhe River mouth, such as the eroded steep scarps in front of the bars, the lagoonal clay exposed on the frontal edge of the bars (Plate 5-7) and the roots exposed, inclined and dead bushes on the bars. Due to erosion, some sand islands have become subaqueous bars which submerge during the high water level and emerge during the low water level. And Caofeidian is such an example. Data from Tianjin Historical Museum[1)] shows that as far back as in the early Tang Dynasty (630 A.D.) there were already ships suffered misfortune around Caofeidian. It can thus be concluded that Caofeidian was already present then. Though it is impossible to know the shape of it, the length and width of it were, according to legends, "in an environment 40 *lis* in four directions". Caofeidian was found an isolated island 4 *lis* long and 2 *lis* wide in 1869 (Qing Dyansty) when the on-spot reconnaissance was made for the purpose of setting up Caofeidian light-house. On the island there were temples and stone-tablets built in the years ruled by the emperor Qianlong of Qing Dynasty. In the Mountain-and-Water Chapter of the Luanxian County Annals in 1890 is attached a map of Caofeidian, on which, the exact locations of its temples, stone-tablets, water wells, fish markets and grass lumps were clearly marked (Fig. 5-57). The repeated attack of tsunamis, spring tide and storm waves resulted in the gradual destruction of the island and in 1978 when the onspot reconnaissance was made again, the temples, houses, stone-tablets and grass lumps could no longer be found. With everything on it gone with the sea water, Caofeidian had become a subaqueous bar emerging only during the low water level. From the above it can be seen that the erosion process of the Caofeidian area is very clear: Caofeidian was once a deltaic body uplifted out of the sea, which evolved into an island later, then a barrier bar and finally a lagoon.

(v) *Abandoned river channel*

A river or rivers may form a delta, and consequently, a abandoned dalta must have its corresponding ancient river channel or channels. Satellite images and historical literature show that there were a series of ancient river channels existing in the Luanhe Fan Delta area. The actual water-flowing time for each ancient channel is given in Fig. 5-51. With the crevassing points respectively

1) Tianjin Historical Museum, 1977, Earthquakes in North China Plain.

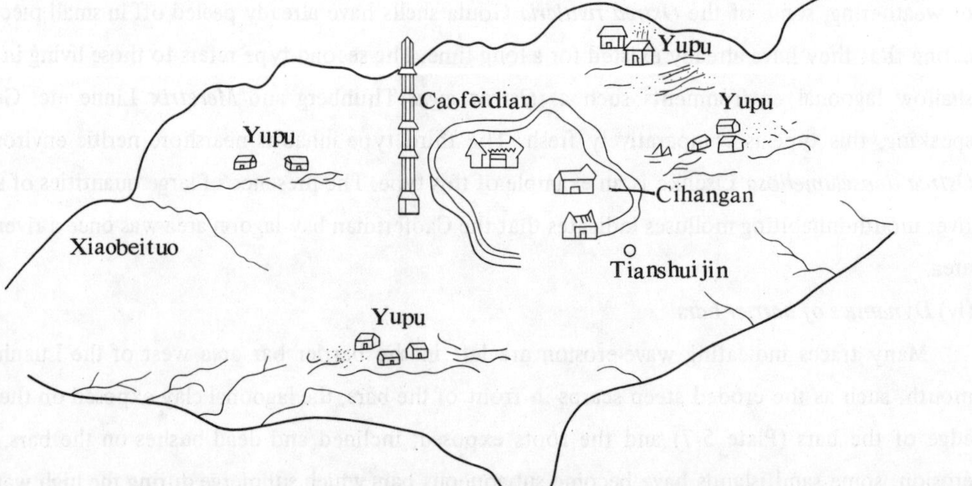

Fig. 5-57. The Caofeidian Island (after the Luanxian County Annals in 1890).

located in Macheng and Shawepu, the river channel kept migrating from west to east until it reached its present location. The ancient river channel 2000 yr ago may correspond to the ancient Caofeidian Delta; while that before 14th century before, to the ancient Dongkengjianzi Delta; that before 18th century to the ancient Daqing River Delta; and that before 19th century, to the ancient Laomigou-Hulinkou Delta. This shows that the abandonment of the deltas also proceeded from west to east and one by one.

(vi) *Arrangement of the barrier bars*

The barrier bars from Caofeidian to the Luanhe River mouth are such arranged that the northeast tail end of the frontal bar, or the extensional line of its trend always locates in the inner side of the bar behind it. A most obvious example of this kind of arrangement can be gained from the relationship between Dawanggang in the Daqing River mouth and the Hulinkou-Laomigou bar. The way in which the barrier bars are arranged indicates that the frontal bars always formed earlier than the ones behind them and hence, the bars of different stages lag behind one another in formation time stage by stage from Caofeidian to the Luanhe River mouth, coinciding with the eastward migration of the ancient river channels. This kind of arrangement of the bars is the result of erosion and reworking of the ancient deltas after their successive abandonment.

(vii) *Abandoned mouth bar*

In some lagoons are still preserved ancient mouth bars, which further demonstrates that these bar – lagoon systems are from abandoned deltas as the result of reworking. Shijiutuo, an isolated island in a lagoon, can be used as such an example, which is 3 km long, 0.6 km wide and 2 km away from the coast. The water depth there varies between 1.0 and 1.5 m, the trend of it intersects the coast obliquely and the sediment of it is characteristic of mouth bars. All this shows that Shijiutuo may be a mouth bar formed when the Luanhe River entered the sea through the Daqing River. Presently Shijiutuo is under erosion and so tends to decrease in size.

(d) *Developing model of the delta*

After entering the piedmont, the Luanhe River channel shifts repeatedly due to the frequent flow diversion. When the Luanhe River emptied into the Bohai Sea, a delta formed and when the river changed its route, the detla abandoned and the abandoned delta, reworked by waves with the result that sheet-like sands occurred and nestled against the edge of the deltaic sedimentary system. With a series of abandoned deltas nestling in turn against it, the deltaic sedimentary system prograded towards the sea. When the river changed its route again, it had to form a delta in a further seaward place (Fig. 5-58). This kind of developing process is comparable to the minor and major cycles in the developing process of the Huanghe Delta.

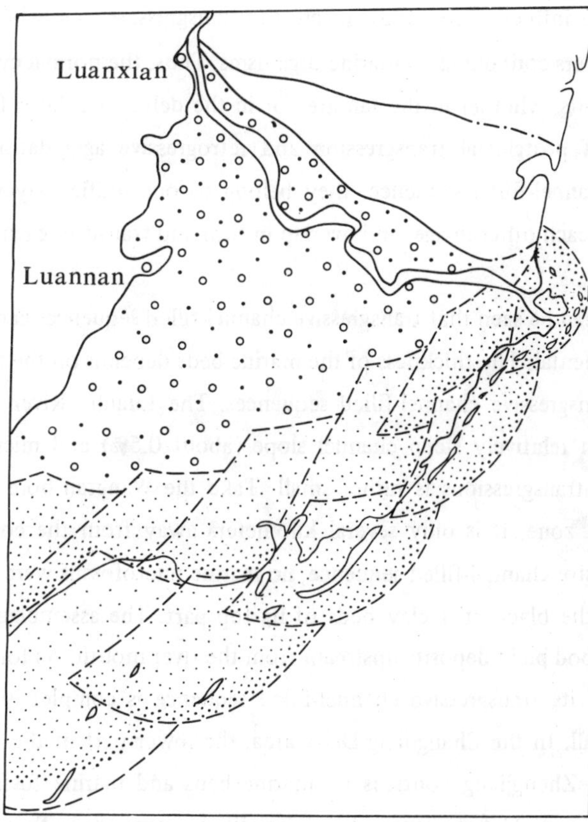

Fig. 5-58. Development model of Luanhe Fan-Delta during postglacial period.

The migration of the Luanhe River channel from west to east in the historical periods mentioned above, during which the abandoned subdeltas nestled successively against the edge of the deltaic sedimentary system, is only the last major cycle or the latest episode.

The sheet-like sands formed by the Luanhe Delta system are capped either by the bar – lagoon system, or by the tidal flat deposits.

(4) ***Genesis relations between the alluvial Fan and the Delta***

The postglacial sea level rising, seawater intrusion and the resulting retrogressive aggradation would inevitably lead to the filling of the channels and hence, the formation of the transgressive channel-filled sequences. The formation process of the transgressive channel-filt sequences is similar to that in the Changjiang and Zhujiang delta areas. In the process of the transgressions, the channels served as the major water-flowing channel alternatively so that sandy and gravelly lenses formed in the channel floors. Now and then there were flows rushing out from the channel fills so that sandy and gravelly lenses intercalated in sand beds were formed.

In the process of the postglacial transgressions, as the retrogressive aggradation distance was much longer than the distance influenced by seawater, marine fossils, as shown by the Wangtan bore (Fig. 5-54), would occur in the transgressive channel-fill sequences formed by retrogressive aggradation in the areas sea-water influence could reach. In those areas such as Machenh that retrogressive aggradation could, but seawater influence could not reach, the transgressive channel-filled sequences would be typical terrigeous layers containing no marine organisms. Thus, the normal cycles in the lower part of the postglacial deposits, whether in the fan area or in the delta area, both formed in the unified depositional process of postglacial transgression and retrogressive aggradation. Representing one unified transgressive channel-filled sequence, they belong to one unified layer in stratigraphy, are roughly comparable to each other in age, and overlap in turn and transit to each other in sedimentary facies.

From the above it can be seen that transgressive channel-filled sequences can contain marine beds or can be totally continental. The thickness of the marine beds depends on the type of the deltas and the location of the transgressive channel-filled sequences. The Luanhe River is a small or at most medium-sized river with relatively steep channel slope (about 0.5%) and micro-tide, and hence its distance influenced by transgressions is rather small. Take the Wangtan bore as an example, being located in the marginal zone, it is only several kilometers away from the coast, yet, in the lower member of its transgressive channel-filled sequence, no marine microfossils were found, and foraminifera presented only in the black stiff clay beds of its top part. The assemblage of the foraminifera resembles that in the flood plain deposits upstream from the river mouth. As for the upstream area of the Luanhe Fan-Delta, its transgressive channel-filled sequence is completely terrigenous with no marine fossils in it at all. In the Changjiang Delta area, the lower part of the transgressive channel-filled sequence around Zhengjiang contains no marine beds and marine fossils occur only in the muddy beds of the top part. The structure of the deposits corresponds to that in the nearshore zone of the Luanhe Delta. However, in the nearshore zone of the Changjiang Delta, the postglacial transgressive channel-filled sequence contains marine microfossils not only in its top muddy beds but in its lower sand beds as well. This may result from the fact that the Changjiang River is with small channel slope, large tidal range and its distance influenced by sea water is relatively long. Thus, based on the extent influenced by sea water in the transgressions and the presence or absence of marine-fossil-bearing beds in the postglacial transgressive channel-filled sequence and their thickness, the different zones of the Luanhe River or the Changjiang River can be arranged into one series: the series from the proximal part of the Luanhe Fan-Delta to the marginal part of the Luanhe Delta, or the proximal part of the Changjiang Delta and then to its marginal part. Following the series, the marine-fossil-bearing beds in the postglacial transgressive channel-filled sequence change from absence to presence and from thin to thick (Fig. 5-59). Drilling has revealed that, in the postglacial transgressive channel-filled sequence of the Luanhe Delta area, the part containing similar marine microfossils as the marginal zone of the Changjiang Delta situates in the offshore neritic area1).

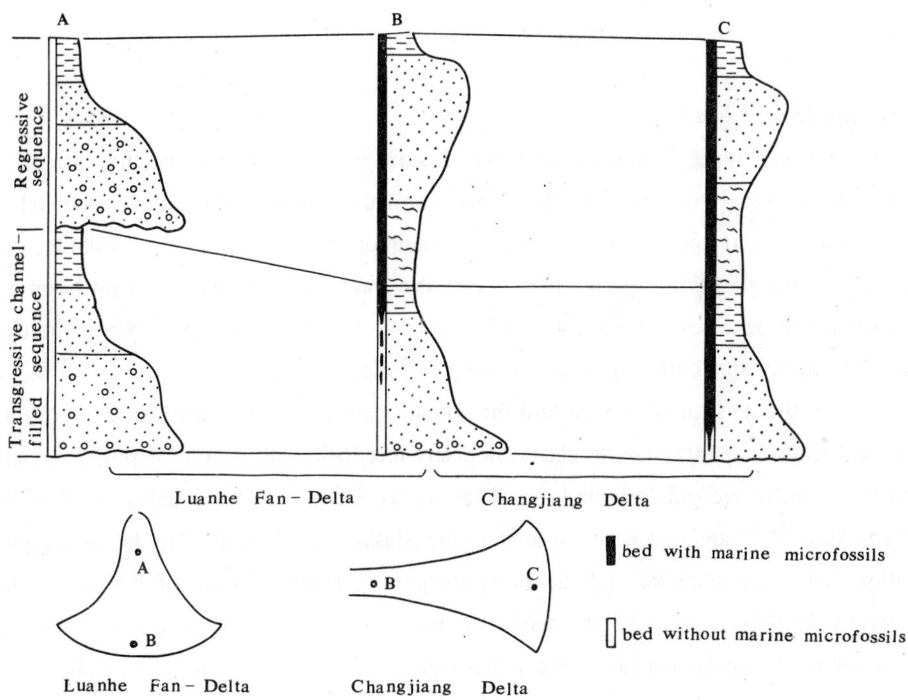

Fig. 5-59. Correlation between the marine-fossil-bearing beds in the transgressive channel-filled sequence in the Luanhe Fan-Delta and Changjiang Delta areas.

Besides, it should be pointed out that transgressive channel-filled sequences can only occur and be preserved in the area where the river debouches from the mountain and enters the plain. In the further inland areas, because of the strong scour-and-siltation, only normal fluvial sequences but no transgressive channel-filled sequences can be preserved. Such is the case for the areas further landward

1) Dagang Oil Field, Bureau of Marine Petroleum Exploration and Tongji University, 1982 Study of the sedimentary system of the Luanhe Fan-Delta.

from the Luanxian County.

In the Luanhe Fan-Delta area, the lower sedimentary cycle of the postglacial deposits, whether having been subjected to seawater influence or not, was formed in the process of transgressions, while the upper one, deposited during regressions. Combining together, the two cycles constitute the complete postglacial sedimentary set, the sediments of which turn upward from çoarse to fine and then to coarse again. The granule gravel layers appear in a two-storyed form. The intervening muddy layer thins landward from the sea and finally tapers out. Where the intervening muddy layer disappears, the upper granule gravel layer will lie directly over the lower one, thus forming a continuous granule gravel layer with much greater thickness (Fig. 5-53). All this is similar to the construction of the postglacial deposits of the Changjiang Delta. However, the muddy layer in the postglacial deposits of the Luanhe Fan-Delta not only thickens seaward, but also grades seaward into marine-continental-transitional facies and marine facies, which is where the postglacial deposits of the Luanhe Fan-Delta differs from that of the Changjiang Delta.

(5) ***Type of the Luanhe Fan-Delta***

Though of many types, fan-deltas have one feature in common, that is, they are all against mountains and near seas. In other words, they are all formed in the area where a river debouches from a mountain or an ancient uplift and flows into the piedmont and then empties into the receiving sea basin. Based on the elevation difference between the mountain and the receiving basin and the piedmont slope, fan-deltas can be divided into three types: (i) the Alaska type. When the elevation difference is relatively small and the slope, relatively gentle, this type appears, in which the alluvial fan is adjacent to the delta in plane view and the contact between the two is jigsaw-like both in plane view and vertically. In strata, coarse clastic deposits interlock with marine deposits, which is an important criterion for recognizing this type of fan-deltas. This type of fan-deltas was first found in the southern Alaska[84] and hence, was termed as the Alaska type[86]. (ii) The Jamaica type. When the elevation difference increases and the slope steepens, the frontal edge of the alluvial fan will extend directly into the receiving basin with the marine factors such as waves and tidal currents directly reworking the coarse clastic deposits. This type of fan-deltas is called the Jamaica type[86], in which, with the progradation of the fan-delta, coarse clastic materials will overlap fine clastic deposits in succession, thus forming the coarsening upward sequence. The sequence is similar to the normal deltaic sequence except that it is composed of second-order normal cycles. As alluvial fan deltas are formed by unconstant rivers, scour planes are veiy common in their coarse clastic deposits. Moreover, the scouring effects have not only given birth to these scour-planes in coarse clastic deposits, but also exerted influences on the underlying fine-grained materials or even the clay layer with the result that the coarse clastic deposits lie uncomformably over the deepwater muddy layer. This is another point in which the sequence differs from the deltaic sequences formed by large rivers. (iii) The Miyang type. When the elevation difference is rather obvious and the slope, very steep, the river goes directly into the deep water area as soom as it debouches from the moutain. This type of alluvial fans is called the Miyang type, which is completely under water and with no subaerial part at all. Judging by the sedimetnary characteristics, the Luanhe Fan Delta belongs to the first type, i.e., the Alaska type. In tectonically-active piedmont zones, the fan-delta in one area can transform from one type into another as the elevation difference between the mountain and the basin and the piedmont slope change with time.

III. Common and Specific Features of Different Types of Deltas

It can be seen from analizing and comparing the major types of deltas in our country that deltaic sedimentary bodies, no matter what types they are of, all possess some general characteristics, which are decided by the depositional and developing regularities abided by then jointly, and based on which, they can be distinguished from the other sedimentary bodies. However, as different types of deltas situate in different tectonic and depositional settings, and as the areas with deltas differ sharply from one another in dynamics and its relative intensity, different types of deltas also possess their own specific characteristics, and these specific characteristics find expression mainly in the associations of deltaic sand bodies. Consequently, the distribution and association of deltaic sand bodies are of key significance in the research of ancient deltas and the exploration of oil and gas.

1. Generality of Deltas

(1) *Distribution of mouth sediments and vertical sequences of deltas*

Outside river mouths, river flows wane gradually and become less competent. In addition, the water depth increases seaward and the wave effects on sea floor weaken correspondingly. All this will result in the differentiated settling of the sedimens along their transportation routes, the grain size of which turns finer outward. Such is the case both for large rivers like Changjiang and Huanghe and for small rivers like Dafengjiang and Fangchenghe. Mud spots occur even outside those small river mouths with strong wave action such as the Yanghe River mouth and some canal mouths along the coast of East Hebei, etc. The sediments in the mouths of the Mississippi River[30], Niger River[32], and Orinoco River[96] etc, all distribute in this way without even one exception. With the delta prograding rapidly and the mouth sedimens overlapping one another in succession, the deltaic sequence will form. The vertical sequences of different types of deltas, though different from one another in their details[11], all compose three facies (from top to bottom: delta plain, delta front and prodelta) and their basic sedimentary constructions are the same. The deltaic sequence composed of the above three facies is the concentrical reflection of the generality of deltas and the major criterion for identifying deltas. Nowadays, deltaic sequences can also be used to determine the extent and boundary of deltas. It should be pointed out, however, that the deltaic sequence in the subaqueous part of a delta is incompletely developed, lacking the delta plain facies. Besides, in those ancient deltas having experienced many times of tectonic variations and destructions, the facies of delta plain is always the part most severely destructed. Therefore, the most basic parts of a deltaic sequence are the delta front facies and the prodelta facies. If only these two sedimenary units are found in its strata, the sedimentary body can be determined as a deltaic sedimentary body.

(2) ***Transgressive and regressive sequences in delta areas***

Deltas are offlapping sedimentary bodies. Before the formation of deltas transgressions always happen. And so transgressions can be considered the prelude to the formation of deltas. Postglacial transgressions are the prelude to the formation of modern deltas, during which, transgressive channel-filled sequences occurred in the ancient channels of delta areas. The transgressive channel-filled sequences formed by large river usually contain marine fossils or marine authigenic minerals, the content, foraminifera diversity and mariness of which all tend to increase in the ascending order. Such is the case for the Changjiang and Zhujiang deltas. Transgressive channel-filled sequences of this type have already been found in the ancient strata and what's more, they are very rich in gas resources[1)]. The transgressive channel-filled sequences in fan-delta areas with small tidal range and steep slope, however, can be entirely continental with no marine indicators at all. But even in this case, they can still be correlated with the downstream transgressive channel-filled sequences containing marine organisms in stratigraphy, and their formation processes are also the same. All this shows that the two types of transgressive channel-filled sequences are correlative with each other in genesis. The Luanhe Fan-Delta can be used as such a typical example. The transgressive sedimentary sequence formed in the interfluve between its ancient channels can be correltated with transgressive channel-filled sequences in stratigraphy though they differ from each other in sediment construction.

The transgressive sequences are overlain by deltaic sequences. Being correlative with each other in genesis, the two form the unified set of postglacial deposits. The development of the transgressive and regressive sequences has given birth to the two-storeyed distrubution of the sand bodies and peat layers in delta areas. There are postglacial deposits composed of transgressive and regressive sequences developing in all the major river delta areas of our country. And their characteristics are shown in Table 5-25.

(3) ***Delta and estuary***

As above mentioned, deltas always take transgressions as a prelude in their formation and development. In the process of the maximum postglacial transgression when postglacial deltas had not yet been built up, the locations of all the major river delta areas of our country were occupied by estuaries. The estuaries of the Changjiang and Luanhe Rivers appeared at that time in a trumpet form while that of the Zhujiang River, in a funnel shape because of the widespread distribution of bedrock islands. In this period, many river delta areas of the world were also occupied by estuaries, such as the delta areas of the Mississippi, Niger and Amazon rivers etc. What should be specially mentioned is the Danube River emptying into the tide-free Black Sea, the delta area of which was also

1) Luo Qihou, 1982 Transgressive channel-filled sequence of Xiangxi group, Sichuan Province & its oil-andgas prospect.

Table 5-25 Characteristics of the transgressive and regressive sequences in detla areas

River	Pre-postglacial Deposit	Transgressive sequence		Regressive sequence		Source of Data
		Ancient river valley	Ancient interfluve	Ancient river valley	Ancient interfluve	
Changjiang River	Terrigenous dark-green mild clay layer	Lower Part: grey fine sand, with about 10 cm of gravel-bearing sand beds in the basal part Upper part: littoral flood-plain sand caly	Lower Part: littoral sand bar-bay lagoonal system Upper Part: nearshore shallow-water muddy facies deposits	Deltaic sequence composed from bottom to top of prodelta-neritic marine facies, delta front facies and delta plain facies	Lower Part: littoral neritic marine muddy deposits Upper Part: littoral plain-littaral bar (radiating bar) deposits	Li Congxian, 1980
Zhujiang River	Red bed-rock weathering-crust and bed rock	Lower Part: channel gravel-bearing sand beds, with 30 cm of gravel beds in the basal part Upper Part: flood plain-littoral marsh greyish-yellow muddy deposits, the top part being black rotten wood silt mud beds		Deltaic sequence composed from bottom to top of prodelta-neritic marine facies, delta front facies and delta plain facies, with rotten-wood silt mud beds developing in the top part		Li Chunchu, 1981 Wu Zhongwen, 1981
Hanjiang River	Granite and its weathering crust	Lower Part: yellow granule gravel beds Upper Part: greyish-green clay-containing shells		Lower Part: greyish-black silt mud with shells Upper Part: grey and greyish-black fine sand and medium sand with shells		Geography Department, Nanjing University, 1976; South China Sea Institute of Oceanology, Academy of Sciences, 1976

(*to be continued*)

Table 5-25 (*continued*)

River	Pre-postglacial Deposit	Transgressive sequence		Regressive sequence		Source of Data
		Ancient river valley	Ancient interfluve	Ancient river valley	Ancient interfluve	
Fangchenghe River	Bed-rock weathering crust		Muddy silt and very fine sand; upward, brackish-water organisms decrease marine organisms increase		Lower Part: prodelta neritic marine muddy silt and caly Upper Part: coarse and medium to fine sand, with the top locally covered by muddy silt	Institute of Oceanology Academy of Sciences, China 1976,
Luanhe River	Yellowish-brown, brownish-red clay, sand clay muddy	Lower Part: mottled granule gravel beds, greyish-white gravel-bearing sand beds and sand beds Upper Part: grey and greyish-black sandy clay or soft silt mud	Lower Part: greyish-white sand beds Upper Part: black sandy silt mud	Lower Part: mottled granule gravel beds gravel-bearing sand beds and sand beds Upper Part: yellowish-brown and greyish-yellow sandy clay	Lower Part: grey clayey silt and sandy clay with organisms Upper Part: yellowish-brown and ashen-yellow sandy clay, greyish-black lagoonal sand clay	Li congxian et al., 1982,

an estuary during the maximum postglacial transgression[94]. Judging by their present tidal ranges, the tidal ranges of the different estuaries during the maximum postglacial transgression vary and close to their present ones. It is due to this that, based on the tidal range (large or small, present or absent), the estuaries can be roughly divided into three types: the estuaries (i) with strong tide, (ii) with weak tide, and (iii) with no tide. The common character of these estuaries is that the sediment discharges of their rivers are all relatively large. It is the rapid accumulation of the sediments that has converted these estuaries into deltas. That some estuaries can survive up till now is only due to the fact that the rivers of these estuaries are small in sediment discharge and hence the sediment amount received by them was too small to convert them into deltas. It is generally believed that the scouring of tidal currents should bear the main responsibility for the formation of estuaries. In fact, however, ever since the postglacial period, nearly all the estuaries have experienced, to different extents, accumulation and filling-up[97]. Take the Qiantang Estuary for example. It is one of the world-famous estuaries with strong tide, But in the last several thousand years, it has been subjected to silting-up and so its area has kept on decreasing[1)]. It can thus be seen that estuaries are not the outcome of strong-tide action. On the contrary, it is the shape of estuaries leading to the deformation of tidal waves and hence the increase of tidal ranges and tidal currents in the estuaries. The real reason that allows the survival of some estuaries is the insufficiency of the sediments carried into the estuaries. If, from now on, the sea level remains roughly constant, the present estuaries will be bound to be filled up and converted into deltas. The deltas of the Changjiang, Huanghe, Zhujiang and Luanhe rivers etc. are all deltas converted from estuaries. And the still surviving estuaries such as those of the Qiantang and Zhujiang rivers will also be converted into deltas sooner or later. This demonstrates that estuaries are geological phenomena with short life span, and that the things that will be preserved in strata are deltaic deposits and deltaic sequences.

3. Individuality of Deltas

(1) *Development of deltas*

In studying the Mississippi Delta, Scruton (1960) pointed out that the development of the delta includes two phases: the constructional and the destructional phases[55], which has be acceptea as nearly the common rule that all deltas have to aoide by in their development. Facts, however, do not always follow this conclusion. Take the major river deltas of our country as an example. In the development of these deltas, sometimes these two phases expressed themselves very clearly such as in the case of the Huanghe and Luanhe deltas and sometimes not very clearly such as in the case of the Changjiang Delta.

The accumulation of large amount of sediments in river mouth areas will inevitably result in the formation and progradation of deltas, and the larger the river sediment discharges, the less competent the marine reworking factors, the faster the deltas will prograde. As is known to all, the Huanghe

1) After Lang Hongru.

River ranks first in the world in sediment discharge. This, combined with another factor that the Bohai Gulf, into which the Huanghe River empties, is a semi-closed bay, has made the Huanghe Delta prograde very rapidly, the maximum prograding velocity reaching as high as 10 km/yr. As for those deltas with small river sediment discharge and competent wave reworking factors, their prograding velocity is only 10–30 m/yr. On the other hand, the flow diversion of rivers is bound to cause the abandonment and destruction of deltas, and for a delta, the larger its river sediment discharge, the finer its sediments, the faster it will be destructed once it is abandoned. The Huanghe Delta can be used as such an example. The desturction velocity of the River can reach 150–200 m/yr and the maximum one, over 600 m/yr. As for those deltas with small river sediment discharge, coarse sediments and low prograding velocity, they will not be destructed very rapidly once abandoned. From the above it can be seen very clearly that the river sediment discharge has much to do with the progradation or retreatment of deltas. The destruction of deltas can occur in response not only to the flow diversion of the rivers, but also to the annual decrease or seasonal variations of the input sediment amounts, which can result in the scouring on the edge of the deltas, the development of wave-cut scraps or the exposing of lagoonal facies on the frontal edge of barrier bars. Such is the case for the Luanhe and Futuanhe deltas.

Though the development of deltas includes the constructional and destructional phases, their general developing trend is the gradual enlargement of their areas. For those rivers with large sediment discharge and frequent flow diversions, the fan shape of their deltaic systems owes its origin to the development of multi-stages of subdeltas. After a deltaic sedimentary system has formed into a complete fan-shaped body, the reflow-diversion of the river will lead to the further extension of the channel so that a new subdelta will be formed further far ahead and a new megacycle, commence. It can thus be seen that deltaic systems usually experience not only megacycles but also microcycles in their development. The river-dominated Huanghe Delta and the Luanhe Fan-Delta are among the deltas possessing of such kind of a developing model.

The development of the Changjiang Delta is characterized by apparent stages. In the postglacial period, the delta experienced 6 major stages with the result that 6 stages of subdeltas were formed. However, because of the strong tidal current effects, the Coriolis Force could always give full play to its influence. And so, in the development of the delta, it was always the south distributary that got enlarged with the mouth bar becoming the core and main body of the next stage of subdelta. The north distributary would, on the other hand, gradually wane and finally get silted up, making the subdelta merge into the north bank. The different stages of subdeltas successively merging into the bank has inevitably resulted in the shrinking of the Changjiang Estuary, the narrowing of the orifice and the enlarging of the delta. And that is why no distinct constructional and destructional phases can be distinguished in the development of the Changjiang Delta.

(2) ***Deltaic sand body***

The factors that can produce sand bodies in delta areas are mouth flows, waves and tidal currents and the outcomes of them are respectively mouth sand bodies, litteral sand bodies and tidedominated sand bodies. Besides, channel sand bodies and tide-dominated sand bodies, can also occur in response to the progradation of deltas and the extension and shifting of main river channels. Deltaic channel sand bodies are different from normal channel sand bodies and the differences between them lie in that the former usually contains marine fossils and the salinity of its sediment tends to increase. Deltaic channel sand bodies also differ from transgressive channel-filled sand bodies in that the marine organisms in them decrease upward in content, and that they are overlain by terrigenous layers. Thus, there are altogether 4 types of sand bodies, in delta areas: mouth sand bodies, littoral sand bodies, tide-dominated sand bodies and channel sand bodies. A delta can at the same time has all the 4 types of sand bodies but with one dominating. And the dominating one will decided the shape and form of the delta. Some deltas can lack one or two types of sand bodies. And the sand bodies are absent either because they have never been built up or becasue they have been reworked and converted into other types. Being large in area and complicated in environment, the Changjiang Delta possess all the 4 types of sand bodies, among which, the tidedominated one is important in body-shape and wide in distribution. Here, even those typical mouth sand bodies are elongated due to the tidal current effects. All this shows that the Changjiang Estuary area is characterized by large tidal range and strong tidal curretn. Being narrow and with thin sand beds, the littoral sand bodies of the Changjiang Delta usually take the elongated form and intermittently distribute in muddy layers. This indicates that, though the conditions favourable to the development of littoral sand bodies do exist here, the wave action is relatively weak. The sand-body association of the Changjiang Delta suggests that the Changjiang Delta is a typical tide-dominated delta, which is coincident with the actual tidal situation of the present delta area. As for the Huanghe and Zhujiang deltas, their mouth bars are inlaid within muddy deposits and distributed in a digital shape. When the two deltas are in the process of progradation, their littoral sand bodies could hardly be seen and their tidedominated sand bodies are not very distinct either. The sand-body associations of the two deltas are characteristic of river-dominated deltas, which is consistent with the hydrology of the two deltas. Wave-dominated deltas are usually typified by littoral sand bodies and their mouth bars tend to be very quickly reworked and converted into littoral bars, or even though they can be preserved, their shape and deposits will bear some features of littoral sand bodies. As for the shape of deltaic littoral sand bodies, some are elongated and parallel to the coasts such as the littoral sand bodies of the Hanjiang Delta; some others are sheet-like such as those of the Dafengjiang and Fangchenghe deltas. There are also some deltas whose littoral sand bodies seem to be elongated in shape because of the alternating distribution of bars and lagoons, but whose sand bodies, when looked as a whole, are still sheet-like because all the lagoons are located in the low and depressed areas between the littoral sand bodies and also because the underlying layer of the lagoonal deposits is still beach sands. Such is the case for the deltas of the Futuanhe and Xiaojiangzihe rivers, Shandong Province.

In recent years, based on the data of dozens of deltas in the world, Coleman et al. (1975) have raised 6 sand-body distribution models[11]. These models, though reflect the efforts made in recent years for studying deltaic sand bodies and yielding correlation models for the exploration of sedimentary deposits, can not be considered as an exclusive sand-body association model. Take the

Changjiang Delta as an example, in which, the 4 types of sand bodies coexist and the main one, the mouth bars are arranged in echelon. This type of deltas has not been included in the Coleman's models. Besides, the sand-body distribution of some of the wave-dominated deltas in our country is also more or less different from the San Francisco and the Senegal deltas. Take the Futuanhe and Xiaojianzihe and the Senegal deltas as examples. It is true that their sand bodies all occur in response to the longshore movement of sedimens. However, in the south sides of the Futuanhe and Xiaojiangzihe river mouths, not only bars distribute alternatively with lagoons, but also the lagoons will disappear southward far away from the mouth so that the sand bodies will merge into a whole one. In the mouth of the Senegal River, on the other hand, as the sand spits will not merge into a whole one even after they have extended southward over 480 km, the littoral sand bodies always distribute alternatively with the fluvial deposits[11].

(3) *Deltas and their adjacent river mouth deposits*

Coastal sediments come mainly from coastal erosion and river transportation. Statistics shows that 99% of the coastal sediments are fluvial sediments while the rest 1%, materials derived from coasts[98]. Therefore, river-transported sediments are the main source of coastal sediments. The accumulation of large amounts of sediments in mouth areas will give birth to the formation of deltas, the seaward convexing of coastal lines and the thickening of deposits. Even along the coast sections under erosion, the inputting of rivers and the adding of sediments will also ease up or even balance the erosional effects. As for the deltas formed by sediment-rich large rivers such as the Huanghe, Changjiang and Zhujiang, they are of profound and many-sided influences on the adjacent coastal sections.

The sediments of the large river deltas of our country all have the tendancy to move directionally along coasts. Take the Changjiang Delta as an example. Its fine-grained sediments tend to move southward so that relatively thick silt-mud deposits have been formed not only subaqueously but also in the coastal plain of Zhejiang and Fujian Provinces. According to statistics, the total sediment discharge of the rivers entering the sea from Zhejiang and Fujiang Provinces is 20×10^6 tons, among which clay and fine silt make up about 30%, that is, about 6×10^6 or 7×10^6 tons[99]. With such an amount of fine-grained materials alone distributed along the coast of the two provinces, it is impossible to form very thick sedimentary layers. However, since the Holocene, thick silt-mud deposits have been formed not only in the nearshore zones along the coasts of the tow provinces[100], but also in their coastal plains. For example, in the Xiaoshao, Wenhuang and Wenrui plains, the muddy deposits can reach as thick as 40–50 m and the thickest one, 80 m[101]. From this it can be seen that the actual amount of silt and mud here is far more than that the rivers entering the sea from the two provinces can supply. Thus, it can be concluded that most of the fine-grained materials have arrived here through the longshore transportation of the Changjiang River. According to reports, of the fine-grained sediments distributed in the Leqing Bay of Zhejiang Province, 90% are from the Changjiang River, and only 10%, from the local muddy materials[102]. Therefore it is the continuous supply of fine-grained materials from the Changjiang River that has resulted in the filling up of the gulf between bedrock points by muddy deposits. In the postglacial period, as transgressions and regressions took place alternatively in the Changjiang Delta area, abrupt facies variations and all the other unusual phenomena also occurred accordingly. For example, in the Leqing Bay and Zhoushan Islands area, tidal silt-muds were found directly covering wave-cut platforms and beach gravels. As is well-known, wave-cut platforms and beach gravels are the outcome of high-energy coasts while silt-mud layers, of low-energy environments. That the two directly contact with each other is a reflection of the postglacial developing process of the Changjiang River. During the maximum post-glacial transgression, the Changjiang River mouth retreated to the place near Zhengjiang and Yang-zhou. At this point, the sediments of the Changjiang River were mainly used to fill its own estuary and there were no spare materials that could be transported into the open sea. Later, as the prograda-tion of the Changjiang Delta resulted in the filling up of the estuary, large amounts of materials began to enter the open sea and, under the influences of nearshore currents, moved southward. The con-tinuous supply of fine-grained materials, in turn, led to the accumulation of large amounts of muddy materials along the coast of Zhejiang and Fujian Provinces, and the weakening of the erosion effects. Finally, the wave-cut platforms would evlove from its destructional phase into its construc-tional phase with the result that silt-muds began to lie directly over the beach gravels. From this it can be seen that the facies variations of the deposits along the coast of Zhejiang and Fujian Provinces are essentially a record of the postglacial transgressions and regressions in the Changjiang Delta area[1)]

(4) ***Fan-delta***

The tectonically-active piedmont zones are the environments most favourable for the develop-ment of fan-deltas. Fan-deltas are a kind of "against-mountain-and-near-water" sedimentary system formed by near-source rivers, inside which coarse clastic deposits interlock with marine deposits (the continental-and-marine transitional facies, included– both horizontally and verticlaly, and the sand bodies are comparatively low in maturity. Fan-deltas resemble the deltas of large rivers such as Changjiang and Huanghe rivers in the process of formation and development of their postglacial deposits, but differ from them in their sedimentary constructions and so they alone can be grouped into one type. Fan-deltas containing rich oil and gas resources[2)] widely develop in many of the sedimentary basins of our country. That is why the study of the modern fan-delta systems must be strengthened for the purpose of yielding correlation models identify ancient analogues and prospect-

References

[1] Li Qinyuan, 1935, On the emergence-submergence of coastline in China, Acta Geographica Sinica, V.2, No.2, pp 129–166. (in Chinese)

---

1) Qian Liping, 1982 Coastal deposits in the Zhoushan Area & transgression and regression of the Changjiang Delta.

2) Zheng Changming, 1981 Geological Character of the oil and gas resources in the rift-basin of the Liaohe River.

[2] Simonov, A. E., 1960, Regularities of the change in speed and dischange of the flow in the delta front and processes companing with them in the prodelta of the River Cuban, USSR. GOIN Publications, V. 49, pp 19–37. (in Russian)

[3] Bates, C. C.,1953, Rational theory of delta fromation, AAPC. V. 39, pp 2119–2162.

[4] Pang Jiazhen et al., 1980, Evalution of the Yellow River Estuary, Proceedings of the International Symposium on River Sedimentation, Beijing, China, Guanghua Press.

[5] Inman D. L. et al., 1971, On the tectonic and morphologic classification of coasts, Jour. Geol., V. 79, No. 1, pp 1–21.

[6] Editing Committee of Phisical Geography of China, 1981, Phisical Geography of China, Science Press. (in Chinese)

[7] Moore G. T., 1969, Interaction of rivers and oceans: Paleocene petroleum potential AAPG, V. 53, No. 12.

[8] Davies I. L., 1964, A morphogenetic approach to world shoreline, Zeits fur Geomorph., V. 8, pp 127–142.

[9] Sun Xiangping et al., 1981, Outline of the hydrology and meteorology in the China coast, Science Press. (in Chinese)

[10] Wang Pinxian et al., 1975, Discovery of Cenozoic brackish-water foraminiferal fauna in some basins of East China and its significance, Professional Papers of Stratigraphy and Paleontology, V. 2 pp 1–35, Beijing Geological Publishing House. (in Chinese)

[11] Coleman J. M. and Wright L. D., 1975, Modern river deltas: variability of processes and sand bodies, In "Deltas: the model for exploitation".ed. Broussard M'L., pp 99–149.

[12] Komar P. D., 1976, Beach processes and sedimentation, Prentice-Hall Inc., England Cliffs, New Jersey.

[13] Coleman J. M. et al., 1972, Sedimentation in Malaysian high tide tropical delta, 1972, National Conference on Earth Science, pp 61–87.

[14] Lee J. S., 1959, Development of the geological works in China since last 10 years, Kexue Tongbao, No. 18. (in Chinese)

[15] Wang Pingxian et al., 1980, A preliminary note on the Late Quaternary microfauna of the Gulf of Tonkin and its bearing on sea-level changes, Papers on Marine Micropaleontology. pp 14–145, China Ocean Press, Beijing. (in Chinese)

[16] Fisher W. L. et al., 1969, Delta systems on the exploration for oil and gas, Bur. Econ. Geol., Austin, Texas. s

[17] Galloway W. E., 1975, Process framework for describing the morphologic and stratigraphic evolution of deltaic depositional systems. In "Deltas: the models for exploitation" ed. Brousserd M. L., pp 87–98.

[18] Wissmann H., 1941, On geographical problems of lower reach of River Yangtse, Geography, V.1, No. 1–4. (in Chinese)

[19] School D. W., 1963, Sedimentation in modern coastal swamps, Southwestern Florida, AAPG V. 47, No. 8, 1581–1603.

[20] Wang Pingxian et al., 1980, Characteristics of the foraminifieral and ostracoda thanatocoenoses from some river mouths of China and their geological significance, Papers on Marine Micropaleontology, pp 101–111, China Ocean Press, Beijing. (in Chinese)

[21] Wang Kaifa et al., 1982, The sporo-pollen and algae assemblages from the surface layer sediments of the Changjiang Delta, Acta Geographica Sinica, V. 37, No. 3, pp 261–271. (in Chinese)

[22] Le Blanc R. J., 1975, Studies of modern and ancient deltaic sediments, in "Deltas, models for exploration", et. Broussard M. L., pp 13–85.

[23] Eppo Oomkens, 1974, Lithofacies relations in the late Quaternary Niger Delta complex, Sedimentology, V. 21, No. 2.

[24] Li Congxian et al., 1982, Sediments and sand bodies on the tidal flats, Oceanologia et Limnologia Sinica, V. 13, N. 1, pp 48–59. (in Chinese)

[25] Dai Zeheng et al., 1980, Evalution and hamess of riverbed of Qiantangjiang Estuary, Proceedings of the International Symposium on River Sedimentation, Beijing, China, Guanghua Press.

[26] Chen Jiyu et al., 1964, Formation and evolution of the big sand bar inside the Qiantangjiang Estuary. Acta Geographica Sinica, V. 30, No. 2, pp 109–123. (in Chinese)

[27] Qian Ning et al., 1964, The fluvial processes of the big sand bar inside the Qiantangjiang Estuary, *ibid,* V. 30, No. 2, pp 124–142.

[28] Dai Zeheng et al., 1958, Erosion and acummulation in the channel of Qiantangjiang Estuary and the transportation of the suspended load, Fluvial Load Research, V. 4, No. 3. (in Chinese)

[29] Wang Pinxian et al., 1980, Micropaleontologia characteristics of relict sediments of the East China Sea, Acta Oceanologica Sinica, V. 2, No. 1, pp 67–78. (in Chinese)

[30] Gould H. R., 1970, The Mississippi Delta complex, in "Deltaic sedimentation: modern and ancient", pp 1–30.

[31] Wang Kaifa et al., 1981, Spore-pollen and algal assemblages from the peat beds in continental shelf of the East China Sea and the sea level changes, Marine Geological Research, V.1, No.1 pp 57–67. (in Chinese)

[32] Liu Jingling et al., 1977, Studies on the Quaternary sporo-pollen assemblages from Shanghai and Zhejiang with palaeoclimatic significance, Acta Palaeontologica Sinica, V. 16, No. 1, pp 1–11. (in Chinese)

[33] Li Congxian and Ming Qiubao, 1981, The time of Holocene transgression and sea level changes in apical area of Changjiang Delta, Journal of Tongji University, No. 3, pp 104–108. (in Chinese)

[34] Li Congxian et al., 1982, Postglacial marine beds in the coastal and deltaic areas in East China, In "Quaternary Geology and Environment of China", ed. Liu Tung-sheng, China Ocean Press.

[35] Sun Shuncai, 1981, Deltaic sediments of the Changjiang Delta and its characteristics, Acta Oceanologica Sinica, V.3, No. 1, pp 97–113. (in Chinese)

[36] Wang Jingtai et al., 1980, Relationship between sealevel changes and climatic fluctuations in East China since Late Pleistocene, Acta Geologica Sinica, V. 35, No. 4, pp 299–312. (in Chinese)

[37] Fairbridge R. W., Eustatic changes in sea-level, Physics and Chemistry of Earth, V. 4 pp 85–99.

[38] McFarlan E. Jr., 1961, Radiocarbon dating of late Quaternary deposits, South Louisiana Bull., Am. Geol. Soc., V. 72, No. 1.

[39] Shepard F. P., 1963, Submarine geology, 2–ed, Harper and Row Publishers, New York, Evanston, San Francisco, London.

[40] Yi Huanzhang et al., 1962, Some cognitions on Neolithic curtures in Taihu Lake area, Jiangsu Province, Kaogu, No. 3. (in Chinese)

[41] Kraft J. C. et al., 1975, Transgressive and regressive sedimentary lithosome at the edge of a Late Holocene marine transgression, 9me Congres International De Sedimentologie Theme 6.

[42] Li Congxian, 1981, Rates of delta sedimentation and its geological significance, Marine Sciences, No. 3. (in Chinese)

[43] Tan Qixiang, 1973, Development and evolution of land and sea in Sanghai area, Kaogu No. 1. (in Chinese)

[44] Huang Xuanpei et al., 1976, Ages of the land formation in Shanghai and development of Shanghai Port during the history, Wenwu (Cultural Relics), No. 11. (in Chinese)

[45] Xia Nai, 1977, Archaeology before the History of China and ^{14}C dating, Kaogu, No. 4. (in Chinese)

[46] Tai Qixiang, 1972, On changes of the land and sea in Shanghai during the history again, Weihuibao, Aug. 29. (in Chinese)

[47] Kolf C. R. et al., 1975, The Mississippi and Mekong deltas a comparison, in "Deltas: models for explotations" ed. Broussard M. L.

[48] Allen, J. R. L. 1965, Late Quaternary Niger Delta and adjacent areas: sedimentary environments and lithofacies, AAPG, V. 49, No. 5, pp 547–600.

[49] Guiyang Geochemistry Institute, Geochemistry, V. 4, No. 3. (in Chinese)

[50] Kolf, C. R. et al., 1966, Depositional environment of the Mississippi River deltaic plain Southeastern Louisiana, in "Deltas and their geologic framework", Houston Geol. Soc., pp 17–61.

[51] Wang Jingqin et al., Acta Geologica Sinica, V. 55, No. 1. (in Chinese)

[52] Ting, V. K., 1919, Report on the geology of the Yangtse Valley below Wuhu, Wangpoo Conservation Board.

[53] Heidenstan, H. V. 1922, The growth of the Yangtze Delta, Jour, North-China Branch of the Royal Asiatic Society, V. LIII, pp 21–36.

[54] Gressy, G. B., 1928, The Geology of Shanghai, The China Jour. Sci and Arts, V. 8–9,

[55] Scruton P. C., 1960, Delta building and deltaic sequence, in "Recent sed. NW. Gulf of Mexico", pp 82–102.

[56] Eppo Oomkens, 1970, Depositional sequences and sand distribution in the postglacial Rhone Delta complex, in "Deltaic sedimentation: modern and ancient", ed. Morgan J. M., pp 198–212.

[57] Nelson, B. W., 1970, Hydrography sediment dispersal and recent historical development of the Po River Delta, Italy, In "Deltaic sedimentation: modern and ancient", ed. Morgan J. M. Soc, Econ. Ploont lineral, Spec, Pub., No. 15.

[58] Meckel L. D., 1975, Holocene sand bodies on the Colorado Delta area, Northern Gulf of California, in "Deltas: models for explotation", ed. Broussard M. L., pp 239–265.

[59] Li Chunchu, 1981, The depositional characteristics and some problems of the process of development of the Zhujiang Delta, Oceanologia et Limnologia Sinica (supplement), Science Press. pp 115–122. (in Chinese)

[60] Wikon son, B. H. et al., 1977, Lavaca Bay-Transgressive deltaic sedimentation, in Central Texas Estuary, AAPG, V. 61, No. 4, pp 527–545.

[61] Wikison, B. H. et al., 1978, Late Holocene history of the Central Texas coast from Gelaveston Island to poss Gavallo, Geol, Soc. Am. Bull., V. 89, No. 10, pp 1592–1600.

[62] Ren Mei-e et al., 1980, Late Quaternary continental shelf of East China, Acta Oceanologica Sinica, V. 2, No. 2, pp 94–111. (in Chinese)

[63] Yang Guangfu, 1985, On sea level changes and transgressive sediments in the continental shelf of East China Sea since Late Pleistocene, Oceanologia et Limnologia Sinica, (in Chinese)

[64] Pang Jiazhen et al., 1980, Fluvial process of the Huanghe Estuary: I. hydrographical character and the region of the sediment silting, Oceanologia Sinica, V. 11, No. 4, pp 295–305. (in Chinese)

[65] Lisitzin, A. P., 1972, Sedimentation in the world ocean, Soc. Econ. Paleot. Mineral Spec. Pub., No. 17, 218.

[66] Qin Yunshan et al., 1982, Study on the suspended matter of the seawater of the Bohai Gulf, Acta Ocean-

ologica Sinica, V. 4, No. 2, pp 191–200. (in Chinese)

[67] Chen Lirong et al., 1980, Mineral assemblages and distribution patterns in the sediments of the Bohai Gulf, *ibid,* V. 11, No. 1 pp 46–64.

[68] Gao Mingdao et al., 1980, Study on the source of mud-sand in the port of Tangu, Studia Marine Sincia, No. 16, pp 33–57. (in Chinese)

[69] Wu Shiying and Fang Zecheng, 1982, The characterisitics of caliches in the Haizhou Bay, Huanghai Sea and its geological implications, Scientia Geologica Sinica, No. 2, pp 207–214. (in Chinese)

[70] Wang Shaohong, 1979, Apreliminary study of marine bed and its depositional conditions in Late Quaternary off the west coast of Laizhou Bay, Oceanologia et Limnologia Sinica, V. 10, No. 1, pp 9–23. (in Chinese)

[71] Zhou Zhide, 1980, The development of the coastline of the Huanghe Delta and its effects on the upstream channels, *ibid,* V. 11, No. 3, pp 211–219.

[72] Cen Zhongmian,1957, Evolution history of the Huanghe River, People Press. (in Chinese)

[73] Pang Jiazhen et al., 1979, Evolution of the Huanghe Estuary: evlaution in the history, Oceanologia et Limnologia Sinica, V.10, No. 2. (in Chinese)

[74] Frazier D. E., 1967, Recent deltaic deposits of the Mississippi River: their development and chronology, Gulf Coast Assoc. Geol. Soc. Trans. V. 17, pp 287–315.

[75] Yuan Yingru et al. 1981, Mineral assembly characteristics of the sediments and its distribution pattern in the NW part of South Huanghai Sea, Oceanologia et Limnologia Sinica, V. 12, No. 6, pp 512–521. (in Chinese)

[76] Xu Junming, 1973, Zhujiang Delta, Guangdong People Press. (in Chinese)

[77] Ren Mei-e, 1964, Features of dynamic geomorphology and exploration of tidal flat, Jour. of Nanjiang University, V. 8, No.1. (in Chinese)

[78] Huang Zhenguo et al., 1982, Formation and development of the Zhujiang Delta, Science Popularization Press. (in Chinese)

[79] Wu Wenshong, 1981, The characteristics of sedimentation in the Zhujiang Delta and the origin of manure water, Nanhai Studia Marina Sinica, V. 2, pp 49–59, Science Press. (in Chinese)

[80] Zhao Huanting et al., 1980, Physiographic development of Shizi Yang (Lion Channel) at the Zhujiang River mouth *ibid,* V. 1, pp 17–34.

[81] Fish H. N., 1961, Bar-finger sands of Mississippi Delta, in "Geometry of sandstone bodies) a symposium", ed. Peterson J. A. and Osmond J. C., Tulsa, Okla, Am. Assoc, Potrol. Geol., pp 240.

[82] Sedimentary Group of Marine Geology Department, South China Sea Institute of Oceanology, Academia Sinica, 1980, Characteristics of the surface sediments on the North Shelf of the South China Sea, Nanhai Studia Marina Sinica V. 1, pp. 35–50.

[83] Makkabeev N. I., Experimental geomorphology, Moscow State University Press. (in Russian)

[84] Galloway W. E., 1976, Sediments and stratigraphic framework of the Copper River fan-delta, Alaska, Jour. Sed. Petrol., V. 46, No. 3.

[85] Shen, A., 1979, Late Pleistocene fan-delta along the Dead Sea rift, J. Sed. Petrol., V. 49, pp 541–552.

[86] Wescott W. A. et al., 1980, Fan-delta sedimentology and tectonic setting-Yallahs Fan-Delta, Southeast Jamaica AAPG, V. 64, No. 3, pp 374–399.

[87] Kotton, S. P. 1982, Pennsylvanian Fan-Delta and carbonate deposition Mobeete Field, Texas Pannandle AAPG, V. 64. No. 4.

[89] Guo Shunmin et al., 1977, Discussion on the regional structural background and the seismogenic model of the Tangshan Earthquake, Scientitia Geologica Sinica, No. 4, pp 305–320. (in Chinese)

[90] Pei Wenshong, 1959, Exavation of Quaternary mammalian faunas in Qianan County, Province Hebei, Papers on archaeology in Province Hebei, pp 1–19, Hebei People Press. (in Chinese)

[91] Schumm S. A., 1977, The fluvial systems, John Wiley and Sons, New York, London, Sydney, Toronto.

[92] Reineck, H. E. and Singh I. B., 1978,Depositional sedimentary environments, 2-ed., New York-Heidelberg-Berlin, Springer-Verlag.

[93] Zhao Xitao et al., 1979, Sea level change of the Eastern China during the past 20000 years, Acta Oceanologica Sinica, V. 1 No. 2, pp 269–281. (in Chinese)

[94] Zenkovich V. P., 1962, Processes of coastal development, Academia Press, USSR. (in Russian)

[95] Li Congxian and Chen Gang, 1982, Sedimentation and development of the barrier-lagoon systems along the coastal zone of East China, Marine Geological Research, V. 2, No. 1, pp 47–57. (in Chinese)

[96] Andel Van, 1967, The Orinoco Delta, Jour, Sed. Pet., V. 37, No. 2.

[97] Stanley D. J. et al., 1976, Marine sediment transport and environment management.

[99] Cai Aizhi, 1982, Diffusion of sediments of the Changjiang River Dischanging into the sea, Acta Oceanologica Sinica, V.4, No.1, pp 78–88. (in Chinese)

[100] Qin Yunshan, 1963, Preliminary study of the topography and sediment types on continental shelvies of China, Oceanologia et Limnologia Sinica, V. 5, No. 1, pp 71–86. (in Chinese)

[101] Feng Zian, 1980, Some features of tidal flat profires, Province Zhejiang, Marine Sciences, No. 2. (in Chinese)

[102] Qian Liping et al., 1983, Sedimentation in the Zhoushan coastal zone and transgression-regression of the Changjiang Delta, Marine Science Bulletin, V. 2, No. 3, pp 37–44. (in Chinese)

by Li Congxian
Tongji University, Shanghai

Chapter Six
Submarine Sand Ridges

I Introduction

The submarine sand ridges are a sort of particular and common erosion-accumulation geomorphological body at the inner shelf of shallow sea. They consist of a series of linear sand bodies parallel to the direction of flow and the troughs between them. Their plane shapes appear parallel or fingered. They are formed by the strong currents. According to their cause of formation, the submarine sand ridges can be classified into two types–the storm current ridges and the tidal current ridges, which are all the products of modern marine hydrodynamical conditions.

This chapter will make a preliminary summary of sedimentary characteristics, growth conditions, distribution patterns and formation mechanism of China's submarine sand ridges on the basis of data obtained in the surveys of the neighbouring seas of China in recent years.

II. General Characteristics of Tidal Current Ridges

The tidal current ridges are composed of a series of linear sand bodies running parallel to the direction of tidal currents, and the trenches and troughs among them. By comparing the morphology of horizontal distribution of most of the tidal current ridges , they may be classified into the parallel and the fingered types. Most of the tidal sand ridges are composed of sandy sediments. According to the nomenclature of river sand and wave sand, they may be called the tidal current sands. In studying the geomorphology of the tidal current ridges, the authors preliminarily have studied their characteristics based on the grain size, minerals, chemical and spore-pollen analyses of the tidal current sand in offshore area of Jiangsu and those in the Bay of Korea (samples were collected only from the west of 124°34′E). A comparison with the shallow beach sediments of Liaodong shows that the tidal current sand is neither identical to the wave sand nor to the river sand, but has sedimentary feature of the tidal current that is a unique agent to form them.

1. Morphological features

Due to the differences of tidal current velocities and their directions, substance supply and geomorphic position result in great difference in the scale of current ridge sand bodies. Generally, the distance between ridges is 1–10 kilometers, a single sand body is 2–100 kilometers long, and the height difference between the furrow and the ridges is 7–30 meters. The width of sand bodies commonly is proportional to their height; the higher the sand body is, the wider the width will be.

The plane configuration of tidal current sand ridges is controlled by their geomorphic position and hydrodynamic condition. Off (1963) considered strict parallel arrangement of the sand bodies as an important characteristics of the tidal current ridges.

The most typical tidal current ridges in the world, such as the tidal current ridges of the western Bay of Korea, the northern Huanghai Sea and those of the southern part of the North Sea, all belong to parallel tidal current ridges. In fact, most of stormy current ridge sand bodies are approximately parallel, such as the linear sand bodies on the continental shelf of Maryland off the eastern coast of U. S. A. The tidal current ridges have also fingered patterns besides parallel ones. The flows spread at outlets of rivers or channels make the plane patterns of the sand bodies fingered such as the tidal current ridges at the east and west outlets of the Qiongzhou Straits and those of Jiangsu. When tidal current velocity is not large enough and its directon is not strictly in reciprocating movement, the sand bodies often develop badly, which show crescent or "S" pattern, have poor stability and vary easily, for instance, the tidal current ridges on the Taiwan Shoal at the south mouth of the Taiwan Strait of China. In the area affected strongly by wave and coastal currents or the area having abundant external substances, the configurations of most sand ridges are irregular.

The profile configurations of the tidal current ridges are not fully alike. Some are humped, such as the tidal current ridges of the western Bay of Korea and those of the Liaodong Shoal; some are sineshaped, such as the tidal current ridges at the mouth of the Qiongzhou Straits (Fig. 6-1).

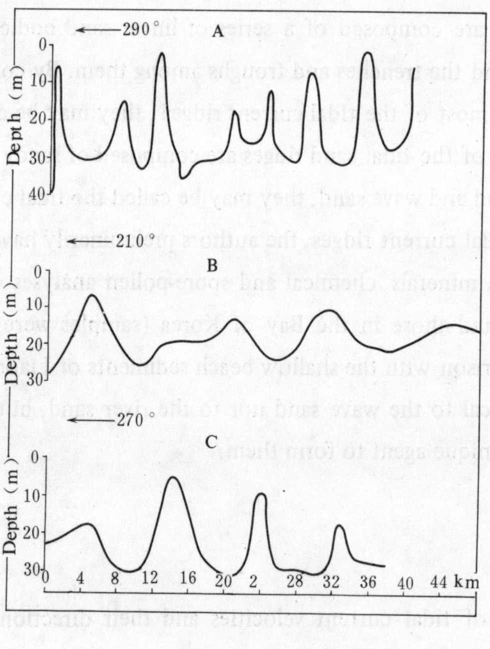

Fig. 6 1. Profiles of the tidal current ridges.
A–Korea Gulf; B–The west mouth of Qiongzhou Straits; C–Liaodong Shoal.

2. Grain size characteristics

Tidal current sand is dominated by sandy sediments, and its size range is 2–4ϕ. Because the ebb and flood tides repeatdly transport and wash sediments, the coarse-grained substance is left and fine-grained substance is carried away, so the size of sediments is homogeneous and wellsorted.

The probability accumulation curve of tidal current sand mainly consists of jumping component which accounts for over 80 percent and even 95 percent; suspended component ranks the second whose content is less than 20 percent; rolling component is a little lacking. With regard to sorting, the jumping component is better than the suspended and rolling components and consists of two sections of straight lines of slightly different sorting. The sorting of its first jumping component is superior to that of the second jumping component, and belongs to normal partial distribution (Fig. 6-2). This curve reflects that the sediments are controlled by two currents with different directions and velocities. Tidal current sand outwardly looks like beach sand, but inwardly the former is different

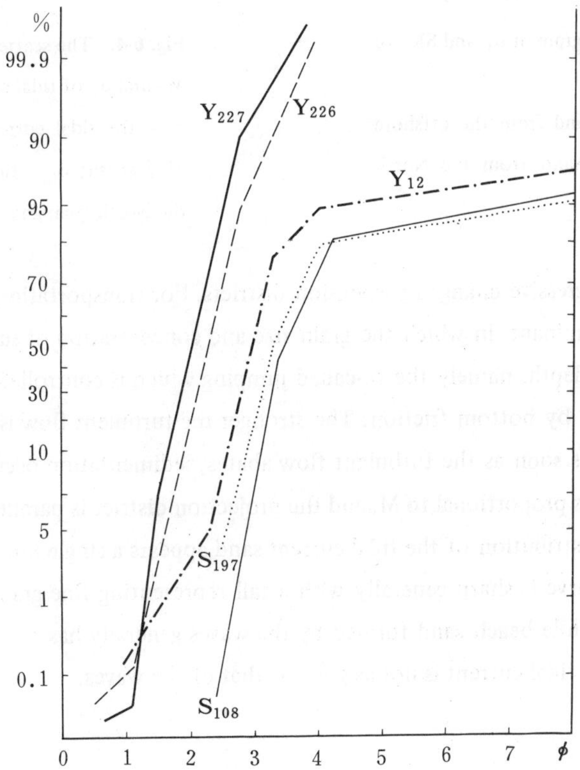

Fig. 6-2. The probability accumulation curve of tidal current sand.

Y_{227} 124°E 38°20′N In the North Huanghai Sea;

Y_{226} 123°50′E 38°20′N In the North Huanghai Sea;

Y_{12} 123°27′E 39°20′N In the North Huanghai Sea;

S_{108} 122°07′30″E 33°40′N In the offshore of Jiangsu;

S_{197} 121°08′E 33°00′15″N in the offshore of Jiangsu.

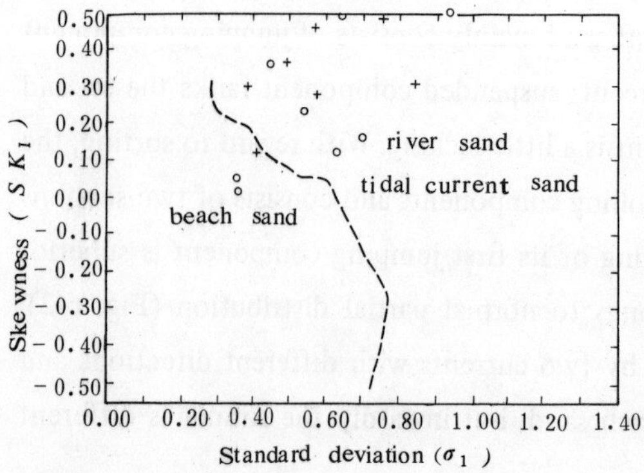

Fig. 6-3. The scatter diagram of σ_1 and SK_1 of the tidal current sand.
+ – the tidal current sand from the offshore of Jiangsu; o – the sand from the North Huanghai Sea.

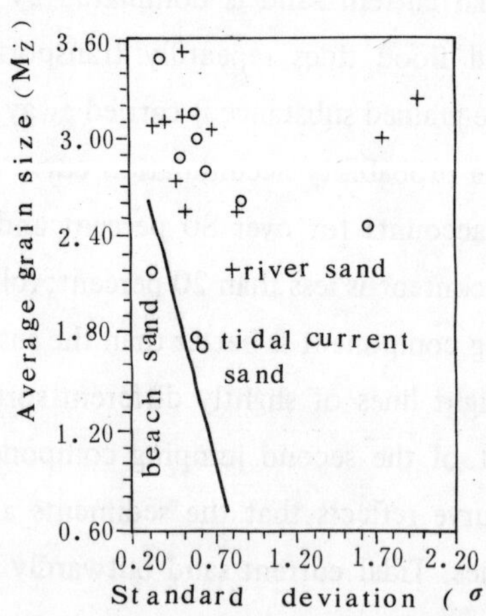

Fig. 6-4. The scatter diagram of the parameters M_Z and σ_1 of tidal current sand.
+ – the tidal current sand from the offshore of Jiangsu; o – the tidal current sand from the North Huanghai Sea.

(VI) close to the progressive changed suspension districts. For transportation way, suspension changed progressively is dominant, in which the grain size and concentration of suspended matter increase regularly with water depth, namely the so-called jumping which is controlled and lifted to sorting by turbulent flow caused by bottom friction. The stronger the turbulent flow is, the coarser the suspended particle will be; As soon as the turbulent flow abates, sedimentation occurs. The most important properties are that C is proportional to M, and the projection district is parallel to C=M line (Fig. 6-5).

The frequency distribution of the tidal current sand appears a single humped pattern, its content is concentrated, its curve is sharp generally with a tail representing fine grain, and its mode is within 2 to 4 ϕ (Fig. 6-6), while beach sand formed by the waves generally has not this tail. Thus, it is clear that the sorting of the tidal current is not as good as that of the waves.

3. Mineral characteristics

Repeated washing and sorting made by the ebb and flood tidal currents results in relative enrichment of heavy minerals and stable minerals, and the former one accounts for over 5 percent of the total amount of light minerals and heavy minerals in tidal current sand. The unbreakable granular minerals with heavy specific gravity, high hardness and non-cleavage, and dominant. In tidal current sand of the western bay of Korea, diastatite, garnet, and gregorite rank the first place; in that of northern Jiangsu, diastafite and acanthconite rank the first place, metallic minerals (including magnetite, gregorite, brown hematite and hydroferrite) and garnet the second, the average content of heavy from the latter. Although beach sand also forms two jumping components under the effects of wave's washing and returning current, the sorting of its second jumping component is better than that of the first jumping component, and belongs to negative partial distribution, and this differs from tidal current sand.The probability curve of sediments of the Liaodong Shoal also shows that the sediments are tidal current sands instead of wave sands.

The probability accumulation curve of tidal current sand is similar to that of estuarine bar, for instance, the probability curve of Jiangsu's tidal current sand precisely bears a resemblance to that of bar sand of the modern estuary of the Changjiang River. This is because the bar sand of this kind of estuary has properties of tidal current sand body and their formation mechanisms are alike– in trough water course and affected by movement of reciprocating current.

Friedman (1967) put the parameters of grain size of river sand and beach sand in scatter-dot diagrams of normal difference to partial distribution and mean grain size by using statistical method and got their rough distribution positions, so as to show their different environments during formation. We also calculated the parameters of a part of tidal current sand through diagrams by Folk-Ward formula (Table 6-1).

Table 6-1 Parameters of diagrammatic measurements of tidal current sand offshore Jiangsu

No.	150	178	190	214	225	293	312	318	320	327	149
M_Z	3.103	3.536	3.083	3.060	2.733	3.003	2.556	2.563	2.210	3.256	3.137
σ_1	0.405	0.587	0.395	0.761	0.498	1.782	0.892	0.543	0.846	1.947	0.578
SK_1	0.113	0.284	0.298	0.481	0.360	0.800	0.606	0.583	0.300	0.852	0.450
KG	1.079	1.180	1.305	2.160	1.977	8.400	3.712	3.582	3.263	4.744	1.164

It is interesting to note that in the above-mentioned scatter diagram most of the tidal current sands nearly lie in the range of river sand area (Figs. 6-3 and 6-4), and this just explains that the formation mechanism of tidal current sand is similar to that of river sand. Relative to beach sand, both tidal current sand and river sand are restricted by flow movement of trough water course controlled by the boundaries. The difference of flows is that the river is one-way current, but the tidal current is reciprocating current. Thus, the probability curve of river sand only has one jumping component, but that of tidal current sand has two jumping components. The variance in amount of two jumping components is a reflection of difference of the ebb and flood tidal currents in speed and time.

M_Z of sediments of the Liaodong Shoal is 2.7 to 3.2 ϕ, their σ_1 is 0.41 to 0.54, the maximum σ_1 is 1.6, and SK is –0.04 to –0.13 and 0.01 to 0.59; other group is basically in range of river sand area and tallies with characteristics of tidal current sand except a group in negative SK.

In C–M grain-size graph, tidal current sand mainly distributes in the suspension district (V) where medium turbulent flow changes progressively, and a few tidal current sands in the suspension district (IV), where strong turbulent flow changes progressively and the even suspension district

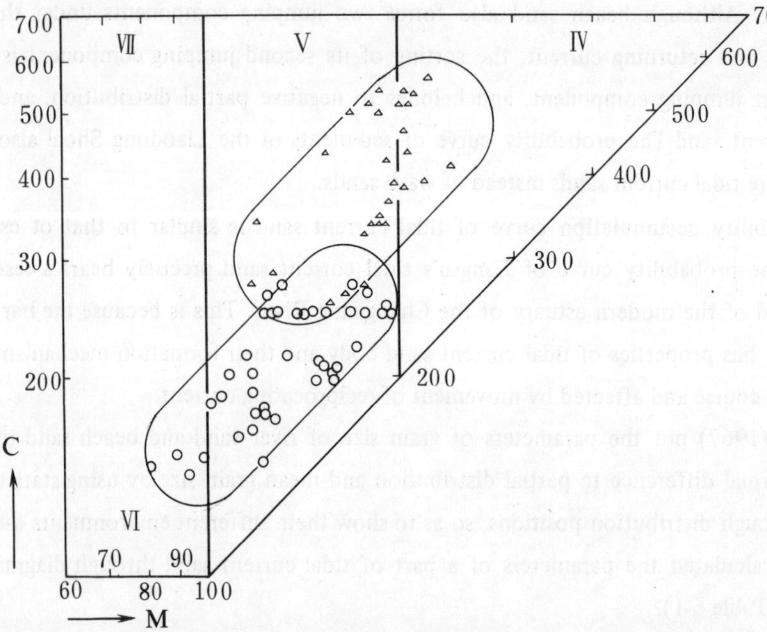

Fig. 6-5. The C=M grain size pattern of the tidal current sand.

Δ – tidal current sand from the offshore of Jiangsu; O – tidal current sand of the North Huanghai Sea.

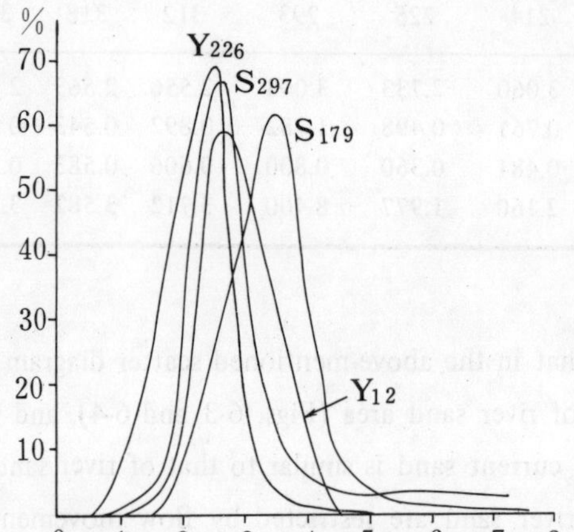

Fig. 6-6. Frequency curves of tidal current sand.

Y_{226} – In the North Huanghai Sea 38°20′N, 124°E;

Y_{12} – In the North Huanghai Sea 39°20′N, 123° 27′E;

S_{179} – In the offshore of Jiangsu 33°10′N, 121°37′30″E;

S_{297} – In the offshore of Jiangsu 32° 30′N, 121°53′ 30″E.

minerals being 5.5 percent. diastatite and acanthconite, in relativly light specific gravity, rank the first; in trough, fine grain garnet and metallic minerals, in heavy specific gravity, rank the first (Table 6-2). The content of heavy minerals in sediments of the Liaodong Shoal is higher than that in its neighbouring sea areas, even reaches 32.2 percent at some stations . The content of stable minerals, such as gregorite, garnet, zircon and so on, can reach 68% in heavy minerals, but that of unstable minerals, common diastatite, acanthconite, etc. decrease to below 30 percent.

Table 6-2 Comparison of heavy mineral contents of crests and troughs of tidal current ridges in the offshore area of Jiangsu

Mineral	Metal		Hornblende	
Geomorphologic location	crest	trough	crest	trough
Average content (%)	11.2	9.3	41.9	48.8
Number of samples	52	41	28	31

4. Chemical properties

The tidal current sand is mainly sandy sediment, whose porosity is high and permeability is good. So it mostly belongs to the oxidizing environments, where, in general, the Eh value is more than 100 millivolts and the ratio between high and low valent irons is more than 1, but the content of organic matters is less than 0.5 percent. The Eh value of tidal current sand of Jiangsu is within 200 to 400 millivolts, and its maximum value even reaches 590 millivolts; the average ratio between Fe^{3+} and Fe^{2+} is 1.15, and the maximum is 3.09; the average content of organic matters is 0.29 percent. The Eh value of tidal current sand of the western Bay of Korea is more than 200 millivolts, and the maximum is 364 millivolts, and the ratio between Fe^{3+} and Fe^{2+} also is more than 2, and up to 44.5; The content of organic matters are less than 0.5 percent and down to 0.05 percent.

5. Sporo-pollen characteristics

The sporo-pollen content is generally lacking in tidal current sand. Most of sporo-pollen bodies are carried far away under stronger hydrodynamic effects, and only a few bigger ones remained. Even so, the remained sporo-pollen bodies usually are worn and broken, and their surfaces are folded. *Picea aspereta,* fir, spore of bryophyte and others can often be observed in tidal current sand offshore Jiangsu, the single body is commonly more than 80 microns in size and they belong to a *Chrysanthemum-fern-artemisia* group. It is very interesting to note that though the spore-pollen content is less, spore-pollen of *Picea asperata* and fir, which reflect cold and moist climate, are relatively more, even up to 4.8 percent. Sporo-pollens of sediments of the Liaodong Shoal reflect that vegetation was mainly conifer dominated by pine interspersed by oak and willow and other broad-leaf forest and that climate at that time was cooler than that of present. Based on the conclusion that sporo-pollen combination of this tidal current ridge area reflects that climate at that time was cooler or colder than that of present. It can thus be speculated that material source of tidal current sand is older and may belong to the strata of the Late Pleistocene or the early Holocene.

III. Distribution of Tidal Current Ridges and Conditions for Their Development

As a submarine geomorphic body in continental sea, tidal current ridges develop at specified geomorphic sites and in hydrodynamic environment. Usually they appear in the neritic regions of the wide continental shelves. Owing to the tidal range and tidal current velocity varying directly with the width of the shelf, the wider the shelf, the greater the tidal range, the stronger the tidal current, the more favorable for the fomation of tidal current ridges. In the Huanghai Sea, East China Sea and South China Sea, the east coast of the U.S. and the Arafura Sea between Australia and New Guinea, there is a good positive correlation between the width of the shelf and the tidal range (Fig. 6-7) (Ren Mei-e, 1980). Off's study also shows that the tidal current ridges are generally distributed in the shallow sea area where the shelf is wider and the tidal range is more than 3 m.

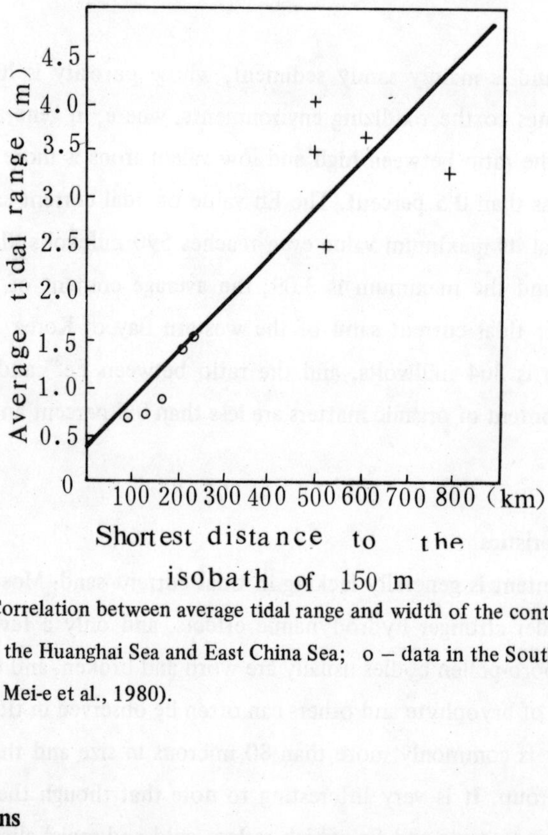

Fig. 6-7. Correlation between average tidal range and width of the continental shelf.
+ – data in the Huanghai Sea and East China Sea; o – data in the South China Sea.
(From Ren Mei-e et al., 1980).

1. Morphologic locations

The morphologic locations of the development of tidal current ridges are: bays, estuaries, mouths of channels and the offshore areas where tidal ranges are fairly great and water depths are mostly less than 35 m. Tidal current ridges can develop at the tops , in the middle portion and at the mouth of bays, depending on the shape of the bay as well as the conditions of the tidal current. At the top of fairly wide bays, tidal waves transmitting from outer seas are influenced by the coastal boundaries and landform, producing harmonic resonance and sea water rises to the top of the bays to cause sudden increase in tidal range and reinforcement of tidal currents. The world-famous big tidal range regions are almost located at the tops of bays such as the Bay of Fundy of Canada and the eastern Bay of Korea etc. Bays often have rivers emptying into them, and rivers rush down large amount of mud and sand into bays, thus providing abundant materials for the development of tidal current ridges. The tidal current ridges in the bays often run parallel to the axes of the bays and intersect coast lines, such as the western Bay of Korea, the northern part of the Persian Gulf, the Bay of Bengal, the Kanbei Bay of India, the Papuan Bay of New Guinea, the Delaware Bay and the Bay of Nushagak of Alaska of the United States of America, in all of which tidal current ridges in various scales have developed. In fairly enclosed bays, tidal current ridges are often formed inside and outside their mouths, as is the case for those in the mouth of the Jiaozhou Bay.

In some estuaries where tidal current action is fairly strong, especially when the direction of estuaries is the same as that of the tidal current, distinct bidirectional currents are formed, and tidal current ridges may possibly appear both inside and outside estuaries, mainly depending on the ratio of runoff volume to tidal prism. In estuaries where runoff plays the leading role, the net transport of sand is directed outward. Thus, tidal current ridges are developed in estuaries such as those of the Oder River (Australia), the Thames, the Amazon, the Ganges-Brahmaputra, the Zhujiang, etc. In estuaries where tidal current plays the leading role, tidal waves, after entering estuaries, become more deformed with the increase of upstream distance, the front slopes becoming steep, whereas the back slope slant, thus resulting in shortening the phase of flood tide and prolonging the phase of ebb tide and causing the average current velocity of flood tide to greatly surpass that of ebb tide. As a result, there exists a net upstream shear stress in the river bottom, and there is more sand transport at flood tide than at ebb tide and tidal current ridges develop inside the estuaries such as those of the Oder River, the north branch of the Changjiang River and the Qiantang River. In some seasonal rivers, the annual runoff is very unevenly distributed. During flood seasons the dominant factor is the run-off; whereas during dry seasons, tidal current predominates. Thus, tidal current ridges develop both inside and outside estuaries. The Qiantang River is famous in China for the bores at its mouth. Those who have seen the Qiantang bores know that the tides rise with the force of an avalanche, the incoming sand from the sea surpassing the silt discharge of runoff. According to the data from a certain observation station, the average silt content at flood tide was 8.8 kg/m^3, and that at ebb tide only 2.34 kg/m^3, and the results of many years showed that sand and mud moved inwards. Thus, no sand ridges were formed outside the funel-shaped estuary, but the massive (leveed) shoals in the upper river course of Ganpuzhen town may be considered as tidal current ridges within the river mouth.

In natural world, some bay tops which are often the places where rivers enter the sea are more favourable for the development of tidal current ridges. For examples, the bay top of the west of Korea Bay is the mouth of the Qingchuanjiang River, that of Bay of Bengal is the mouth the Ganges, that of Delaware Bay is the mouth of the Delaware River and Lingdingyang is the mouth of the Zhujiang River.

When tidal currents pass through the strait areas, the profile of passing water becomes reduced, whereas the current velocity inereases and the effect of Coriolis force causes the flood and ebb tides to have relatively fixed channels. Practically these channels are similar to river beds. The current velocity within the channels may be as great as over 5 knots and, the channel outlets being like river outlets, the current velocity decreases because of the dispersion of the current, and accumulations occurred, often forming at the outlets of straits fingered tidal current ridges which are called tidal deltas by some people. For example, the tidal current ridges of the Liaodong Shoals are located just at the Laotieshan channel of the Bohai Straits, the fingered tidal current ridges at the east and west outlets of the Qiongzhou Straits are also of this type.

In the inshore areas where the tidal range is great, so long as there are relatively strong and oriented tidal movement and abundant materials, tidal current ridges can be formed, such as the finger-shaped tidal current ridges of Jiangsu Province and the tidal current ridges to the east of the Jiaozhou Bay.

2. Tidal action

A strong and oriented tidal current action is one of the prerequisite conditions for the development of tidal current ridges. Tidal waves transmitted to the inshore areas will be affected by coastline and landform, which will cause the tidal currents to have the nature of oriented reciprocal currents and, when having a considerable velocity, will probably develop tidal current ridges. Belderson et al. (1966) considered that, with tidal current velocities changing from great to small, the basal morphology will consist of a succession of the following morphological provinces: 1. bedrock or gravels, 2. sand ribbons, 3. sand waves, 4. smooth sand , and 5. sand patches (Fig. 6-8). Kenyon (1970) considered that on mid-tidal level the average current velocity of 2.5 to 3 knots may form sand ribbons which have the same direction with that of the tidal currents. Off considered that tidal current velocities of 1 to 4 or 5 knots are most favorable for the development of tidal current ridges. In accordance with the conditions for the development of tidal current ridges along the coast of China, tidal currents with velocities ranging from 1.0 to 3.5 knots can all develop into sand ridges, especially those of 2–3 knots being the most favorable (Table 6-3). If the tidal current velocity is too great, its action will be mainly that of erosion, whereas if too small, sand waves with smaller scale and their direction vertical to that of the tidal currents will be formed. In nature, sand waves with secondary wavy relief can often be seen on the sand ridges. This is formed under different times, weather conditions and tidal current velocities.

The above-mentioned relation in which the submarine relief changes with the current velocity is basically suited to the conditions in China. The succession from the channels of Laotieshan to Liaodong Shoal and up to the northern wavy landform completely corresponds to the progressive decrease in tidal current velocities. The Sino-American cooperative investigation of sediment dynamics outside the Changjiang River mouth found that outside the tidal current ridges of Jiangsu Province, there are also sand wave fields formed by tidal currents, the area of the fields amounting up to 20,000 km^2, wave lengths from 5–12 m, and wave height less than 1 m. On the outer side of the tidal current ridges at the Lingtingyang there are also sand wave fields owing to the decrease in tidal current speed (Feng Wenke et al., 1982). In general, with the change in tidal current speed, a series of different geomorphological feature are moulded on the sea floor, and the tidal current ridge is only an important and specific link of this chain.

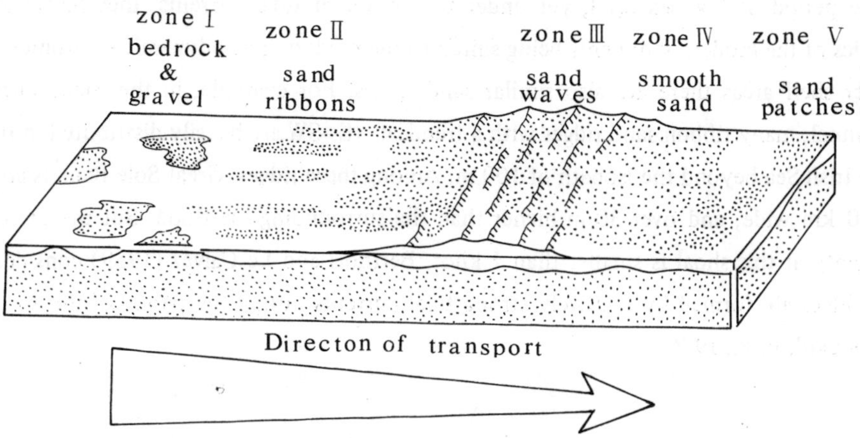

Fig. 6-8. Succession of morphology along a tidal transport. Based on Belderson et al., (1970).

Table 6-3 Tidal current velocities of coastal waters in China

Location	Jiangsu offshore area	W. Korea Bay	Taiwan Straits	Liaodong Shoal
Velocity (knots)	2–3	2–3	2–3	2–3
Location	mouth of Qiongzhou Straits	Jiaozhou Bay	North branch of Changjiang Estuary	Lingdingyang Bay
Velocity (knots)	2–3.5	1–2	2–3	1–1.5

3. Material basis

Tidal current ridges generally occur in sandy shoal seas and modern or old river deltas. In these regions, there are either residual loose deposits or abundant supply of modern materials. Only under this condition can the erosion-accumulation action of tidal currents prevail. Hence, abundant loose deposit is another prerequisite condition for the development of tidal current ridges. For instance, the tidal current ridges of Jiangsu Province developed on the underwater deltas of the abandoned Changjiang River and the abandoned Huanghe River. In high-latitude areas glacial debris and fluvioglacial deposits are often the sources of underwater sand ridge materials. Examples are the area in the North Sea, George Shoals of New England, etc.

Intense mechanical weathering during the Pleistocene glacial age as well as the retreat of the eroded coast during the course of the rise of sea level gave rise to a large amount of sand deposits of the fossil sea shore on the continental shelf of today. This provided the material basis for the development of tidal current ridges. Although most of the materials of tidal current ridges were those in the period of low sea level, yet under the action of tidal currents, they have shown the characteristics of the modern sediments being suited to the modern hydrodynamic environment.

In outer shelf areas there are also similar sand ridges. For example, in the island outside the English Channel, many ribbon sand ridges with a trend of NE–SW are heavily distributed in the outer shelf of the Irish Sea beyond the isobath of 100 m. Among these ridges, Great Sole Bank is about 150 km long, 10 km wide, and over 30 m higher than the surrounding sea flood. But the present tidal current velocity at this shoal is smaller than 1 knot. Berthois and Le Calvez (1959) considered that these sand ridges are formed by relatively strong tidal currents during the period of the lowering of sea level (Shepard, F. P., 1976).

IV. The Tidal Current Ridges in China's Coastal Waters

The tidal current ridges in China's coastal waters not only are widespread with many types, but also have their own evolutionary law. A lot of submarine sand ridges with different sizes and shapes develop at the Liaodong Shoal, the Liugu Estuary, the Bay of Korea, the Jiaozhou Bay, the shoal of Jiangsu, the north branch of the Changjiang River Estuary, the Qiantang River Estuary, the Oujiang Estuary, the Taiwan shoal, the Lingdingyang Bay and the eastern and western mouths of the channel of the Qiongzhou Straits in vast sea area from the Liaodong Bay to the Beibu Gulf (Fig. 6-9). In accordance with their different geomorphic locations, they can be subdivided into the channel type, the bay type and the other types parallel to coast. From the point of view of hydrodynamics, they are generally controlled by the reciprocating flow. Somewhere the flood flow is dominated and others the ebb current, but the fingered tidal current ridges of the Jiangsu offshore area are formed by the cross action of two different directional tidal currents. In the characteristics of evolution, they have a trend of siltation, heightening and gradually being put together with coast. They are different from the evolutionary characteristics of submarine ridges of the east coast of America, where the storm current ridges even more separate from the coast along with the rise of sea level during the Holocene Epoch. In this section we discuss the distribution, properties and hydrodynamic environment of these tidal current ridges separately.

1. The fingered tidal current ridges of the Jiangsu offshore area

The large tidal current ridges attracting worldwide attention have grown in the Jiangsu offshore area, which is about 200 km long from north to south and about 90 km wide from east to west, and is composed of more than seventy big or small sand bodies. A single ridge is dozens or even a hundred more kilometres long and 5 to 10 km wide.Taking Qianggang as a center and spreading

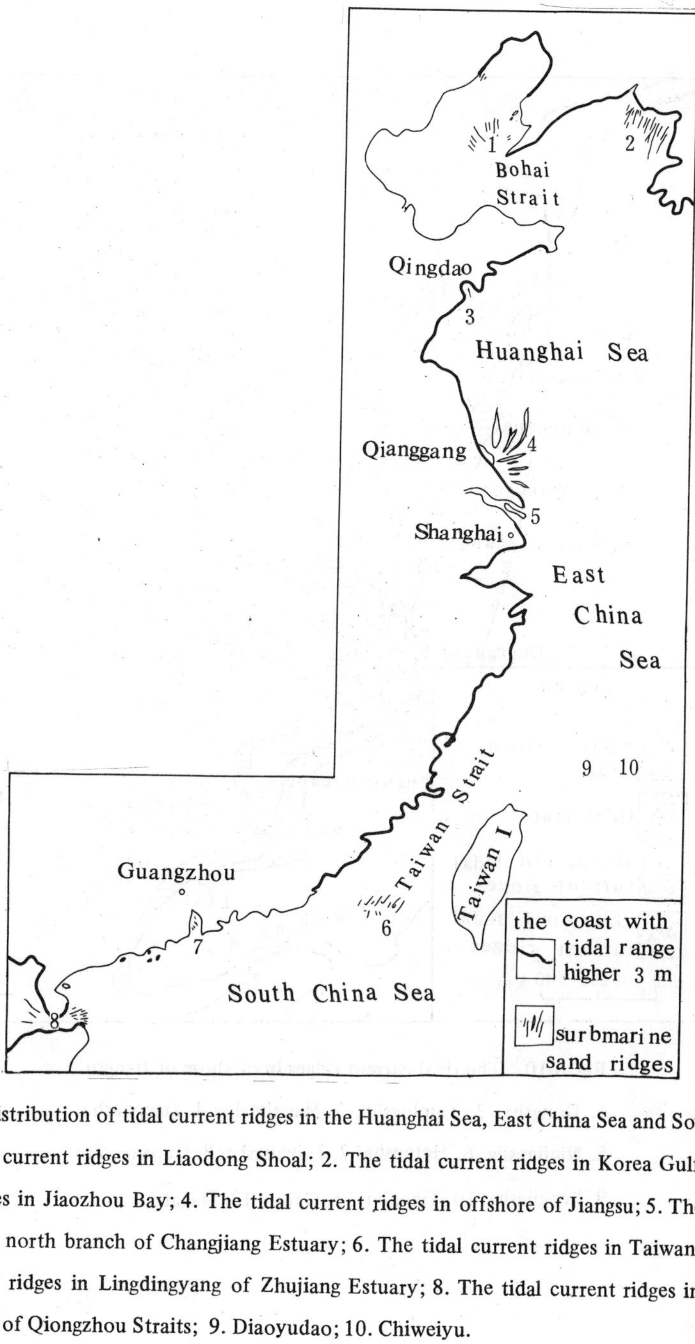

Fig. 6-9. Distribution of tidal current ridges in the Huanghai Sea, East China Sea and South China Sea. 1. The tidal current ridges in Liaodong Shoal; 2. The tidal current ridges in Korea Gulf; 3. The tidal current ridges in Jiaozhou Bay; 4. The tidal current ridges in offshore of Jiangsu; 5. The tidal current ridges in the north branch of Changjiang Estuary; 6. The tidal current ridges in Taiwan Shoal; 7. The tidal current ridges in Lingdingyang of Zhujiang Estuary; 8. The tidal current ridges in the east and west mouths of Qiongzhou Straits; 9. Diaoyudao; 10. Chiweiyu.

outwards, in fingered shape, they are Dongsha, Taipingsha, Maozhusha, Jiangjiasha, Niujiaosha, Hetunsha, Taiyangsha, Huoxingsha, Lengjiasha and Yiaosha sands clockwise arrayed from north to south. There still exists Tiaozini outside Qianggang (Fig. 6-10).The furrows between ridges are generally 10 m or so deep. In the center the crests are high, the furrows are shallow, and the depths are less than 15 meters. The outer furrows are over 30 meters deep, Dongsha Sands, whose emerged area is more than 600 square kilometers during low tide, is the largest one of these ridges.

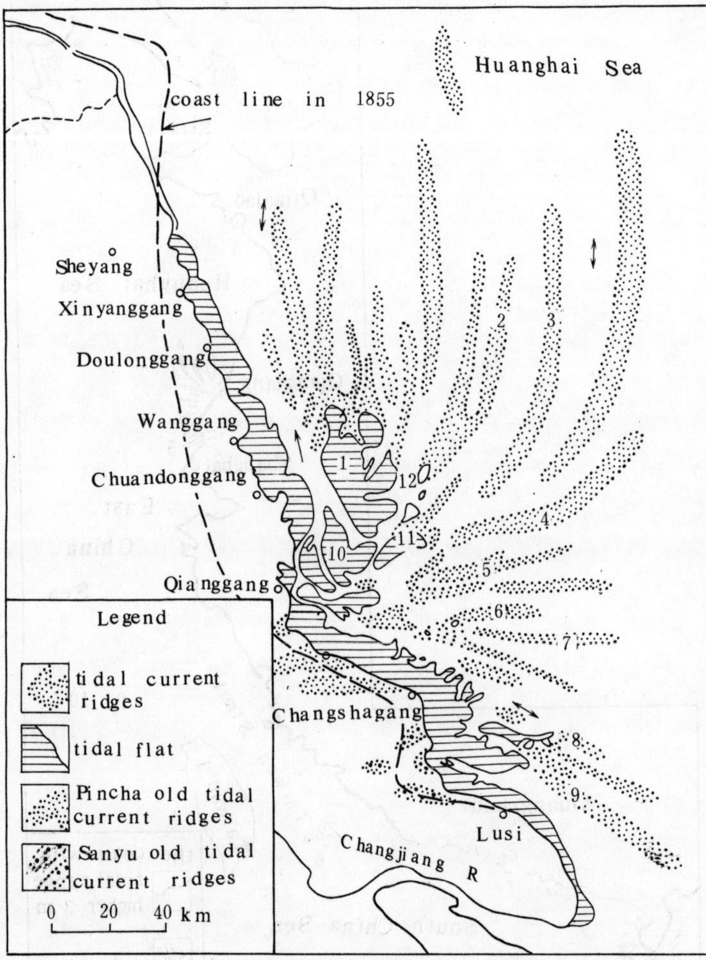

Fig. 6-10. The tidal current ridges in offshore of Jiangsu.

1. Dongsha; 2. Taipingsha; 3. Maozhusha; 4. Jiangjiasha; 5. Niujiaosha; 6. Hetensha; 7. Taiyangsha; 8. Huoxingsha; 9. Lengjiasha; 10. Tiaozini; 11. Gaoni; 12. Macai.

(a) *Material source*

Material of the tidal current ridges of the Jiangsu offshore area is mainly from the submarine deltas of the ancient Changjiang River and the ancient Huanghe River. It is well known that the Huanghe River and the Changjiang River entered the sea here being a decisive factor for the mould of the Jiangsu coast for a long time. Average annual silt discharge of the Huanghe River is 1.1 billion tons, which ranks second in the world. Migrating southward or northward in the long geological periods, the Huanghe River entered the sea through the Jiangsu area many times. The last time was from 1128 to 1855, when it moulded a huge estuary delta in the Jiangsu offshore area. The average annual silt discharge of the Changjiang River is 0.48 billion tons, which ranks second in China. Since the Holocene, the estuary of the Changjiang River has continually moved southward, so the old deltas were left in these old estuaries. The Qianggang was the old Changjiang entrance to the sea (Zhong Deling et al., 1983). It can be considered that the Jiangsu offshore area including some littoral plains was the dumping place of the silt of the old Changjiang River and the Huanghe River, where enormous loose sediments provide materials for the development of the tidal current ridges.

Undoubtedly, the coastal currents from north or south also provide part of material for the development of these ridges. According to in situ measurements and drift bottle data, during the flood season, the coastal current northward from the Changjiang mouth reaches the Changsha port, around the Lengjiasha (sands), goes southeastward, then joins with a southward current from the northern Jiangsu coast. The average velocity of the Changjiang northward coastal current is 0.2 to 0.4 knot. The maximum value is 1 knot, and the current width is 15 nautical miles. In the rest time, the Changjiang River discharge is mainly southward. Moreover, after fresh water of the Changjiang River flows through the estuary, the water area widens and gradient decreases, the current velocity cuts down, the coarse substance accumulates near the mouth, and only fine-grained substance goes following the current. So this northward coastal current is only a seasonal current. Its velocity and extent are limited. By comparison, the southward coastal current from the old Huanghe River mouth is stronger, and affects whole sea area within 50 m deep, whose velocity is generally less than 0.5 knot. Since the Huanghe River migrated northward, the source of sediment here has been cut off, and under the erosion of wave and tidal current the coast has constantly moved back to the land and erosive material of some parts of coast moves southward following the coastal current of northern Jiangsu. By comparing the present Jiangsu coast line with that of 1855, it can be found that the coastline direction has changed from former near N–S into NNW–SSE. The coast withdraws in north and grows outward in south. The old Huanghe River mouth moves back by an average of 150 m a year, and the extent of increase of Dongtai's outside land is the biggest; in the Wanggang area the coast grows outward at an annual average of 50 meters (Ren Mei-e, 1983). Thus it can be seen that the effect of this coastal current carrying sand and mud is very strong. Nowadays, because of building dykes and strengthening coast along the old Huanghe River mouth, the rate of erosion and retreat has been cut down. The sand amount carried by this southward coastal current has thus decreased considerably.

In general, the current velocity at furrow and trough is higher than that at crest, so normally the substances at crest would be finer than those at furrow and trough. But the tidal current ridges in the Jiangsu offshore area show an opposite distribution pattern (Fig. 6-11). This is due to the facts that the tidal current ridges of the Jiangsu offshore area develop on the submarine deltas of the old Changjiang River and the old Huanghe River, the basal substances are rather fine. The furrows and troughs have already been cut, and the base of the old submarine deltas has emerged, so the distribution pattern, coarse at crest and fine at trough, appears. The results of sporo-pollen, analysis indicate that the sediments contain *Picea asperata,* fir and other sporo-pollens reflecting cold climate, and that the base consists of those sediments of cold climate during the Late Pleistocene or the Early

Holocene. Off the coast of Qianggang, all ridges are linked together, water depth shallows, and hydrodynamic condition is rather active. In this area, the sediments in the south mainly are fine sands and just a little coarser than those very fine sands in the north, which is closely related to different sources of substances from north and south and hydrodynamic condition.

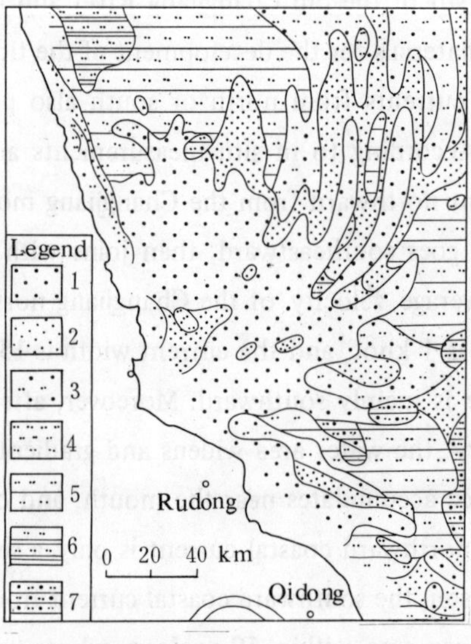

Fig. 6-11. Distribution of sediment in the offshore of Jiangsu.

1. Fine sand; 2. Silty sand; 3. Sandy silt; 4. Silt; 5. Clayey silt; 6. Silty clay; 7. Sand-silty-clay.

(b) *Hydrodynamic condition*

The specific tidal environment near the Jiangsu offshore area produced such tidal current ridges with strange shape and broad in scale. According to the data of hydrologic investigation, the Qianggang offshore area is the place where two kinds of the tidal waves meet. The area to the north of Qianggang is controlled by the rotating tidal wave system of the west of the Southern Huanghai Sea, whose crest line goes from north to south, and this shows a rotating current in the individual area lying to the east of the distribution area of the sand ridges, and a reciprocating current in the rest area. To the west of the Dongsha Sands, the main current is parallel to the coastline on the whole, which is a reciprocating current in NNW–SSE direction, and the period of flood is shorter than that of ebb, and the average velocity at flood tide is higher than that at ebb tide. To the east of the Dongsha Sands, tidal current direction is NNE–SSW, whose differences of period and velocity between flood current and ebb current are smaller. To the south of Qianggang, the Pacific tidal waves go through the East China Sea into the Huanghai Sea in a form of forward wave. To the north of the Lengjiasha Sands the tidal current direction is NWW–SEE and from the south of the Lengjiasha Sands to the north bank of the Changjiang River it is NW–SE. The flood current velocity of some troughs is higher than the ebb current velocity, and the case of other troughs is contrary to this. In a word, the period and velocity differences between flood and ebb currents are not great. These two groups of different directional tidal currents meet in the Qianggang offshore area, the time of their strongest flood and ebb tidal field is in keeping fairly well with the time when current changes direction, so that tidal energy is gathered to form high tidal range and strong tidal current environment. The average high tidal current velocity is 2.5 knots in the sea area north to the Qianggang offshore area, while 3 knots south to the Qianggang offshore area. Under the repeated effects of the flood and ebb currents, the furrows and troughs are eroded, sediments are accumulated at the crests, the undulation of submarine relief is gradually increased, and then the geomorphologic patterns of tidal current ridges of crest alternating with furrow are constructed. It can be considered that the tidal current ridges of the Jiangsu offshore area are the results of intersection of two groups of the fingered ridges following tidal current in the area of Qianggang offshore area. The Dongsha, Taipingsha, and Maozhusha sands, etc. in the north are a group of fingered ridges in nearly N–S direction. All ridges in the south of the Jiangjiasha Sands are a group of fingered ridges in S–E direction. With erosion and accumulation action of tidal current, local substance and that carried by the coastal current obviously are differentiated. Coarser fine sand remained to form tidal current ridges; clay silt and other fine-grained substance are transported to the seashore to participate in the siltation, or to the open sea. Thus a tidal flat over ten kilometers wide has developed on the inner side of the ridges.

Secondly, the effects of wind-driven wave cannot be ignored either. This sea area is significantly affected by monsoon. In summer season, this area is controlled by subtropical high pressure, southeast wind is dominant, and the wind speed is not high except in typhoons. In winter season, this area is controlled by high pressure of strong cold current from Mongolia, and northerly high wind prevails. Waves mainly are mixed wave taking wind-driven wave as dominant role. Normal waves are northerly. The direction of strong wave is northeastern or southeastern. Because the northern part of the sea area is extensive, northerly or northeasterly wind goes straight to the area, so the coast is serious destroyed. In the middle area–the tidal current ridges area, wind-driven waves undergo breakage twice, so that its erosion to the coast weakens. But its disturbance on the sea bottom and action setting sands in motion are very obvious. Judged from the distribution of suspended sediments of the Jiangsu offshore area , during winter and summer, wind-driven waves have serious influence over this area. The content of suspended sediments in winter is often many times than that in summer or even more, so the area of muddy water in winter is also much larger than that in summer (Fig. 6-12). Storm effects on this area are also important, especially in summer, typhoons often affect this area. According to statistical data from 1953 to 1975, 75 typhoons affected this area, the average being three times annually. Especially when storm meets spring tide, destructive power is much bigger. On 1, 2, and 3, September, 1981, when Typhoon 8114 passed through this area and happened to meet spring tide, the highest water level at all hydrological stations of Jiangsu Province broke the historical records and the wave height reached more than 9 meters; after typhoon had passed this area, branching stream were changed beyond recognition.

Based on the theory that water depth affected by wave is less than half wavelength, all part less than 25 meters deep in this area is affected by wave, and this makes originally complicated tidal current ridges be more complicated. The irregular shape outside the Dongsha sands is related to strong northeast wind. Not only storm disturbs the sea bottom violently, but also wind current makes some small sand bodies remove. But the wind current cannot remove other bigger sand bodies, thus they have a relative stability.

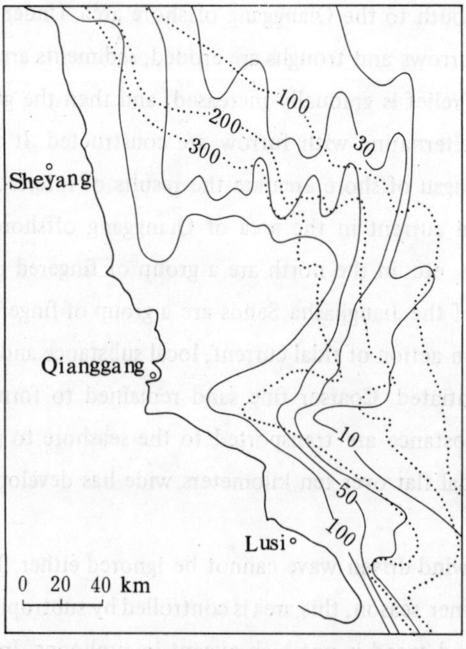

Fig. 6-12. Concentration of suspended material in the offshore of Jiangsu (mg/1). Solid lines and dots lines are, respectively, the data in July and December 1980 (From Zhang Jinian).

(c) *Tendency of development and evolution*

Formation of the tidal current ridges of the Jiangsu offshore area is an inevitable result that hydrodynamics and substance interact on each other in particular environment. The research indicates that the basic configuration of present coast was formed about seven thousand years ago. Since then, there has hardly been any difference of marine hydrodynamic condition of the Jiangsu offshore area as compared with that of present . Because the Pacific tidal wave and rotating tidal wave of the western part of the South Huanghai Sea meet here, it is inevitable to form fingered tidal current ridges; because the flood current velocity is higher than ebb, net displacement of substance points to the convergence part of the sand ridges to make them gradually silted and rise. Finally they will join with coast to form land. Since the Holocene, three stages of fingered tidal current ridges have successively formed. According to the study made at the Tongji University (Li Congxian, 1979), the oldest stage, the Pincha radial shoal, formed 1000 to 2000 years ago; the second stage is the Sanyü radial shoal; the third stage is the Qianggang radial shoal, namely the Jiangsu tidal current ridges. It can be inferred that from formation, growth, siltation to joining together with coast, one stage of the sand ridges lasts less than one thousand years.

The final result of evolution of the tidal current ridges in the Jiangsu offshore area will be that they join together with coast, but the levels of development of each part are different. In the sea area north to Qianggang offshore area, flood current velocity is higher than ebb tidal current. Tidal wave is dominated by standing wave. That is to say, during the middle tide, the velocity is the highest; during the high and low tides, the velocity decreases. This kind of tidal wave is favourable for sedimentation. The current velocity is low at the beginning of flood tide, and then increases with tidal level's rising, thus tidal current has great ability to transport sand and mud and to erode furrow and trough. At high tide level the current velocity decreases again, so more sand and mud accumulates near the high tidal line. After that, the ebb tide begins, and because of smaller current velocity, the tidal current cannot carry away the sand and mud which were brought in at the flood tide, so coarser substances remain to heighten sand ridges and flat. The above-mentioned facts are the reasons why Dongsha sands and flat from Xingyanggang to Qianggang silt and rise quickly. So the sea area north of Qianggang is in the process of washing and deepening furrows and troughs, accumulating sand ridges and silting flat. For example, ten-meters deep furrow of Pingtuyang deepened to 15 m; the depth of Xiaoyangkou increased from 17.2 m in 1964 to 22 meters; the deepest site of trough east of Nirouhang deepened from 16 m to more than 20 m. Conversely, the depth of Gaoni sands decreased from 3 to 1 meters at high tide, the height of Dongsha sands increased from 3.7 m to 5.8 m , and the height of Niruohang sands increased from 2.7 m to 3 m.

The trend of evolution of these ridges can be observed through comparing landsat images of 1973, 1978 and 1979. Tiaozini sands at the joint of two groups of tidal wave have been developing to the advanced stage of sand ridges, which silted and grew quickly, joined with coast gradually and linked with the end of Jiangjiasha sands. The area of Tizozini's shoal above the lowtide line nearly increased by one time as compared with that of six years ago. But at Qianggang where the tidal range is big and the tidal current is strong, because of violent washing of the tidal current at high tide, barrier was destroyed, coast collapsed and flat advanced. At present, vigorous measures are being taken to prevent barrier and coast from further collapsing. Although Dongsha sands is still in the process of silting and growing, its joining together with land cannot be expected soon because of higher velocity of tidal current at Xiyang Trough between Dongsha sands and land, such as near Wanggang the great velocity reaches more than 9 knots, and some areas are at washing stage. Gaoni Sands, Macaihang Sands and other small sand bodies at the end of Taipingsha Sands also silt and grow quickly, which will first bring out merging and then rise and silt.

In the sea area south of Qianggang, in spite of great effect of tidal current, the size of sand ridges is smaller and accumulation is not very obvious because tidal wave is forward wave. That is to say, the current velocity is the greatest at the high and low tides and the smallest at the middle tide, and the difference of duration and that of current velocity between the flood and ebb tides are not great, which are disadvantageous to the silting and growing of sand ridges.

In a word, evolutional law of Jiangsu's tidal current ridges is silting and joining together with coast, and this long process has been going on from west to east stage by stage.

2. The tidal current ridges of West Korea Bay

The most typical parallel tidal current ridges in the world are developed in the submarine area of more than ten thousand square kilometers from the Yalu River mouth to the Dartong River mouth (Fig. 6-13).

Dozens sand bodies constitute broad subfluvial comby sand ridges in scale, which extend in NNE–SSW direction as tidal current. Their relative height difference is 7–25 m and average height is about 20 m ; a single sand body is above 80 km long; the trough between two ridges is more narrow at the head of bay and broad at the mouth of bay; interval distance of ridges is 1.4–8 km. The shapes of the above tidal current ridges can be seen clearly on satellite images, too.

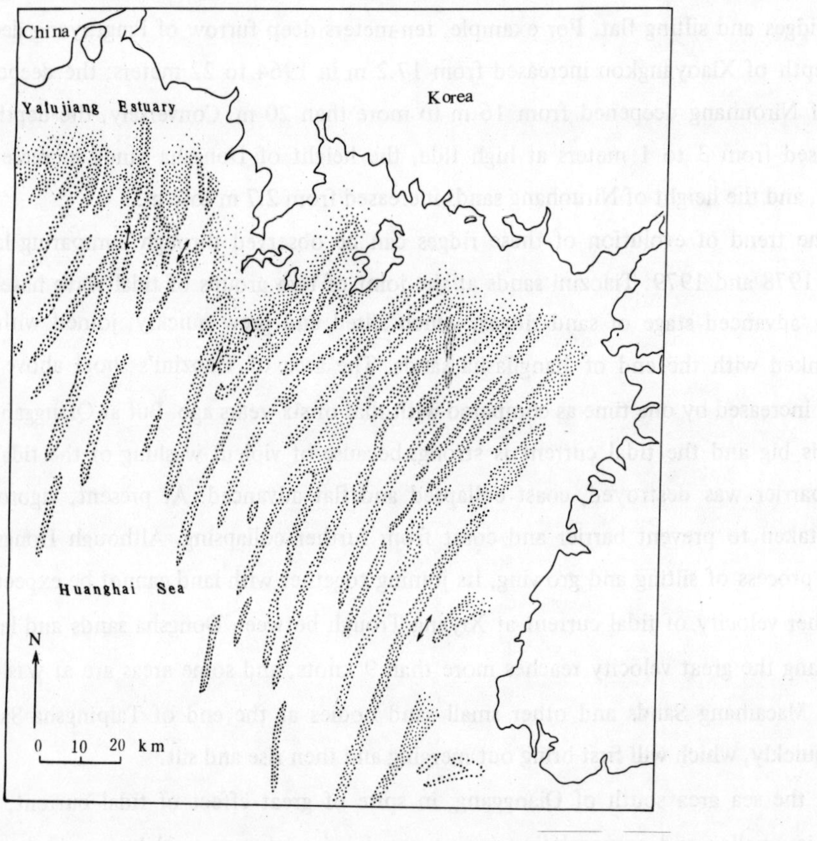

Fig. 6-13. The tidal current ridges in Korea Gulf.

The constituent substance of tidal current ridges of the West Korea Bay is mainly dark grey or brownish-grey medium-fine sand. Though samples of sediments belong to the edge of the West Korea Bay's tidal current ridges, analytical results of grain size, mineral and chemistry indicate that these medium-fine sands have typical properties of tidal current sands.

Because of effect of Coriolis force, tidal range of eastern part of the Huanghai Sea is generally greater than that of western part of the Huanghai Sea. The tidal range of the West Korea Bay reaches above 8 meters and the current velocity is 2–3 knots. In addition, standard linear sand ridges are developed there due to water and sand coming from high gradient rivers.

3. The tidal current ridges in the Liaodong shoal

The tidal current ridges in the Liaodong shoal lie to the western side of the Liaodong Peninsula and to the north of the Laotieshan channel. They belong to the channel type of the tidal current ridges with a finger pattern. In the area of about three thousand square kilometers, six large sand ridges and some small ones have been growing, and four of them parallel with each other and extending in S–N direction, in the middle are the most typical. The east sand ridge are in NNE direction. The west one is the largest in NNW direction; its southern part is subdivided into two branches, which join together in the north apex and appear to be an upside-down hook (Fig. 6-14). By making a comprehensive survey of the plane distribution of all sand ridges, it can be seen that the sand ridges approximately spread in a finger pattern from the mouth of the strait to NWW–NNE direction. A single sand ridge is about 2–4 km wide and about 10–36 km long. The height difference between its gully and keel is about 10–24 m. The interval between the two keels is 10 km or so. Near the strait, the troughs of sand ridges are narrow. The topographic relief is large. But after sand ridges protrude into the bay and spread out, the topographic relief becomes gentle gradually, and is finally substituted by smooth submarine relief.

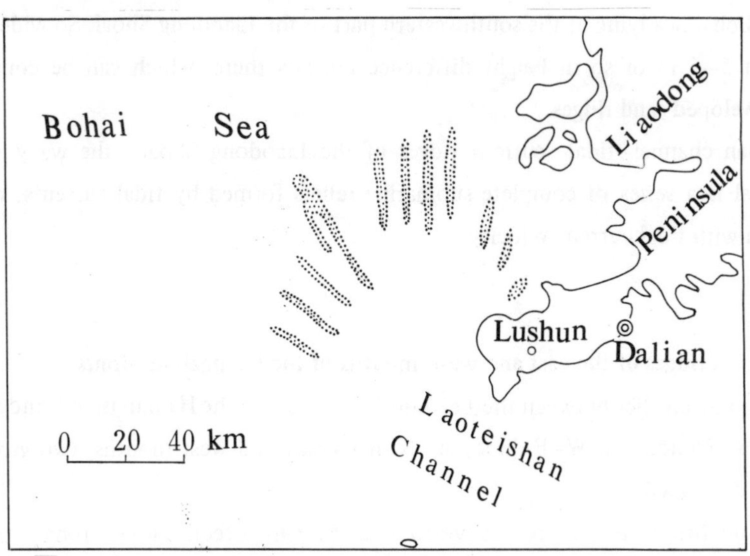

Fig. 6-14. The tidal current ridges in Liaodong Shoal.

The sediments of sand ridges area are yellowish-brown fine sands or the finest sands, bearing a small amount of pieces of fresh-water-shells of littora facies. They have the typical characteristics of tidal current sand based on the results of the grain size analysis. The mineral composition indicates that the substance comes from output of the Liaohe River, the Dalinghe River and other rivers. But the fine sand substance cannot be transported into the Liaodong Shoal by the Liaohe River and the Dalinghe River under the current condition. Sporo-pollen analysis data and shell bodies of mollusc show that the substance has some characteristics of relict sand during low sea level. The relict sand at that period provided a good substantial conditions for the development of the tidal current ridges of the Liaodong Shoal.

Consistency between trend of the sand ridges of the Liaodong Shoal and that of the current reflects that the sand ridges obviously depend on the effects of the modern tidal currents. The tidal current velocity at the Laotieshan channel amounts to five knots or so. Such a strong current has brought a violent scouring, so that the modern sediments have hardly remained there, and the bind beds have emerged in some areas. After tidal water passes the strait into the Bohai Sea, because of the spread of flow, the current velocity decreases and reaches about two knots or so at the Liaodong Shoal. In addition, there are abundant loose sediments at the sea bottom. These two factors provide perfect condition to form sand ridges. The tidal current has gradually transformed the original relief into the typical sand ridge relief. In the process of moulding sand ridge relief, the tidal current fully washed the original finer sediments from the Liaohe River and the Dalinghe River and made them coarsen. After the tidal flow passes the Liaodong shoal, the current velocity is further reduced to below one knot, so that the tidal current never has enough energy to accumulate materials and form sand ridge relief. The current only forms moderate sand ripple, wave mark and other geomorphological patterns.

Because of lower tidal current velocity, the current cannot form typical sand ridge at the shoal in the center of the Bohai Sea lying at the southwestern part of the Liaodong Shoal. So wide and smooth ridge relief about 5–6 m or so in height difference appears there, which can be considered as a kind of poorly developed sand ridges.

The Laotieshan channel–tidal current ridges of the Liaodong Shoal– the wavy relief of the north of the shoal is a series of complete submarine reliefs formed by tidal currents, which shows a good correlation with tidal current velocity.

4. The tidal current ridges of the east and west mouths of the Qiongzhou Straits

The Qiongzhou Straits lies between the Leizhou Peninsula and the Hainan Island, and is controlled by fault structures extending in W–E direction. Near its east and west mouths, two groups of tidal current ridges are developed.

The Qiongzhou Straits is the area in which tidal current affects more strongly than in any other areas of the South China Sea. Because of effect of relief, the tidal current velocity reaches 4 to 6 knots. The tidal current violently erodes sea bottom to form rough abrasion trough, where the deepest depth is 120 m. The tidal current velocity is slowly reduced to about 2–3.5 knots off the Straits. Its flow direction in the Straits is the same as the trend of the Straits, moving back and forth in W–E direction. The flood tidal current flows from east to west, while the ebb current flows in the opposite direction. Off the mouth, the direction of the current is fingered.

The favourable condition of hydrodynamics results in developing two groups of tidal current ridges with fingered pattern at the east and the west mouths of the Qiongzhou Straits. A group of tidal current ridges at the west mouth is controlled by the flood tidal current and the other one at the east mouth by the ebb current (Fig. 6-15). There are more than ten large and small sand bodies altogether, and someone named them tidal deltas. Their morphological features are listed in Table 6-4 (Jin Bo et al., 1982).

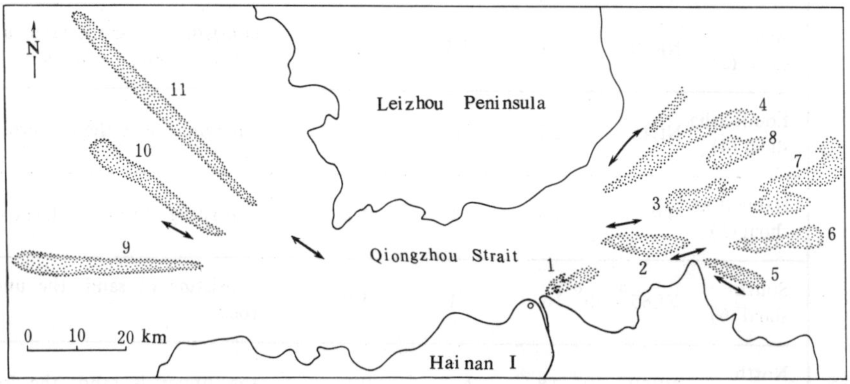

Fig. 6-15. The tidal current ridges in the east and west mouths of Qiongzhou Straits.

The tidal current ridges at the west mouth of the Qiongzhou Straits have a few individuals, longer in length and complete and typical in morphology. But the tidal current ridges at the east mouth have many individuals, smaller in size and untypical in morphology. This is due to the effects of stronger wave and coastal current at the east mouth. There exists a coastal current from north toward south, whose maximum velocity can reach 0.8 knot. There exists a circulation in the counter-clockwise rotation in the winter half-year at the Gulf of Guangzhou. The wave effects of the east mouth are also much stronger than those of the west mouth. In the winter northeastern waves are dominant while in the summer southeastern waves are predominant. The maximum wave height amounts to 5 m. The total effects make the materials from the land and the open sea be moved to the east mouth, so that the tidal current ridges are fragmentary, short and small.

The distribution characteristics of substance constituting the tidal current ridges are: coarse at furrow and fine at keel; coarse near the mouth; the farther away from the mouth, the finer; medium-coarse sand and gravel are dominant in furrow near the mouth; medium-coarse sand is dominant at keel. All these are related to strong tidal current and rocky base. The substance changes into medium-fine sand gradually outward.

Even now, the tidal current has constantly been digging and eroding the straits and the furrows between ridges. In addition, material is continuously provided by mud and sand carried by the rivers entering into the sea, for the development of the tidal current ridges, thus heightening sand ridges uninterruptedly.

Comparison between the bathygraphic chart measured in 1972 and that measured additionally

Table 6-4 Characteristics of tidal current ridges in the west and east mouths of Qiongzhou Straits

Location	Name	Direction	Length (km)	Width (km)	Depth (m)	Characteristics
The east mouth	Baisha shoal (1)	WE	15	1.5	0.4	
	Southwest shoal (2)	WE	8	2	1.7	consisting of sand; large undulation of the underlying bed rock
	West shoal (3)	NE70°	11	2.5	1.8	consisting of sand; large undulation of the underlying bed rock
	Loudousha shoal (4)	NE45°	32	6		emerging above the sea level
	Hainantou shoal (5)	WE	20	4	0.2	consisting of four small shoals
	South shoal (6)	NE85°–90°	16	1	1.9	consisting of sand; the underlying bed rock
	North shoal (7)	NE40°–50°	18	2–3	0.5	the upper is sand; the lower is bind
	Northwest shoal (8)	NE	12	3–6.5	2	consisting of medium-fine sand
The west mouth	Southwest Bank (9)	WE	68		7.4	ribbon
	Middle Bank (10)	NW305°	35		6.5	ribbon
	Northwest Bank (11)	NW318°	67		11	ribbon

from 1955 to 1962 shows that the tops of sand ridges have heightened about one meter and their furrows have been cut down about one meter.

5. The tidal current ridges of the Taiwan shoal

The tidal current ridges of the Taiwan shoal are distributed near the south mouth of the Taiwan Strait, covering about 8,800 square kilometers in area. There exist several tidal sand bodies in NE–SW direction. The maximum length of sand body is about 5 km . The height difference between gully and keel is 6–20 m (Fig. 6-16).

Substances constituting the sand ridges are more coarse, most of which are well sorted yellowish-brown medium-coarse sand. The better psephicity of sand reflects that littoral environment has long been affected by stronger hydrodynamics. The average content of heavy minerals is about 2.1 per-

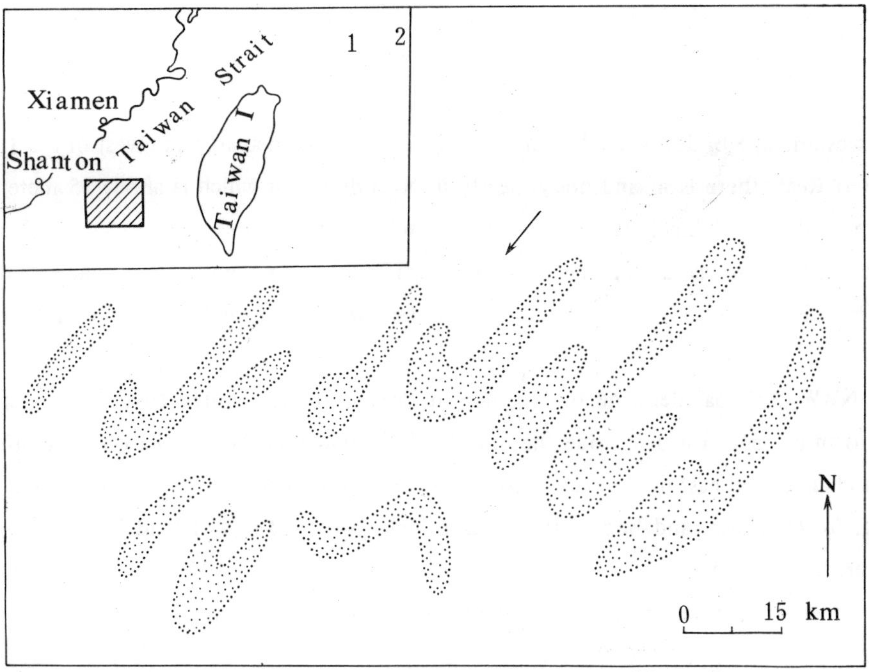

Fig. 6-16. The tidal current ridges in Taiwan Shoal.
1. Diaoyudao; 2. Chiweiyu.

cent. The characteristics minerals are ilmenite gregorite and andalusite. The mineral composition are hornblendite, schorga, ilmenite gregorite, chlorite, aimant magnetite, andalusite and acorite (Shen Shunxi et al., 1980).

The deposits contain beachy and intertidal and even fresh water environmental biodetritus, such as *Ostrea Pestigris, Cardita* sp., *Solellina* sp., *Corbicual* sp. etc., which reflect the characteristics of relict deposits (Zhen Tiemin et al., 1980).

The subfluvial relief of the Taiwan Shoal has obvious feature of line spreading, with no difference in direction with ebb current. It can be seen that the development of sand ridges is mainly controlled by the ebb current. In this area, the direction of the flood tidal current is northward, while that of the ebb current is sorthwestward. The ebb current velocity is about 2 to 3 knots, and this is an ideal velocity for the development of tidal current ridges. Because flood and ebb tidal currents do not move strictly in the form of reciprocating flow, the development of sand ridges is not fully typical, and their plane pattern is "S"-shaped, or crescent. Moreover, the sand ridges often develop with a bedrock knob as the core, so in the group of sand ridges there are often some bed rocks towering on sand ridges.

6. The tidal current ridges of the mouth of the Jiaozhou Bay

The Jiaozhou Bay is a well-known tide-funnel-type and semiclosed bay in China. Its outlet is narrow and the water area inside the bay is broad. Because of narrow at the bay mouth, the velocity of periodic reciprocating tidal current passing in and out the bay is increased to 2–4 knots. The bay mouth is similar to a small channel, so some small-scale tidal current ridges develop at the exit and entrance of the bay.

At the western side of deep-trough channel which is about 1 km apart from west of the Tuandao Island-Matijiao Reef, there is a sand body, nearly in N–S direction which is about 15 meters high, 0.5 kilometer wide and 4 km long. Its constitution substance is coarse-medium sand. Along with northward extension, the sand ridge becomes wide gradually and its constituent changes into silty fine sand. There is a sand body in N–W direction to the northeast of the Huangdao Island inside the bay as well.

Several NWW–SEE parallel sand bodies, named Beisha, are distributed to the south of Yaner Island and north of deep trough outside the mouth of the Jiaozhou Bay. In the waters south of the deep trough channel and east of the Daqiao Island, there is a sand body of the same direction, named Nansha (Fig. 6-17). These sand bodies are 6–20 m high, 100–300 m wide and 1–3 km long. The size is smaller, the substance is coarser, and the base of the sand body is bedrock. They are formed through the effect of reciprocating current which flows at a small angle with coast.

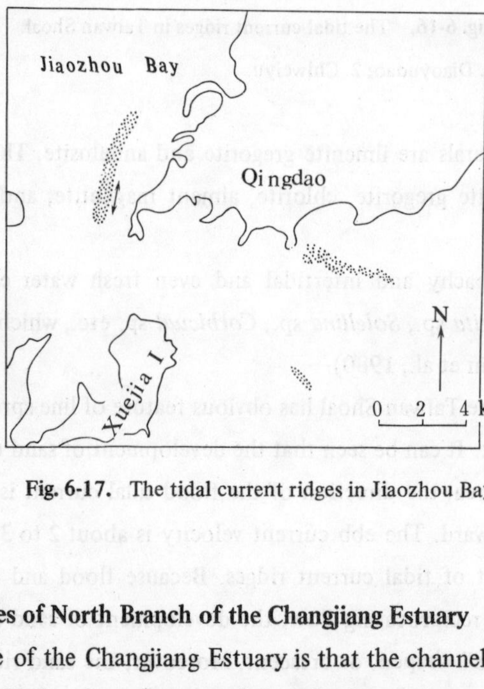

Fig. 6-17. The tidal current ridges in Jiaozhou Bay.

7. The tidal current ridges of North Branch of the Changjiang Estuary

A remarkable feature of the Changjiang Estuary is that the channel is divided into two branches, the south and the north. Under the influence of Coriolis force, the river flows and the ebb currents mainly flow down through the south branch, while the flood currents mainly go up the north branch. Because of increase of discharge and current velocity caused by superposition of the river flow to the ebb currents, the river channel of the south branch is scoured and deepened. Conversely, because of decrease of dynamic effects caused by the river flow going against the flood currents, the river channel in the north branch is silted and shallowed. From 1915 through 1978, the capacity of the north branch of the river decreased from 26.31 billion cubic meters to 10.44 billion cubic meters, three fifth of the river channel was silted. During the period from 1958 through 1970, the average rate of silting was 0.433 hundred million cubic meters per year; during the period from 1970 through

1978, it was 0.22 hundred million cubic meters per year.

Table 6-5 Characteristics of the tidal current ridges of the North Branch of the Changjiang Estuary

Location	Length (km)	Width (km)	Height difference (m)	Shape
East of Xinglongsha Sands	30	0.5–1	3–8	ribbon
East of Santiaogang Port	8	0.3–1	3–7	ribbon
West of Daxinggang Port	6	0.3–1	2(south)–17(north)	ribbon
West of Lingdiangang Port	4.5	0.3–0.4	5–6	ribbon
West of Xinglongsha Sands	11		5–9	crescent

From 1915 to 1959, the percentage that the river flow of the north branch took up the whole amount of the river flow was reduced from 25 percent to 1 percent. In the 1970's, it was changed into a negative value. That is to say, in the north branch, the flood tidal currents dominated gradually instead of the ebb currents. According to the measured data made in September 1971, the amount of the flood tide of the north branch took up 21.5 percent of the whole tidal amount, while that of the ebb current only 8.8 percent. The reverse trend that mud and sand in the north branch filled reversely into the south branch occurred, and thus the north branch evolved as a tidal type estuary.

At present, the average flow velocity of spring tide of the north branch is 2–3 knots, its maximum tidal range is 5–6 m, and mean tidal range is 3 m or so. In addition, because of the existence of abundant mud and sand provided by upper reaches of the Changjiang River and the sea area, several long narrow and crescent sand bodies have developed in the estuary (Fig. 6-18). Their characteristics are that topographic relief is gentle, the height difference between crest and trough is small, and some sand bodies, such as the Yonglongsha, have been silted into shoal to incorporate into the coast, others, such as the Xinglongsha, have gradually emerged from water. Main morphological elements of each sand body are listed in Table 6-5.

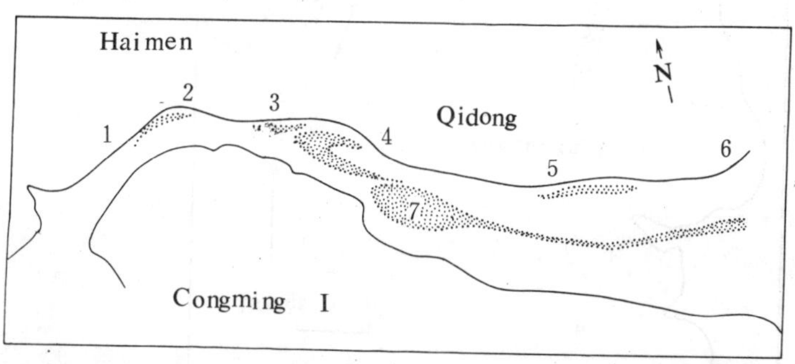

Fig. 6-18. The tidal current ridges in the north branch of Changjiang Estuary.
1. Qinglonggang; 2. Daxingang; 3. Lingdiangyang; 4. Sanhegang;
5. Santiaogang; 6. Lianxinggang; 7. Xinglongsha.

8. The Tidal Current Ridges of Lingdingyang Bay (Zhujiang Estuary)

The tidal current ridges in the estuary bay are the results of the common effects of tidal current and the river flow. The Lingdingyang tidal current ridges off the Zhujiang Estuary is a good example (Fig. 6-19). The Lingdingyang bay, nearly in S–N direction is a funnel-shaped estuary, being a submerged valley formed by transgression during postaglacial period. The silt discharge of the Zhujiang River is 83.36 million tons per year, most of which go into the Lingdingyang Bay. Besides, in the dry season, some sands come from sea area, and the important marks are that sediments contain a small amount of marine authigenic minerals and marine biodetritus, such as authigenic brazil, glauconite, foraminiferal shell, sponge bone needle, diatom, coral fragments and others. Particularly in the eastern part, the above minerals and biodetritus are considerably distributed (Zhao Huating, 1981). The tide in the area belongs to the type of irregular half-day tide. Tidal range increases with the distance from the mouth of estuary to the apex of estuary, resulting from the estuary's funnel boundary. Although tidal range is not large, because of obvious seasonal change in

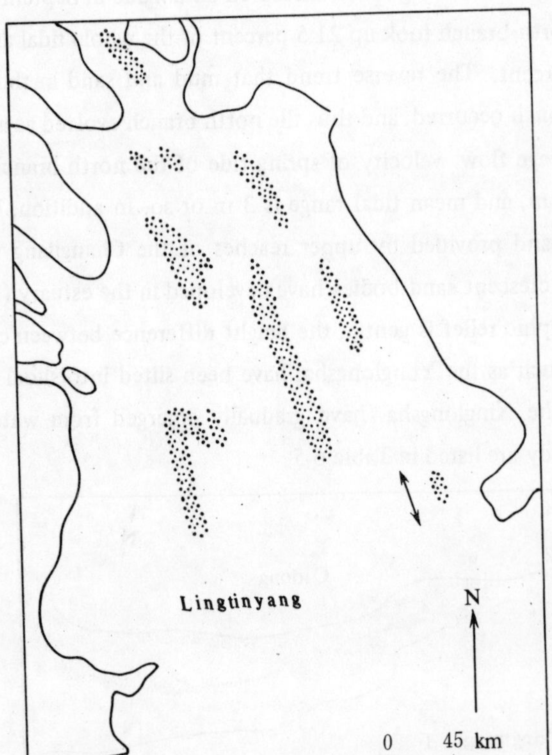

Fig. 6-19. The tidal current ridges in Lingdingyang of Z ıujiang Estuary.

the discharge of the Zhujiang River, tidal current action is comparatively remarkable, its velocity ranges from 1 to 1.5 knots, and the maximum velocity is 2 knots. With regard to the whole section of the Lingdingyang Bay, the discharge of the ebb current is larger than that of the flood flow. Development characteristics of the tidal current ridges in the Lingdingyang Bay depend on the above facts. Several sand ridges with the Zhujiang River mouth as apex, such as Jiaoyi Sands, Gong Sands, Ruo Sands etc., were formed, whose trend is approximately in NW–SE direction and which are some to ten more kilometers long and about 2 km wide, and the height difference between crest and trough is 5–6 m. Because of quite small velocity, the development of the tidal current ridges is not typical. At present, the west trough is still being silted. Because of strong tidal currents, the east trough tends to further scouring. Beyond doubt, it is important to study the development processes of the tidal current ridges and the channel between crests in the estuary for maintaining the depth of navigable channels in the estuary and for reclaiming land from delta.

V. Storm Current Ridges

The sand ridges by storms do not develop widely in the offshore areas in China. So far, the study on the storm current ridges has not yet been begun in China, thus we have less understanding of their distribution characteristics and development regularities. But the shelf of the South China Sea is rather extensive and the prevailing typhoons in summer are much stronger than the storms in the eastern shelf of North America. According to statistics, from 1949 to 1969 typhoons and strong typhoons attacked the South China Sea 196 times. The annual average was about 9 times, and there was a maximum of 18 times in 1964 (Sun Xiangping et al., 1981). In Weigel's opinion (1964), the velocity of wind-driven current is about one twentieth to one fiftieth of that of surface wind. Because the velocity of a strong typhoon is higher than 32.6 m per second, it can be calculated that a strong typhoon will drive a wind drift with a speed of 65.2–163 cm per second (about 1.3–3.2 knots) or more, and even a common typhoon with a speed of 17.2–32.6 m per second will also drive a wind drift with a maximum speed of 86–163 cm per second (about 1.5–3.2 knots). This value of wind drift speed comparatively approaches the ideal current velocity for the formation of submarine sand ridges, as advanced by N. H. Kenyon (1970). So the shelf of the South China Sea provides favorable hydrodynamical conditions for the formation of storm current ridges, and is a potential area for the distribution of storm current ridges.

During the glacial epoch the Zhujiang River, the Hanjiang River and other rivers pouring into the shelf of the South China Sea and carried a lot of sediments to accumulate there. The sediments of these old channels and deltas were submerged below water surface following the rise of sea level during post-glacial period so as to form many sand bars and shoals, which are substance foundation for the development of the submarine sand ridges.

The meteorological data of many years indicate that most of typhoon tracks in the South China Sea are approximately in W–E direction and only one or two typhoons formed in the South China Sea blow from south to north toward the Zhujiang River mouth. The data obtained by the First

Brigade of the Bureau of Marine Geology Survey using lateral sonar indicate that sand ridges and sand waves appear in groups in the offshore area of the Zhujiang River mouth, the Guangzhou Bay and the sea area east of the Leizhou Peninsula in the South China Sea, and that sand ridges develop at sea bottom of 10–15 meters deep and generally run parallel to the coast in W–E direction. The height difference between crest and trough is about 3–5 meters, the ridges are hundreds to thousands meters long, and the distance between crests of the heighbouring sand ridges is about 20–30 meters (Feng Wenke et al., 1982). It can be seen from this that the trend of sand ridges is identical with the direction of the main typhoon. It is considered that submarine sand ridges constructed by storm currents can be called storm current ridges. Because of the tidal range less than 3 m, and small tidal current velocity, the tidal current ridges cannot be formed there. A few sand bodies perpendicular to the coast are distributed in the submarine delta of the offshore area of the Zhujiang Estuary, and this may be related to the activities of typhoons in N–S direction.

VI. Conclusions

The submarine sand ridges widely developed in the shallow sea areas on the continental shelves is a kind of erosion-accumulation geomorphologic system. According to the hydrodynamic conditions of their formation, the ridges are mainly divided into two kinds, the storm current ridges and the tidal current ridges. China adjoins a vast shallow sea of the continental shelf that is coverd by abundant and loose relict sediments or modern sediments. The coastal relief is complicated, the tidal range is large and the tidal current is strong. In the South China Sea the storm currents caused by strong typhoon often occur. Therefore, many submarine sand ridges have been developing in the sea area adjacent to China. Their distribution and characteristics are concluded as follows:

1. The tidal current ridges mainly develop in the following morphological locations: the heads of the open bay, the mouth of the strait or the bay, the tidal estuary, as well as the offshore areas with the strong currents. There are two types of the plane shapes: the parellel and the fingered.

2. The ridges are made up of sand with well sorting, which is called tidal current sand. There are two saltation populations in the log-probability curves of the tidal current sand. Sorting of the first saltation population is better than that of the second one, and the curves have positive skewness. In the C–M grain-size pattern the tidal current sand is mainly in graded suspension segment and the projective range parallels to the straight line of C-M. The characteristics of mineral are that stable mineral with high hardness and large density concentrate. The chemical feature shows oxidation environment, Fe^{+++}/Fe^{++} and Eh values are high, and concentration of organic matters is low. Their biological assemblage is characteristic of relict deposit.

3. The conditions for forming the tidal current ridges are that the velocities of reciprocating tidal current range from 1 to 3.5 knots, that there is the supply of abundant sediments and that water depth is generally less than 35 m.

4. The basic dynamics of the formation and the development of the ridges is that the tidal current in trough moves forward in double helical flows, which is similar to the flows in a straight

river channel. That is to say, the flow in the center of trough converges down and erodes the trough deeply; the flow on the sides of trough diverges up and accumulates the ridges until equilibrium marine dynamic condition attained.

5. There are many tidal current ridges on the continental shelves of the Huanghai Sea, the East China Sea and the South China Sea, such as the Liaodong Shoal , the West Korea Bay, the Jiaozhou Bay, offshore area of Jiangsu, the north branch of the Changjiang Estuary, the Taiwan Shoal, the Lingdingyang Bay and the east and west mouths of the Qiongzhou Straits .

6. There are a few storm current ridges in the South China Sea where the strong typhoon with steady direction has caused strong storm current, which may construct a few small-scale storm current ridges.

References

[1] Belderson, R. H. and A. H. Stride, 1966, Tidal current fashioning of a basal bed, Mar. Geol., V. 4, pp 237–257.

[2] Feng Wenke, Bao Caiwang, Chen Junren and Zhao Xitao, 1982, Preliminary study on submarine relief of the northern South China Sea, Acta Oceanologica Sinica, V. 4, pp 465–466. (in Chinese)

[3] Feng Wenke and Bao Caiwang, 1982, Topographic and geomorphological characteristics of the South China Sea, Marine Geological Research, V. 4. No. 4, pp 80–93. (in Chinese)

[4] Friedman, G. M., 1967, Dynamic processes and statistical parameters compared for size frequency distribution of beach and river sand, J. Sedim. Petro., V. 39, pp 327–354.

[5] Geng Xiushan, 1981, The geomorphological features and forming factors of submarine relief in the Bohai Sea and the Huanghai Sea, Acta Geographica Sinica, V. 4, pp 424–433. (in Chinese)

[6] Houbolt, J, J. H. C., 1968, Recent sediments in the southern bight of the North Sea, Geol. Mijnbouw, V. 4, pp 245–273.

[7] Jin Bo, Bao Caiwang and Lin Jisheng, 1982, Geomorphic feature and origin of the deltas at the east and west mouths of Qiongzhou Straits, Marine Geological Research, V. 4, No. 4, pp 95–99. (in Chinese)

[8] Kenyon, N. H., 1970, Sand ribbons of european tidal seas, Mor. Geol., V. 9, pp 25–39.

[9] Li Chengzhi and Li Benchuan, 1981, Studies on the formation of Subei sand clays, Oceanologia et Limnologia Sinica, V. 12, pp 321–330. (in Chinese)

[10] Li Chunchu, 1982, The characteristics of estuary delta affected mainly by tidal, Marine Sciences, No. 1, pp 39–42. (in chinese)

[11] Li Congxian, Guo Xumin, Xu Shiyuan, Wang Jingtai and Li Ping, 1979, The characteristics and distribution of Holocene sand bodies in Changjiang Delta area, Acta Oceanologica Sinica, V. 1, No., pp 252–267. (in Chinese)

[12] Lin Zhenxia and Xia Dongxing, 1983, A preliminary study of tidal current ridges, Oceanologia et Limnologia Sinica, V. 14, pp 286–296. (in Chinese)

[13] Off, T., 1963, Rhythmic linear sand bodies caused by tidal currents, AAPG, V. 47, pp 327–337.

[14] Ren Mei-e and Zeng Chenkai, 1980, Late quaternary continental shelf of East China, Acta Oceanologica Sinica, V. 2, No.2, pp 94–106. (in Chinese)

[15] Ren Mei-e, Chang Zhenshang, Yang Juhai, 1983, Sedimentation on tidal mudflat of Wanggang area, Jiangsu China, Collected Oceanic Works, V. 6, No. 2, pp 103. (in Chinese)

[16] Sanders, J. E., 1963. North-South trending submarine ridges composed of coarse sand off False Cape, Virgina. Abs., Amer. Assoc. Petrol. Geol. Bull., V. 46, pp 278.

[17] Shepard, F. P., 1973, Submarine Geology, third edition, pp 287.

[18] Stride, A. H., R. H. Belderson and N. H. Kenyon, 1972. Longitudinal furrows and depositional sand bodies of the English Channel. Mem. Bur. Rech. Geol. Min., No. 79, pp 233–244.

[19] Stubblefield, W. L., J. W. Lavelle, T. F. Mckinney, and D. J. P. Swift, 1975, Sediment response to the Present hydraulic regime on the Central New Jersey Shelf. J. Sedim. Petrol., No. 45, pp 337–358.

[20] Shen Shunxi, et al., 1980, Heavy mineral composition and distribution characteristics in Taiwan Shoal, Huanghai Sea and East China Sea Geology, pp 98–104. (in Chinese)

[21] Sun Xiangping, et. al., 1981., A survey of hydrometer in the littoral area of China, pp 191. (in Chinese)

[22] Swidt, D. J. P., 1975, Tidal sand ridges and shoal retreat massifs. Mar. Geol., V. 18, pp 105–134.

[23] Swift, D. J. P., 1976, Continental shelf sedimentation. In: Marine Sediment Transport and Environmental Management, pp 311–350.

[24] Swift, D. J. P., G. L. Freeland and R. A. Young, 1979, Time and space distribution of megeripples and associated bedforms, Middle Atlantic Bright, North American Atlantic Shelf, Sedimentology, V. 26, pp 389–406.

[25] Swift, D. J. P., and Field, M. F., 1981, Evolution of a classic sand ridge field: Maryland Sector, North America Inner Shelf, Sedimentology, V. 28, pp 461–482.

[26] Weigal, R. L., 1964, Oceanographical enginnering, prentice-Hall, Englewood, Cliffs, New Jersey., pp 532.

[27] Wright, L. P., 1978, River deltas, Coastal Sedimentary Environments, pp 55–68.

[28] Xia Dongxing and Liu Zhenxia, 1984, The formation mechanism and devolopmental conditions of tidal current ridges, Acta Oceanologica Sinica, V. 6, No. 3. (in Chinese)

[29] Zhao Huanting, 1981, Bottom relief of lingdingyang of the Zhujiang Estuary, Acta Oceanologica Sinica, V. 3, pp 255–274. (in Chinese)

[30] Zheng Tieming and Zhang Junyuan, 1980, A preliminary study of submarine relief and sedimentary characteristics in the Taiwan Shoal and Its Neighbourhood, Huanghai Sea and East China Sea, Geology,pp 52–66 (in Chinese)

[31] Zhong Delin, Shen Xianzhong, Xia Dongxing and Liu Zhenxia, 1983, Explanation of satellitic photography of the ancient Changjiang Delta region in Early Holocene, Marine Sciences, No. 2, pp 16–17. (in Chinese)

[32] Zhou Changzheng and Sun Giasong, 1981, On the genesis of the shoal off North Jiangsu, Marine Geological Research, V. 1, No.1, pp 83–91. (in Chinese)

by Liu Zhenxia and Xia Dongxing

First Insitute of Oceanography, National Bureau of Oceanography, Qingdao

Chapter Seven

Coral Reef Sediments

Among various types of coasts of China, coral reef coast is important and unique. With the exception of one or two islands (e.g. Gaojianshi of Xisha Islands, consisting of tuff agglomerate-lava with a K-Ar dating of 2,050,000 years BP; Plate 7-1), China's islands in the South China Sea all consist of coral reefs, and are called coral islands. From the viewpoint of sedimentology, it is the general name of lime-sand (gravel) islands and reef islands.

Coral reefs of China are distributed in four regions, namely, Taiwan, coasts of South China, Hainan Island, and South China Sea Islands. The South China Sea Islands include Dongsha Islands, Zhongsha Islands, Xisha Islands, Nansha Islands, and Huangyan Island, 160 nautical miles east by south of the Zhongsha Islands.

Dongsha Islands are situated on the continental slope of the northeastern South China Sea, mainly consisting of Dongsha Island, Beiwei Bank, and Nanwei Bank. Dongsha Island, situated in the western part of an atoll flat, is about 2 km^2 in area with a topography of high northeast and low south-west[1]. The configuration of this island is crescent and so fishermen call it "Crescent Island".

Xisha Islands, consisting of 34 island, banks, reefs and shoals, are developed on the step-like continental slope. According to distribution, the islands are divided into 2 groups: the eastern group is named Xuande Islands (Xuande horse-shoe reefs), of which the largest one (1.85 km^2 in area) is Yongxing Island, which is also the largest one among the South China Sea Islands. In the eastern group there is a highest island–Shidao Island, with an elevation of 16.1 m; it is also the highest one among all the South China Sea Islands.

Zhongsha Islands are virtually a completely submerged large atoll. It is the largest one among all atolls of the South China Sea Islands. It extends in an ellipse way with a northeast-oriented long axis up to 76 nautical miles long and 33 nautical miles wide. At its periphery there are about 20 groups of shoals and reef banks, most of them are submerged about 12–26 m deep and the shallowest one is 9.1 m deep (Manbu Shoal). In the central lagoon region of the northeastern part, there are also 6 shoals and reef banks, virtually patch reefs in the lagoon of the atoll. Most parts of the lagoon have a water depth of 60–80 m; the southwestern part is the deepest (up to 109 m). Except the northeastern edge slope, edge slopes in other places of the Zhongsha large atoll are all fairly steep[2].

Nansha Islands are situated in the South China Sea basin. This is an island group with the largest number of islands, reefs, shoals and banks in the South China Sea Islands and with the widest distribution. Its islands, reefs, shoals, and banks large or small, number 230 odd with a distributional scope of over 500 nautical miles long from south to north, and of more than 400 nautical miles wide from east to west. Of these, there are 25 single islands (the larger ones number 10, the largest one is Taiping Island with an area of only 0.85 km^2), 77 shoals, 128 banks and reefs[1].

Huangyan Island, situated in the eastern part of the South China Sea basin, is 3500 m higher than the basin. It is a typical atoll, like an isosceles triangle. Large reef blocks on the central reef flat of the atoll number a hundred, and are 0.31–1.5 m higher than the sea surface. Its lagoon has an area of about 150 km^2[3].

As indicated above, the present chapter deals with not only coastal sediments on coral reefs, but also the lime-sand (gravel) island sediments and island reef sediments.

Viewing from biotic reef, carbonate sediment and diagenesis, though sedimentation is indeed a prelude, diagenesis seems to be an accompanying process, i.e. syngenetic (hypergene) diagenesis. Hence, in the present chapter, rock types and cement types of reef rocks should be mentioned, but diagenesis includes not only cementation, but also transformation, replacement, dissolution, and filling. However, these problems the author and his colleagues have already dealt with in other papers[4–9] and so, they will be merely quoted here for discussion. It should be stated that for coral reef sedimentation, we only deal with characteristics of the facies or sub-facies zones, but are not going to sum up with the popular sedimentation "model" concepts. This is because though many other authors have their own sedimentation "models", there is still not a consistent opinion and so, they will not be discussed here.

Since 1962, the author has been engaged in coral reef research. He is grateful for the concerns and direction of his teacher Prof. Ren Mei-e, who also is the revisor of the present chapter. Zhu Yuanzhi, Zhong Jinliang, Ne Baofu, Sha Qing'an, Wang Guodong, Song Chaojing, Xu Zhongfan and Guo Lifen have taken part in the field work and in the laboratory work. The author is much obliged to Zou Renlin, Xie Yixuan, Liu Zhaoshu, Wong Chong, Zhang Tiangang, Kong Jinxiang, Zhou Mingfu, Xu Zukang, Mao Shuzhen, Shi Xiaoyuan, Luo Kaifu, Xu Jinzhi, Yu Shan, Yang Shifeng, Ye Longfei, Huang Chengfa, Chen Shaomou, Zhao Huanting, Zhang Chang'an, Liu Zengyi, Dong Weiguang, Jiang Guoyuan, Wang Qinyuan, Huang Shuren, You Piaoxiang, Chou Dezhong, Chen Honglai, Huang Guangwei, Chen Suhua, Lin Xiuzhen, Chen Lihong, Zeng Zhaoxuan, Zeng Wenzhou, Chen Shijian and Luo Hui for thier helps. Part of the material and viewpoint of the present chapter has been disccussed in various occasions with the following foreign colleagues: D.R. Stoddart (Cambridge Univ., UK), J. Chappell (Australian National University), B. Thom (N.S.W.University, Australia), L.D. Wright (Virginia Ocean Science Institute, U.S.A.), O.K. Emery (retired, U.S.A.), P. Toreora (Rome University, Italy), and D.R. Choi (Miami University, U.S.A.). The author would like to take this chance to express his thanks to them.

I. Major Reef-Building Organisms

In view of material composition, coral reefs in China, either primary or secondary, or loose accumulatants, are all mainly of Hermatypic corals. Other biotic components in order are molluscs, calcareous algae, foraminifera, echinoderms, crustaceans, *Millepora, Alcyonacra*, bryozoans, etc.

1. Hermatypic corals

Reef-building corals refer to the shallow water hermatypic corals of Anthozoa. In China 45 genera (Table 7-1), 179 species and sub-species[1)] have been recorded, all belonging to Indo-Pacific fauna.

According to the many-year investigations by researchers of the South China Sea Institute of Oceanology, Academia Sinica, of the hermatypic corals found in reef areas of China, *Stylophora* and *Seriatopora* of Pocilloporidae can only be collected in the areas of the South China Sea Islands. Distribution of Pocilloporidae, either in quantity or in species, is decreasing from south to north. Distribution of Fungiidae is similar to that of Pocilloporidae. In the Xisha Islands area, there are not less than eight species of Pocilloporidae, namely, *Fungia actiniformis, F. actiniformis Palawensis, F. danai, F. echimata, F. fieldi, F. fungites, F. repanda, F. scutaria;* but in the Hainan Island area, only three species can be seen, i.e. *F. fungites, F. echinata* and *F. paumotensis.* However, distribution of *Turbinaria* of Dendrophyllidae is just the opposite, that is, it shows a tendency of increase from south to north either in species or in quantity. For example, in the Xisha Islands area, only 2 species (*Turbinaria irregularis* and *T. crater*) can be seen, but in the Hainan Island area there are five species (*T. agaricia, T. mantonae, T. peltata, T. undata* Bernard, and *T. crater*), whereas along the coast of South China, they amount to 10 species (*T. contorta, T. elegans, T. foliosa, T. globalaris, T. mesenterina, T. titizimaensis, T. undata, T. peltata, T. agaricia* and *T. crater*).

According to the developing morphology, hermatypic corals may be divided into massive and branching types. The former, such as *Porites, Favites, Favia, Platygyra* and *Symphyllia,* are dominant in the high energy fore-reef zone due to their strong wave-resistance, and they are also the major builders of the primary reefs on reef flats. The latter, such as *Acropora* (Plate 7-2), *Montipora* (Plate 7-3), grow rapidly, sometimes with a growth rate many-fold of that of the massive ones, even to a magnitude order. However, as the latter has a poor wave-resistance, their fragments usually are the important components of secondary reefs or loose accumulatants, shch as sand bars and sand islands. Branching corals are mostly distributed in the low energy environments, e.g. hinter-reef lagoon or in negative topography on reef flats.

2. Molluscs

There is a great variety of mollusc species in reef areas of China. In Xisha Islands area, over 500 speices of marine shellfishes have been found and their shells take part in the reef-building. According to the statistics of the biomass on atoll reef flats, except that of the hermatypic corals, it is up to

---

1) Zou Renlin, Chen Youzhang, 1983, A preliminary study on the geographical distribution of shallow-water scleractinia corals from China, Nanhai Studia Marina Sinica, 4:89–96.

Table 7-1 Genera and distribution of major hermatypic corals in China

Genus	Distribution			
	Taiwan	Along the coast of South China	Hainan Island	South China Sea Islands
Psammocora	+	+	+	+
Acropora	+	+	+	+
Montipora	+	+	+	+
Pavona	+	+	+	+
Goniopora	+	+	+	+
Porites	+	+	+	+
Favia	+	+	+	+
Favites	+	+	+	+
Goniastrea	+	+	+	+
Platygyra	+	+	+	+
Hydnophora	+	+	+	+
Leptastrea	+	+	+	+
Cyphastrea	+	+	+	+
Galaxea	+	+	+	+
Merulina	+	+	+	+
Turbinaris	+	+	+	+
Pocillopora			+	+
Astreopora			+	+
Pachyseris	+		+	+
Fungia			+	+
Polyphyllia			+	+
Parahalomitya			+	+
Caulastrea			+	+
Echinopora			+	+
Diploastrea			+	+
Lobophyllia			+	+
Symphyllia			+	+
Acanthastrca			+	+
Pectinia			+	+
Stylophora				+
Seriatopora				+
Herpolitha				+
Halomitra				+
Euphyllia				+
Alveopora	+			+
Acrhelia	+			+
Xishasiderastrea				+
Ancropora		+		
Pseudosiderastrea		+		
Leptoseris			+	
Podabacia			+	
Echinophyllia			+	
Scapophyllia			+	
Catalaphyllia			+	
Plesiastrea		+		

$1997 g/m^2$, accounting for 25–50% of the total biomass. In loose sediments on reef flats, fragments of mollusc shells account for, in general, 15–27% of the grains.

Turbo, Strombus, Lambis chiragra, Lambis lambis, and various genera of *Cypraeidae* are commonly seen in the reef flat area. They are characterized by rapid calcification of shells, a short life cycle, and a high transformation rate of population; thus, they provide a certain amount of material for the coral reefbuilding in China's waters.

Bivalves taking part in the reef-building in this region are mainly *Tridacna* and *Ostrea,* especially *Tridacna.* There are six species of *Tridacna* all over the world, which can all be found in China's coral reef areas. *Tridacna* are symbiotic with zooxanthellae. In fringing reef areas, sometimes oyster reefs may be formed with *Ostrea* as the dominant species, which can be seen, for example, in the remotest corners of the Hainan Island.

3. Calcareous algae

Calcareous algae directly join in the reef-building processes, e.g. coralline algae of the red algal phylum, of which some genera are widely distributed in different climatic zones; those common in all oceans are *Lithothamniun, Jania,* and *Corallina;* others are only limited in tropical waters, such as *Porolithon, Neogoniolithon;* these two are typical tropic species. All genera mentioned above are commonly seen in the reef areas or on reef rocks in the country. Though these algae have not yet formed "marine algal ridges", on the fore-flats of these reef areas, genetic enclosure and cementation resulting from these algae are not negligible.

The green algae *Udalea, Halimeda,* etc. are also commonly seen in reef areas in the country and the aragonitic segments of *Halimeda* are a kind of special fragments in sediments of these areas. Drillings reveal that in the Tertiary reef rocks in the country, *Halimeda* segment debris have transformed into microcrystalline calcite, or into dolomite due to dolomitization. *Halimeda* are seen in wave action areas in reef area or in the rear-reef lagoons.

4. Foraminifera

There are over 100 genera of foraminifera, mainly benthic, in reef areas in the country. Viewing from the species composition, the foraminifera fauna in reef areas of the country belongs to the Indo-Pacific tropic neritic fauna. They have obvious reef-building significance. For example, Shidao Island in the eastern group (Yongle Islands) of Xisha Islands is zooxanthellae limestone consisting of calcareous algae and foraminifera and forming an elevated reef 16.1 m above sea level.

In the loose sediment of reef areas, there are lots of foraminifera and their debris which may account for 15% and more of the grains in the specimen[10]. On reef flats of the fringing reef of the Hainan Island, they may reach 7.32%[1)]. *Amphistegina, Calearina, Cymbaloporetta, Elphidium, Heter-*

1) Gao Zhiwen, 1982, Sedimentation characteristics of sediments on the modern coralline shore reef in Luhuitou, Hainan Island. (To be published).

bina Rotalia and others, altogether 20 genera, are commonly seen[1)].

Viewing from the shell wall types of foraminifera, the calcareous porous type and calcareous non-porous type are dominant, while the sandy cemented type is proportionally scarce[11]. Typical benthic foraminifera of calcareous porous type are Rotaliidae and Calcarinidae; and foraminifera of calcareous non-porous type are mainly *Quinqueloculina* and *Triloculina* of Miliolidae; typical foraminifera of sandy cemented type are Textulariidae. The shell walls of the above-mentioned foraminifera (including those of sandy cemented type) all are calcareous, and hence, they are very important in reef-building.

5. Crustaceans

Crustaceans in the reef areas of the country include barnacles, ostracods, shrimps and crabs.

Barnacles are only collected at the depth shallower than –7m along the coast of Gaojianshi Island, Xisha Islands[2)]. They attach to the tuff agglomerate lava. In loose accumulatants and reef rocks in the reef areas of the country, barnacle shells and debris have not yet been found, therefore, it may be suggested that they are not significant in reef-building.

Ostracoda may be found in loose accumulatants and reef rocks in the reef areas of the country. So far over 20 genera have been known, of these the neritic genera and species, such as *Triebelina, Xestoleberis* and *Quadracythereis,* are commonly seen.

There are plenty of shrimps and crabs in the coral reef area. They may live in the beach caves, or hide in coral colonies or holes of reefs. They may use mollusc shells as their habitats, some even lead a life that seems separating from the host, and in a symbiotic relationship with the host. At branches of some branching corals, such as *Acropora* and *Stylophora,* "small tumours" are often seen; if they are broken, small crabs may be found inside. In Xisha Islands area there are 40 species belonging to Xanthidae alone[12]. Some shrimps are also in a symbiotic relationship with their hosts, e.g. the well-known "living all life in the same cave" describes *Spongicola venusta* are symbiotic with sponge as their host. So, it can be seen that shrimps and crabs in the country are also significant in reef-building.

Investigations on ecology of reef flats indicate that besides reef-building corals, crustaceans may account for over 20% of the total biomass; their internal shells (mainly carapaces and appendages) consisting of calcified scales take part in reef-building and their fragments are easily distinguished under a microscope.

6. Alcyonacra

Of Alcyonarians only *Coenothecalina* produces calcareous skeletons, so it is obviously significant

1) By courtsey of Ms. Tu Xia for identification.

2) The specimen is deposited in the Laboratory of Marine Biology of the South China Sea Institute of Oceanology, Academia Sinica.

in reef-building, but it has never been found in reef areas of the country. However, *Heliopora* and *Tubipora* are commonly seen in the reef areas, especially in atoll areas in the country. Other Alcyonarians take part in reef-building with their spines secreted from the middle gelatinous coating. Hence, Alcyonarians are considered as "soft, plastic and viscously springy"[13].

On top of the elevated primary reef (with ^{14}C dating age of 4930±185–5180±190 years BP) along Luhuitou on the south coast of Hainan Island, there is an overlying limestone consisting of soft-coral spines with a thickness of 40cm[2]. In loose accumulatants in reef areas of the country, particularly in the samples collected from Xisha Islands, red *Tubipora* debris have been found scattering within, which indicates that *Tubipora* are widely distributed in the reef areas and take part in reef-building. In a word Alcyonarians actively take part in reef-building.

In addition, some reefphilic organisms also provide materials for reef-building. For example, debris of bryozoa may be found not only in reef rocks, but also in loose accumulatants in the reef areas. Debris of sea urchins and polysporous hydranths are also commonly seen.

Summing up, reef-building organisms in the country obviously are dominated with hermatypic corals and so it is very suitable to name the reefs as coral reefs[14]; yet, other reef-building organisms and part of reefphilic organisms also take part in reef-building. Thereby, coral reefs in China are comprised by components of both reef-building organisms and reefphilic organisms.

II. Coral Reef Sediments

Briefly speaking, coral reef sediments are a kind of marine carbonate sediments. Owing to the ecological adaptation of the "thermal-philic nature" of reef-building corals, they are roughly distributed in the waters between 24°S–24°N. According to a rough estimation, coral reef in the world provide carbonate sediments of 3×10^9 tons per year[1)] and distributed in an area of over 6×10^5 km^2.

According to sedimentary environments and sedimentation processes of coral reefs, four major facies, i.e., fore-reef sediment, reef-flat sediment, supratidal sediment, and lagoon (rear-reef) sediment, may be distinguished as follows:

1. Fore-reef sediment

Vertically, the fore-reef sediments may usually be divided into two sub-facies zones with the base-level of waves as their boundary, namely, the high energy zone above the base level and the low energy zone below the base level. In reef areas of China, the base level of waves is roughly at a water depth of 20 m. A concise description of the two sub-facies sediments is as follows:

Fore-reef sediments below the base level of waves: according to the analysis of samples, sediments

1) D.R. Stoddart, 1981, Scientific and technical data of geography (23), 63p. The Department of Geography, Nanjing University.

here are "mixed accumulation" of reef blocks and debris of organism skeletons. Here, the sedimentary processes are obviously under the effects of gravity, underwater cementation, and activities of benthic organisms, rather than action of waves. Under the splash of waves, coral colonies and algae above the base level of waves are partly pushed onto reef-flats, partly broken and fall below the base level of waves, and accumulate into rock debris heaps (layers). Debris of other organisms also shift downward to the underwater slope, where they can remain in a gravity equilibrium. Plankton living in the upper water layer, particularly the planktonic foraminifera (*Globigerina, Globigerinoides,* and *Globrotalia,* etc.). sink to the bottom of the sea below the base level of waves, and their shells become one of the major components of the bottom debris. There are not a few benthic organisms living below the water depth of 200 m, e.g., some deep sea benthic foraminifera (*Siphonina* and *Marginulina*), and certain worms that secrete calcareous tubeshells, as well as some algae.

In the sub-facies sediments, with the exception of reef blocks that consist of corals and algae, according to microscope identification of the composition of fragmental (grained) organisms, the component of foraminifera is dominant (of these, the planktonic species are dominant), next comes calcareous algae, and followed by mollusc shells and coral fragments (Plate 7-4); besides, there are some spines of organisms and other components. Among the other components, there are a few (or extremely scarce) grains that are non-calcareous dark brown fine grains.

The analysis of the debris sample collected at a water depth of 240 m from the southern slope of Xianfa Shoal, Zhongsha Islands, indicates that those with a sorting coefficient of 1.39 and grain size of 0.25–1 mm account for over 77% of the total (Table 7-2, Fig. 1).

Table 7-2 Grain size and biotic composition of the fore-reef sediments below the benchmar level of waves of Xianfa Shoal, Xisha Islands

Grain size (mm)	Number of grains	Of the total grains (%)	Biotic composition (number of grains)					
			Foraminifera	Mollusc shells	Corals	Calcareous algae	Spines	Others
>2	10	0.56	2	1		2	2	3
2–1	204	11.50	52	43	21	70	12	6
1–0.5	739	41.06	279	107	107	206	26	24
0.5–0.25	647	36.45	219	122	113	184	3	6
0.25–0.1	102	5.74	48	12	23	16		3
<0.1	73	4.69	25	7	7	29	2	3
Total	1775	100.00	635	292	271	507	45	45

Besides on the biotic components listed in Table 7-2, observations on the reef blocks with a diameter of 5–8 cm show that the block surface is dark brown and coral fragments are cemented or enclosed by algae. But fractures are greyish white in colour and the surface is porous. Sometimes the biotic debris are arranged in layers.

Identifications of fore-reef sediments below the wave base level with X-rays diffraction and dyes show that the mineral composition is aragonite and magnesium calcite.

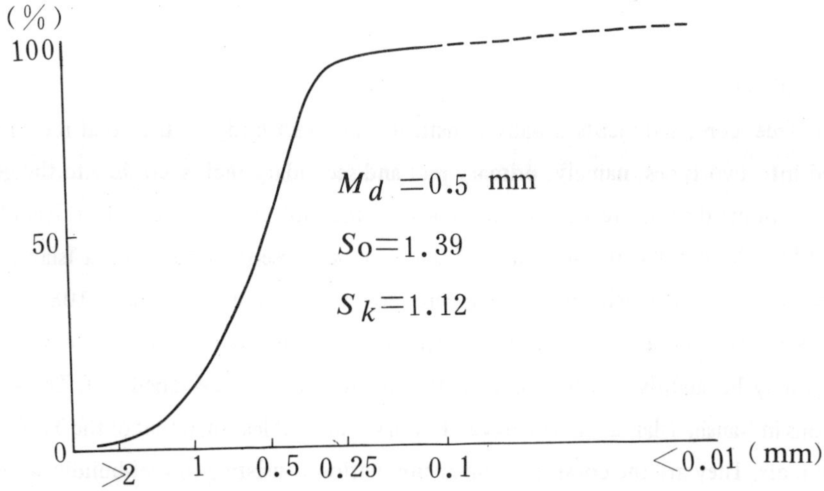

Fig. 7-1. The curve of grain size accumulation of the fore-reef sediments below the benchmark level of waves about Xianfa Shoal, Zhongsha Islands.

Topographically, the fore-reef sedimentary sub-facies below the wave base level is usually with steep slopes and some slope sections are vertical wall-like, e.g. the fore-reefs of the Coral Island, Xisha Islands, and the eastern fore-reef of the Great Atoll, Zhongsha Islands.

It must be pointed out that there is a transitional relationship between the continental slope of the fore-reef sedimentary sub-facies below the wave base level and the deep sea plain. For example, at the station DA20 in the central South China Sea (14°57′05″N, 110°03′00″E) with a water depth of 449 m, it is a continental slope environment[15]. Here, sea bottom is clayey silt, the biotic components are foraminifera and ostracoda with the percentage content (individuals/50 g) of 98100 and 118 respectively, and no trace of corals and algae can be found; while at the Station DA24 (14°55′ 36″N, 113°56′36″E), it is a deep sea plain environment with a water depth of 4220 m. Here, sea bottom is silty clay, the percentage content of biotic components (individuals/50g) consists of 624 radiolarians, 54 foraminifera, and one ostracod. The above-data indicate that water depth of the lower limit of the fore-reef sub-facies below the wave base level is less than 500 m. Considering from sample analyses and taking underwater topography into account, it is very likely that the lower limit depth is about 250 m.

Fore-reef sediments above the wave base level. This sub-facies is in the high energy zone with a changable relief. Here, it is difficult for ships to manoeuver and not easy for boats and diving. The several observations and samplings indicated that there are wave-resistant *Porites, Faviinae (Favia, Favites), Symphyllia,* and *Platygyra* growing here. And, *Porolithon, Lithothamnium* and *Hydrolithon* are commonly found. Space between reefs and wave-eroded trenches cut the region in question. The trenches are 2–5 m deep, 1–3 m wide and several tens of meters long. They cut upward into the reef-flat, usually extend radially from the reef-flat to the fore-reef. Major sediments in this sub-facies are coral fragments which can be commonly seen on the trench bottom. But here, they are only temporarily accumulated; they are transported and grinded by waves and currents, and migrating up and down.

2. Reef-flat sediments

Reef-flat (reef-core) sediments usually constitute the main body of the coral reef, and can be distinguished into two types, namely, primary reef and secondary reef, according to the genesis and rock character. Some drillings reveal that thickness of the Holocene reef in Hainan Island is greater than 20 m[14]. The coral reef formation of the Xuande horse-shoe reef, Xisha Islands, is over a thousand meters thick, of which Miocene-Pliocene dolomized coral limestone is 1082 m thick and the total thickness of Quaternary primary and secondary reefs is 169 m; thickness of Yongluo Atoll, Xisha Islands may be slightly less than that of the Xuande horse-shoe-shaped reef. Thickness of the reef formations in Nansha Islands and Huangyan Island should not less than that of the Yongluo Atoll.

Primary reefs. They are the coral skeleton accumulation consisting of reef-building corals that are sessile to the substrate, and then grown vertically and laterally. It is known so far that the base of the China's atolls developed on the continental slope is genissic granite dated 627 million years BP, and tuff agglomerate-lava dated 2.05 million years BP[24]; the base of the China's atolls developed on sea basins, such as that of Huangyan Island, is basalt. The base of fringing reefs in the country is even more complicated. Taking the fringing reef of Hainan Island as an example, the base is either Pre-Devonian quartzite or metamorphic rock system, or granite and basalt, or even mud[14]. Bedding of primary reefs is not clear, often massive, with apparent coral structure and well-developed pores. Of the biotic components, coral skeletons are dominant; and the reef-philic organisms mostly present a characteristic of accumulation in situ, e.g., mollusc shells are mostly complete individuals buried in situ. Individuals of boring organisms, such as boring algae, *Lima lima,* and *Margilus sp.* and holes and tunnels they bored are all completely preserved in the reefs.

Mineral components of the primary reef are mainly carbonatic minerals, such as aragonite, calcite and magnesium-calcite (Figs. 7-2, 7-3). And the carbonates are mainly calcium carbonate and magnesium carbonate. According to the author's determination of 27 samples collected from Hainan Island, on an average, content of calcium carbonate accounts for 89.25% of the total; of these 27 samples, 18 are with the content of over 90%, the maximum reaches 97.33% (from Tanmen reef-flat), and the minimum is still 61.03% (Linwu reef-flat). According to the analysis of the reef-flat samples collected from Xisha Islands, content of calcium carbonate is 57.60–95.00% and that of magnesium carbonate 4.30–4.60%. Although content of magnesium carbonate is larger than that of calcium carbonate in sea water, quantity of calcium carbonate taken by reef-building corals from sea water is larger than that of magnesium carbonate and such peculiar biochemical function results in the concentration of content of calcium carbonate in primary reefs. It should also be pointed out that there is a relationship of rise and fall (and vice verse) between the contents of calcium carbonate and magnesium carbonate in primary reefs, i.e., when content of calcium carbonate is high, that of magnesium carbonate is low, but when content of calcium carbonate decreases, that of magnesium carbonate increases. This is mainly dependent on the algae in the primary reef samples. Usually, high content of magnesium carbonate is found in the primary reef that is in cohesion with or coated by algae.

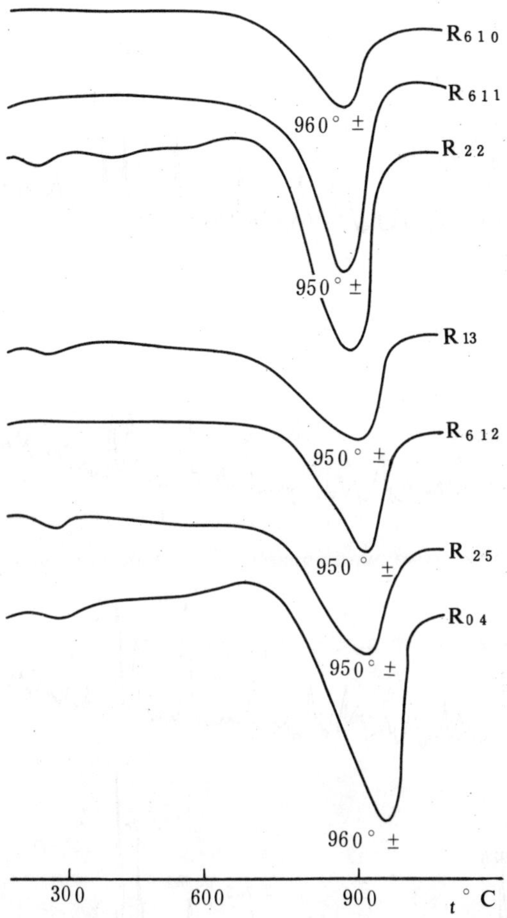

Fig. 7-2. Differential thermal curve of carbonatic mineral in primary reefs.

Reef-flats consisting of primary reefs are several meters to 3 km wide. Usually, reef-flats of fringing reefs are relatively narrow, while those of atolls are broader. Relief of a reef-flat is rugged with well-developed micro-karst landscape[5]. The planar form of a reef-flat is shown in Fig. 7-4.

Secondary reefs. Under sea erosion, coral colonies and primary reefs are destroyed and transported elsewhere, become heterotopical biotic debris and reef fragment heaps, which undergo diagenesis and form coralline gravel (sand) debris limestone. It bears the general features of littoral deposits, such as the well-developed horizontal bedding and cross-bedding. Deposits are obviously mechanically differentiated, and rounded and sorted to a certain extent. According to grain size, the composition may be classified as reef block, coral-shell-gravel and coral-shell-sand debris.

According to energy, topography, geomorphology and the sedimentary diagenetic environment, a reef-flat from sea to land (island) may be divided into three zones, namely, outer reef-flat high energy zone, middle reef-flat reef block deposited zone, and inner reef-flat low energy zone (Fig. 7-5).

The outer reef-flat high energy zone is always washed by high seas; here, the reef-building coral colonies grow flourishingly at the water depths of 3–5 m (fringing reef) and shallower than 8 m (atoll and platform reef), and are mainly wave-resistant genera and species, such as *Porites, Platygyra, sym-*

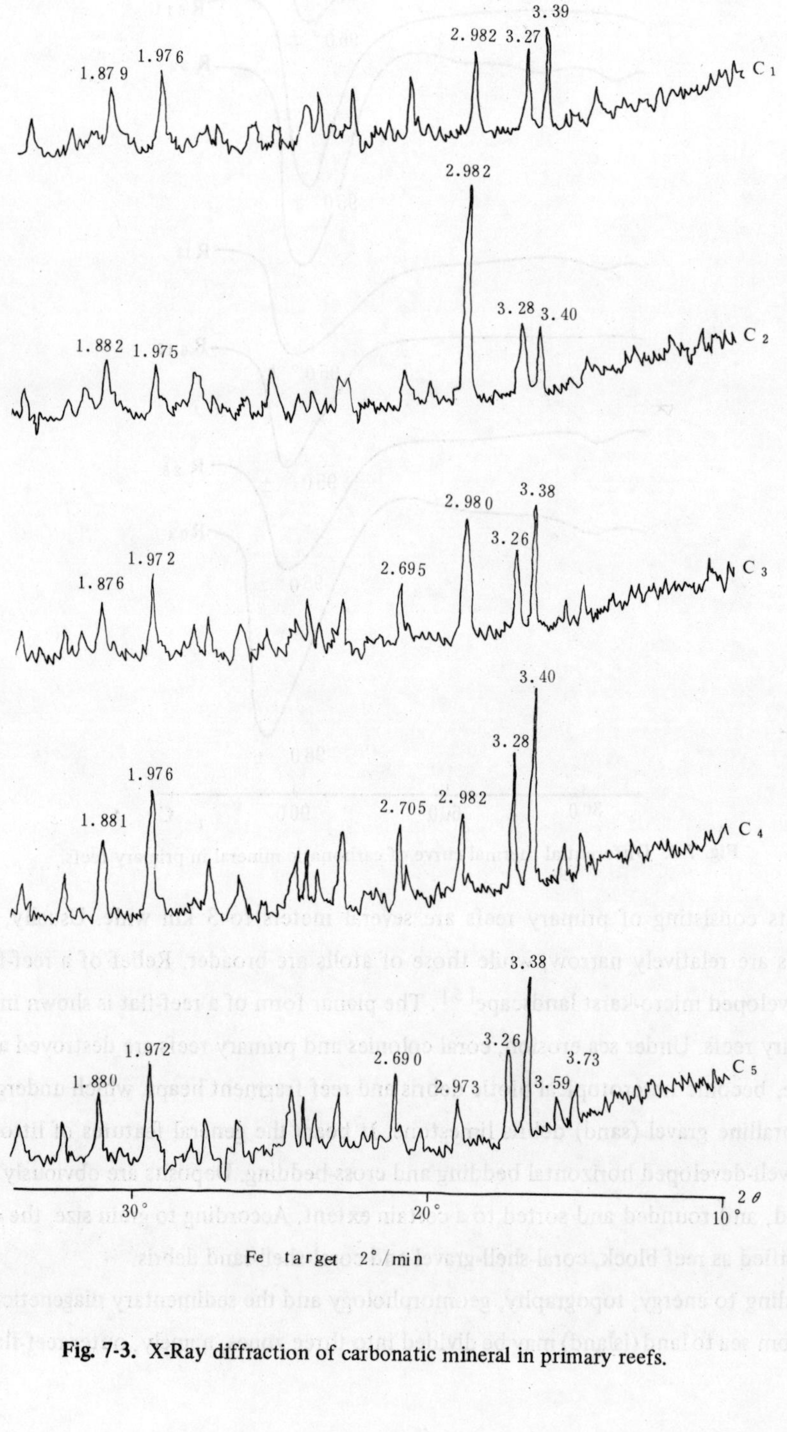

Fig. 7-3. X-Ray diffraction of carbonatic mineral in primary reefs.

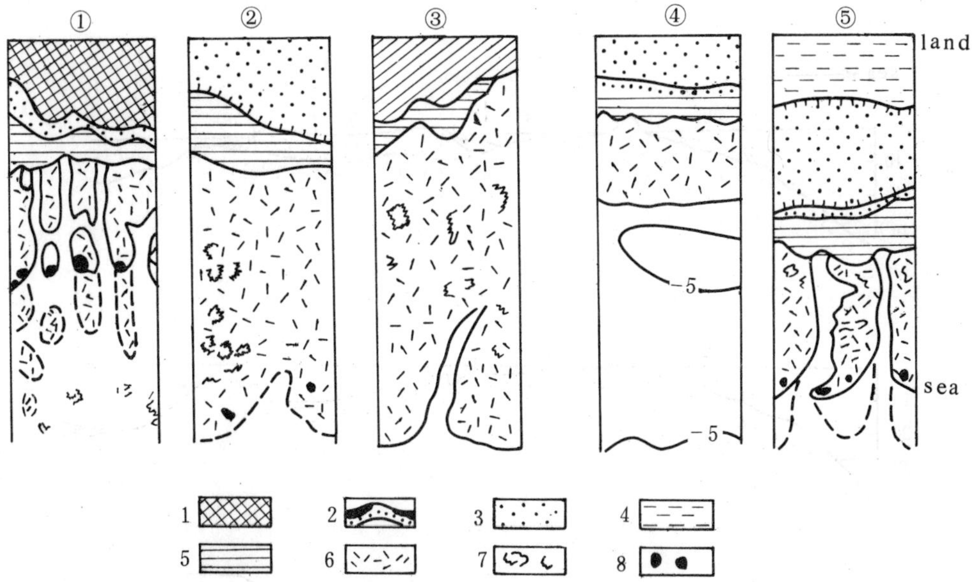

Fig. 7-4. Planar form of the reef-flat.

(1) Dongmao Island, Hainan Island; (2) Bangtang of Hainan Island;

(3) Baohu Bay of Hainan Island; (4) Basuo Bay of Hainan Island·

(5) Jinqing Island, Xisha Islands.

1– Bed rock; 2– Beach & beachrock; 3– Sand bar or sand cay; 4– Atoll lagoon;

5– Reef embankment (Karst basin); 6– Reef flat; 7– Karst hoodous ("lapies");

8– Reef lumps.

phyllia, Favia and *Goniastrea;* some branched corals may also exist, such as *Acropora humilis* (Dana), *A. bruggemanni uncinata* (Brook) and *Pocillopra damicornis* (Linnaeus) etc. In this zone, the phenomena of cohesion and coating by algae are relatively apparent, which reflect the wave-resistance of the reef-building organisms, but geomorphologically, "algal ridges" have not been formed yet. Slope of this zone is about 5°–10° or steeper; trenches developed perpendicularly to the coast give this zone a saw-toothed surface and loose sediments only temporarily stay in the trenches.

The middle reef-flat reef block deposited zone is situated in the highest position of the reef-flat. On the reef-flat that is near low tide level, (the reef block accumulation may be emerged 1–3 m or so above low tide level) the living corals can only better develop in the trenches (Plate 7-5), the horizontal extention of *Acropors prostrata* (Dana) covers parts of the trench. Though some middle reef-flat have not emerged, biotic fragments covered and coated by algae may also by found, e.g. the middle reef-flat on the northeast coast of Shu Island, Xisha Islands. However, most of the middle reef-flats on the South China Sea Islands are reef block concentrated areas, e.g., on the middle reef-flat of the Huangyan Atoll, reef blocks number hundreds[4]. On these middle reef-flats, diameter

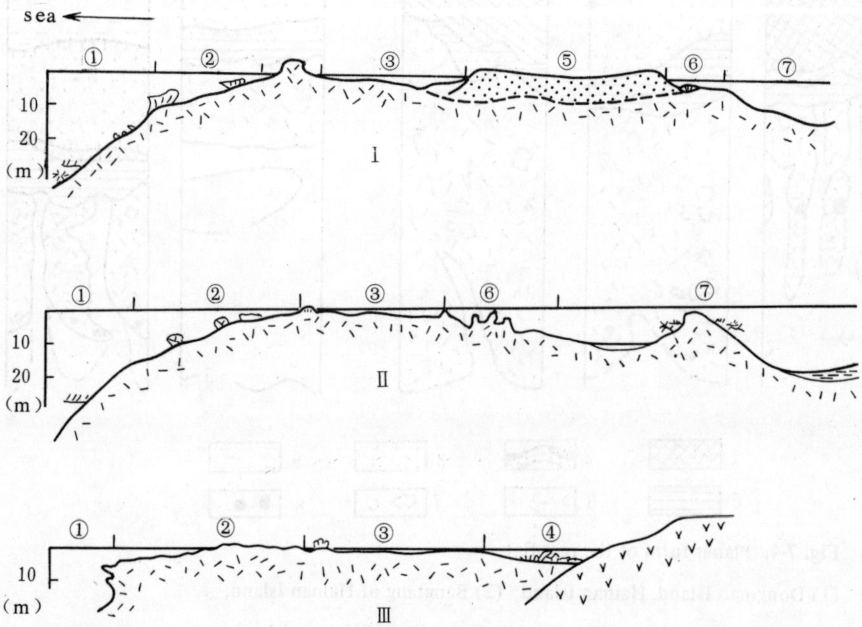

Fig. 7-5. A sketch of the reef-flat zonation.

I– Reef-flat in atoll with sand cay;

II– Reef-flat in atoll without sand cay;

III–Reef-flat in fringing reef.

(1) Fore-reef; (2) Oute reef-flat; (3) Middle reef-flat;

(4) Inner reef-flat or reef embankment; (5) Sand cay or reef-rock island;

(6) Lagoon slope; (7) Lagoon bottom and patch reef.

of large blocks reaches 3.5 m (on the reef-flat of Dongmao Island by the south coast of Hainan Island). It is suggested that middle reef-flats are energy-dissipated accumulation areas which, comparatively speaking, are the most oligobiotic zone.

The inner reef-flat low energy zone is situated between the middle reef-flat and the lagoon in atoll areas or between the middle reef-flat and the beach in fringing reef areas; whereas on platform reef-flats, this facies zone is absent. Geomorphologically, this inner low energy zone that constitutes the reef-flat of fringing reef may be called reefpool or dissolved basin. It is resulted from dissolution-sea erosion through the continuous percolation of freshwater which mixes with seawater. On this negative relief, sediments are the mixed accumulation of terrigenous debris and fragments of marine organisms (mainly reef-building organisms) with the grain size of medium sand-coarse silt. Mangroves grow here (seen in Heishidi of Hainan Island). Some are covered with communities of sea turtle weeds (seen on the south coast of Hainan Island). Inner reef-flats of atolls may have either sedimentary characteristics: one is that of the inner reef-flat on leeside, such as those on the northeast coasts of the Coral Island and Jinqing Island, Xisha Islands; on these flats filled with biotic-sand debris, sea urchins densely concentrate (<10 individuals/m^2) and ophiuroids are not seldom seen, but corals do not grow here. The other one is that of the inner reef-flat facing the prevailing wind, where wind-wave currents and the flood tide current carry plenty of food and therefore, not only corals grow here, but there is also a variety of reef-philic organisms.

Growth rate of coral reefs on the reef-flat of fringing reef is 1 mm/year[14], that on atolls and platform reefs– 2–3mm/year[10,14].

3. Supratidal sediments

Supratidal sediments in the coral reef areas in China usually form sand bars, carbonate islands, including sand (or gravel) cay consisting of loose deposits, and islands consisting of reef rocks (Plate. 7-6,7). In the fringing reef area, besides sand bars consisting of loose deposits, coral reef terraces consisting of raised reefs may also be seen (e.g. those seen on the south coasts of Taiwan and Hainan Island).

For the sand bar in a fringing reef area, the sediment composition is mainly characterized by the terrigenous debris. Take the Luhuitou fringing reef area on the south coast of Hainan Island as an example: content of terrigenous debris accounts for 3.99% of the total, but grain debris of carbonates, such as shell and coral, amounts to over 93% grain size parameters of the sediments are Md 1.51 mm, σ_1 0.80, SK –0.16, KG 1.03. In the Shalao fringing reef area on the east coast of Hainan Island, except for the percentage content of terrigenous debris that is increased, the rest are similar to those in the Luhitou fringing reef area.

For coral reef terraces in fringing reef areas, as they are dealt with in other papers [5,16,24], they are not mentioned here.

Reef rock islands. Shiyu and Shi Island of Xisha Islands may be cited as representatives. Shiyu Island (Jinsha Island) is the only reef rock island consisting of coral reef limestone (coral sand-gravel debris limestone) in the west group (Yongluo Atoll) of Xisha Islands. It is on the same reef-flat with Jinqing Island to the south, 2.3 m above sea level. The reef rock displays a feature of sedimentary anticline (Fig. 7-6). Age of the reef rock dated by ^{14}C is 2630±200 years BP. The reef rock is cemented by carbonate mud and no trace of freshwater processes has been found so far, which indicates that its diagenesis occurred in an intertidal environment, that is, the loose deposits of the tidal sand bar consolidated into rocks and was in the supratidal environment not long ago. The northeast coast of Shiyu Island faces the northeast monsoon-induced wave high energy area, and cliffs are formed by sea erosion.

Another reef rock island is Shi Island. It is a component of the east group (Xuande Islands) of Xisha Islands and is situated on the same reef-flat with Yongxing Island, the largest sand cay of the country (with an area of 1.85 km^2)[1]. The distance between these two is 1130 m and is wadable when it is low tide. Shi Island is 16.1 m above sea level

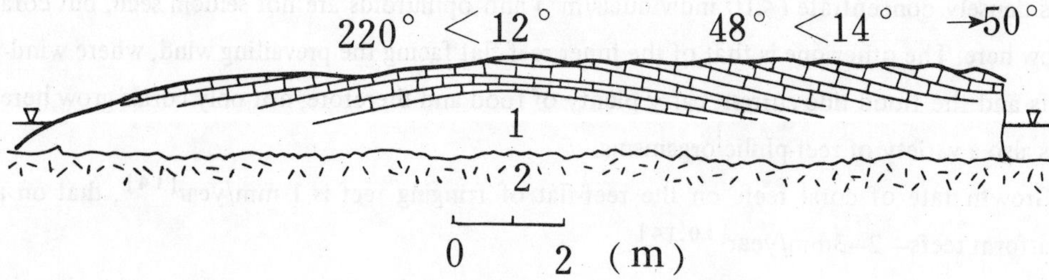

Fig. 7-6. Shiyu Island (Jinsha Island) geologic profiles.

1– Coral sand & gravel rubbish limestone, froming sedimentary anticline;

2– Connate reef froming reef flat.

when it is low tide. Shi Island is 16.1 m above sea level1), being the highest reef rock island in the country. The Shi I. reef rock consists of biotic-sand debris limestone; its biotic composition is mainly coralline algae, next comes foraminifera, then, lamellibranches, corals, echinoderms and gastropoda. Cement consists of grained calcite. According to the biotic composition, the reef rock may be called "foraminifera-algae limestone"[7]. Percentage contents of the chemical components are as floolws: CaO 50.22–52.22, MgO 1.75–2.47, Al 0.03–0.3, B 0.08–0.10, Cu 0.001–0.004, Fe 0.1–0.3, Mn 0.005–0.01, Na 0.03–0.05, P 0.1–0.6, Si 0.1–0.3, Ti 0.003–0.1.

The foraminifera-algae limestone is with clear bedding (Plate 7-8). From top to bottom, it may be divided into four layers:

That the supratidal sediment in reef areas in the country is considered as more significant is due to the fact that it form, snad cay, commonly seen on the reef flats of atolls and table-reefs.

Characteristics of grain size of the sand cay sediments. Analyses of samples indicate that sand cay sediments in the country are mainly composed by fine gravel and coarse sand, which may account for 62.5% of the number of the samples. It is generally accepted that if sediments possess a sorting coefficient So=1–1.58, then, their sorting is good. According to this standard, majority of the sand cay sediments in the country are of good sorting, with the exception of the sediments on the southwest coast of Chuanfu Island (So=1.90), on the northwest coast of Shenhang Island (So=2.02), and on the southwest coast of Zhaoshu Island (So=1.90). Hence, generally speaking, waves are in full play here. This is similar to the sand cay sediments on the atolls of the Pacific Ocean(17). The sand cay sediments in the country are also with a relatively good roundness (0.5–0.7 degree), but a poorer sphericity[10]. Sediments in different azimuths of a sand cay are with different grain sizes. For example, on Chuanfu Island, which is located in the northern part of Yongluo Atoll, sediments on the north coast are with the largest grain size (Md=7.8 mm), indicating the obvious effect of the strong northerly wind-induced waves; next comes that on the southwest coast (Md=1.04 mm), refrecting the effect

1) Remeasured in 1975.

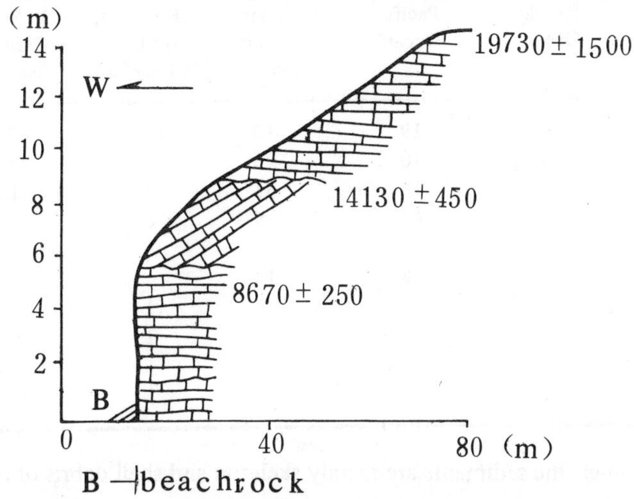

Fig. 7-7. Geologic profiles of Shi Island foraminifera-algae limestone.

Bed 4: 4.5m thick, thin-bedding, a single bed is 1–3cm thick, occurrence nearly horizontal, consisting of medium sand, good sorting, rocks greyish white-yellowish brown in color, dissolved eluvial root tubes commonly seen with a hole in the center of the tube. Its age is 19730±1500 years BP by ^{14}C dating.

Bed 3: 3–5 m thick, thin-bedding, occurrence 250° L38°–318°L15°, consisting of medium sand layer containing fine sand, good sorting, yellowish brown in color. Its age is 14130±450 years BP. by ^{14}C dating.

Bed 2: consists of medium-fine sand, thin-layered, occurrence 200°L33°–230°L27°, the bed as a whole 4.5 m thick, rocks greyish white and yellowish brown in color. Its age is 8670±250 years BP by ^{14}C dating.

Bed 1: consists of fine sand layer containing medium sand, thin-layered, occurrence nearly horizontal, rocks yellowish brow in color, visible thickness 2 m.

of the strong southwesterly wave in summer; while on the east and south coasts, sediments are with finer grain sizes of 0.65 and 0.78, respectively. As the southern part of Yongluo Atoll is under the effect of the strong southwesterly, so the southwestern part and the south part of Shenhang Island face the strong wave; thereby, coral gravel barriers have been accumulated and grain size of the gravels reaches 100–200 mm; while on the northwest of the island, sediments are with Md=0.7 mm. It can be easily seen that there is a great difference between the two. In addition, in different relief positions of a sand cay, sediments are also with different grain sizes; generally, grains near the high tidal line are coarser, whereas on top of the shore barrier, where the wind strengthens, grains are finer.

Biotic composition of the sand cay sediments. Sediments on sand cays of the country are basically of marine nature, no terrigenous materials have ever been found within, which is consistent with the biotic composition of the sand cay sediments in other reef areas of the world (Table 7-3).

It is shown in Table 7-3 that in China's sand cays either developed on atoll reef-flats (e.g. Dongdao I.), or formed on table (platform) reef flats (e.g. Zhonjian I.), the sediment components are mainly corals, next comes mollusc shells, and then, calcareous algae and foraminifera.Whereas in sand cays

Table 7-3 Biotic composition of the sand cay sediments on coral reefs

Place / Com-Position Sea	Perlas Arch., Caribbean	Hawaii I. Pacific Ocean	Midway I. Pacific Ocean	Caingal, Indian Ocean Sea	Dongdao I. (Hewu I.), South China Sea	Panshiyu I., South China Sea	Zhongjian I. South China
Hermatypic	25	9	19	10	60	67	73.7
Molluscs		33	10	6	16	23	21.6
Foraminifera	15	21	19	51	15	1.5	0.7
Coralline algae		32	47	9	6	7	2.5
Halimeda	60		5	14			
Bryozoa					1	0.5	0.1
Other calc. debris		5	2	3	1	0.5	1.0
Diatoms					1	0.5	trace quant.

on the oceanic atolls, though the sediments are mainly skeleton and shell debris of marine organisms, they are with different content percentages. For example, in sand cays on the Pacific atolls, sediments are mainly calcareous algae, next foraminifera, and coral skeleton fragments taking the third place. In sand cays in Caribbean Sea, sediments are still mainly calcareous algae, but coral frágments the second, and foraminifera the third. And in sand cays of the Indian Ocean, sediments are mainly foraminifera, next calcareous algae, and corals the third. Therefore, in different reef areas, contents of biotic components of sand cay sediments are different, but they are all carbonates. In the sand cay sediments of the South China Sea, there is a trace quantity of non-calcareous debris (diatoms and siliceous sponges etc.).

Coral fragments from the sand cays in the country are of Hexacorallia (Hermatypic), their grain number generally accounts for 55–80% of the total grain number. There are red *Tubipora* fragments scattering in each sample, which indicates that *Tubipora* are distributed in all reef-flats; it is consistent with the character of the Indo-Pacific type, as there are no *Tubipora* in the tropic Atlantic type[18]. Mineral composition of coral fragments is aragonite with fiber-like structure.

Shell fragments are of broken mollusc shells, generally account for 15% of the total number of the grains, ronging from 27% to 5%. Viewing as a whole, shell fragments take the second place in the total amount of loose deposits. Mineral composition of the shell beds is either totally aragonite, or the intercalated beds of aragonite and calcite, but high-and low-magnesium calcite may appear.

Foraminifera shells are largely rather complete; they all belong to benthic calcareous foraminifera. The shells are secreted by cellular plasma; most genera secrete shells of calcite or aragonite[19]. Majority of the outer shells is lens-like, very few are cone-like or disc-like; their size is mostly 0.5–1 mm, that of 0.5–0.25 in the second place, the third is the larger ones of 1–2 mm. Part or most radial spines of the foraminifera have been ground out, and some small swollen rooms have been leveled. In the samples collected from the bottom of reef pools, many primitive forms of hard bodies of foraminifera are well-preserved. Almost every sample contains foraminifera shells but with quite different number of individuals. In loose sediments of Zhaoshu Island, Dong (Hewu) I., Jinqingdao I., Yongxing I., Coral Island and Shi Island, there are more foraminifera shells, accounting for 6–15% of the total grain number in the samples, while those in the samples collected from Panshiyu and Zhongjian, less foraminifera shells are found, only accounting for 1.1% of the total.

Nearly every sample contains calcareous algae, but their grain sizes are different and their number is larger than that of bryozoa. Their apparent forms are either smooth, or rough with a single irregularly porous plane, or branchletlike, corticated or grain-like. Generally, they are white in color, but with gloomy luster. R.G.C. Bathurst (1971) pointed out that "various species of calcareous algae are distinguishable by the forms and sizes of their respective calcareous parts"[20]. They are either calcified into calcite (generally high-magnesium calcite) or aragonite, or precipitated in the cells and presented as calcified cell walls (red algae), or precipitated outside the cells, generally inside or around certain tissues, as a by-product of photosynthesis of plants (green algae). Grains of dead calcareous algae are coarser than those of living ones. Quantity of calcareous algae is often larger than that of foraminifera, generally accounting for 3–6% of the total number of grains, the most up to 14–22%, the least less than 1%.

Nearly every sample contains bryozoa hard bodies, but quantity is less than that of foraminifera, generally accounting for 1–2% of the total number of grains, the most up to 8% (e.g., in the northeastern part of Yongxing I.), the least not to 1% or without a single sign. Mineral composition of bryozoa hard bodies is high in magnesium calcite or aragonite, or the mixture of these two[20].

In the samples, no standard oolite grains are found; only some white calcareous flocculates that are easily separated can be seen under a microscope.

The non-carbonatic grains are mainly some non-calcareous sponge fragments and non-calcareous algal (e.g., planktonic diatoms) particles.

Trends of the sand cay sediments. For the sand cays with vegetation, particularly with high forests, such as Yongxing I. and Coral Island, the accumulation is relatively stable, because roots of the plants fix the sand and the plants proper can trap wind-winnowed sands to accrete the ground surface; such cays may be 5–8 m above sea level, e.g., Coral Island, 5.1 m above sea level, and Ganquan Is., 8.3 m. As for the cays without vegetation, their forms are not stable and with lower elevation, e.g., Quanfu, only 2.2 m above sea level, and Zhongjiandao, also merely 2.6 m. In 1953, Zhongjian was kettle-shaped and might be drowned under typhoon storm surges, but now it has become a shallow disc and with an obviously increased area. Fishermen there retailed that during 1977–1979, Zhongjian has changed a great deal, particularly the shallow lagoon in the central part has rapidly shrunk and got shallower and its water depth measured by the author is merely 50 cm.

For sand cays, typhoon induced-waves and storm surges are either constructive process or destructive force. The constructive process is usually present at the stage when the typhoon-induced waves are weakened. For example, in 1972, after the attack of the typhoon No. 20 on Xisha Islands, a new sand bar appeared at the southmost end of the northeast atoll of the Xuande Reef which is in a horesshoe shape; the sand bar is 3 m above sea level. The destructive force is not only present in the shifting of sand cays but also in the destruction of the sand cays covered with vegetation. It can break or even draw out high trees without roots left in the earth. Such process is especially obvious in the early stage of typhoon-induced high waves.

The statistics on directions of accumulation of over 20 sand cays (including sand bars) show that the accumulation mostly expands towards south-southwest[10], which is resulted from the process of prevailing wind-driven waves and undoubtedly, also from the process of sea currents.

According to the calculation of the unearthed relics from the sand cays, in recent 650 years, accumulation rate of the sediments, on an average, is 0.15 cm/year, while the growth rate of coral reefs in the sand cay areas is 0.2–0.3 cm/year[10]. This not only indicates that the growth rate of reefs is faster than the accumulation rate of sand cay sediments, but also demonstrates that sand cays in the reef areas of China are rich in resources of carbonite materials.

4. Lagoon sedimnets

Lagoons are neritic low energy environments. According to their water salinity, lagoons may be classified into three kinds, namely, normal lagoon, brackish water lagoon and super-saline lagoon.

The present text only deals with normal lagoons where water salinity is close to that in the offshore area. Most lagoons on the fringing reefs and atolls of China belong to this type. Water in the lagoon can exchange rapidly with water from the offshore area. Which first flows through the windward reef or tidal channel into the lagoon, and then withdraws through leeward reefs or tidal channels from the lagoon and so that the water in the lagoon can be renewed. Such lagoons are widespread in the South China Sea and are also seen along the coast of Hainan Island.

Sedimentary facies of lagoons with fringing reef. Guilcher (1954) suggested that coral reefs avoided lagoons on the sea shore, as there was not enough dissolved oxygen in the lagoon water and water temperature and salinity were too variable[21]. However, along the coast of Hainan Island, coral reefs are developed. Reefs preserved along the coast of a lagoon may be named as lagoon fringing reefs[14] in some lagoons. Actually, they are a variety of fringing reefs, their planar form varies with the lagoon form, but their environment differs from the common fringing reef environment.

Lagoon fringing reefs of the South China Sea are seen in the Xinchun Port and Yulin Port on the south coast of Hainan Island. Salinity of such lagoons is usually about 34‰, close to that of the offshore water. Their water temperature is also suitable for the growth of reef-building corals. Lagoon water exchanges with the offshore water through tidal channels, bringing in enough food needed by corals. Depth of lagoon water is shallower than 10m and wave is very weak. Hence, reef-building corals here are mainly branching corals of weak wave-resistance, and the *Fungia* that can avoid the precipitation of sand and mud. Taking the lagoon fringing reefs in Xinchun Port as an example, in the samples collected by divers, there are altogether 52 species and an unidentified species, such as *Psammocora contigua* (Esper), *Pocillopora damicornis* (Linnaeus), *Acropora formosa* (Dana), *A. pulchra stricta* Brook, *Montipora ramosa* Bernard, *Pavona decussata* Dans, *Fungia echinata* Pallas, *Porites andrewsi* Vanghan, *P. pukoensis* Vaughan, *Favia pentagona* (Esper), *Goniastrea aspera* Verril, *Turbinaria mantonae* Crossland, etc.

In lagoons, the above-mentioned corals grow in a water depth of –1 – –4m but individuals may grow at the water depth of –5 m[1].

Reef-flats of the lagoon fringing reef consist mainly of primary reefs, with an age of about 5000 years BP, Its surface is rather even, often covered by thin sedimentary layers of sand and mud debris and nearly no wave-eroded and dissolution eroded forms can be seen. At the rear margin of the reef-flat (the nearshore part), there are often mangrove swamp; and the fore margin of the reef-flat (towads the lagoon) slopes to the center of the lagoon with an extremely gentle slope.

The sedimentary facies of this kind of lagoons bears an obvious zonation (Fig. 8). Usually, at the lagoon mouth and in the tidal channel, sediments become coarse and coarse sand, coarse and medium sand debris are accumulated. In the lagoon there are still two samll patches of sediments of this kind, which may be the relic of an ancient channel. These coarse sediments reflect the transportation of strong currents. According to the in situ measurements by the South China Sea Institute of Oceanology, Academia Sinica, in the spring tide, the maximal current velocity of the surface layer in the middle part of the Xinchun lagoon mouth is 98 cm/sec and current direction is 114° in flood tide, and 110 cm/sec, 283° in ebb tide. Current velocity in the bottom layer is close to that in the surface layer. It shows that the velocity of the ebb tidal current is larger than that of the flood tidal current and so coarse sediments are accumulated off the lagoon mouth like a tongue. Inside the lagoon, sediments are distributed in ring-zonation; from the lagoon shore to the lagoon center, there are the medium-fine sand zone, the fine sand zone, and finally the silt zone in the centre (about –5m or deeper).

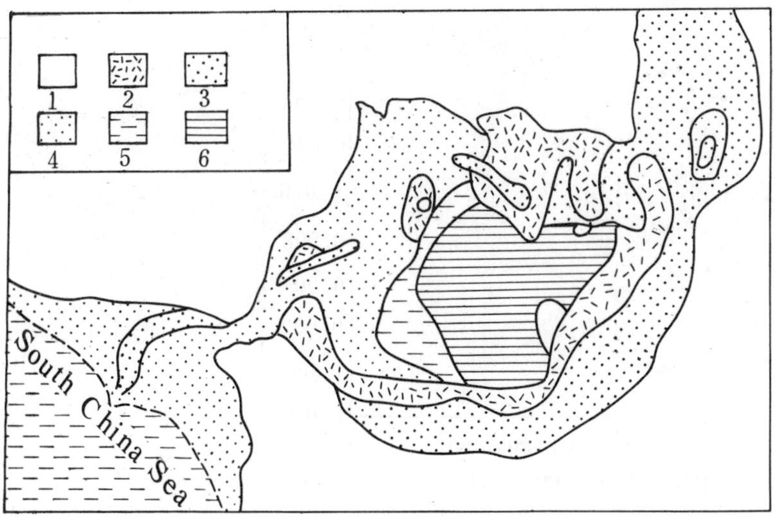

Fig. 7-8. The Xinchun lagoon fringing reefs and the substrates.

1– Land; 2– Reef flat; 3– Coarse sand; 4– Medium sand; 5– Fine sand; 6– Silt.

---

1) Huang Jinsen, Wang Guodong, Huang Shuren, Zhu Yuanzhi, Nie Baofu and Xu Zukang, 1975, An investigation report on the coral reef coast of Hainan Island, Proceedings of Coastal Geomorphology of the South China Sea. (2), pp. 47–168.

It should be pointed out that in the last decade, the ecological environment of the Xinchun lagoon where fringing reefs are developed has changed a great deal. This is because of the port construction and the mining of corals. In the early 1960's, on the underwater shore slope of the lagoon, coral fringing reefs were developed flourishingly; but now (grain data are from the samples collected in 1978) the slope is covered by medium-fine or fine sand, even by silt. The residual corals grow poorly; they are mainly *Acropora, Montipora ramosa* Bernard, and *Porites* in small lumps. The human destruction leads to changes of the ecological environment and to the extinction of coral fringing reefs, as well as to the variation of grain size of sediments.

By analyzing ϕ_{25} – Md$_\phi$ –ϕ_{75} indexes of the sediment of the Xinchun lagoon fringing reefs (Fig. 7-9), it can be clearly seen that in different sedimentary environments the grain size is different. On the lagoon bottom it is mainly the silt, with Md$_\phi$ 5.5–9.1, i.e., from medium silt to clay. Most of the quartiles bear an obvious skewness. With the exception of individuals that are nearly symmetric, most of the Sk$_\phi$ values are of positive or negative skewness. The coefficients of sorting are all poor or very poor (2.03–3.96), except for individuals that are intermediate (Fig. 7-9).

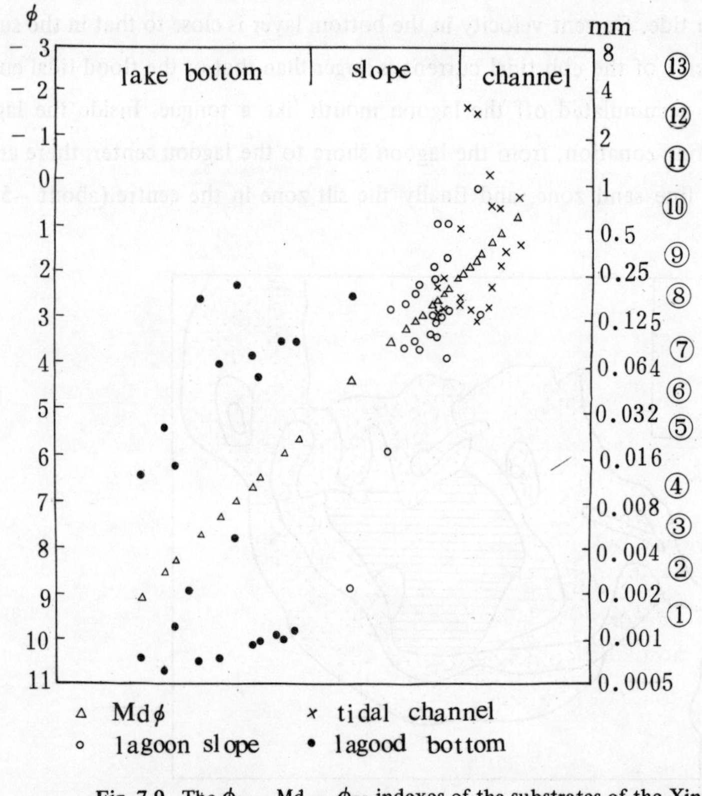

Fig. 7-9. The ϕ_{25} –Md$_\phi$ –ϕ_{75} indexes of the substrates of the Xinchun lagoon fringing reefs. 1 Clay; 2 Extremely fine silt; 3 Fine silt; 4 Medium silt; 5 Coarse silt; 6 Extreme coarse; 7 Extreme fine; 8 Fine sand; 9 Medium sand; 10 Coarse sand; 11 Extremely fine gravel; 12 Extremely fine gravel; 13 Fine gravel.

Parts of the grain sizes of the bottom sediments in the channel and on the lagoon slope appear to be close to one another. Md of sediments in the channel is mostly between medium sand and coarse sand, except for two samples. Sediments in the channel are of good sorting. Most QD_ϕ values are 0.26–0.91, a few are of negative or positive skewness, even extremely negative. Md_ϕ of sediments on the lagoon slope are 2–3.5, individuals–1.5 or 4.35, that is, they are mostly fine sand, a few are medium sand or extremely coarse silt. They have a good sorting, half or the QD_ϕ values are 0.41–0.51, i.e., pretty well sorted; a small amount are 0.72–1.20, i.e., fairly well sorted; the rest are 1.48–1.55, i.e., intermediately sorted; one or two is 3.21, i.e., very poorly sorted. Skewness is largely close to symmetry, one or two is of extremely negative or extremely positive skewness.

Composition of the above-mentioned grains consists of terrigenous and biogenous materials. The terrigenous materials are mainly quartz, also a small amount of feldspar and mica; content of kaolinite increases with the decrease of the grain sizes; biogenous materials are mianly fragments of coral skeleton and mollusc shell, and fecal pellets can be seen on beaches.

The sedimentary facies of atoll lagoons:

There is a lot of atolls in the South China Sea. They are one of the major types of coral reefs in China[22]. Coral atolls may be divided into two types: one is the atoll of oceanic type, e.g., Maldive Atoll in the Indian Ocean and the Minzhu Reef (Huangyan I.) in the South China Sea[3,23]; the others are atolls on the continental slope terrace (submarine plateau), e.g., Yongluo Atoll of Xisha Islands and Zhongsha Atoll[2].

In Huangyan Atoll, there are 12 families, 20 genera, and 46 species and subspecies (2 unidentified species) of Hermatypic corals1). In addition, Huangyan Atoll is rich in reef-habitants, e.g., *Turbinaria ornata* (brown alga) grows in patches on the inner reef-flat; *Turbo chrysostomus* (mollusc) is of rather high production here and falls one of the major preys to the Chinese fishermen; there are also not less *Holothuria atra* (echinoderm), which may reach 20–30 cm long, and so on. After a preliminary identification, the nomenated reef-philic organisms here amount to 84 species, belonging to nine phyla. Of these, there are 22 species of sea algae, belonging to 18 genera, 12 families, 4 phyla; the most are brown algae with 7 genera, 9 species; next comes red algae with 7 genera, 7 species; then, green algae, 3 genera, 5 species; blue algae, 1 genus, 1 species.There are 2 genera, 2 species of soft corals of Coelenterata; 3 genera, 3 species of polychaetes; 24 genera, 39 species of molluscs; 3 genera, 5 species of crustaceans; 10 genera, 13 species of echinoderms2). In the lagoon of Huangyan Atoll, 55 species of plankton were collected from nine sampling stations3). Of these there are 46 species of phytoplankton; the most are diatoms, next comes dinoflagellates, then the *Trichodesmium;* there are 9 species of zooplankton, mainly copepods of small size. Besides, there are the planktonic

1) Zou Renlin, Chen Youzhang, 1983, A preliminary study on the geographic distribution of hermatypic corals in China. Nanhai Studia Marina Sinica, 4:89–96.

2) Benthos Study Group of Marine Biology Laboratory, South China Sea Institute of Oceanology, Academia Sinica, 1977, A preliminary investigation on the organisms on the reef-flat of Huangyan I. List of benthic animal species. Reports of investigations on the adjacent waters of Zhougsha, Xisha Islands in the South China Sea, 23–27.

3) Ye Jiasong, Chen Qingchao, 1977, A preliminary investigation on the plankton Huangyan I. Ditto, 13–17.

larvae of benthic animals. Viewing from the planktonic composition, of the zooplankton, most are small-sized coastal and pelagic surficial species; most of the phytoplankton are coastal species, but some are species of intertidal zone. Therefore, it may be suggested that in the lagoon of Huangyan Island, the plakton consist mainly of surficial species outside the reef and the speices living along the lagoon shore and near the bottom inside the lagoon, most of them are eurythermal species of tropic and temperate zones.

The above-mentioned organisms, particularly the hermatypic corals, calcareous algae and molluscs with their skeleton and shell debris, serve as biotic sources for the lagoon sediments of Huangyan Atoll. Hence, in the loose deposits in the lagoon of Huangyan Atoll, components with grain sizes of >2–0.1 mm are mainly hermatypic coral fragments, next comes calcareous algal debris, then shell fragments, and a small smount of foraminifera, bryozoa, spines of Alcyonarians and the ambulacral plate debris of sea urchins, a trace amount of non-calcareous debris is also a component of biotic sources, e.g., diatoms. It can be seen that sediments of such atoll lagoons are without any terrigenous components, which are similar to those of the atoll lagoons in the Pacific Ocean and Caribbean Sea (Table 7-4). Table 7-4 shows that loose deposits of the atoll lagoons in the Pacific Ocean and Caribbean Sea are with more calcareous algal debris, the content of which may reach 33–54%, next comes the hermatypic coral fragments; whereas in sediments of Huangyan Atoll lagoon in the South China Sea, content of hermatypic coral fragments may reach 81%, next comes clacareous algal debris, and content of shell fragments equals to that of the Serrana Atoll lagoon in the Caribbean Sea (5%).

Foraminifera shell debris are of apparent significance in reef-building. In the loose sediments in reef areas of the South China Sea, content of foraminifera takes the fourth place, next to corals, calcareous algae and shells. Particularly, they are mainly of the calcareous non-porous type and calcareous porous type species. Those of sand-cemented type account for a small amount. The uplifted reef of Shi Island, Xisha Islands, with an age of about 20,000 yr BP, consists of foraminifera and calcareous algae debris and thus, the rock is named foraminifera-algae limestone. Taking waters around Xisha Islands as an example, there are 175 species (including sub-species), belonging to 39 families, 90 genera, (of these 3 genera and 7 species are new to science) of foraminifera[11]. Of these 175 species, except for 13 species of plankton, the rest 162 speices are all benthic foraminifera.

According to the statistics, depth of an atoll lagoon is usually related to the diameter of the atoll. Lagoon of a small atoll or a bracelet reef is generally with shallower water depth, e.g., Huangyan Atoll is –19.5 m deep; the Huanguang Reef of Xisha Islands is with a diameter of 30 nautical miles from east to west and the lagoon is only 20 and odd meters deep[1]; lagoon of a large atoll is usually deeper, e.g., the Zhongsha Atoll is with a lagoon of –80 m or so deep and the maximal depth is –109 m[2]; the lagoon of Yongluo Atoll, Xisha Islands is –50 m deep[24]; Eniwetok Atoll in the Pacific Ocean is with a lagoon of –70 m deep[25]. However, it should also be pointed out that atolls in Caribbean Sea, large or small, are all with lagoons of –5 – –115 m deep[20]. Actually, the surface at the water depth of –15 m was the low sea level of the reef area in 7000 yr BP[1)].

---

1) Stoddart, D.R., 1977, Coral reefs of the recent two million years. Record of the academic lecture presented at Guangdong Hall of Science on Aug. 17, 1977.

Table 7-4 Biotic composition of sediments in some atoll lagoons(%)

Composition \ Atoll	Caribbean Sea				Pacific Ocean			South China Sea
	Courtown Cays	Albuquer -que Cays	Roncador Bank	Serrrana Bank	Ifaluk Cays	Midway Island	Kure Cays	Huangyan Island
Hermatypic coral	35	30	25	33	36	25	15	81
Mollusc shells	10	9	15	5		12	10	5
Calci-algae	49	53	41	41	33	41	54	9
Foraminifera	3	2	3	3	23	14	10	3
Bryozoa								<1
Mixed shells		1	1	1	9	8	9	~2
Cryptocrystalline		2		3				
Cumularspharolith								
Non-shell mixture			11	10				
Non-calcareous Debris								Trace quantity

An atoll lagoon usually includes three geomorphological types, namely, lagoon slope, lagoon bottom and point reef (patch reef, Fig. 7-10).

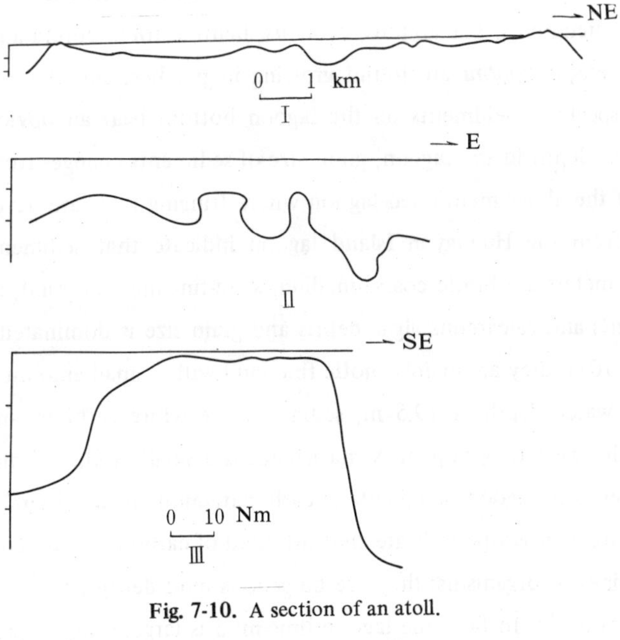

Fig. 7-10. A section of an atoll.

I– Huangyan Island Atoll;

II– Bei Reef Atoll;

III– Great Atoll of the Zhongsha archipelago.

Lagoon slope. It is also the inner slope of a reef-flat, usually with a steep gradient, sometimes approximate to the angle of repose of the sediment accumulation. Hermatypic corals are well-developed in the zone, but are mainly branching and fragile species and genera, such as *Acropora* and *Montipora.* In this zone, some algae, such as *Turbinaria turbinata* (L.) J. Ag. (brown alga) grows in patches, covering a large area; there are quite a lot of *Turbo chrysostomus* and *Trochus pyramis* Born; point reefs are also able to develop on the lagoon slope.

Patchy reefs. Such coral reefs are developed in atoll lagoons. In the South China Sea, patchy reefs are widely distributed in atoll lagoons, some are small and low coral heads or coral hills (e.g. those seen in the Bei Reef lagoon, Xisha Islands), some are large massive reefs (e.g. those seen in Huanguang Reef lagoon, Xisha Islands), others grow up close to the low water level, over 10 m higher than the lagoon bottom (e.g. the spot reef in the atoll lagoon of Huangyan Island). There are even spot reefs on which coral and shell gravel and debris are accumulated (lime-gravel cays), for example, Xianshe and Yagong cays in the Yongluo Atoll lagoon, Xisha Islands.

Patchy reefs can either provide material sources for the bottom sediment of a lagoon, or affect the movements of water and debris. At the periphery and on top of a patchy reef, branching corals grow. *Tubipora* (Alcyonacea, red-coloured) grow at a water depth of over 4 m in the periphery of the patchy reef. In addition, algae and various reef-habitated animals grow at the patchy reef.

Lagoon bottom. It refers to the portion of lagoon bottom without patchy reefs. Though water depth there is larger than that where patchy reefs are, its environment is the neritic low energy one. Populations of sea cucumbers, sea urchins, *Trochus*, benthic foraminifera and ostracoda, as well as green algae grow there; *Acropora* are found growing in patches; and there is not a few individual *Fungia.* Generally speaking, seidments on the lagoon bottom bear an obvious zonation: with the increase of the water depth in the lagoon, grain size of sediments changes from coarse to fine, which is similar to that of the above-mentioned lagoon where fringing reefs are developed. Analyses of the samples collected from the Huangyan Island lagoon indicate that sediments at the water depth shallower than five meters are biotic coarse-medium sand-fine-medium sand; at the water depth of 8 m are coral fragments and calcareous algae debris and grain size is dominated by fine sand grade; at the water depth of 10 m they are mainly biotic fine sand with a small amount of silt; at the center of the lagoon, where water depth is 19.5 m, sediments are white carbonate-mud (silt) (Table 7-5). Lingyang Reef in the western group of Xisha Islands is a small atoll, and may be called a bracelet reef[1]. In the center of its lagoon, sediments of carbonate-mud are also found. Observations by using an electronic scanning microscope indicate that this kind of carbonate mud consists mainly of corals, foraminifera and spines of organisms; they are biogenous mud debris (Plate 7-9) rather than product of chemical precipitation1). In fact, the lagoon lime-mud is largely biogenous, e.g. the lagoon lime-mud in Alaklan Reef and Belize[17].

It should be pointed out that material sources of sediments on the lagoon bottom are mainly the reef-flats around the lagoon and patchy reefs. When water from outside sea flows in over the atoll reef-flat or through the tidal channel, it may carry into the lagoon some exo-habitated or reef-habi-

1) Huang Jinsen, 1979, Diagenesis of marine carbonate lime-mud, Proceedings of South China Sea Oceanography (Abstracts).

Table 7-5. Characteristics of grain size of the sediments in the atoll lagoon of Huangyan Island

No.	Hu-2	Hu-3	Hu-7	Hu-1	Hu-4	Hu-5
Relief position	Upper margin of lagoon slope	Upper margin of lagoon slope	Inside channel mouth of lagoon	Lagoon slope	Lower section of the lagoon slope	Bottom of the lagoon center
Water depth (m)	4.0	3.4	3.0	8.6	10.8	19.5
$\phi 25$	0	0.6	0	3.10	0.13	3.14
$\phi 50$	0.54	1	1	3.43	3.48	4.80
$\phi 75$	1.1	1.7	1.2	3.73	3.85	5.72
QD_ϕ	0.55	0.55	0.6	0.32	0.36	0.99
SK_ϕ	0.01	0.15	−0.4	−0.02	0.01	−0.07
Nominated	coarse-medium sand	fine-medium sand	coarse-medium sand	fine-sand	fine sand with coarse silt	coarse silt with fine sand (sand-mud)

tated fauna and strongly filtering feeder, as well as their skeleton and shells. Therefore, it is clear from sediments collected from the lagoon bottom that the lagoon biotic populations are without any special species and genera.

Accumulations of biotic sand debris on the lagoon bottom often form sand ridges with a relative height of 2–6 m. It is not only found in the atoll lagoons of the South China Sea (e.g. the Bei Reef lagoon, Xisha Islands), but also in the atoll lagoons of Caribbean Sea; where relative height of sand ridges is only 0.5 m[17]. The sand ridge continuously expands and heightens and may form sand bars in the lagoon. e.g. Lizhou (the Inner Bar) in the Dongsha Atoll lagoon in the South China Sea.

Analyses of the grain size of sediments collected from the atoll lagoon bottom of Huangyan Island indicate that in the area close to the lagoon channel, where current is strong, sediments become coarser–coarse-medium sand with a small amount of fine gravel. In addition, they have a rather steep ϕ value accumulated curve, reflecting that grain sizes are concentrated in certain grades (Fig. 7-11).

As for the sediment sorting of the Huangyan lagoon, most QD_ϕ values are smaller than 0.6, indicating the sorting is good. As for sediments on the bottom of the lagoon center, QD_ϕ=0.99, it also is of good sorting. They have a very small skewness, most are nearly symmetric, only a sample collected near the reef-flat (Hu-3) is of positive skewness, and another sample collected near the channel mouth (Hu-7) is of extremely negative skewness (Table 7-5).

Chemical analyses and the analyses with atomic absorption spectro-photometry indicate that chemical composition of the above-mentioned carbonate lime-mud is mainly Ca, and Mg accounting for 1.85%, while Sr only 0.35%.

Mineral composition of the lagoon sediments is aragonite and magnesium calcite.

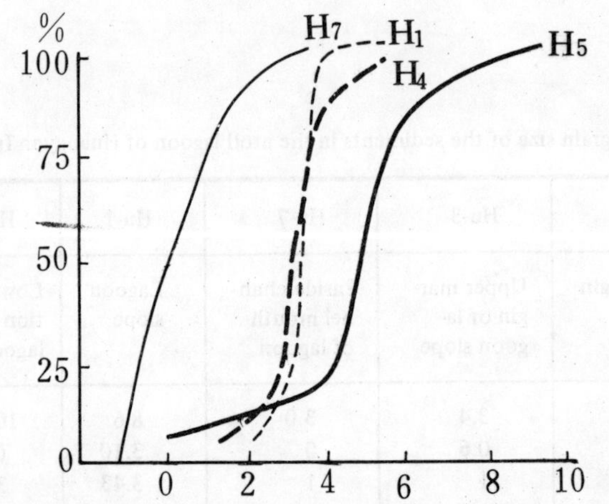

Fig. 7-11. Accumulated curves of substrates of the atoll lagoon of Huangyan Island. H_1 & H_4 – Lagoon slope; H_5 – Lagoon center; H_7 – Inside of a tidal inlet.

III. Coral Reef-Rocks and Their Types of Cementation

Coral reef rocks are formed of hermatypic corals and parts of reef-prech biological bone shells, which are cemented by carbonates and given the general name of rock, commonly also called coral limestone or coral reef limestone. It is obvious that they are also soluble carbonate rocks.

First of all we will discuss the petrographic types of coral reef-rocks in China, according to their reef formation effects, in particular the cementation effects in the three different environments of the subtidal, intertidal and supratidal zones.

1. Subtidal environment

The hermatypic corals are adapted to shallow sea environment, and not found in any big ocean of the world, as their growth is limited to the depth of 70–80 m[25]. The best development of the hermatypic coral in fringing reef in Hainan Island, is in a depth of 7–8 m, but in a depth of 3–5 m, coral is very well grown[14]. While in the lagoon of the Yongle Atoll in the Xisha Islands, there are Echinophyllia, Acropora, Heliopora, Porites, Seriatopora, Lobophyllia, Turbimaria, Glaxea, Leptoseris and Portes rigeseens.

Hermatypic coral has developed holes, and always preforated by drill-opening organisms (Plate 7-10). Coral skeletons are composed of lime, accreted by the coral polyps. These limestone skeletons, accumulated in situ, through diagenesis, form coral reef limestone. These rocks are called skeleton limestone[26].

coral reef limestone is a dominant species of coral reef rock in China and forms a firm carbonate rock structure. During coral reef formation, some of the inner and outer empty pores of the hermatypic coral were without cement or mineral crystals (e.g. as seen in SEM Photoes of the subtidal samples from Coral and Zhongjian Island of Xisha Islands, Plate 7-11); some are cemented by granular calcite or aragonite needle (Plates 7-12, 13); but most of them are wrapped by algae or cemented by algae (Plate 7-14). From literature, we know that in the subtidal of other regions, there are carbonate cement of high magnesium calcite[17]. In China, subtidal reef rock have also cementation of high magnesium calcite (Plate 7-15). but usually, in the subtidal environment, cementation is a little slower than in other environments. In shallow water area, they are mostly cemented by aragonite and high magnesium calcite; in deep water condition, cement is mostly of high and low magnesium calcite. Evidently, the development of aragonite crystals are affected by water depth and temperature.

The development and structural types of subtidal carbonate mineral crystals are shown in Fig. 7-12.

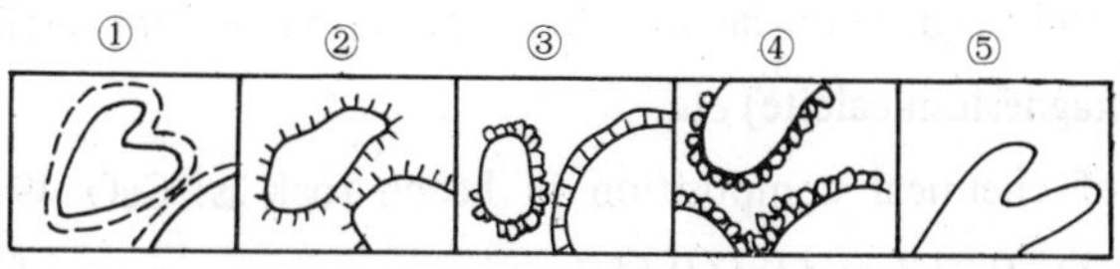

Fig. 7-12. Diagram of subtidal reef-rock cementation and structural types.

(1) – Algae cementation and shell wrapping;

(2) – Aragonite needle;

(3) – High magnessium calcite;

(4) – Granular calcite;

(5) – Coral framework, without cement.

By the way, we have to point out that there are many reports about the old coral reef limestones of China, e.g. the Upper Permian cryatal algae coral reef limestone of Datang, Guizhou province, formed a nucleus of the reef but secondary change of coral reef limestone was very great.

The chemical composition and content percentage of coral reef limestone in China, are: CaO 53.05–55.70%, MgO 0.08–0.14%, SiO_2 0.19–0.42%, Fe_2O_3 0.07–0.27%, and SrO 1%. Porosity of the coral reef limestone is 22.15–47.18%, while porosity of microsection is 34–10%, and permeability is 479–9110 darcy. That is the reason why coral reef limestone is called porous carbonate rock. At Xianfa Ansha (Constitution Hidden Shoal) ^{14}C age of reef limestone is 22,300±1000 yr BP.

Algae cohesive rock. They are formed by magnesium-calcitic algae cemented with hermatypic coral and reef biological bone-shells, sometimes consisting of inter layers of coral and coralline algae. Usually the algae layer is thin while the coral layer is thick, because the growth of corals is faster than of algae.

In a stable or lowering sea level, the subtidal reef rocks increase vertically, and form a dangerous area for Navigation. Even in a condition of sea level rising, if speed of vertical development of the subtidal rock is faster than that of sea level rise, it will all the same creates an unfavourable area for navigation. On the other hand, a reef rock of a certain depth makes a "cover layer" for the submarine guyot, and provides a natural testing ground for many uses, which of coures is favourable for men.

2. Intertidal environment

The formation of rocks by carbonate mineral cement in the intertidal environment is called beach rock (Plate 7-17), named according to their morphological position.

In China, beach rock is widely distributed. As its rock formation is fast glass, pieces of iron instruments and Hongwu currency of the Ming Dynasty may by cemented in beach rock. Besides, there is also beack rock of the Holocene period, e.g. the ^{14}C dating of the Yingghai beach rock at Hainan Island is 6250±210 years BP.

Biodetritus of beach rock includes coral (aragonite), Tubipora (magnesium-calcite), coralline algae (magnesium-calcite), Halimeda (aragonite), foraminifera (magnesium-calcite and calcite), lamellibranchiate (aragonite and magnesium-calcite), Bryozoans (magnesium-calcite), Millepora (aragonite), and Echinodermate (magnesium-calcite) etc.

The percentage of chemical composition in beach rock is: CaO 49.89–52.01%, MgO 1.36–3.20%, SiO_2 0.62%, P_2O_5 0.21% and SrO 0.65%.

Comparatively speaking, the porosity of beach rock is 11.71–40.23%, and porosity of microsection is 7–24%, lower than that of coral reef limestone. Its permeability is also poorer than that of coral reef limestone, being 106–3247‰ darcy. But compared with old limestone, porosity or permeability of beach rock is higher, because the old limestone was formed long ago and has undergone strong secondary changes.

In the reef areas of Ching, beach rock cement is mainly of aragonite needles and aragonite of mud-crystals (latent crystals), and also of columnar high magnesium-calcite and high magnesium-calcite of mud-crystals. According to their different combination, they may be divided into the 5 following cementation types.

1. Aragonite mud and aragonite needles (scicular aragonite) rim cementation: first of all a turbid and dark layer of aragonite wrapping is formed outside the particles, while aragonite mud crystalisation is unnoticable, or perhaps because there were organic matter mixed in it (?); its thickness generally is 0.02-0.03 mm. In SEM Photo., aragonite mud have varied sizes and forms (Plate 7-18). Outside this aragonite mud layer, aragonite needles develop and form a rim, which is arranged according to the aragonite vertical direction of the particle surticle. The length and size of these rims are even, with a normal length of 0.02–0.05 mm, while outside the particles and aragonite mud a thick skin is formed. This type of cementation can be observed at Coral Island, Ganquan Island and the Jinyin (Plates 7-19, 20) and Jinqing Island of Xisha Islands, and also at Dongmao and Ximao Island in the Sanya Bay of Hainan Island. This type of cementation is also reported from the coral reef island of the world, e.g. in Belize (formerly British Honduras[27]), the Bahamas[28], the Great Barrier Reef of Australia[29], and Cayman Island in the Carribean Sea[30], etc.

This type of aragonite spiculis crystals have varied patterns; observation by scanning electric microscope shows fine needle shapes (those seen in Jinyin Island), plate shapes (those in Bèi Island)

and thick blunt shapes (seen in Ganquan Island, Plate 7-21), etc. Sometimes the aragonite mud layer is not discovered, and then aragonite needles grow directly on the particle surface; but sometimes several interlayers of aragonite mud and aragonite needle with appearances of several cementation periods are seen (Plate 7-22).

2. Aragonite needle rim + aragonite and high magnesium-calcite mud with silt detritus cementation: they are formed on the foundation of the first type cementation by bioclastic silt and mud sedimentation. As silt and mud are composed of biological hone-shell detritus, so their mineral constituents are also aragonite and high magnesium-calcite. We generally see large amounts of biological bone-shell consists of aragonite matter, with the result that basically their silt and mud detritus are also of aragonite. Because silt and mud detritus are mechanically deposited, so that distribution and thickness are uneven. Sometimes they more or less display the direction of deposition, i.e. concentrated in one side of the intergranular pores; sometimes filling in between the particles, forming a "bridge form" cementation. This type of cementation can be represented by beach rock of Bei Island in the Xisha Islands (Plate 7-23). The katar peninsula beach rock of the Persian Gulf also belongs to this type of cementation[31].

3. Cementation of aragonite with high magnesium-calcite mud and silt detritus + aragonite needle: this type is just the opposite of the above mentioned cementation, namely a part or the whole of outer of the particle has an uneven thickness of aragonite or high magnesium-calcite biosilt or mud detritus on which aragonite needles grow (Plate 7-24). This type of cementation can be observed at Zhaoshi Island, Yongxing and Quanfu Island in Xisha Islands. Cementation at Yongxing Island is inferior to that of Zhaoshi Island; while at Quanfu Island only biosilt detritus filled up the intergranular pores, with comparatively only a little biological mud detritus (Plate 7-25), and aragonite needles only grow on silt detritus, but are very sparse and small. Perhaps, this is due to the fact that intertidal zone of this island is narrow and steep, and also because loose sediments were formed rather lately.

According to the above, aragonite rim + aragonite, with high magensium-calcite mud and silt detritus cementation belongs to the same type as aragonite, with high magnesium-calcite mud and silt detritus + aragonite needle cementation. The type of cementation depends upon depositional environment at that time. If the environment is favourable to growth and precipitation of aragonite needle, cementation of aragonite needle rim + aragonite, with high magnesium-calcite mud and silt detritus will appear. On the contrary, aragonite with high magnesium-calcite and silt detritus + aragonite needle cementation would occur. No matter which type of cementation it is, their mud and silt fillings are very even, with the result that on the outside of the particles, aragonite needles directly grow on the particle surface.

4. High magnesium-calcite mud cementation: at present this type can only be observed in the beach rocks of the Tianyahaijiao in Hainan Island. The beach rock is composed of biological bone-shell detritus and quartz sand, the intergranular matrix being wholly made up of high magnesium-calcite mud (Plate 7-26), which according to estimation of darklight adjustment in colouring process, the contents of $MgCO_3$ can not be less than 15% (mg). Observation under the micrescope

indicates that the structure of this high magnesium-calcite is even without any indication of crystalisation, but under the scanning electric microscope, crystals of 5 microns, like rhombusses or flat triangles, could be observed. This type of cementation has also been reported abroad, e.g. in the west coast of Italy[13]; at St Croix Island, east of Florida[17] and in Belize[27] etc. According to SEM photographs of rock samples taken from these areas, they are very similar, with the same size and distinct crystal characteristics, which are easy to differentiate.

5. Columnar high magnesium-calcite cementation: at present this type can only be observed at the hidden beach rocks of Yinggehai in Hainan Island, which are generally flooded by seawater now. Grain detritus of the rocks are mainly made of quartz sand containing a little biological boneshell detritus. The intergranular pore is filled with high magnesium-calcite cement (Plate 7-27), which is shaped into columns and grows out of the periphery of the grains. The top of these columnar crystals is often shaped into flat triangles. Sometimes, high magnesium-calcite mud and spherulite fillings are formed in the intergranular pores which belong to bioclastic detritus when they are observed under the scanning electronic microscope. According to chemical analysis and X-ray identification by crystal powder of the columnar high magnesium-calcite, $MgCO_3$ is 19% (mg). This type of cementation is often observed at Katar Peninsula in the Persian Gulf[31].

The above 5 types of beach rock in China can be seen in Fig. 7-13, and are compiled according to a comprehensive observation by microscope and scanning electronic microscope.

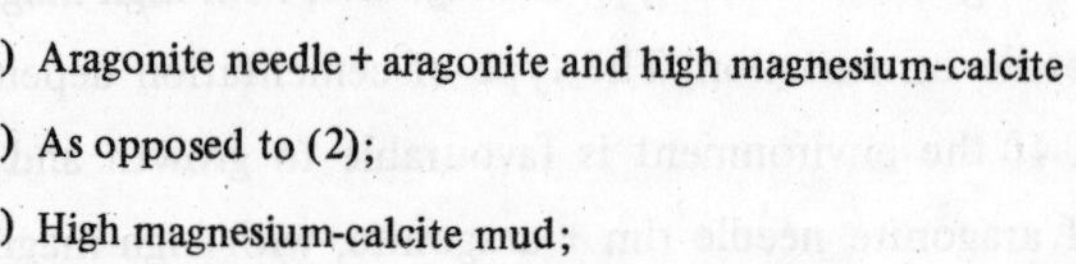

Fig. 7-13. Diagram of the beach rock cementation types in China.

(1) Aragonite needle rim;

(2) Aragonite needle + aragonite and high magnesium-calcite silt and mud;

(3) As opposed to (2);

(4) High magnesium-calcite mud;

(5) Columnar high magnesium-calcite.

From the available data, We can see that in the beach rock cementation of China, aragonite is mainly produced on the beaches of the coral reef area, while high magnesium-calcite appears on the beaches of the south weatern coast in Hainan Island. Although the Tianyahaijiao lies not far from Dongmao and Ximao Island, but their beach rock cementation is clearly not the same.

Carbonate cement in beach rock is different from cement produced under the effect of subaerial fresh water. Under the effect of meteoritic and fresh water, grained calcite (low magnesium) is formed.

The intertidal area is a zone which is periodicly immersed by sea water and exposed under the effects of atmosphere and fresh water, but generally speaking, sea water effect is dominant. In such a special environment, there is a possibility that the rich $CaCO_3$ sea water becomes supersaturared because of the loss of CO_2, so that the carbonate minerals crystalise and grow on the rim of the grain detritus or in the intergranular pores. Perhaps this is also the reason why cementation of beach sediment is fast and widespred. These biosilt and mud detritus fillings of the intergranular pores are obviously the result of transportation and deposition of the wave and currents. Growth of this cement (or matrix) depends on the existence of this type of material at that locality and during that time, and also on the existence of the negative topography. Of course, we have also to consider that carbonate mud on the rims of gtains of beach roch can not rule out the possibility of the effects of biological actions (e.g. of algae).

The intertidal zone is a significant area where the effects of sedimentation and diagensis are rather complicaed. At present, the sea at geological periods known by us are mostly epeiric sea where intertidal zone is not only well-developed but is also wide and therefore, old beach rocks are certainly common and wide spread. To be able to distinguish the environment of rock formation the cementation type and their characteristics are very important marks–they indicate a particular facies. Although aragonite and high magnesium-calcite are unstable and substable minerals, which will change into calcite with the change of time and environment, they will finally become calcite. Especially as the change of aragonite into calcite will be accompanied by changes in crystal stucture this will make distinguishing of old beach rock cementation types rather difficult. But as long as we really know how to distinguish beach rock facies and the characteristics of the rock cementation types, they will enlighten us in rightly analysing the old coast zone–this is the principle of realism in geological science.

It must also be pointed out that beach rock has certain wave resistance, so that in coastal engineering, especially in coastal engineering of the coral reef area, we must, consider the value of beach rock in coastsl protection and make full use of this natural wall and not excavate or damage them. If the natural dynamic equilibrium of beach profile is damaged, the speed of erosion or accumulation in the projected coast section will be increased.

The low lying beach rocks can be used as an ideal breeding ground for aquaculture. In foreign countries, the question of breeding grounds for coasts without beaches is solved by creating man-made reefs.

3. Supratidal Environment

The supratidal zone is effected by atmosphere and fresh water, where sodium, magnesium and associated anions are rather low and effect of metabolism of the microorganisms and of physical and chemical precipitation is strong and large quantities of calcite mud cryatals can be formed on the surface. In vadose zone, because recrystallisation is very slow, calcite is grobular bright-crystal. Deep underground, crystal water, salt water and atmospheric water are mixed and concentration of sodium anion is generally greater than at of normal sea water, but concentration of magnesium anion is low because of clay absortption and dolomitization. Under low Mg/Ca, decrease of magnesium anions eliminates control of these anions in calcite type. Thus, in the mixed zone, there is not only the possibility of dolomite formation, but also of aragonite and high magnesium-calcite changing into calcite which will be characteristic spar calcite with the false apprarance of primitive minerals. Of course, different speed of crystal growth can influence mineral type.

Generally, reef rocks exposed in the supratidal zone form emergence reefs. We have already seen coral reef limestone, marine fresh water origin limestone, phosphate coral limestone, sand dune rocks, boneneedle limestone, and ostrea reef limestone, etc. in China.

In the formation of emergence coral reef limestone, aragonite in the coral skeleton is already partly or wholly changed into calcite, with cementation of granular spar calcite. Its chemical composition is: CaO 50.22%, MgO 2–3%, SiO_2 1–2%, P_2O_5H 0.03%, Fe_2O_3 0.015%, SrO 0.35%. Porosity of microsection is 55–15%. This rock can be observed at Taiwan, Hainan Island and Islands of South China Sea; and also at New Guinea[33].

The other rock of emerged coral-reef is marine fresh-water origin limestone, which is deposited under marine environments and later elevated (or descending of sea level) to the supratidal zone, and formed rock by the processes of freshwater diagenesis (5,7). Its chemical composition is CaO 50.52%, MgO 1.75–2.47%, SiO_2 0.66%, P_2O_5 0.87%, Fe_2O_3 0.10%, SrO 0.26%. Rock cement is granular calcite with porosity of 22.65–47.23%. Porosity of microsection is 14–9.8% and permeability is 10457% darcy.

In the supratidal reef rocks, effects of dissolution and filling are rather distinct, e.g. in the foraminifera coralline limestone of Shi Island in Xisha Islands, some aragonite bone-shell detritus were already wholly dissolvde, with only the mud skin of outer shell left, and some were already partly or wholly filled up with granular calcite (Plate 7-31). In this type of filling up, calcite crystalisation can sometimes be very automorphous (Plate 7-32), a phenomon which can also be observed in old limestone. Dissolution and filling of aragonite bone-shell detritus are also very well developed in the supratidal carbonate rocks of Wuman and Tandun of Hainan Island.

Under the conditions of the subaerial vadose zone in the supratidal environment, their change, dissolution, filling up and cementation do not occur uniformly in the sediments, because the vadose zone are not equally immersed in water. Thus, crescent and gravity comentation can be observed in the supratidal vadose zone. These cementation types can also be seen in the intertidal (Plate 7-33). But from the viewpoint of ground water hydraulics, their dissolution must be stronger than that in the vadose zone.

If dissolution and filling generally occur in the aragonite boneshell or aragonite oolitic grains of the rocks, we may consider that they have exposed on the land surface, and have the possibility to produce large amounts of dissolution porosity, thus forming good oil-gas reservoir layers.

Whether in the vadose zone or phreatic zone, their $CaCO_3$ source is limited. In a given time, sediments in the subtidal environment may be wholly cemented, but in the vadose zone of the supratidal environment, cementation at most only partly occur, preserving the original porosity and secondary dissolution porosity. In the old limestone of vadose zone, cementation is "crescent", with the result that it contains many intergranular porosity. Preservement of the intragrannular porosity (produced during rock formation in the vadose or phreatic zone) is also commonly observed.

Thus in comparison, the sediments (or rocks), which in the early had exposed on the land surface), will have more and better chance in producing large amount of porosity pores than the sediments which had not passed through this process of diagenesis. In research, recognition of this type of limestone can help in finding facies zones and layers which had experienced subaerial diagencsis during petroleum exploration. According to this knowledge, we will also be able to furnish the foundation for correct comparison of the known oil-gas layers in this type of oil-gas field.

In short, knowledge of the marine fresh water origin limestone will give us a deeper knowledge of old limestone and help us in applying it in geological work.

Among rocks constituting the uplifted reefs, there is also the phosphate coral limestone which is formed by replacement of coral aragonitic skeleton or bone detrital by phosphate of guano with cement mainly of collophane (Plate 7-35) and a little cryptomerous estramadurite. Its P_2O_5 content is 5.20–32.00%. Sometimes they are formed of interlocked phosphate and calcium of many generations. This type of rock is typical in Dong Island of Xisha Islands and its ^{14}C dating reveals an age of 3630±150 years BP.

Bone needle limestone is also one type of rock forming emerged reef, and can be seen at the west coast of Luhuitou in Hainan Island, being deposited in thin layers (with a thickness of 20–40 cm). Biodetritus make up 70% of the rocks, and are wholly composed of Alçyonacea bone needle (which include bone needle of the crown and base). Alcyonacea bone needle is of low magnesium-calcite, cemented by carbonate mud which are infiltrated into rock (Plate 7-36). Bone needle limestone is brown yellowish indicating that rock formation occured in the supratidal environment. The mixed and disorder arrangement of bone needle detrital (vertically, obliquely, horizontally), and mixture of crown and base bone needle with the large amount of carbonate mud indicate that they were deposited in the tidal flat environment. The significance of bone needle in reef formation is already acknowledged by the marine geologists[17]. From sedimentation and diagenesis, the bone needle limestone in Hainan Island belongs to the category of marine freshwater origin limestone. According ^{14}C dating of their upper and lower layers, its age is between 4900–4400 years BP.

Ostrea reef limestone or ostrea detrital limestone can be seen at Tianya haijiao in Hainan Island (Plate 7-37), where they grow at the base of coastal granite pillars formed by wave erosion.The ostrea shell is made of calcite with only 15% mg of $MgCO_3$, with granular calcite with a little carbonate mud filling as cement. Its age is 4850±265 years BP. It is considered that the ostrea reef which lies in between 64°N and 44°S is the most widely distributed carbonate reef[17].

Sand dune rocks (dunites) are seen in Hainan Island and along the coasts of South China. The aeolian dunes cover the tops of coastal barriers. In the windward areas, they may be deposited on the middle part of slope of low hills. These dunes are already partly petrified into rocks and called dune rocks. Their age is about 2000 years BP.

The constituents of the dune rocks are mainly of silt quartz which contain little shell and coral detrital with granular calcite cementation. Those rocks seen at Qingzhou of Guangdong and Mulantou of the Hainan Island are often cemented into arcs with concaved surface and composed of one calcite orystal grain. In some rocks, the inter-granular porosity is wholly filling up. Generally, in the vadose region, water flows down through the intergranular pores, with the result that it can only come into contact with grains in or nearby the intergranular pores. As water is adhered to the walls of grains, it assumes an arc surface. With continuous supply of water with limited $CaCO_3$ content, calcite first gets crystalised and forms the crystalised surface of an arc. This is called "crescent" cementation [34]. Sometimes there is another type of cementation in the supratidal environment, namely that in some large bone shell detritus there is no or very little cement, while on others they develop very well. This obviously reflects the effect and influence of gravity which is a special cementation type in the vadose zone. In the vadose zone, cementation generally begins from contact with or nearby the grains and afterwards spreads out slowly, but does not happen in many places at the same time.

The phreatic zone in the supratidal environment is often immersed in water, water will fill up the porosity, when the condition is favourable to calcite crystallization, calcite will grow around or on the rim of the hole-pores, or the intergranular porosity will wholly be filled up with granular calcite crystals. This is different from cementation process and structure in the vadose zone.

According to the above data, carbonate cementation minerals and structural types in the supratidal environment are shown in Fig. 7-14.

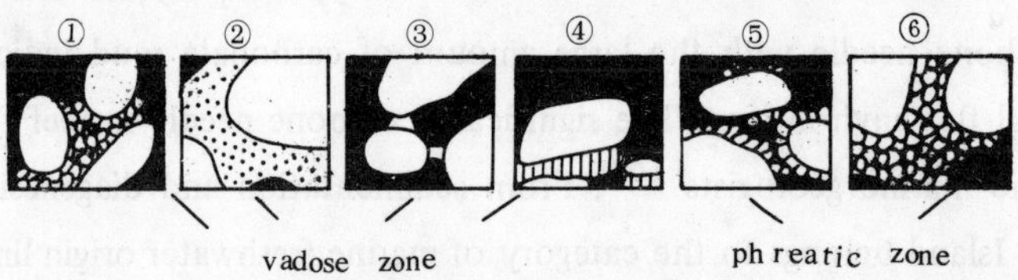

Fig. 7-14. Diagram of the carbonate mineral cementation in the supratidal environment.

(1) Granular calcite; (2) Carbonate mud; (3) Granular calcite, crescent type; (4) Granular calcite of gravity effect; (5) Granular calcite rim; (6) Granular calcite filling pores.

The above cementation shows that the sediments of ancient meteoritic-fresh water cementation, can be recognized by their cementation type (e.g. "crescent" or gravity cementation, etc.). In samples from the top of the Cambrian peroid oolitic limestone in Quyang of Hebei Province, "gravity cementation" is typical, indicating that the top of the oolite layer had been under the vadose zone of a supratidal environment where gravity cementation was produced (Primary generation). Afterwards the remains of intergranular porosity continued to be filled up by calcite (secondary generation). Cementation of these two generations is very clear. Also, the crescent cementation structure of some oolite limestone in Jurassic system, Mexico maintains their primary intergrnular pores, forming very good oil-gas reserveoirs.

Under the condition of the vadose and phreatic zone in the supratidal environment, the effects of atmosphere and fresh water produces karst landscape[5] (Plates 7-38,39). First of all, the unchanged aragonite biological bone-shell will be dissolved, forming holes. At the some time, this will supply part of $CaCO_3$ for the nearby sediments. If the dissolved surface reserves a great amount of secondary dissolved holes, then they too would make good oil-gas reservoirs. That is why in the study of reefs and rocks, knowledge of dissolution, filling up and cementation not only have significance in theory, but also have great practical importance.

References

[1] Huang Jinsen, 1974, Science Popularization, No. 10, pp 25–26. (in Chinese)

[2] Huang Jinsen et al., 1982, Symposium on Research Reports on the Sea Area of South China Sea, (1), 39–67, Science Press. (in Chinese)

[3] Huang Jinsen, 1982, Acta Oceanologica Sinica, V.1, pp94–106.

[4] Huang Jisen et al., 1978, Scientia Geologica Sinica, No. 4, pp 358–394. (in Chinese)

[5] Huang Jinsen, 1982, Tropic Oceanology, V. 1, No. 1, pp 12–20. (in Chinese)

[6] Sha Qingan, 1977, Scientia Geologica Sinica, No. 2, pp 172–178. (in Chinese)

[7] Sha Qingan et al., 1982, Studies on the Seidmentary Petrology (symposium), 226–242, Science Press. (in Chinese)

[8] Huang Jinsen, 1982, Marine Sciences, No. 5. pp 64. (in Chinese)

[9] Huang Jinsen, 1982, TCQ Symposium, 34–36, Science Press. (in Chinese)

[10] Zhong Jinliang and Huang Jinsen, 1979, Oceanologia et Limnologia Sinica, V. 10, pp125–135. (in Chinese)

[11] Zheng Zhizhong and Zheng Shouyi, 1978, Studia Marina Sinica, No.12, pp 149–166. (in Chinese)

[12] Lan Jinyun, 1982, Symposium on Research Reports on the Sea Area of South China Sea, (1), 241–250, Science Press. (in Chinese)

[13] Alexanderson, T., 1969, Sedimentology, V. 12, pp 47–61.

[14] Huang Jinsen, 1965, Science Bull, pp 85–87. (in Chinese)

[15] Chen Shaomou, 1982, Symposium on Research Reports on the Sea Area of South China Sea, (1), 69–98, Science Press. (in Chinese)

[16] Peng Tsunghung et al., 1977, Memo. Geol. Soci. China, No. 2, pp 57–69.

[17] Milliman, J. D., 1974, Marine Carbonate, Springer-Verlang Berlin-Heideberg New York.

[18] Wells, J. W., 1957, Treatise on Marine Ecology and Paleoecology, V. 1, pp.609–631.

[19] Jeans, O. J., 1959, An Introduction to Microfossil, 158–184, Science Press.

[20] Bathurst, R. G. C., 1971, Carbonate Sediments and their Diagenesis, Development in Sedimentology, 12, Elsevier Publ., New Yirk.

[21] Guilcher, A., 1954, Morphologic Littorale et Sousinarine, Paris.

[22] Huang Jinsen, 1979, Third Symposium (Abstract) on QRAC, 79–80.

[23] Xu Zongfan and Zhong Jinliang, 1980, Nanhai Studia Marina Sinica, No.1, pp 11–16. (in Chinese)

[24] Huang Jinsen et al., 1981, Symposium on Geomorphological of GSC in 1977, 108–115, Science Press. (in Chinese)

[25] Kuenen, Ph. H., 1950, Marine Geology, John Wiley and Sons. Inc., New York.

[26] Nelson, H. F., 1982, Symposium on Carbonatite Classify, Chongqing Branches of the Science & Technology Press.

[27] Stoddart, D. R. and Cann, J. R., 1965, J. Sedim. Petrol., V.35, pp 237–243.

[28] Illing, L. H., 1954, Bull. Am. Ass. Petrol. Geol., V. 38, pp1–95.

[29] Davies, P. T. and Kinsey, D. W., 1973, J. Sedim. Petrol.,V.43,pp58–81.

[30] Moore, C. H. Jr., 1971, In Bricker, O. P. (ed), Carbonate Cements, The Johns Hopkins University Studies in Geology,No.19, pp9–12.

[31] Taylor, J. C. M. and Illing, L. V., 1969, Sedimentology, V. 12, pp 69–107.

[32] James, N. P. et al., 1976, J. Sedim. Petrol.,V.46,pp523–524.

[33] Walker, D. and Chappell, J., 1983, Tropic Oceanology,V.2, pp154–162.

[34] Dunham, R. J., 1971, In Bricker,).). (ed), Carbonate Cements, The Johns Hopkins University Studies in Geology, No.19, pp 297–300.

[35] Su Jin et al., 659, A New Chinese Matera Medica, Shanghai Science & Technology Press (Process Plate in 1959.). (in Chinese)

by Huang Jingsen
South China Sea Institute of Oceanology, Academia Sinica, Guangzhou

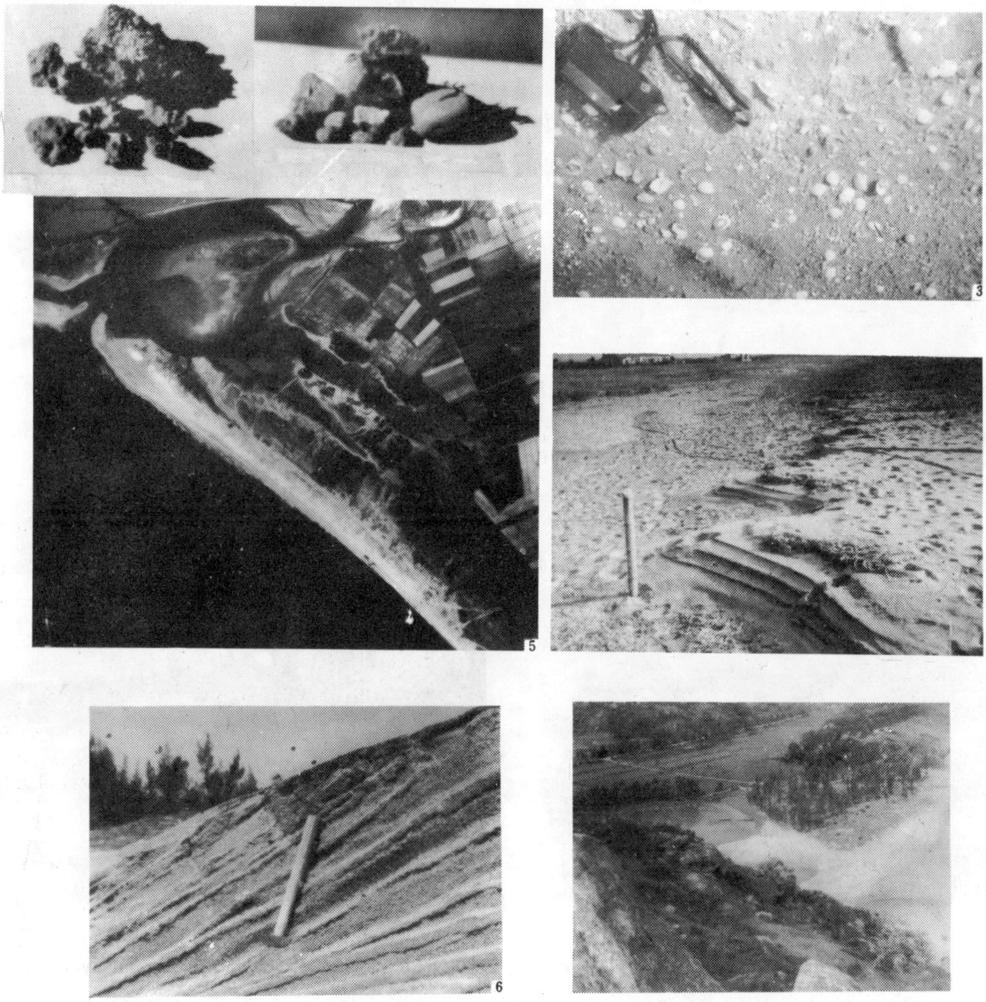

1. Terrigenous concretions in sediment of the central Wutong Bay, Xiamen; 2. Marine molluscs and plants on the surface of the well-rounded gravel, sample taken from the Wutong Bay, Xiamen; 3. Original state of shells in the lagoon of Longkou Tombolo, Shandong; 4. A new beach ridge migrating shorewards on the southern coast of Longkou Tombolo, Shandong; 5. Coastal barrier and lagoon of the Fenjia Estuary, southern Shandong Peninsula (air photo), From [17]; 6. Washover of barrier on the Zhenhai Bay mouth, southern Fujian. Oblique bedding is shown; 7. In Zhenhai Bay mouth, sand is pushed into lagoon by storms.

Plates 2-8–2-13

8

9

10

11

12

13

8. Wind erosion and wind ripples on Jidong dune; 9. Jidong dune is retreating and the lagoon layer is exposed on the shoreface; 10. Coastal dune of Jiantian, eastern Chanle, Fujian; 11. Bedding and ripple of Jiantian dune in Chanle, Fujian; 12. Wind sand of Zhenhai passes over the ridge and fills into the back wind slope; 13. A small dune in Dongshan, a slowly accumulating coastal dune.

Plates 3-1–3-5

Plates 3-6–3-10

7

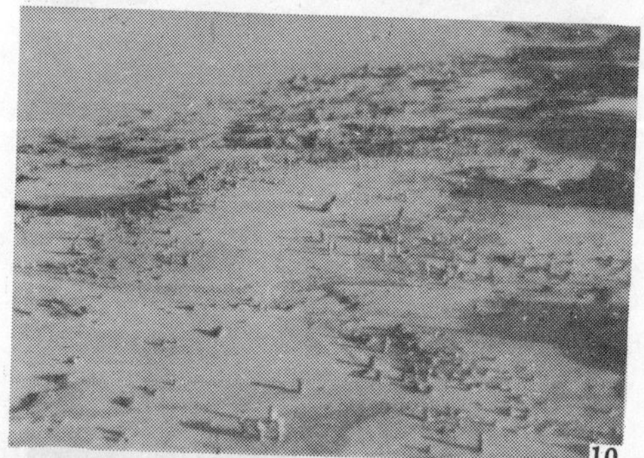

10

6

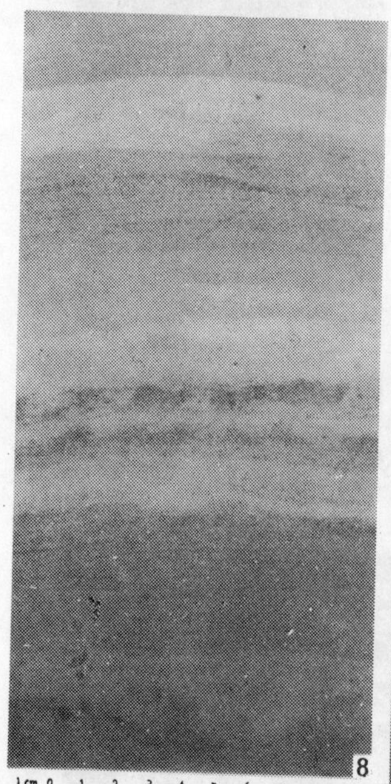

8

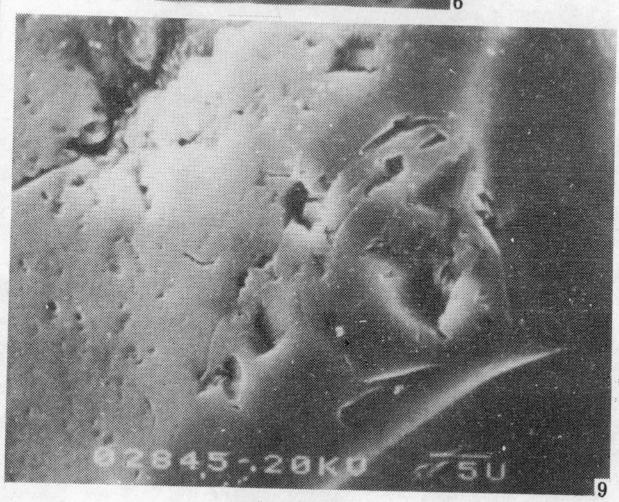

9

Plates 3-11–3-15

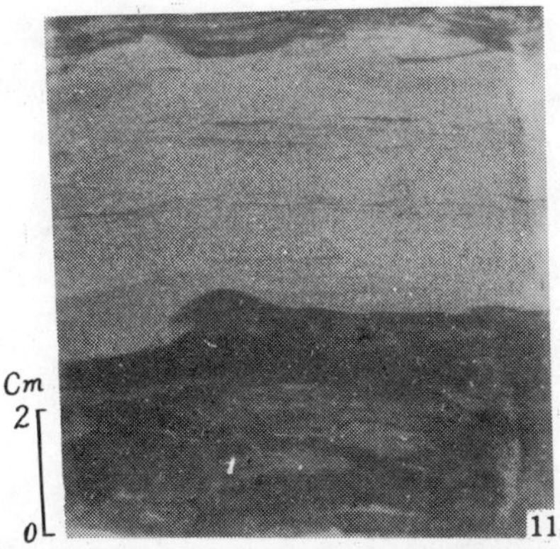

11

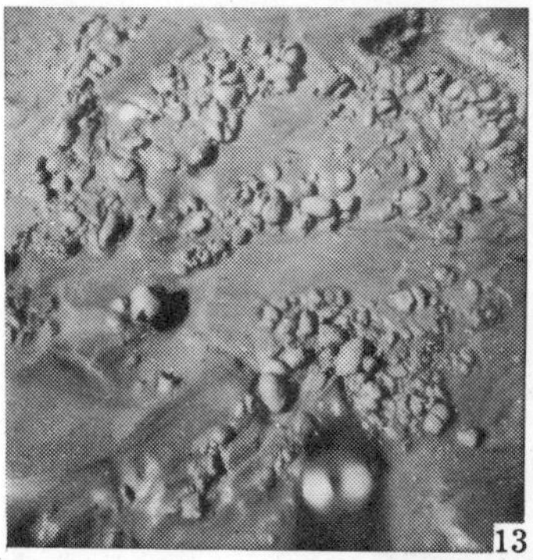

13

12

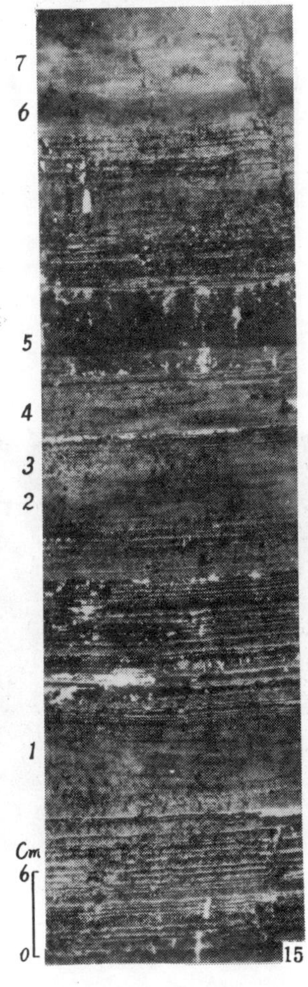

15

14

Plates 3-16–3-19

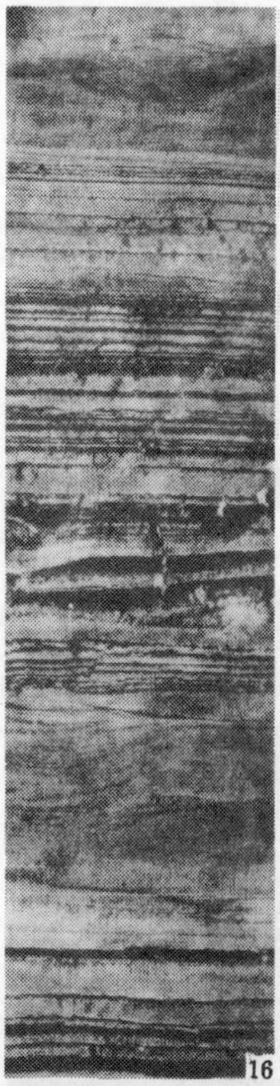

16

18

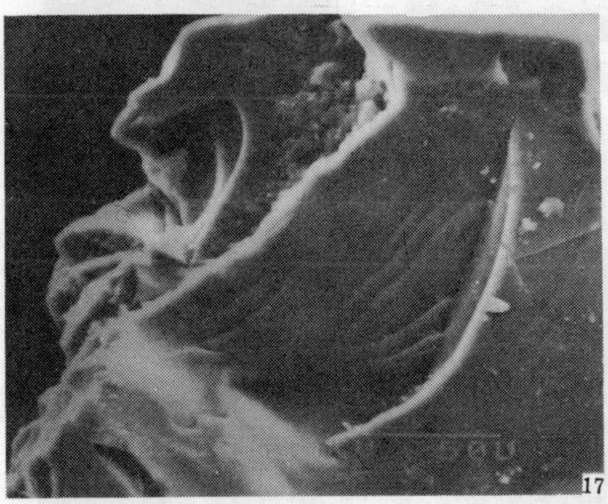

17

19

20

21

22

23

24

25

Plates 4-1–4-8

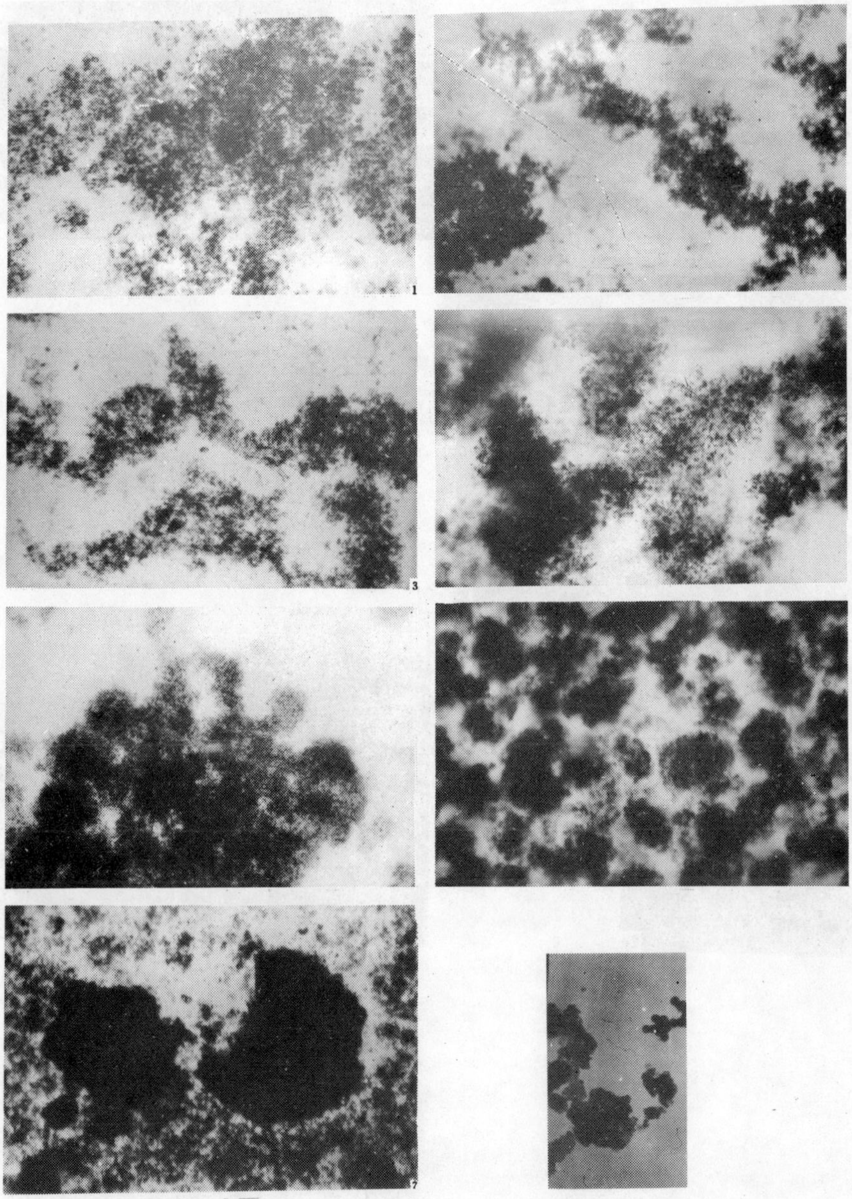

1. Flocculation colony (X375) 20‰ *S* 5 g/L; 2. Single catenary structure (X150) 25‰ *S* 20 g/L; 3. Double catenary structure (X150) 25‰ *S* 2 g/L; 4. X-type structure (X375) 5‰ *S* 30 g/L; 5. Honeycomb Structure (X150) 5‰ *S* 2 g/L (loss calcareous); 6. Medium aggregates (X375) 20‰ *S* 150 g/L; 7. Large network (X150) 25‰ *S* 30 g/L, 8. Illite and hematite (Black Ball at Right Upside) (X4700) St. 7.

Plates 4-9—4-17

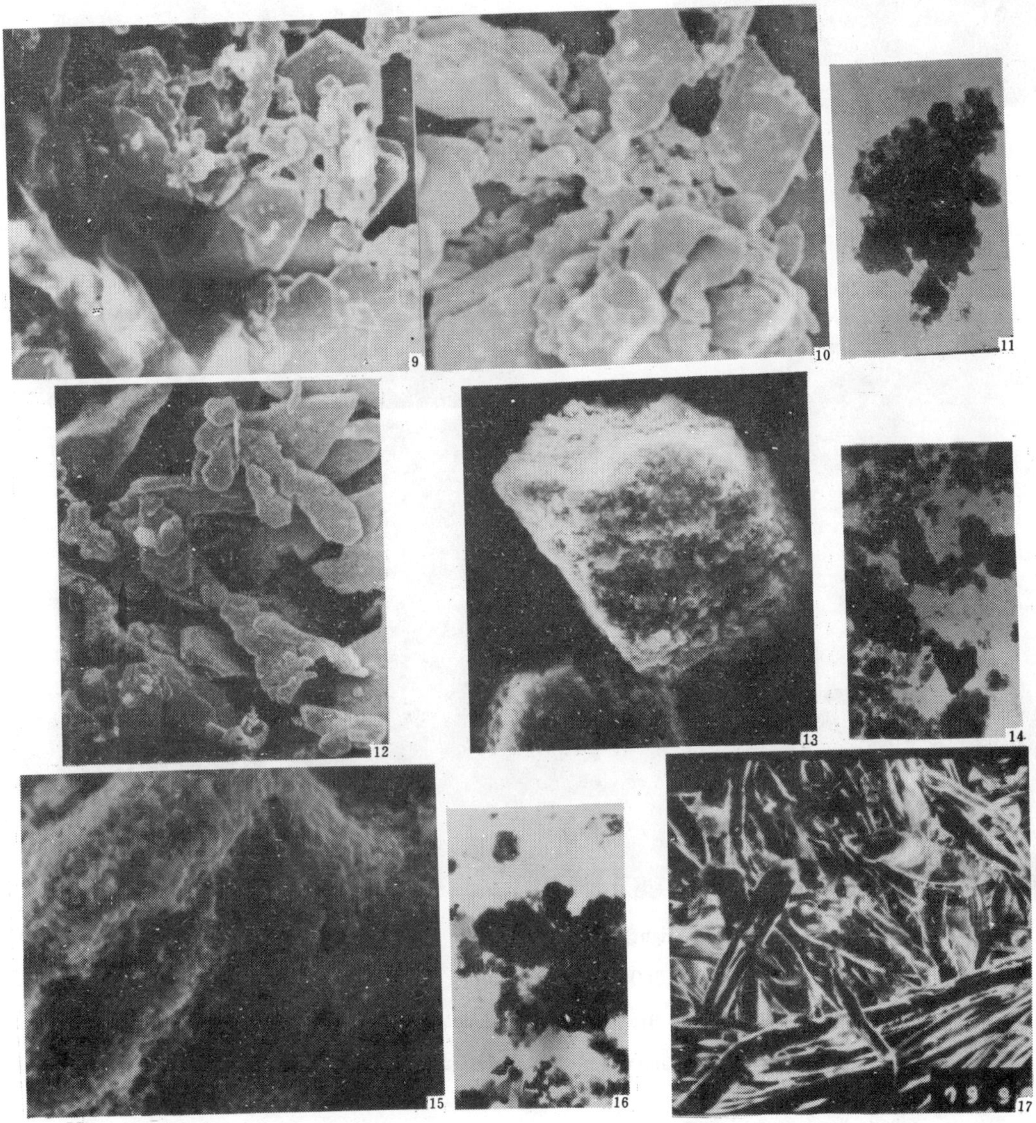

9. Hexagon kaolinite (X400) No. 5; 10. Platy illite (X500) St. 5; 11. Montmorillonite with vague margin (X5000); 12. Slabbing illite (X500) St. 5; 13. Flocculation form of illite (X500) St. 5; 14. Chlorite with straight edge and pointed outline (X5000) St. 10; 15. Microstructure of chlorite surface. It is similar to insect burrows St. 10; 16. Kaolinite with perfact hexagon outline (at right upside) (X5000) St. 13; 17. Crimping pipe halloysite (X500) St. 13.

Plates 4-18–4-22

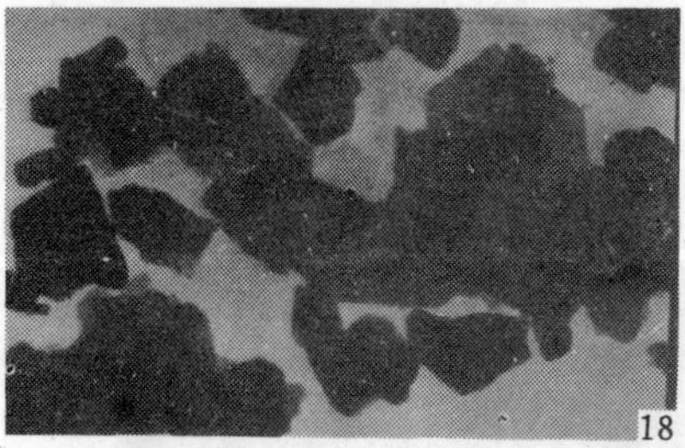

18

19

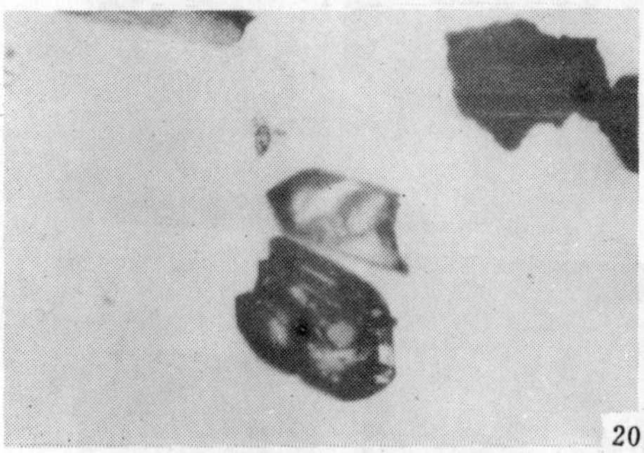

20

21

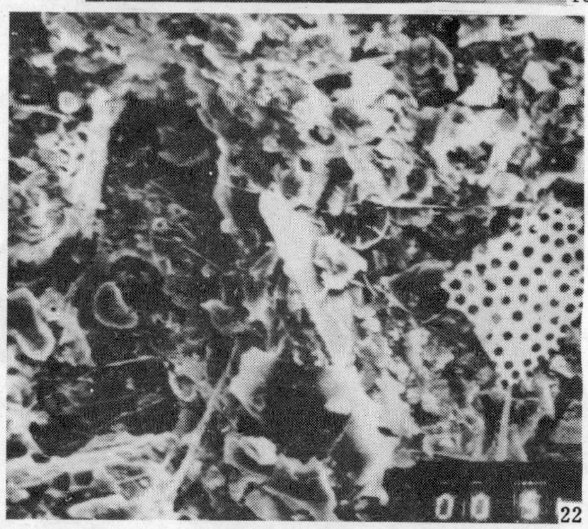

22

18. Chappy vermukulite (middle, right) chlorite and illite (X6000) St. 13; 19. Montmorillonite: fine scale of polymers (X1200); 20. Tetragonal dipyramidal and short columnar singly polarized light (X120); 21. Breaking zircon in tidal bore area singly polarized light (X120); 22. Resuspended diatom and other detrital matter of bottom layer in the offsore of Changjiang Estuary (X600).

1

2

1. Tidal ridges on Qiantaing Sand Bar (Photo. of NASA ERTS 21 Oct. 1976); 2. Radiating bars in Qianggang Area (Photo. of NASA ERTS 30 Apl. 1975).

Plates 5-3–5-7

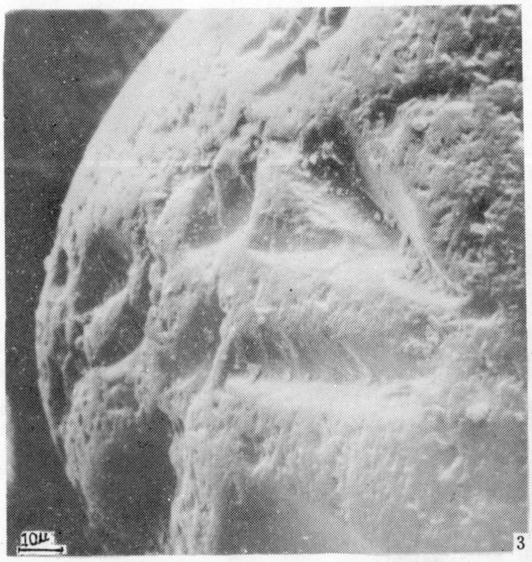

3

5

6

4

7

3. SEM photo. of the quartz surface in Luanhe Delta; illustrating V-indentations (X520); 4. Small cross beddings in flood plain of Luanhe Delta; 5. Lagoonal clay outcroped at front of barrier in Luanhe Delta; 6. Shells in oyster reef of the Galatuo, north of Caofeidian, offshore of Luanhe Delta; 7. Erosion cliff of barrier and lagoon clay at front of the barrier, Luanhe Delta (yuetuo off Daqinghe River mouth).

Plates 7-1–7-5

1. The Gaojianshi Island of Xisha archipelago formed of tuff lava; 2. Acropora; 3. Montipora and Acropora; 4. Foraminifera. algae and lamellibranchiate etc. in fore-reef sediment at –240 meters, Xianfa Ansha (Constitution Hidden Shoal). X40, (+); 5. Biodetritus deposits and trepang. starfish etc. In grooves of reef-flat, west coast of the Jinqing Island, Xisha archipelago.

Plates 7-6–7-13

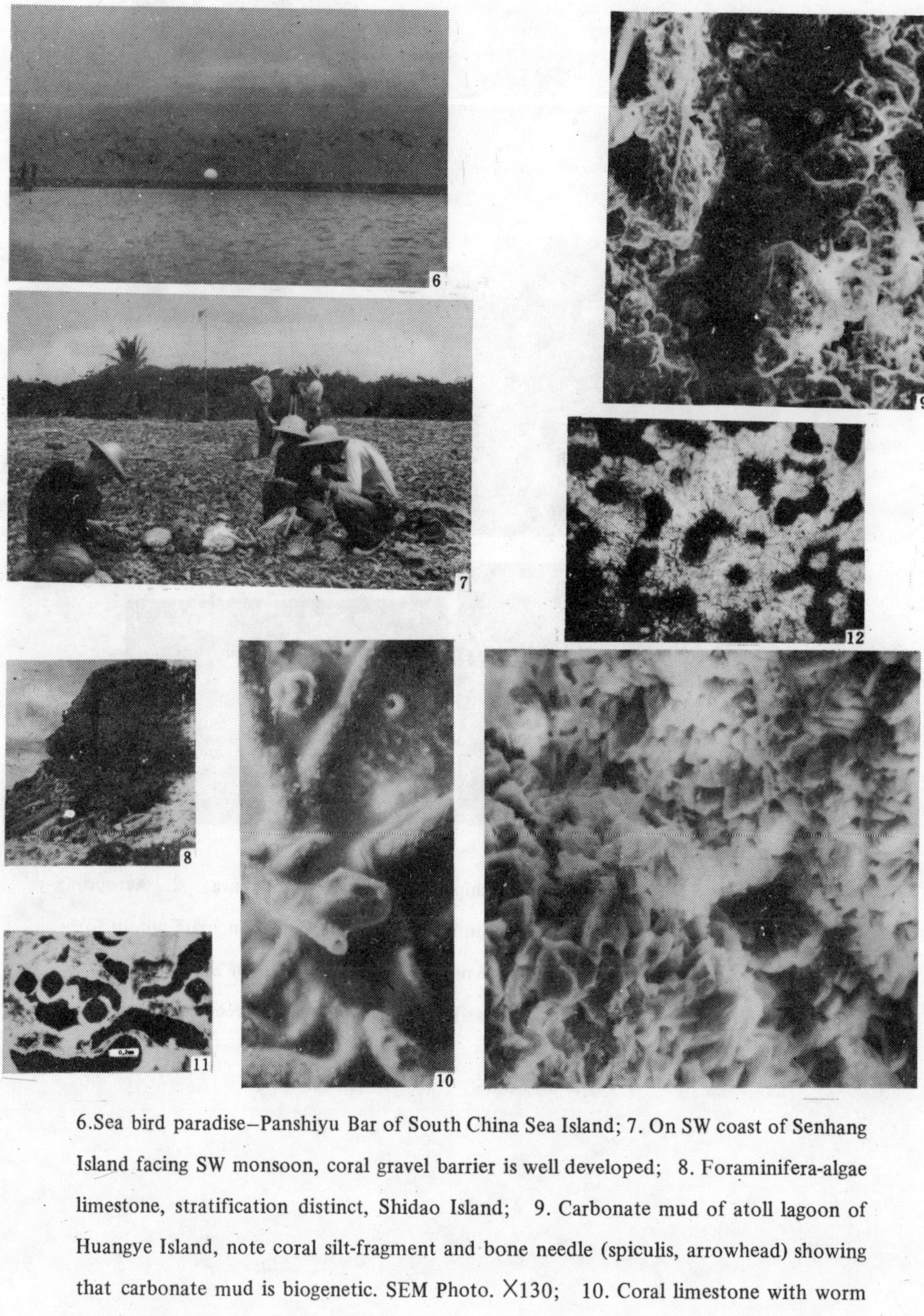

6.Sea bird paradise–Panshiyu Bar of South China Sea Island; 7. On SW coast of Senhang Island facing SW monsoon, coral gravel barrier is well developed; 8. Foraminifera-algae limestone, stratification distinct, Shidao Island; 9. Carbonate mud of atoll lagoon of Huangye Island, note coral silt-fragment and bone needle (spiculis, arrowhead) showing that carbonate mud is biogenetic. SEM Photo. ×130; 10. Coral limestone with worm bore hole, at –90 m, SW coast of Shanhu (Coral) Island, SEM Photo. ×315; 11. Well developed, Coral algae holes Note worm pipelines in holes. Zhongjian Island, –90 m, scale=0.2 mm, (+); 12. Coral limestone, with aragonite needles grown in pore spaces, –2 m, Huangyu Island, ×30, (+); 13. Granular calcite cementation in subtidal coral limestone, Shidao Island, SEM Photo. ×600.

Plates 7-14–7-20

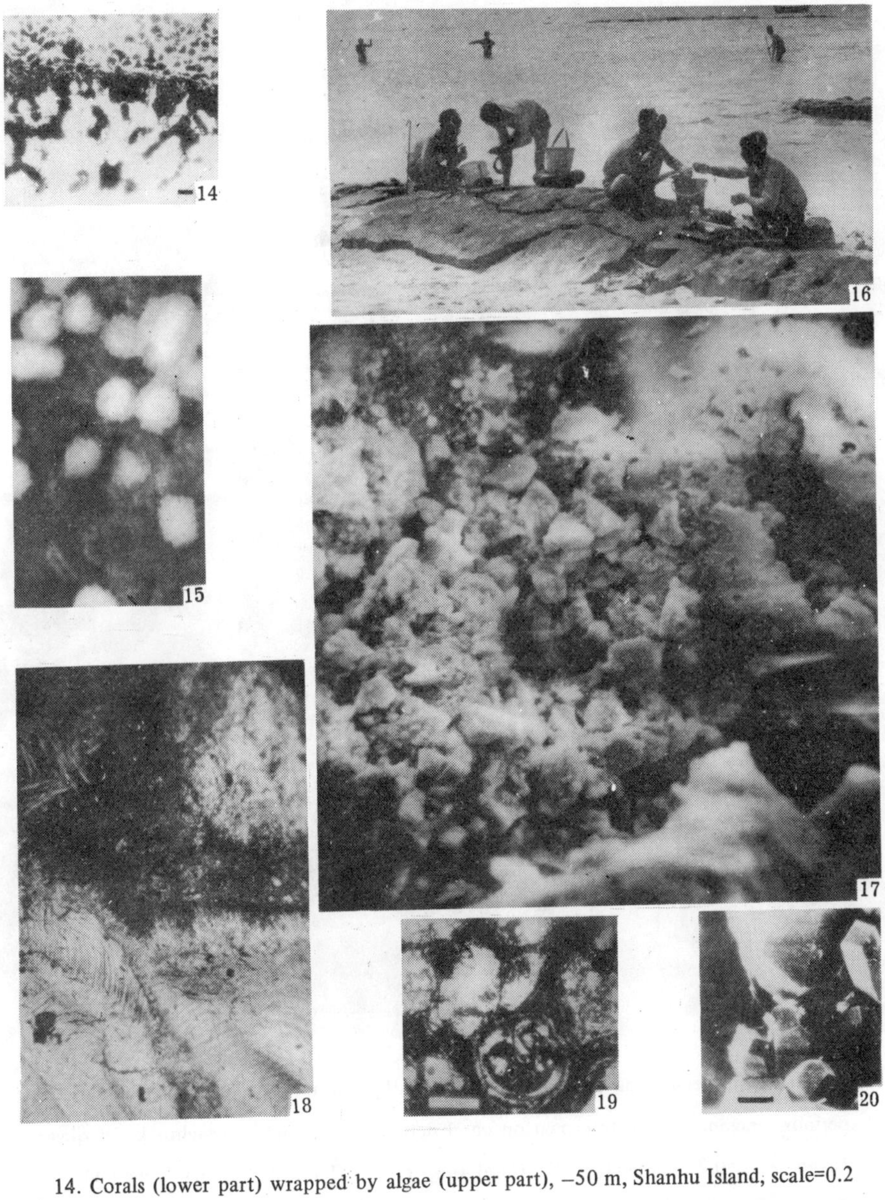

14. Corals (lower part) wrapped by algae (upper part), –50 m, Shanhu Island, scale=0.2 mm, (+); 15. Goniastrea pore spaces filled by high magnesium-calcite foraminifera in coral limestone, Huangyu Island, SEM Photo. X1680; 16. Beachrock on SW coast of Jinqing Island; 17. Aragonite mud cementation in beach rock, Jinyin Island. They differ in size and form, SEM Photo. X800; 18. Aragonite needle rim, this needle is 0.02 mm long, beachrock, Jinyin Island, X 40. (+); 19. Coral clastics, with a Gastropoda in the center, cemented by aragonite mud + aragonite needle rim, beachrock, Jinyin Island, scale=0.2 mm, (+); 20. Columnar high magnesium-calcite cementation, buried beachrock, Yingohai coast of Hainan Island, SEM Photo. scale=10 μ.

Plates 7-21–7-26

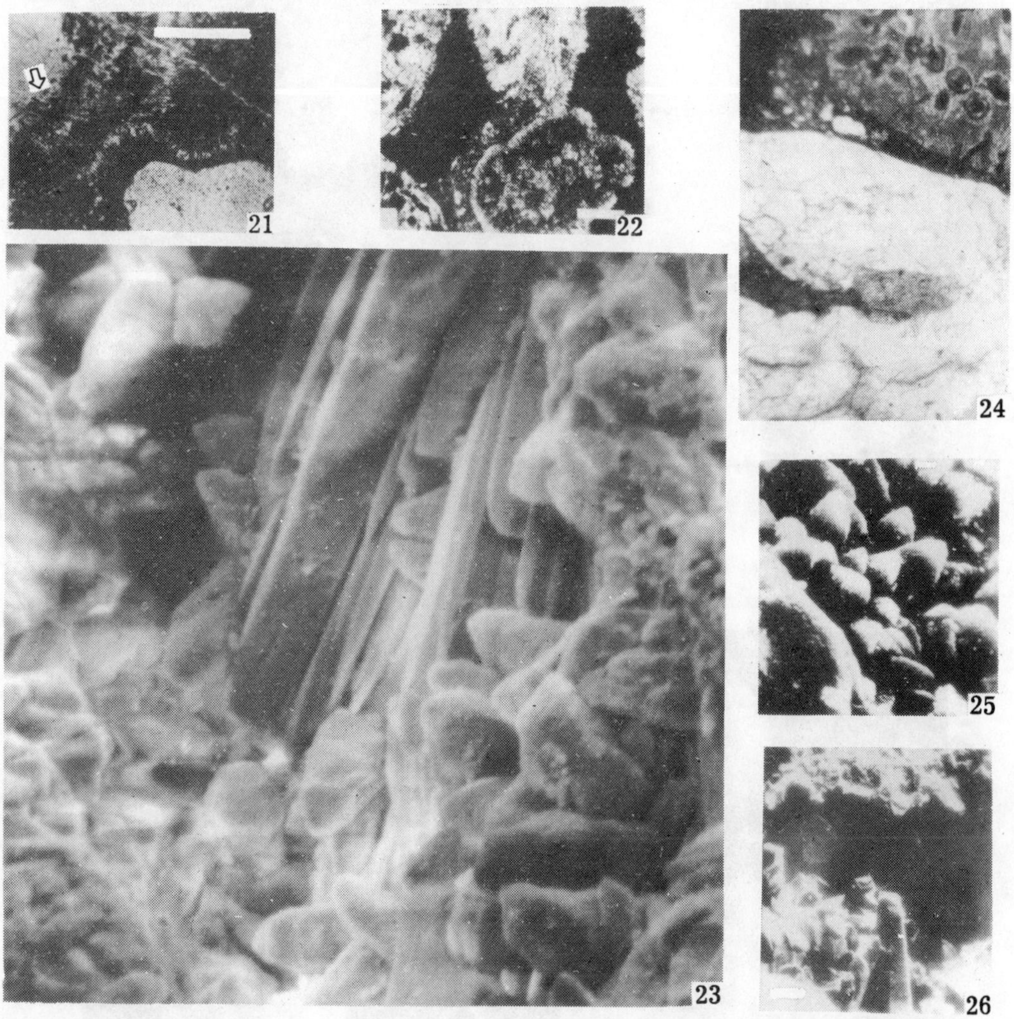

21. Aragonite needle and aragonite mud rim with appearances of several cementation periods. aragonite is easily grown on coral detritus, (arrowhead), beachrock, Jimqing Island, scale=0.2 mm, (+); 22. Coral (upper) and Halimeda (lower), cemented by aragonite needle + bioclastic silt and carbonate mud, beachrock, Beidao Island, scale = 0.2 mm (+); 23. Aragonite needles grown on high magnesium-calcite mud (granular), beachrock, Su Island, SEM Photo. $\times$1900; 24. Algae (upper) and coral (lower), aragonite and high magnesium-calcite bioclastic silt, filling inter-granular pores beachrock of Quanfu Island, $\times$50, (+); 25. High magnesium-calcite mud cementation, beachrock, Ends of the Earth, Hainan Island, SEM Photo. scale=2μ; 26. Columnar high magnesium-calcite cementation in bioclastic sandstone, Yinggehai, Hainan Island, SEM Photo. scale=30μ.

Plates 7-27—7-33

27

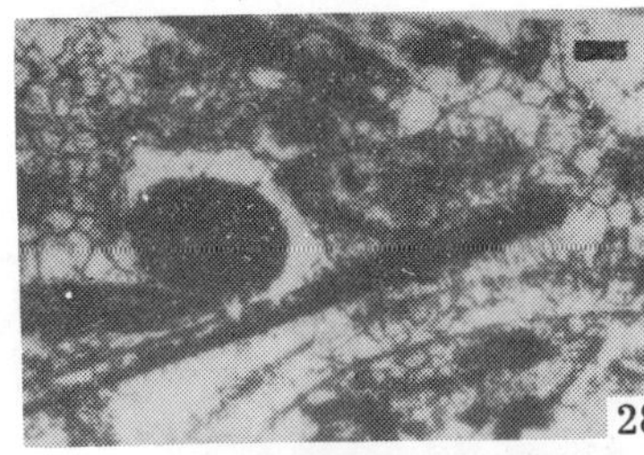

28

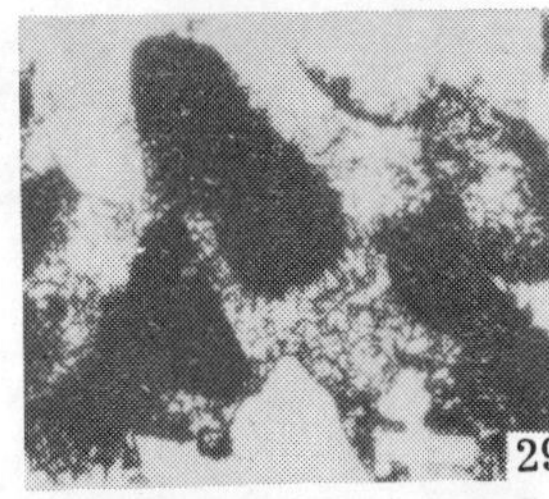

29

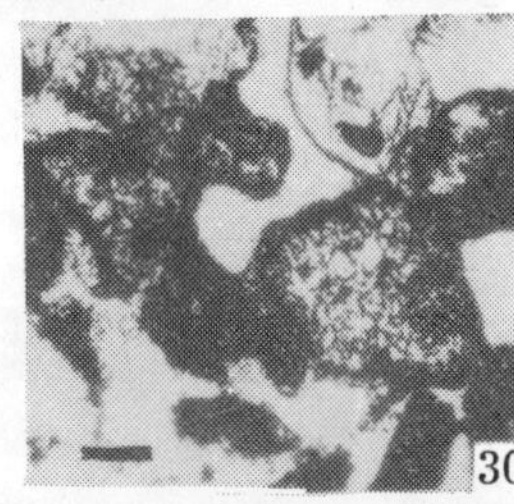

30

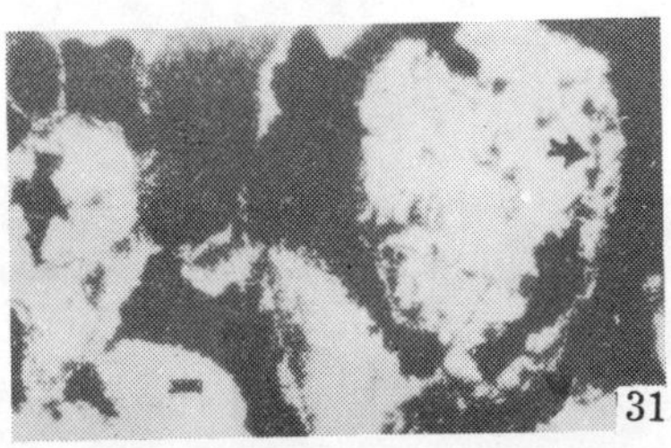

31

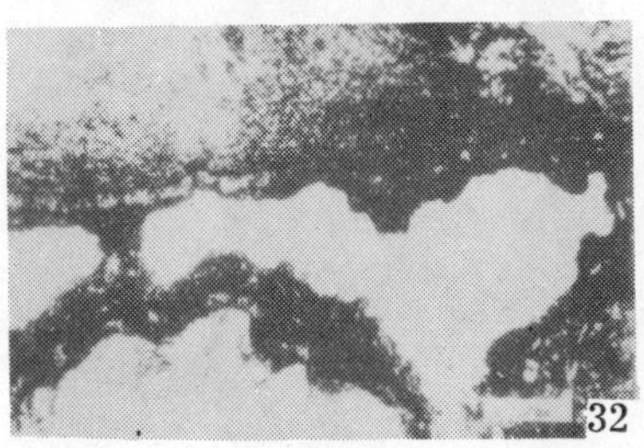

32

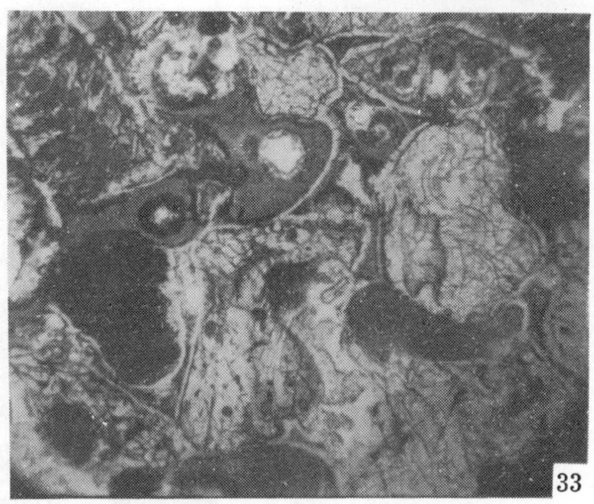

33

27. Raised reefs of Oman in Hainan Island; 28. In the centre are urchin and a long shell, sea urchin has grown up, below sea urchin, Lamellibranchiata shell has changed into calcite, the upper left is intergranular calcite cementation (granular), Xandun, Hainan Island, scale=0.2 mm, (+); 29. Intergranular calcite cementation (granular), Raised reefs of Oman scale=0.3 mm (+); 30. Intergranular calcite cementation (granular), intragranlar dissolution-pore also filled by calcite (granular), foraminifera-algae limestone, Shidao Island, scale=0.2mm (+); 31. Dissolved aragonite bone and shell, fill up by calcite. Individual calcite has distinct eleutheromorph (arrowheed), foraminifera-algae limestone, Shidao Island, scale=0.1 mm, (+); 32. Aragonite needle rim + aragonite and high magnesium-calcite silt between algae (upper) and coral (lower). Note the effect of gravity cementation , Beidao Island, scale=0.2 mm, (+); 33. Bio-detriitus wrapped by collophane cement, phosphate coral limestone , Jinqing Island, X50, (–).

Plates 7-34–7-39

34. Phosphate coral limestone cementation-fine texture of collophane: cotton fibre type Jinqing Island, SEM Photo. X683; 35. Biodetriitus bone and shell with collophane (thickness=0.1 mm, black) & calcite (white), cementations Dongdao Island, X41, (+). 36. Bone needle limestone, cemented by biodetriitus silt, Luhuitoo, Hainan Island, X26, (+); 37. Ostrea reef limestone, the Ends of the Earth, Hainan Island; 38. Karst landscape, Shidao Island; 39. Dissolution zawn in algae Limestone, Shidao Island, Xisha archipelago.